D1252673

Sustainable Groundwater Development

Special Publication reviewing procedures

The Society makes every effort to ensure that the scientific and production quality of its books matches that of its journals. Since 1997, all book proposals have been refereed by specialist reviewers as well as by the Society's Books Editorial Committee. If the referees identify weaknesses in the proposal, these must be addressed before the proposal is accepted.

Once the book is accepted, the Society has a team of Book Editors (listed above) who ensure that the volume editors follow strict guidelines on refereeing and quality control. We insist that individual papers can only be accepted after satisfactory review by two independent referees. The questions on the review forms are similar to those for *Journal of the Geological Society*. The referees' forms and comments must be available to the Society's Book Editors on request.

Although many of the books result from meetings, the editors are expected to commission papers that were not presented at the meeting to ensure that the book provides a balanced coverage of the subject. Being accepted for presentation at the meeting does not guarantee inclusion in the book.

Geological Society Special Publications are included in the ISI Science Citation Index, but they do not have an impact factor, the latter being applicable only to journals.

More information about submitting a proposal and producing a Special Publication can be found on the Society's web site: www.geolsoc.org.uk.

GEOLOGICAL SOCIETY SPECIAL PUBLICATION NO. 193

Sustainable Groundwater Development

EDITED BY

K. M. HISCOCK
University of East Anglia, UK

M. O. RIVETT
University of Birmingham, UK

R. M. DAVISON
University of Sheffield, UK

2002
Published by
The Geological Society
London

THE GEOLOGICAL SOCIETY

The Geological Society of London (GSL) was founded in 1807. It is the oldest national geological society in the world and the largest in Europe. It was incorporated under Royal Charter in 1825 and is Registered Charity 210161.

The Society is the UK national learned and professional society for geology with a worldwide Fellowship (FGS) of 9000. The Society has the power to confer Chartered status on suitably qualified Fellows, and about 2000 of the Fellowship carry the title (CGeol). Chartered Geologists may also obtain the equivalent European title, European Geologist (EurGeol). One fifth of the Society's fellowship resides outside the UK. To find out more about the Society, log on to *www.geolsoc.org.uk*.

The Geological Society Publishing House (Bath, UK) produces the Society's international journals and books, and acts as European distributor for selected publications of the American Association of Petroleum Geologists (AAPG), the American Geological Institute (AGI), the Indonesian Petroleum Association (IPA), the Geological Society of America (GSA), the Society for Sedimentary Geology (SEPM) and the Geologists' Association (GA). Joint marketing agreements ensure that GSL Fellows may purchase these societies' publications at a discount. The Society's online bookshop (accessible from *www.geolsoc.org.uk*) offers secure book purchasing with your credit or debit card.

To find out about joining the Society and benefiting from substantial discounts on publications of GSL and other societies world-wide, consult *www.geolsoc.org.uk*, or contact the Fellowship Department at: The Geological Society, Burlington House, Piccadilly, London W1J 0BG: Tel. +44 (0)20 7434 9944; Fax +44 (0)20 7439 8975; Email: *enquiries@geolsoc.org.uk*.

Published by The Geological Society from:
The Geological Society Publishing House
Unit 7, Brassmill Enterprise Centre
Brassmill Lane
Bath BA1 3JN, UK

(*Orders*: Tel. +44 (0)1225 445046
Fax +44 (0)1225 442836)
Online bookshop: *http://bookshop.geolsoc.org.uk*

British Library Cataloguing in Publication Data
A catalogue record for this book is available from the British Library.

ISBN 1-86239-097-5
ISSN 0305-8719

Typeset by Bath Typesetters, Bath, UK
Printed by Cromwell Press, Trowbridge, UK

Distributors

USA
AAPG Bookstore
PO Box 979
Tulsa
OK 74101-0979
USA
Orders: Tel. + 1 918 584-2555
Fax +1 918 560-2652
E-mail *bookstore@aapg.org*

Australia
Australian Mineral Foundation Bookshop
63 Conyngham Street
Glenside
South Australia 5065
Australia
Orders: Tel. +61 88 379-0444
Fax +61 88 379-4634
E-mail *bookshop@amf.com.au*

India
Affiliated East-West Press PVT Ltd
G-1/16 Ansari Road, Daryaganj,
New Delhi 110 002
India
Orders: Tel. +91 11 327-9113
Fax +91 11 326-0538
E-mail *affiliat@nda.vsnl.net.in*

Japan
Kanda Book Trading Co.
Cityhouse Tama 204
Tsurumaki 1-3-10
Tama-shi
Tokyo 206-0034
Japan
Orders: Tel. +81 (0)423 57-7650
Fax +81 (0)423 57-7651

Contents

Risk assessment methodologies for developing and protecting groundwater resources

Response of aquifers to future climate change

It is recommended that reference to all or part of this book should be made in one of the following ways:

HISCOCK, K. M., RIVETT, M. O. & DAVISON R. M. (eds) 2002. *Sustainable Groundwater Development*. Geological Society, London, Special Publications, **193**.

PRICE, M. 2002. Who needs sustainability? *In*: HISCOCK, K. M., RIVETT, M. O. & DAVISON, R. M. (eds) 2002. *Sustainable Groundwater Development*. Geological Society, London, Special Publications, **193**, 75–81.

Preface

Sustainable groundwater development is a key environmental and social issue for the future. Whereas in the past the management of groundwater resources was based predominantly on the concept of the available renewable resource, today it is necessary to consider protection of springs, river flows and water levels dependent on groundwater discharges, while concurrently maintaining abstractions for water supply and economic benefit. Obtaining this balance between human and environmental needs, and protecting valuable groundwater resources from over-exploitation and pollution, presents a challenge to hydrogeologists that is reflected in the papers contained in this volume.

Following an introductory chapter that provides a discussion and definition of sustainable groundwater development, the remainder of the book comprises 23 papers organized into four sections: (1) approaches to groundwater resources management; (2) assessment and measurement of the impacts of groundwater abstraction on river flows; (3) risk assessment methodologies for developing and protecting groundwater resources; and (4) response of aquifers to future climate change.

In the section discussing approaches to groundwater resources management, the first paper presents an integrated hydrogeological interpretation of current understanding of the geological structure of the Bristol–Bath basin and its relevance to the sustainability of the thermal springs at Bath. The following papers present the views of regulators and academics in debating the current and future direction of groundwater resources management in the UK and Arabian Peninsula, succeeded by case studies that demonstrate experience in managing aquifers to meet environmental and water quality objectives both in the UK and overseas, including North & South America and South Asia.

With the adoption of the EU Water Framework Directive in December 2000, the introduction of the concept of integrated river basin management will require a greater understanding of river–aquifer interaction. Current ideas and methodologies for assessing the effects of groundwater abstraction on surface river flows are discussed in the section on groundwater abstraction and river flows with experience drawn from the UK and Germany.

Contamination of groundwater resources from surface-derived diffuse and point source contaminants is a serious threat to the provision of groundwater supplies and the aquatic environment. On the other hand, for example in urban areas, a more flexible approach to water use for different purposes can enhance the availability of water. The papers in the section on risk assessment for groundwater management demonstrate modelling and risk-based methods for assessing aquifer vulnerability in rural and urban environments. Examples are presented for the urban area of Nottingham in the English Midlands, a grossly contaminated alluvial aquifer situated below oil refineries in Romania and a pulp and paper mill industrial complex in northwest Russia.

One of the greatest challenges for humankind in the 21st century is adapting to global climate change induced by increasing emissions of greenhouse gases. The general scenario for mid-latitudes under increasing average annual temperature is for increased rainfall in winter and drier summers. The translation of these changes into the potential consequences for longer-term management of groundwater resources is discussed in the last section on future climate impacts on groundwater resources. Two papers describe the results of using the scenarios produced by global circulation models (GCMs) on catchment water resources in several European carbonate aquifers and discuss changes in aquifer recharge, water levels, baseflow and groundwater chemistry.

This Special Publication of the Geological Society of London has it origins in the symposium on *Sustainable Groundwater Development* organized by the Hydrogeological Group of the Society and held at the *Geoscience2000* conference at the University of Manchester in April 2000. Following this symposium, additional papers were invited to supplement those papers given as oral presentations. In the process of compiling this volume, the editors are greatly indebted to the time and effort spent by the following referees in providing peer-review of submitted articles and also the input of the Series Editor, Dr Martyn Stoker, and the staff at the Society's Publishing House, including Angharad Hills and Diana Swan.

Kevin Hiscock, Norwich
Mike Rivett, Birmingham
Ruth Davison, Sheffield

October 2001

Referees

The Editors are grateful to the following people for their assistance with their reviewing of papers submitted to this Special Publication

Mr Brian Adams	British Geological Survey, Wallingford
Dr Phillip Aldous	Thames Water Utilities Ltd, Reading
Mr Dave Allen	British Geological Survey, Wallingford
Dr Paul Ashley	Mott McDonald Ltd, Cambridge
Dr Timothy Atkinson	Department of Geological Sciences, University College London & School of Environmental Sciences, University of East Anglia
Prof John Barker	Department of Geological Sciences, University College London
Dr Ron Barker	School of Earth Sciences, University of Birmingham
Dr Mike Barrett	Robens Centre for Public & Environmental Health, University of Surrey
Prof Keith Beven	Department of Environmental Science, University of Lancaster
Dr Phillip Bishop	Thames Water Utilities Ltd, Reading
Dr David Burgess	Environment Agency, Peterborough
Dr Willy Burgess	Department of Geological Sciences, University College London
Dr Mike Carey	Entec (UK) Ltd, Shrewsbury
Dr Richard Carter	Institute of Water & Environment, Cranfield University
Mr John Chilton	British Geological Survey, Wallingford
Dr Neil Chroston	School of Environmental Sciences, University of East Anglia
Dr Dick Cobb	School of Environmental Sciences, University of East Anglia
Dr Luke Connell	Department of Geological Sciences, University College London
Dr Declan Conway	School of Development Studies, University of East Anglia
Jane Dottridge	Komex, London
Dr Dick Downing	Twyford, Berkshire
Prof Mike Edmunds	British Geological Survey, Wallingford
Dr Trevor Elliot	School of Civil Engineering, Queen's University Belfast
Mr Alec Erskine	Montgomery Watson Harza, Edinburgh
Prof Stephen Foster	British Geological Survey, Wallingford
Dr Mark Grout	Environment Agency, Peterborough
Dr Paul Hart	Environment Agency, Peterborough
Dr John Heathcote	Entec (UK) Ltd, Shrewsbury
Dr Alan Herbert	Environmental Simulations International Ltd, Shrewsbury
Mr Adrian Lawrence	British Geological Survey, Wallingford
Mr David Lister	School of Environmental Sciences, University of East Anglia
Dr Rob Low	Water Management Consultants Ltd, Shrewsbury
Mr Phillip Merrin	North-West Water Ltd, Warrington
Mr Bruce Misstear	Department of Civil, Structural & Environmental Engineering, University of Dublin, Trinity College
Mr Brian Morris	British Geological Survey, Wallingford
Dr Mike Owen	Environment Agency, Reading
Mr Mike Price	Postgraduate Research Institute for Sedimentology, University of Reading
Mr Shaminder Puri	Scott Wilson, Water & Environment, Abingdon
Mr Michael Riley	School of Earth Sciences, University of Birmingham
Dr Nick Robins	British Geological Survey, Wallingford
Prof Ken Rushton	School of Civil Engineering, University of Birmingham
Prof Peter Smart	School of Geographical Sciences, University of Bristol
Dr Willie Stanton	Westbury-sub-Mendip, Somerset
Dr John Tellam	School of Earth Sciences, University of Birmingham
Mr David Watkins	Camborne School of Mines, University of Exeter
Emily Whitehead	British Geological Survey, Wallingford
Dr Mark Whiteman	Environment Agency, Peterborough
Dr Janet Whittaker	Entec (UK) Ltd, Shrewsbury
Dr Fred Worrall	Department of Geological Sciences, University of Durham
Prof Paul Younger	Department of Civil Engineering, University of Newcastle-upon-Tyne

Sustainable groundwater development

K. M. HISCOCK[1], M. O. RIVETT[2] & R. M. DAVISON[3]

[1]*School of Environmental Sciences, University of East Anglia, Norwich NR4 7TJ, UK (e-mail: k.hiscock@uea.ac.uk)*

[2]*School of Earth Sciences, University of Birmingham, Edgbaston, Birmingham B15 2TT, UK*

[3]*Groundwater Restoration & Protection Group, Department of Civil & Structural Engineering, University of Sheffield, Mappin Street, Sheffield S1 3JD, UK*

Abstract: Estimated annual water availability per person in 2025 is likely to result in at least 40% of the world's 7.2 billion people facing serious problems with obtaining freshwater for agriculture, industry or human health (Gleick 2001). To meet present and future needs with the currently available surface and groundwater resources, while at the same time preserving terrestrial and aquatic ecosystems, will require a sustainable approach to managing water. This paper discusses the importance of groundwater resources in industrialized and developing countries, and the associated problems of over-abstraction and groundwater pollution, with the objective of defining sustainable groundwater development. It is concluded that sustainable groundwater development at global and local scales is achieved through the maintenance and protection of groundwater resources balanced against economic, environmental and human (social) benefits. This interpretation of sustainable groundwater development is incorporated into the methodologies currently emerging in Europe (the EU Water Framework Directive) and England and Wales (Catchment Abstraction Management Strategies). However, success in achieving future sustainable groundwater development will require a common understanding at the level of the individual based on information and education within a legislatory framework that promotes co-operation and self-responsibility.

According to O'Riordan (2000), the three fundamental principles of sustainable development are to maintain and protect essential ecosystems, to utilize renewable resources to the point of precautionary replenishment and to price the cost of living according to its natural burdens and social disruption. How might these principles be applied to groundwater resources?

Unlike other natural resources such as fossil fuels, water is a renewable resource and most abstractions are strictly sustainable in that abstracted water will, in time, ultimately return to the hydrological cycle. With this in mind, Foster (2000*a*) and Price (2002) argue that there is no fundamental reason why the temporary over-exploitation of aquifer storage for a given economic benefit is an undesirable process as part of a logical water resources management strategy as long as the groundwater system is sufficiently well understood in order to evaluate impacts. Price (2002) identifies a number of examples of non-sustainable use of groundwater that may have had beneficial outcomes. For example, the use of groundwater from the Chalk aquifer of the central London Basin during the nineteenth and early twentieth centuries was not sustainable in the long term but enabled London to develop as a major centre of population and manufacturing.

A dramatic illustration of over-abstraction of groundwater leading to a non-sustainable situation is the High Plains Aquifer in the mid-section of the United States (Dennehy *et al.* 2002). Use of the High Plains Aquifer as a

From: Hiscock, K. M., Rivett, M. O. & Davison, R. M. (eds) *Sustainable Groundwater Development*. Geological Society, London, Special Publications, **193**, 1–14. 0305-8719/02/$15.00

source of irrigation water has transformed the area into one of the major agricultural regions of the world. Substantial pumping from the 1940s up to 1980 resulted in water level declines of more than 30 m in some parts of the region. Intensive arable farming has also resulted in significant increases in nitrate concentrations throughout the Ogallala Formation, the principal geologic unit of the aquifer. Declining water levels are a direct threat to the current way of life of the area and attempts are being made to introduce more water-efficient irrigation and best-management farming practices including, ultimately, a shift away from irrigated agriculture to dry-land farming. Such a shift in agriculture will have far-reaching implications for the regional economy.

Another example, although with environmental considerations, is presented by Anderson *et al.* (2002) of the economically important revenues from copper and gold mining in northern Chile. The demands of the mining industry for both potable and ore-processing water increase the overall demand for water from the Tertiary sedimentary aquifer in this remote, high-altitude and extremely arid environment. Set against the high commercial demand for water is the desire to conserve animal and plant communities developed in the internationally important Tilopozo wetland on the edge of the Salar de Atacama. The combined cost-effective abstraction of groundwater is at least twice the estimated aquifer throughflow but is judged environmentally sustainable in the foreseeable future by virtue of the fact that the aquifer possesses an extremely large volume of groundwater storage (*c.* 10^{10} m^3). A feature of the proposed development strategy is that only a small percentage (*c.* 5%) of the total storage in the aquifer should be mined if overall sustainability criteria are to be fulfilled.

The progressive mining of aquifer storage can potentially have negative long-term consequences for groundwater users, the environment (e.g. the drying up of springs and streams in the London Basin as a result of the lowered Chalk potentiometric surface) and society. Historically, the management of groundwater resources was based on measures of the 'safe yield' of an aquifer, frequently taken to equal the long-term average recharge (Twort *et al.* 1985). This approach had as its primary goal the fulfilment of economically and socially imposed demands for water which was assumed should be satisfied. Under the approach of sustainable development of groundwater resources there has been a shift of emphasis to recognise the needs of the aquatic environment such as groundwater-fed springs and wetlands (Burgess 2002).

To start to understand the implications of sustainability in groundwater development, this paper begins with a global perspective of the sustainability debate and then describes the approaches being adopted at European and national (England and Wales) levels to manage groundwater resources in a wider, integrated framework. Examples are also chosen from developing countries to illustrate the main quantity and quality issues relating to the provision of adequate water supplies.

The main objective of this paper is to give a consistent definition of the meaning of sustainable groundwater development and to highlight the way forward in achieving sustainability. The paper concludes that sustainable groundwater development at global and local scales is not the balancing of available aquifer storage to satisfy a single aim such as meeting water users' demands, but the maintenance and protection of the groundwater resource to balance economic, environmental and human (social) requirements. This interpretation of sustainable groundwater development is illustrated in Figure 1.

Global water demand and Rio 1992

The total amount of water abstracted globally from surface and groundwaters has increased by nine times since 1900 (Fig. 2). Water use per person, however, has only doubled in that time and has even declined slightly in recent years in developed countries due to more efficient use of water (Gleick 2001). Despite this positive trend, improvements in water efficiency may not keep pace with projected population growth. Estimated annual water availability per person in 2025 is likely to result in at least 40% of the world's projected 7.2 billion people facing serious problems with obtaining freshwater for human consumption, industry and agriculture (Gleick 2001).

The United Nations (UN) Development Programme (1998) report into human development provided a scientific audit of the state of the planet and presents sobering statistics. As a result of demographic pressure, 20 countries already suffer from water stress, having less than 1000 m^3 of water per person per year. Furthermore, 30% of the population in developing countries lack access to safe drinking water with two million dying every year from associated diseases. Such evidence of environmental stress and social deprivation is a persuasive argument for a greater sharing of individual opportunities, a topic that has received much attention. A

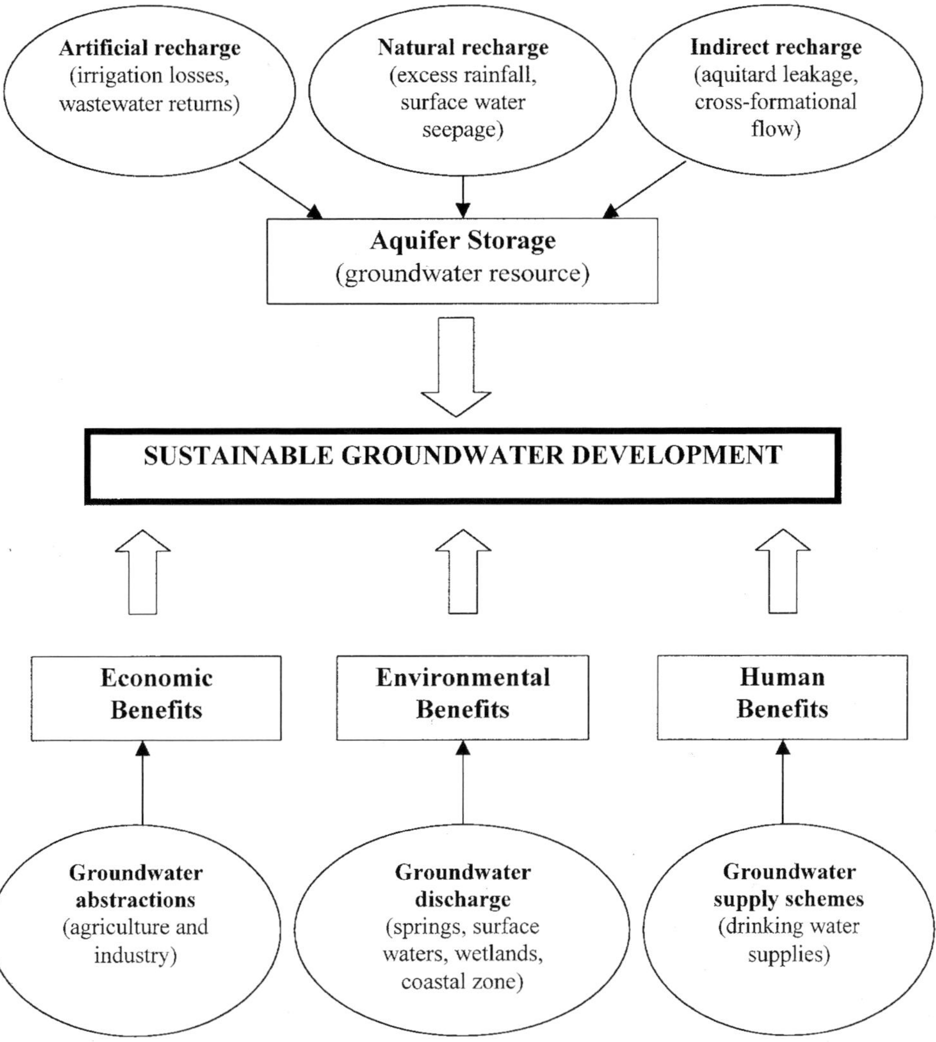

Fig. 1. The achievement of sustainable groundwater development through the balance of recharge inputs to aquifer storage (the groundwater resource) against discharge outputs for economic, environmental and human (social) benefits.

prominent publication in this field is the Brundtland Report (1987) prepared by the World Commission on Environment and Development, established by the UN. This commission was charged with the task of identifying and promoting the cause of sustainable development and discussed the right of all human beings to live in an environment adequate for their health and well-being. This landmark report was followed by the famous UN Conference on Environment and Development: Rio 1992 (also referred to as the UNCED or the Earth Summit).

The UNCED was designed to take stock of the state of the world 20 years after the first major Earth Summit held in Stockholm in 1972. Several agreements were signed at Rio, the centrepiece of which was *Agenda 21*, a 40-chapter report outlining an action plan for sustainable development that integrates environment and development concerns and which is strongly oriented towards bottom-up participa-

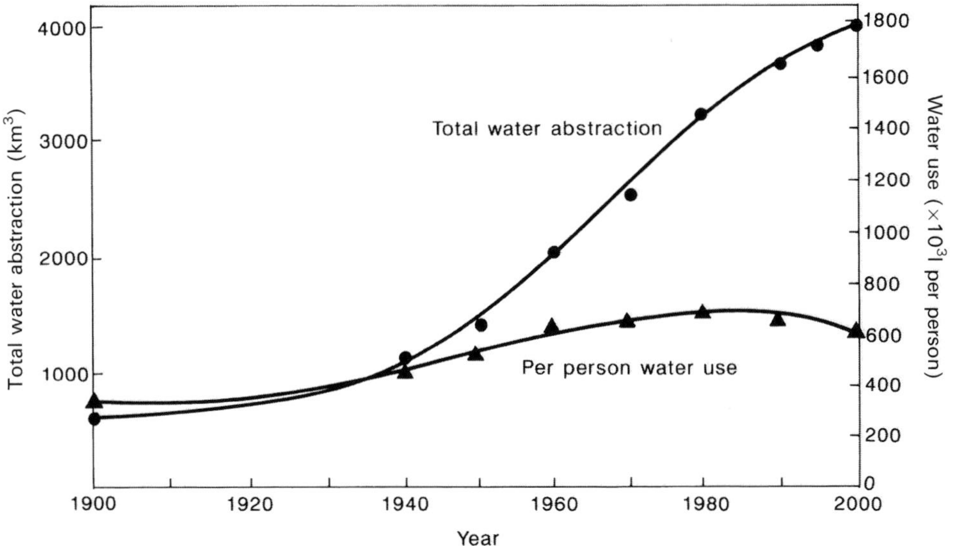

Fig. 2. Annual global water abstractions. Adapted from *Making every drop count* by Peter H. Gleick. Copyright © February 2001 by Scientific American, Inc. All rights reserved.

tion and community-based approaches. *Agenda 21* forms the basis of national sustainable development strategies that are now sent annually by governments to the UN Commission on Sustainable Development. Since the Rio Conference, freshwater depletion has moved into the international spotlight, partly because of persistent warnings that more than a third of the world's population will not have access to sufficient water by 2025 (Gleick 2001). This theme is bound to become more important in future Earth Summits, as demonstrated by the year 2003 being designated the UN International Year of Freshwater.

Sustainable groundwater use in Europe

Groundwater resources have historically provided a local and inexpensive source of drinking water for domestic supplies in Europe. Of the total water abstracted in the European Union (EU), between about 22% (OECD 1997) and 29% (EEA 1995) is taken from groundwater. Table 1 provides a summary of groundwater abstractions in various European countries. In several countries (e.g. Austria, Denmark and Portugal), a large proportion of total abstractions (>40%) comes from groundwater, compared with less than 10% in Belgium, Finland and The Netherlands.

A significant factor in the management of water resources in Europe is agriculture, particularly the demand for irrigation water. In many EU Member States, there has been a relative decrease in the importance of agriculture in comparison with other economic sectors, although agriculture still accounts for approximately 30% of total water abstractions (EEA 1999). In southern European countries (Greece, Italy, Portugal and Spain) this percentage rises to 62% of total uses, particularly where irrigation water use is prevalent, in marked contrast to northern and eastern European countries where, on average, less than 10% of the water resources is used for irrigation.

Although all European countries potentially have sufficient resources to meet national demands, most are interested in drinking water supply problems at regional or local scales. The greatest demand for water is concentrated in the densely populated urban conurbations. The demand for European water resources has increased from 100 km^3 in 1950 to 551 km^3 in 1990 (EEA 1999) with consequent local over-abstraction of water in relation to the available resources, especially in southern Europe and the industrial centres of the north. An overview of the problem relating to groundwater resources is

Table 1. *Data indicating the percentage of total abstractions and percentage of public water supplies contributed by groundwater in European countries. Data summarised by the EEA (1999)*

Country	Groundwater abstraction as a percentage of total abstraction[1]		Percentage of public water supply supported by groundwater abstractions
	OECD data[2]	EEA data[3]	
Austria	34	53	99.3
Belgium	9	9	51.5
Denmark	25	99	100.0
Finland	10	8	55.6
France	16	16	56.4
Germany	13	13	72.0
Greece	26	28	50.0
Ireland	19	31	50.0
Italy	23		80.3
Luxembourg	46	46	69.0
Netherlands	13	7	68.2
Portugal	42	42	79.9
Spain	9	15	21.4
Sweden	20	20	49.0
UK (England & Wales)	19	19	27.4
Average EU15	22	29	62.0
Czech Republic	18		44.0
Estonia		15	
Hungary	16	16	
Iceland	91	95	84.1
Norway			13.0
Poland	16	16	
Slovenia		22	
Switzerland			82.6

[1]Eurostat (1997) and ETC/IW (1998)
[2]OECD (1997)
[3]EEA (1995)

illustrated in Figure 3 that shows reported cases of groundwater over-exploitation that have resulted in severe drawdowns, saline intrusion and ecological damage. Although this is an incomplete inventory, it is apparent that localized over-exploitation occurs widely throughout the EU. In those countries where a systematic inventory has been made (e.g. in several federal states in Germany and in The Netherlands) the problems were found to be extensive (RIVM & RIZA 1991).

Problems of water shortages are compounded by seasonal or inter-annual variation in the availability of freshwater resources, for example the dry years of the mid-1970s and early 1990s that affected northwest Europe. These problems are exacerbated in certain aquifer types where storage is limited. An example is the hard-rock island aquifer of Guernsey (Robins *et al.* 2002) which is developed as a shallow weathered zone of up to 30 m thick in ancient crystalline metamorphic rocks with an estimated specific yield of only 3%. The aquifer is valued as having a significant resource potential in maintaining baseflows to streams but it is under severe pressure in that water demand is increasing when the long-term rainfall total has been declining locally. Furthermore, intensive use of nitrogenous fertilizer has raised the concentration of nitrate in water supplies to unacceptable levels.

Problems associated with groundwater quantity are often accompanied by threats to quality. Urban, industrial and agricultural activities release point and diffuse sources of pollutants that can contaminate vulnerable groundwater

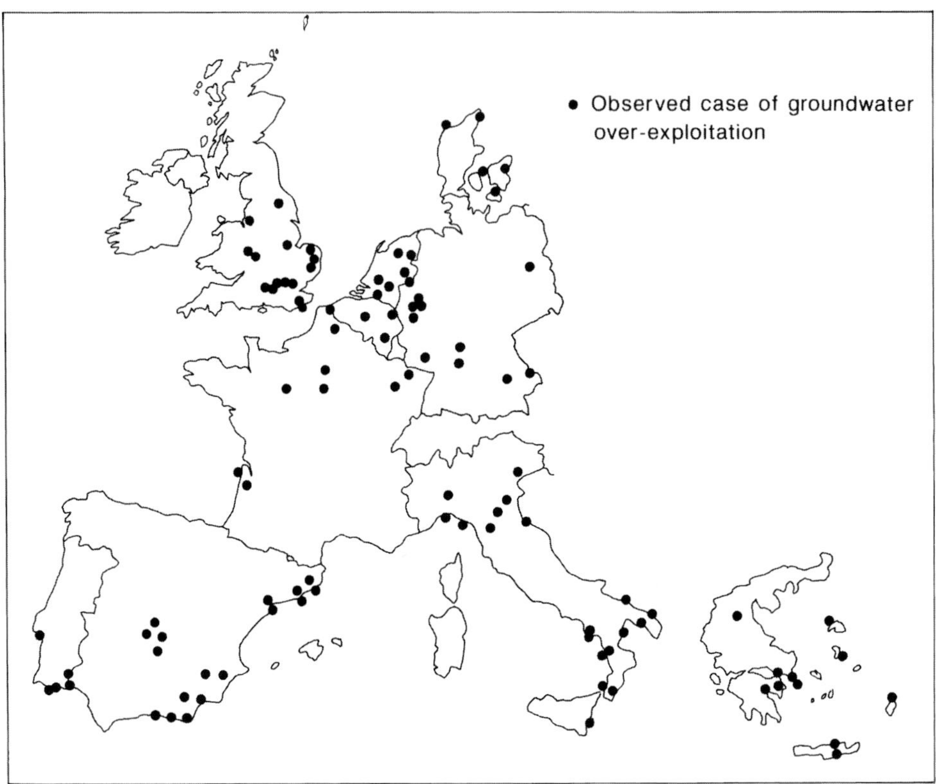

Fig. 3. Locations in Europe of problems encountered as a result of groundwater over-exploitation. Cases include over-development problems, large drawdowns due to abstractions, saline intrusion and ecosystem damage due to the lowering of groundwater levels. Although an incomplete inventory, it is clear that problems associated with groundwater over-exploitation are widespread in the Europe Union. From RIVM & RIZA (1991).

supplies. Widespread problems include diffuse agricultural sources of nitrates and pesticides, landfill disposal and the numerous contaminated urban and industrial sites that are a legacy of economic growth in Europe. Table 2 illustrates the magnitude of contaminated urban and industrialized areas in the EU.

An example of gross industrial contamination from oil refineries is presented by Albu *et al.* (2002) for the area of Ploiesti in Romania. The contamination of groundwater by petroleum products is probably one of the most extensive cases reported to date with one third of the water supplies to the city of Ploiesti affected. The lack of groundwater management has resulted in unacceptable, major impacts on groundwater quality and has jeopardized both the short- and long-term sustainability of the alluvial aquifer resource. Albu *et al.* (2002) outline the measures that have been considered to remediate the existing contaminated areas. An enabling factor in this effort is the privatization of the petroleum industry which is an essential step if Romania is to meet requirements to join the EU. As a result, an EU grant has been made available for the initial stages of investigating the remediation of the area. This example provides an illustration of how by removing institutional obstacles to obtaining financial resources may allow the first steps towards achieving a remediated, more sustainable environment.

The recognition of deteriorating quantity and quality of groundwater resources in Europe came with the ministerial seminar held in The Hague in 1991. The special significance of groundwater, both in the hydrologic cycle and as a source of drinking water, was then recognized by the European Council and led to

Table 2. *Estimation of the magnitude of contaminant source areas in the European Union (from RIVM & RIZA 1991)*

Source	Number	Amount of waste (tonnes)	Area (km^2)	Potentially contaminated area (km^2)
Industrial estates	12×10^6	–	10 000	16 000–40 000
Municipal landfill	$6–12 \times 10^4$	$3–6 \times 10^9$	600–1200	900–7200
Fuel storage tanks	$3–6 \times 10^6$	–	–	250–4000
Mining waste dump sites	$< 1 \times 10^4$	17×10^9	250–500	350–5000
Line sources	–	–	10 000–25 000	1500–7500
Dredged sediment dump sites	$< 1 \times 10^3$	–	–	$< 1 \times 10^3$
Hazardous waste sites	$< 1 \times 10^3$	–	–	$< 1 \times 10^3$
Estimated total contaminated area				$20–60 \times 10^3$

the call for a Community Action. This initiative resulted in a draft proposal for an Action Programme for Integrated Groundwater Protection and Management (GAP) (COM (96) 315 final) which required a programme of actions to be implemented by 2000 at National and Community level, aiming at sustainable management and protection of freshwater resources.

Many of the recommendations in the GAP are now found in the EU Water Framework Directive (WFD) (COM (97) 49 final) that was adopted in December 2000. The overall purpose of the WFD is to establish a framework that will allow for the protection of surface freshwater, estuaries, coastal waters and groundwater that: (1) prevents further deterioration and protects and enhances the status of aquatic ecosystems and, with regard to their water needs, terrestrial ecosystems; and (2) promotes sustainable water consumption based on long-term protection of available water resources; and thereby contributes to the provision of a supply of water of the qualities and in the quantities needed for sustainable use of these resources.

The WFD requires the attainment of good surface water and groundwater status by 2010. Good groundwater status will only be achieved when there is no over-exploitation of aquifers or adverse impacts on groundwater-supported aquatic and terrestrial ecosystems. Thus, the control and management of water quantity will be for the first time a legal requirement across the EU.

One of the significant scientific challenges presented by the WFD is understanding the processes affecting the quantity and quality of water at the groundwater–surface water boundary. Assessing the environmental impacts of groundwater abstractions is a difficult task but analytical solutions to idealizations of the complex river–aquifer interaction can help guide judgements (Kirk & Herbert 2002). In contrast, Rushton (2002) shows that simple analytical solutions are rarely applicable to predicting river–aquifer interaction in fissured Chalk and limestone aquifers and recommends the development of realistic conceptual and mathematical models in order to incorporate non-linear flow behaviour. Uncertainties concerning the hydraulic connection between ground and surface water at a site, for example distribution of areas with different infiltration rates, thickness of sediment layers and hydraulic head gradient, can be reduced by careful fieldwork and laboratory studies as demonstrated by Macheleidt *et al.* (2002).

To prevent further deterioration of groundwater quality and improve the status of aquatic systems under the WFD will require the application of reliable groundwater management tools. Several vulnerability assessment methodologies have been developed in Europe as a means for managing land use activities adjacent to production boreholes (source protection) and also over aquifer outcrops (resource protection). These methods are reviewed by Worrall (2002) who also presents a method for measuring borehole catchment vulnerability to pesticide contamination. The role of the unsaturated zone in affecting contaminant transport is an important consideration in aquifer vulnerability assessment and is one of the hardest to evaluate given the difficulties of establishing relationships between physical (site and contaminant-specific) properties and the mechanisms that affect the transport processes. Connell (2002) recognizes these difficulties and presents a simple analytical solution for solute transport through the unsaturated zone under dynamic variation in water movement that allows for a root zone with distinct properties from the rest of the profile.

In urban and industrial areas, a typical approach is to adopt a tiered risk assessment depending on the amount of available data. For example, Schoenheinz *et al.* (2002) adopted this approach to determine the risk of groundwater pollution arising at a pulp and paper mill site in northwest Russia. In this example, the groundwater pollution risk at individual sample locations across the pulp and paper mill site was obtained by calculation of a toxicological index, or hazard quotient. This approach is relatively crude but does provide a comparative, quantitative assessment of analytical results for different measurement points across a large, complex industrial site, although the assessment is neither source nor target related.

A single conceptual model often cannot represent the complex interaction between an urban or industrialized area and the underlying groundwater due to the inherent uncertainties and limited data. Instead, a probabilistic risk assessment method is appropriate. An example of an advanced probabilistic water management tool applied to an urban aquifer is given by Davison *et al.* (2002) and provides an economically viable approach to the sustainable water management of cities by balancing the need to reduce rising water tables, alleviate water shortage and to reduce the pressure on rural aquifers.

Sustainable groundwater use in England and Wales

The Environment Agency of England and Wales (the Agency) has the duty to conserve, augment, redistribute and secure the proper use of water resources in England and Wales and is the central body with responsibility for long-term water resources planning. The Agency has published its national and regional water resources management strategy for the next 25 years (Environment Agency 2001*a*) where consideration is given to both the environment and society's need for water, and associated uncertainties about future water demand and availability. The strategy adopted follows a twin-track approach that seeks the efficient use of water whilst bringing forward proposals for resource development where appropriate. The underpinning principle to the strategy is sustainable development. The Agency follows the UK Government's approach that sustainable development should ensure a better quality of life for everyone, now and for generations to come (DETR 1999). In practice this means meeting four objectives simultaneously:

(1) social progress which recognizes the needs of everyone;
(2) effective protection of the environment;
(3) prudent use of natural resources; and
(4) maintenance of high and stable levels of economic growth and employment.

These objectives provide a framework against which strategic options can be tested. For example, the Agency uses a technique known as *sustainability appraisal* to measure the contribution of its strategies to sustainable development in which each of the above four sustainability themes is sub-divided into criteria against which each water resources option may be assessed. The appraisal is carried out using specified questions to produce a comprehensive coverage of possible impacts and risks. The appraisal is applied at each stage of the strategy formulation process including strategy objectives, strategy options and policies.

One of the keys to successful adoption of the strategy is to ensure that it is flexible enough to deal not only with current scenarios but at least to some extent with events resulting from future scenarios. For example, in the context of the water resources strategy, the Agency favours schemes that improve the management of water use or developments that can be phased (e.g. groundwater development schemes) over schemes that are inflexible (e.g. new surface water reservoir schemes).

Catchment Abstraction Management Strategies (CAMS) (Environment Agency 2001*b*) are a recent initiative to which the water resources strategy (Environment Agency 2001*a*) contributes by providing information on water demands. The CAMS, in turn, provide information to future regional strategies about the availability of resources on a local scale and about pressures on the aquatic environment. It is envisaged that CAMS will provide a consistent strategy for achieving the sustainable management of water resources within a catchment or group of catchments. With a national programme of sequential implementation and a six-year review cycle, CAMS will have the following objectives:

(1) to provide a consistent and structured approach to local water resources management, recognizing both the reasonable needs of abstractors for water and environmental needs;
(2) to provide the opportunity for greater public involvement in the process of managing

abstraction at a catchment level;
(3) to provide a framework for managing time-limited licences; and
(4) to facilitate licence trading where appropriate.

CAMS will also introduce sustainability appraisal for the first time in the context of a water resources licensing policy with a largely qualitative, two-tier sustainability appraisal process. Tier 1 aims to define the resource availability status that should or could be achieved in each six-year period. If the catchment resource is determined as being over-licensed or over-abstracted then it is necessary to implement Tier 2 that allows for recovery of some water resources as the current situation is unsustainable. Tier 2 assessment defines resource recovery options and appraises the environmental, economic, social and resource implications of these options. Each tier assessment is undertaken by the Agency with the participation of a group of key stakeholders.

Furthermore, the CAMS process provides the Agency with a mechanism for meeting not only parts of the WFD in providing the information required to formulate River Basin Management Plans, but also in meeting other EU legislation, such as the Birds and Habitats Directives (79/409/EEC and 92/43/EEC, respectively) that require the Agency to undertake resource assessments to protect the water requirements of designated sites.

The increasing use of groundwater models can be expected with the introduction of CAMS and the consideration of surface and groundwaters within integrated river basin management planning. In order to be able to select and apply the correct tool, the Environment Agency is currently undertaking a review of its groundwater modelling capability (Hulme *et al.* 2002) and highlights the investment needed to provide a good conceptual understanding of the hydrological system being evaluated in order to develop and improve the use of distributed numerical groundwater models.

The achievement of sustainable development within the Agency's framework for managing water resources will entail greater sophistication in the way in which catchment resources are integrated. Experience already exists from the past development of conjunctive use and artificial recharge schemes (Downing 1993; Burgess 2002) and Hudson (2002) provides a current example of an integrated approach to water resources management in the East Midlands region of England. The East Midlands Resource Zone is supported by a combination of surface and groundwater sources supplying nearly three million people. There is a significant reliance on the Nottinghamshire Sherwood Sandstone aquifer such that the aquifer is over-licensed. In order to meet the existing resource shortfall while at the same time reducing abstraction licences in areas where rivers are impacted by abstractions, the water company is developing groundwater resources beneath the city of Birmingham to support its downstream surface water abstraction on the River Trent. This scheme also shows how urban groundwater resources, which in the past have been considered to be of poor water quality and therefore of little value, can be seen as a valuable future water resource.

The impacts of climate change on groundwater resources and incorporation of predicted effects into water resource plans is in its infancy in hydrogeology. Climate change is expected to lead to an intensification of the global hydrological cycle and have major impacts on regional water resources (IPCC 1997). Yusoff *et al.* (2002) developed a distributed transient groundwater model for the Chalk aquifer in eastern England and show a noticeable and consistent decrease in autumn recharge as a consequence of a smaller predicted amount of summer precipitation and increased autumn potential evapotranspiration.

Younger *et al.* (2002) also adopted a numerical modelling approach to the prediction of the vulnerability of aquifers to future climate change across three different climatic zones in Europe. Taking hypothetical spring catchments representative of major carbonate aquifers, their results indicate increased recharge in the northern maritime zone (e.g. across the UK), a net decrease in aquifer recharge in the continental zone (Germany) and a decrease in recharge in the Mediterranean zone (Spain).

Atkinson & Davison (2002) and McCann *et al.* (2002) demonstrate the need for rigorous geological and geophysical data collection and interpretation if groundwater resources are to be managed sustainably. Both papers study the Bristol–Bath basin in southwest England and the hydrogeological framework of the hot springs at Bath that form part of the regional outflow from the Carboniferous Limestone that underlies the structural basin. The hot springs are the largest natural thermal source in Britain and sustainable use of the waters requires maintenance of their temperature and flow rate. To achieve sustainability, current uncertainties as to the basin's structure need to be resolved, particularly the hydrogeological role of thrust faults that may cut the limestone at depth.

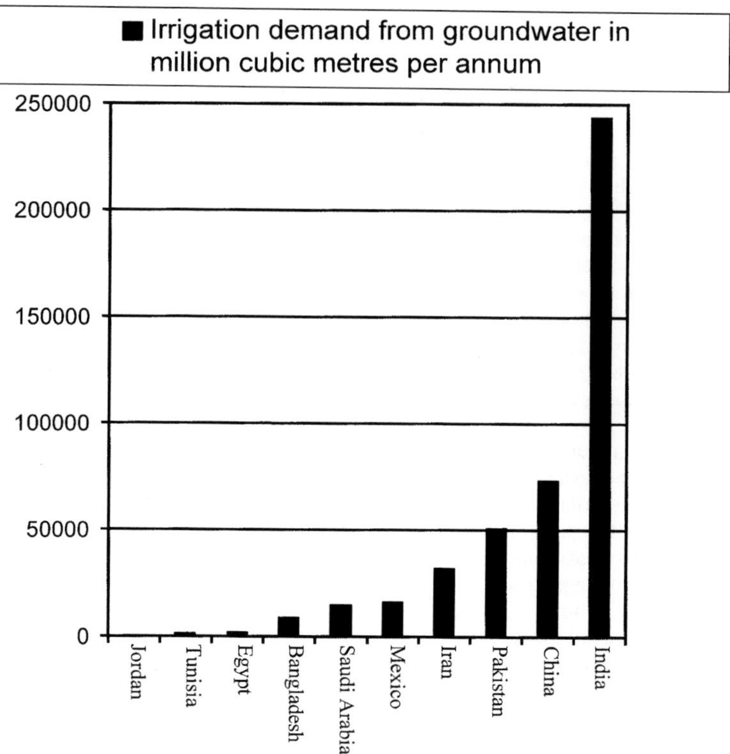

Fig. 4. Agricultural irrigation demand from groundwater ($\times 10^6$ m^3 a^{-1}) for selected countries and representative of the period 1990–1997 (after Foster 2000*b*).

Continuity of the thrust faults has possible implications for the impact of dewatering by large quarries at the limestone outcrops north and south of Bath.

Sustainable groundwater use in developing countries

Provision of groundwater resources has allowed the economical and rapid development of more reliable and better quality water supplies for a large proportion of rural communities across extensive areas of Africa, Asia, and Latin America. According to Foster (2000*b*), this crucial but formidable task began to gain momentum in the 1980s and has led to significant improvements in the quality of life for innumerable rural populations. The continuing challenge is to extend basic services, especially in areas with less favourable hydrogeological conditions, and to improve operational sustainability of the many systems already developed.

During the last 20 years, many developing countries have witnessed an enormous increase in the exploitation of groundwater for agricultural irrigation (Fig. 4). Groundwater resources have enabled the 'green revolution' in agriculture across many Asian nations, and have also permitted cultivation of high-value crops in a significant number of arid regions. In many cases, the development of groundwater resources in developing countries has been poorly managed. The proliferation of wells can lead to over-exploitation and quality problems may also arise from low standards of well construction, induced hydrochemical changes or pollutant migration. Clearly, these factors bring into question the sustainability of the associated rural development.

An example of the issues surrounding the sustainable use of groundwater in developing regions is provided by Rajasooriyar *et al.* (2002) in a study of the Valigamam region of the Jaffna Peninsula. The Valigamam region is moderately highly populated and intensive agricultural practices, that are increasingly dependent on chemical fertilizers and pesticides for greater food production, have led to deteriorating groundwater quality in the unconfined Miocene limestone aquifer. Further problems are associated with high nitrate inputs from congested or improperly planned soakaway pits and saline intrusion caused by over-abstraction of coastal wells. To tackle these problems, local educational programmes are required to strengthen public awareness of the reasons for the poor quality and deteriorated nature of groundwater in the Jaffna Peninsula at the same time as introducing measures such as restrictions on the use of agrochemicals and controls on the location and pumping rates of new and existing coastal wells.

Elsewhere in South Asia there is wide concern for the widespread occurrence of arsenic in groundwater of the Holocene alluvial aquifers in West Bengal and Bangladesh. Regional surveys have demonstrated that aquifers of the Ganges, Brahmaputra and Meghna floodplains are all affected in parts. Groundwater with arsenic in excess of the Bangladesh regulatory limit of 50 $\mu g\ l^{-1}$ is now known to be present over most of the south and southwest of Bangladesh and it is estimated that up to 21 million people may be at risk of arsenic poisoning (British Geological Survey & Mott MacDonald 1999). There have been calls for groundwater to be abandoned as a source of bacteriologically-safe drinking water in the affected areas even though the alluvial aquifers provide water for 95% of the population and are widely used for irrigation. In two papers (Burgess *et al.* 2002; Cuthbert *et al.* 2002) it is argued that the sustainable development of the alluvial aquifers may still be possible. The key questions to address concern the mechanisms of arsenic mobilization in groundwater, the distribution of the arsenic source, the patterns of groundwater flow to tubewells and the processes of water-rock interaction which govern arsenic behaviour on its flow path towards the tubewells.

In the states of the Arabian Peninsula, aquifer stress is frequently reported, reflecting the effects of high population growth, industrial development and national policies that encouraged the growth of irrigated agriculture. Of the total water consumption, some 60–95% is accounted for by irrigation (Young 2002). The over-exploitation of groundwater resources had led to internal social conflict and the prospects of international conflict over shared resources. Young (2002) explains that groundwater abstractions are mostly private without volume limitations, rather than being under the control of a government or an irrigation agency. Consequently, restrictions on abstraction are either limited or absent. For greater sustainable use of groundwater resources, Young (2002) advocates a participatory intervention at the aquifer level as offering scope for introducing reform gradually, through education and the engagement of all stakeholders as a presage to changes in national water policies. This view is endorsed by Foster (2000*b*) who highlights the need to address groundwater over-exploitation by requiring water demand management through changes in the individual use of water to reduce total abstraction while working collaboratively with groundwater regulators in setting objectives and formulating strategies. This is of fundamental importance, since in many cases a divide exists between the approaches advocated by governmental authorities and the perceptions of local users.

Conclusions

The prediction that more than a third of the world's population will not have access to sufficient freshwater by 2025 cannot be ignored and Gleick (2001) advocates that, rather than trying to satisfy projected future increases in demand, it is better to meet needs with currently available water resources, while preserving the ecosystems that are part of our well-being. In essence this is a call for sustainable development. From the survey of papers presented in this volume, we conclude that the defining feature of sustainable groundwater development at global and local scales is not the balancing of available aquifer storage to satisfy a single aim such as meeting water users' demands, but the maintenance and protection of the groundwater resource to balance economic, environmental and human (social) requirements. This interpretation of sustainable groundwater development is illustrated in Figure 1 and essentially captures the methodologies currently emerging in Europe (WFD) and England and Wales (CAMS) based around the concept of integrated river basin management.

A number of key points are identified in this volume when seeking to achieve sustainable groundwater development. First, the successful adoption of any strategy must ensure that it is flexible enough to deal not only with current scenarios but at least to some extent with events resulting from future scenarios, including climate change. Second, removing institutional obstacles to obtaining financial resources may significantly influence achievement of a more sustainable environment, particularly in relation to remediating contaminated aquifers. Third, innovative approaches to sustainable groundwater management are necessary in order to balance conflicting demands, for example risk-based methods for predicting groundwater vulnerability and groundwater models built on a sound understanding of the groundwater system. Fourth, and with particular reference to developing countries, there is a need to bridge the divide that exists between the approaches advocated by governmental authorities and the perceptions of local users. To tackle these challenges in both industrialized and developing countries, local educational programmes in collaboration with groundwater regulators are required to strengthen public awareness of the reasons for the often poor quality and deteriorated nature of groundwater.

Finally, as individuals, we cannot rely on legislation alone for delivering sustainability. Each of us should adopt the precautionary principle in which we use water and handle our raw materials and waste in such a way as to reduce demand and restrict the probability of pollution of water resources. Sustainable groundwater development can therefore be achieved by following the principle and spirit of *Agenda 21*. Hence, to find solutions to the challenge of meeting the global freshwater demand will require a more common understanding based on information and education within a legislatory framework that promotes co-operation and self-responsibility. This is not to say that sustainable groundwater development will be easily achieved, especially in the face of a burgeoning global population and current water shortages in semi-arid and arid areas of the world. However, progress can be made if the principle of sustainability is at applied at each stage of groundwater development

The authors are grateful to Dr Tim Atkinson at University College London and the University of East Anglia and Dr Mark Whiteman of the Environment Agency (Anglian Region) for their comments on an earlier draft of this paper. We thank the following for permission to reproduce material: the European Environment Agency for information contained in Table 1; the Institute for Inland Water Management and Waste Water Treatment (RIZA) for Figure 3 and Table 2; and Scientific American, Inc. for Figure 2.

References

ALBU, M. A., MORRIS, L. M., NASH, H. & RIVETT, M. O. 2002. Hydrocarbon contamination of groundwater at Ploiesti, Romania. *In*: HISCOCK, K. M. RIVETT, M. O. & DAVISON, R. M. (eds) *Sustainable Groundwater Development*. Geological Society, London, Special Publications, **193**, 243–301.

ANDERSON, M., LOW, R. & FOOT, S. 2002. Sustainable groundwater development in arid, High Andean basins. *In*: HISCOCK, K. M. RIVETT, M. O. & DAVISON, R. M. (eds) *Sustainable Groundwater Development*. Geological Society, London, Special Publications, **193**, 133–144.

ATKINSON, T. C. & DAVISON, R. M. 2002. Is the water still hot? - Sustainability and the thermal springs at Bath, England. *In*: HISCOCK, K. M. RIVETT, M. O. & DAVISON, R. M. (eds) *Sustainable Groundwater Development*. Geological Society, London, Special Publications, **193**, 15–40.

BRITISH GEOLOGICAL SURVEY & MOTT MACDONALD 1999. *Arsenic in groundwater in Bangladesh, Phase 1. Rapid Investigation*. Report to UK Department for International Development.

BRUNDTLAND, G. H. (Chair) 1987. *Our common future*. Oxford University Press, Oxford.

BURGESS, D. B. 2002. Groundwater resource management in eastern England: a quest for environmentally sustainable development. *In*: HISCOCK, K. M. RIVETT, M. O. & DAVISON, R. M. (eds) *Sustainable Groundwater Development*. Geological Society, London, Special Publications, **193**, 53–62.

BURGESS, W. G., BURREN, M., PERRIN, J., MATHER, S. E. & AHMED, K. M. 2002. Constraints on sustainable development of arsenic-bearing aquifers in southern Bangladesh. Part 1: A conceptual model of arsenic in the aquifer. *In*: HISCOCK, K. M. RIVETT, M. O. & DAVISON, R. M. (eds) *Sustainable Groundwater Development*. Geological Society, London, Special Publications, **193**, 145–163.

CONNELL, L. D. 2002. A simple analytical solution for unsaturated solute migration under dynamic water movement conditions and root zone effects. *In*: HISCOCK, K. M. RIVETT, M. O. & DAVISON, R. M. (eds) *Sustainable Groundwater Development*. Geological Society, London, Special Publications, **193**, 255–264.

CUTHBERT, M. O., BURGESS, W. G. & CONNELL, L. D. 2002. Constraints on sustainable development of arsenic-bearing aquifers in southern Bangladesh. Part 2: Preliminary models of arsenic variability in pumped groundwater. *In*: HISCOCK, K. M. RIVETT, M. O. & DAVISON, R. M. (eds) *Sustainable Groundwater Development*. Geological Society, London, Special Publications, **193**, 165–179.

DAVISON, R. M., PRABNARONG, P., WHITTAKER, J. J. & LERNER, D. N. 2002. A probabilistic management system to optimize the use of urban groundwater. *In*: HISCOCK, K. M. RIVETT, M. O. & DAVISON, R. M. (eds) *Sustainable Groundwater Development*. Geological Society, London, Special Publications, **193**, 265–276.

DENNEHY, K. F., LITKE, D. W. & MCMAHON, P. B. 2002. The High Plains Aquifer, USA: groundwater development and sustainability. *In*: HISCOCK, K. M. RIVETT, M. O. & DAVISON, R. M. (eds) *Sustainable Groundwater Development*. Geological Society, London, Special Publications, **193**, 99–119.

DETR 1999. *A better quality of life: a strategy for*

sustainable development for the United Kingdom. Department of the Environment, Transport and Regions, London.

DOWNING, R. A. 1993. Groundwater resources, their development and management in the UK: an historical perspective. *Quarterly Journal of Engineering Geology*, **26**, 335–358.

ENVIRONMENT AGENCY. 2001*a*. *Water resources for the future: A strategy for Anglian Region.* Environment Agency, Peterborough.

ENVIRONMENT AGENCY 2001*b*. *Managing water abstraction: The Catchment Abstraction Management Strategy process.* Environment Agency, Bristol.

EEA 1995. *Europe's Environment – The Dobris Assessment.* European Environment Agency, Copenhagen.

EEA 1999. *Sustainable water use in Europe. Part 1: Sectoral use of water.* Environmental assessment report no.1, European Environment Agency, Copenhagen.

ETC/IW 1998. *The Reporting Directive; Report on the returns for 1993 to 1995.* (France S) European Topic Centre on Inland Waters Report to DGXI, PO28/97/1.

EUROSTAT 1997. *Water abstractions in Europe. Internal working document, Water/97/5.* Eurostat, Luxembourg.

FOSTER, S. S. D. 2000*a*. Assessing and controlling the impacts of agriculture on groundwater from barley barons to beef bans. *Quarterly Journal of Engineering Geology and Hydrogeology*, **33**, 263–280.

FOSTER, S. 2000*b*. *Sustainable groundwater exploitation for agriculture; current issues and recent initiatives in the developing world.* Uso intensivo de las aguas subterráneas. Aspectos éticos, tecnológicos y económicos. **Series A, no. 6**. Papeles del Proyecto Aguas Subterráneas.

GLEICK, P. 2001. Making every drop count. *Scientific American.* **February 2001**, 28–33.

HUDSON, M. 2002. Groundwater sustainability and water resources planning for the East Midlands Resource Zone. *In*: HISCOCK, K. M. RIVETT, M. O. & DAVISON, R. M. (eds) *Sustainable Groundwater Development.* Geological Society, London, Special Publications, **193**, 91–98.

HULME, P., FLETCHER, S. & BROWN, L. 2002. Incorporation of groundwater modelling in the sustainable management of groundwater resources. *In*: HISCOCK, K. M. RIVETT, M. O. & DAVISON, R. M. (eds) *Sustainable Groundwater Development.* Geological Society, London, Special Publications, **193**, 83–90.

IPCC 1997. The regional impacts of climate change: an assessment of vulnerability. *In*: Watson, R.T., Zinyowera, M.C. & Moss, R.H. (eds) *A special report of the IPCC WGII.* Summary for Policymakers. Intergovernmental Panel on Climate Change. Cambridge University Press.

KIRK, S. & HERBERT, A. W. 2002. Assessing the impact of groundwater abstractions on river flows. *In*: HISCOCK, K. M. RIVETT, M. O. & DAVISON, R. M. (eds) *Sustainable Groundwater Development.* Geological Society, London, Special Publications, **193**, 211–233.

MACHELEIDT, W., NESTLER, W. & GRISCHEK, T. 2002. Determination of hydraulic boundary conditions for the interaction between surface water and groundwater. *In*: HISCOCK, K. M. RIVETT, M. O. & DAVISON, R. M. (eds) *Sustainable Groundwater Development.* Geological Society, London, Special Publications, **193**, 235–243.

MCCANN, C., MANN, A. C., MCCANN, D. M. & KELLAWAY, G. A. 2002. Geophysical investigations of the thermal springs of Bath, England. *In*: HISCOCK, K. M. RIVETT, M. O. & DAVISON, R. M. (eds) *Sustainable Groundwater Development.* Geological Society, London, Special Publications, **193**, 41–52.

O'RIORDAN, T. 2000. The sustainability debate. *In*: O'RIORDAN, T. (ed.) *Environmental Science for Environmental Management (2nd ed.).* Prentice Hall, Harlow, Essex, 29–62.

OECD 1997. *OECD Environmental Data Compendium 1997.* Organisation for Economic Co-operation & Development, Paris.

PRICE, M. 2002. Who needs sustainability? *In*: HISCOCK, K. M. RIVETT, M. O. & DAVISON, R. M. (eds) *Sustainable Groundwater Development.* Geological Society, London, Special Publications, **193**, 75–82.

RAJASOORIYAR, L., MATHAVAN, V., DHARMAGUNEWARDENE, H. A. & NANDAKUMAR, V. 2002. Groundwater quality in the Valigamam region of the Jaffna Peninsula, Sri Lanka. *In*: HISCOCK, K. M. RIVETT, M. O. & DAVISON, R. M. (eds) *Sustainable Groundwater Development.* Geological Society, London, Special Publications, **193**, 181–197.

RIVM & RIZA 1991. *Sustainable use of groundwater: problems and threats in the European Communities.* National Institute of Public Health and Environmental Protection (RIVM) and Institute for Inland Water Management and Waste Water Treatment (RIZA). Report no. 600025001. The Netherlands.

ROBINS, N. S., GRIFFITHS, K. J., MERRIN, P. D. & DARLING, W. G. 2002. Sustainable groundwater resources in a hard-rock island aquifer – the Channel Island of Guernsey. *In*: HISCOCK, K. M. RIVETT, M. O. & DAVISON, R. M. (eds) *Sustainable Groundwater Development.* Geological Society, London, Special Publications, **193**, 121–131.

RUSHTON, K. R. 2002. Will reductions in groundwater abstractions improve low river flows? *In*: HISCOCK, K. M. RIVETT, M. O. & DAVISON, R. M. (eds) *Sustainable Groundwater Development.* Geological Society, London, Special Publications, **193**, 199–210.

SCHOENHEINZ, D., GRISCHEK, T., WORCH, E., BEREZNOY, V., GUTKIN, I., SHEBESTA, A., HISCOCK, K., MACHELEIDT, W. & NESTLER, W. 2002. Groundwater pollution at the pulp and paper mill Sjasstroj near Lake Ladoga, Russia. *In*: HISCOCK, K. M. RIVETT, M. O. & DAVISON, R. M. (eds) *Sustainable Groundwater Development.* Geological Society, London, Special Publications, **193**, 277–291.

TWORT, A. C., LAW, F. M. & CROWLEY, F. W. 1985. *Water supply (3rd ed.).* Edward Arnold Ltd., London.

UNITED NATIONS DEVELOPMENT PROGRAMME. 1998. *Human development report 1998.* Oxford University

Press, Oxford.

WORRALL, F. 2002. Direct assessment of groundwater vulnerability from borehole observations. *In*: HISCOCK, K. M. RIVETT, M. O. & DAVISON, R. M. (eds) *Sustainable Groundwater Development*. Geological Society, London, Special Publications, **193**, 245–254.

YOUNG, M. E. 2002. Institutional development for sustainable groundwater management - an Arabian perspective *In*: HISCOCK, K. M. RIVETT, M. O. & DAVISON, R. M. (eds) *Sustainable Groundwater Development*. Geological Society, London, Special Publications, **193**, 63–74.

YOUNGER, P. L., TEUTSCH, G., CUSTODIO, E., ELLIOT, T., MANZANO, M. & SAUTER, M. 2002. Assessments of the sensitivity to climate change of flow and natural water quality in four major carbonate aquifers of Europe. *In*: HISCOCK, K. M. RIVETT, M. O. & DAVISON, R. M. (eds) *Sustainable Groundwater Development*. Geological Society, London, Special Publications, **193**, 303–323.

YUSOFF, I., HISCOCK, K. M. & CONWAY, D. 2002. Simulation of the impacts of climate change on groundwater resources in eastern England. *In*: HISCOCK, K. M. RIVETT, M. O. & DAVISON, R. M. (eds) *Sustainable Groundwater Development*. Geological Society, London, Special Publications, **193**, 325–344.

Is the water still hot? Sustainability and the thermal springs at Bath, England

T. C. ATKINSON[1,2] & R. M. DAVISON[2,3]

[1] *Groundwater Tracing Unit, Department of Geological Sciences, University College London, Gower Street, London WC1E 6BT, UK (e-mail: t.atkinson@ucl.ac.uk)*
[2] *School of Environmental Sciences, University of East Anglia, Norwich NR4 7TJ, UK*
[3] *Groundwater Protection and Restoration Group, Department of Civil and Structural Engineering, University of Sheffield, Sheffield S1 3JD, UK*

Abstract: The hot springs at Bath are the largest natural thermal source in Britain. Sustainable use of the waters for a spa requires maintenance of their temperature and flow rate. Together with smaller springs at Hotwells, Bristol, they form the outflow from a regional thermal aquifer that occurs where the Carboniferous Limestone is buried at depths >2.7 km in the Bristol–Bath structural basin. The aquifer is recharged via limestone outcrops forming the south and west portions of the basin rim. Current knowledge of the basin's structure is reviewed, and important uncertainties identified concerning the hydrogeological role of thrust faults which may cut the limestone at depth. A simple numerical model is used to determine the possible influence of thrusts upon groundwater flow within the thermal aquifer. Comparison of the modelled flow patterns with geochemical data and structure contours eliminates the hypothesis that thrusts completely disrupt the continuity of the aquifer. The most successful model is used to simulate the possible impact of dewatering by large quarries at the limestone outcrops north and south of Bath. Substantial reductions in modelled flow at Bath result from proposed dewatering in the eastern Mendips, although the steady-state approach adopted has severe limitations in that it does not take account of the incremental staging of actual dewatering, nor allow for partial restitution of groundwater levels. The geological uncertainties highlighted by the modelling could be addressed by future research into the effect of thrusts on the continuity of the Carboniferous Limestone. More refined modelling to predict the timing of possible impacts of quarry dewatering will require measurements of the storativity of the thermal aquifer.

The concept of *sustainability* in the exploitation of geological resources can be simply defined as the use of a resource in such a way that it is passed on to future generations in a usable form without loss of value. This begs the questions, usable for what and for whom? In the case of conflicts of use, how can priorities be decided and the resources managed accordingly? Most exploitation of groundwater is for water supplies and in this case the aim of sustainable development should be to manage resources so as to maintain the flux, amount of storage and quality of water in aquifers. This perspective underpins official strategies for resource development and groundwater protection in most nations. Developing an aquifer by drilling and abstracting from boreholes may involve trade-offs between storage, flux and quality but the long-term objective is usually clear – resources should be managed in such a way as to maintain quality and rates of abstraction for supplies indefinitely into the future. The impact of development on the environment apart from the aquifer itself is also an important criterion of sustainability. This was recognized long ago in the concept of *Safe Yield* (Todd 1976) which is defined as the

From: HISCOCK, K. M., RIVETT, M. O. & DAVISON, R. M. (eds) *Sustainable Groundwater Development*. Geological Society, London, Special Publications, **193**, 15–40. 0305-8719/02/$15.00

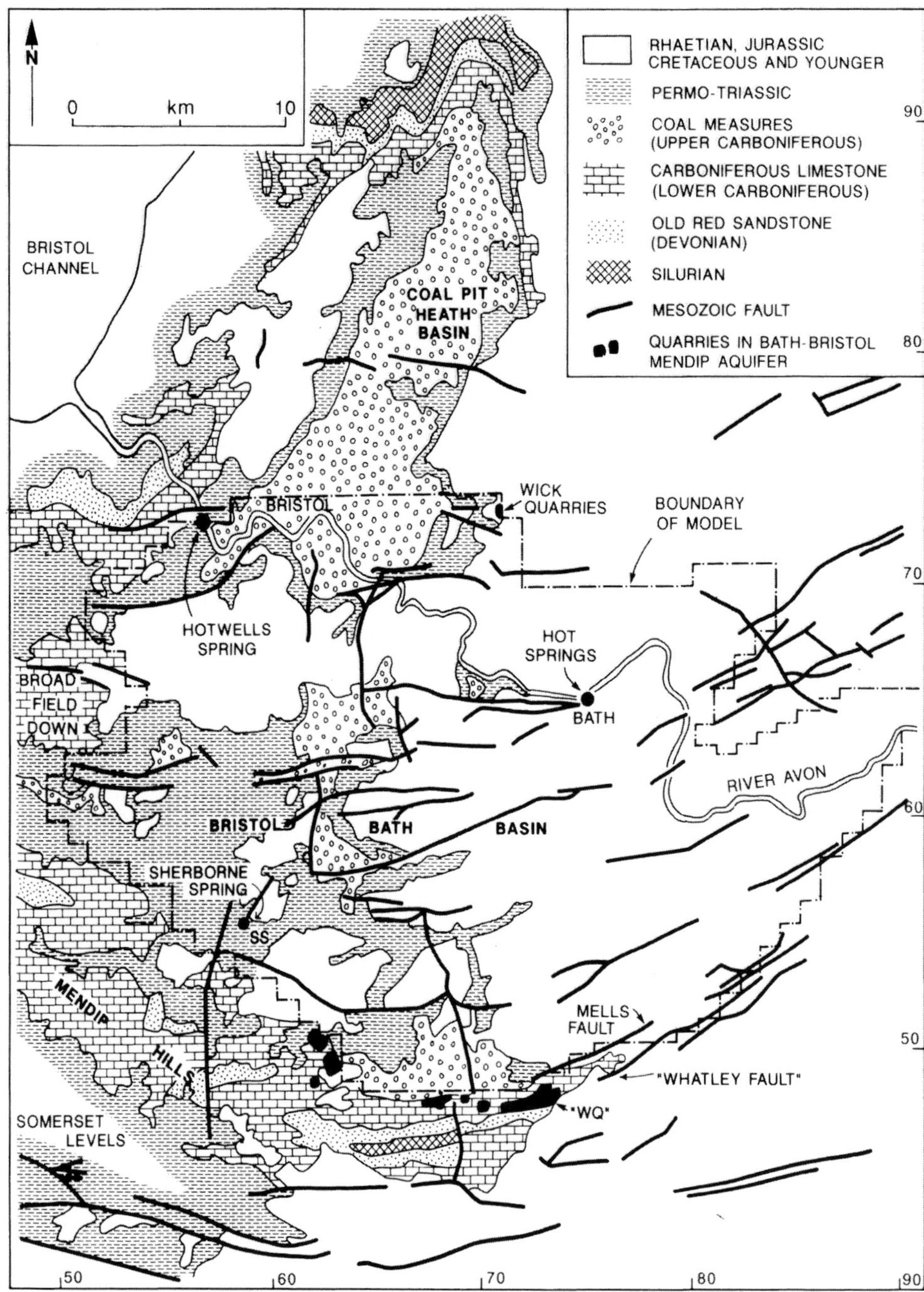

Fig. 1. (a) Geological map of the Bristol and Bath region, showing outcrops, Mesozoic faults, locations of principal springs and quarries, and the boundary of the groundwater model discussed in the text. Tick marks show UK National Grid.

PERIOD	FORMATION	LITHOLOGY	HYDROGEOLOGICAL UNIT	NATURE OF DOMINANT PERMEABILITY
RHAETIAN AND JURASSIC	7 GREAT OOLITE	LIMESTONE	AQUIFER	SECONDARY/KARSTIC
	6 FULLERS EARTH	CLAY	AQUICLUDE	
	5 INFERIOR OOLITE	LIMESTONE	AQUIFER	SECONDARY/KARSTIC
	4 MIDFORD SANDS	SANDS, SILTS	AQUIFER	PRIMARY
	3 LOWER LIAS	CLAY, THIN LIMESTONES	AQUICLUDE, AQUITARD	SECONDARY
PERMO-TRIASSIC	2 KEUPER MARL	CALCAREOUS MUDSTONES	AQUICLUDE	
	1 DOLOMITIC CONGLOMERATES	CALCAREOUS BRECCIA	AQUIFER (LOCAL)	SECONDARY/KARSTIC
UPPER CARBONIFEROUS	UPPER COAL SERIES	SHALES WITH SANDSTONES AND COALS	AQUITARD	SECONDARY (IN SANDSTONES)
	PENNANT SERIES	SANDSTONES	AQUIFER	SECONDARY
	LOWER COAL SERIES	SHALES WITH SANDSTONES AND COALS	AQUITARD	SECONDARY
	QUARTZITIC SANDSTONE GROUP	SANDSTONES	AQUIFER (LOCAL)	SECONDARY
LOWER CARBONIFEROUS	CARBONIFEROUS LIMESTONE	MASSIVE AND WELL-BEDDED LIMESTONES	AQUIFER	SECONDARY, KARSTIC AT OUTCROP
	LOWER LIMESTONE SHALES	SHALES, THIN LIMESTONES	AQUICLUDE, AQUITARD	SECONDARY
DEVONIAN	OLD RED SANDSTONE	SHALES, SANDSTONES, CONGLOMERATES	AQUIFER	SECONDARY
SILURIAN	—	SHALES, SANDSTONES, VOLCANICS	AQUITARD	SECONDARY

1000 METRES

Fig. 1. (b) The stratigraphic succession and principal hydrogeological units in the Bristol and Bath region.

long-term yield that may be abstracted from an aquifer system without adverse effects. Large-scale abstractions usually cause reductions in natural flows of springs and watercourses and limiting such adverse effects to an acceptable level is frequently used as a criterion for deciding upon a 'safe' or 'sustainable' yield.

Thermal spas are a long-established form of groundwater use which, to be sustainable, require the conservation of the discharge rates, temperatures and mineral qualities of their hot springs. A considerable part of the appeal of many spas rests in the springs' status as *natural* phenomena, and maintaining this requires conservation of the thermal aquifer system feeding them. Other forms of groundwater abstraction should be limited to levels that can be demonstrated not to impact upon the thermal springs. In practice this requires scientific understanding and a legal framework for licensing new abstractions. The remainder of this paper examines a case in point, the hot springs at Bath which are Britain's most important thermal source (Barker *et al.* 2000). The springs are known to originate from the Carboniferous Limestone beneath Bath but there is considerable uncertainty as to the extent and structure of the thermal aquifer as a whole. The aim of the paper is to show how simple groundwater flow models can be used to explore the areas of greatest uncertainty, to identify possible threats to the hot springs from dewatering by deep quarries, and to suggest areas of future research that would be most useful for assuring conservation of the Bath springs and sustainability of the spa.

The hot springs at Bath and Bristol

The hot springs at Bath (Fig. 1a) have been in use for at least the past 2000 years. They have temperatures of 44–47°C, with an apparently constant flow of $15 \, l \, s^{-1}$. Three sources, the King's, Cross and Hetling Springs, issue from what were probably once pools on a floodplain terrace of the River Avon, in the centre of Bath. A succession of buildings have been constructed over them or on their margins, beginning with the Roman Baths and temple of the first century A.D. The thermal waters continued to be used until 1978 when the discovery of *Naegleria*, a pathogenic amoeba, in the water and sediments of the King's Spring led to closure of the spa. Extensive geological investigations took place in the following decade, with numerous boreholes being drilled including an inclined bore which intercepts the fissure up which the thermal water rises to the King's Spring. This inclined bore is now the principal source of spa waters which are free of the amoeba. Redevelopment of the spa is in progress at the time of writing.

A second group of thermal springs occurs near Bristol (Fig. 1a). The Hotwells Spring is the warmest (24°C) of a group of small thermal springs located in or near the Avon Gorge, where the River Avon cuts through a ridge of Carboniferous Limestone on the north-west side of Bristol. The history of the Bath and Hotwells springs, detailed accounts of the geological investigations of the 1980s, and much other information are summarized in the book *Hot Springs of Bath* (Kellaway 1991*a*).

The origin of the thermal water: the 'Mendips Model'

The origins of the thermal waters have been the subject of interest and speculation for over two centuries. Early views are summarized by Kellaway (1991*b*). There have been two important phases of modern investigation. One was concentrated in the immediate vicinity of the Bath springs and has already been mentioned as arising from the finding of *Naegleria* and its effect on the spa (Kellaway 1991*c*). An earlier phase was reported by Burgess *et al.* (1980) and Andrews *et al.* (1982) who examined the geochemistry of the thermal waters and other groundwaters in the region, demonstrating that they were all of meteoric origin. The Bath waters were composed of a mixture of several components, the major component having a Ca–Na–Mg, SO_4–Cl–HCO_3 chemistry. The combination of isotopic compositions in the SO_4 and HCO_3 unambiguously indicated a source in the Carboniferous Limestone. The boreholes drilled later in the centre of Bath confirmed that the hot water does indeed rise from this aquifer (Kellaway 1991*c*; Stanton 1991). The radiocarbon age of the King's Spring water is 2000–3000 years, but a fluctuating, minor component of modern water, less than a few decades old, is indicated by traces of tritium that were detected from time to time in 1977–1978. The silica content indicated that the thermal water had attained a maximum temperature of either 64 or 96°C, depending on whether chalcedony or quartz was the solubility controlling phase. Using an estimated geothermal gradient of $20°C \, km^{-1}$, Andrews *et al.* (1982) calculated a circulation depth for the water of between 2.7 and 4.3 km from these temperatures. Burgess *et al.* (1980) estimated the natural groundwater head beneath central Bath to be 40 m OD but more recent investigation has shown that it is less, 27–28 m

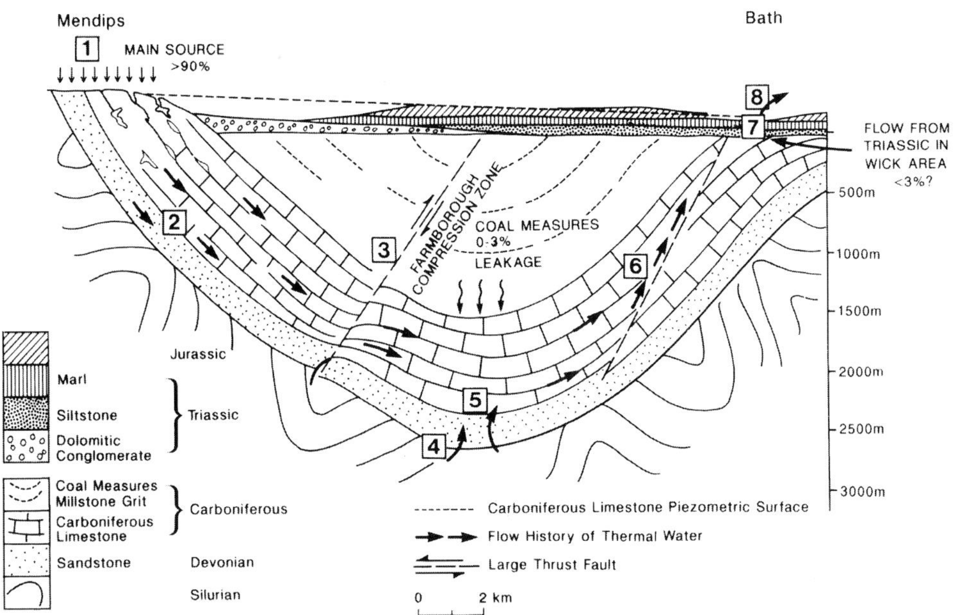

Fig. 2. Conceptual model for the origin of the Bath Hot Springs, modified after Andrews *et al.* (1982). 1, Recharge (9–10°C) at the Carboniferous Limestone/Devonian sandstone outcrop on the Mendip Hills; 2, flow down dip and down gradient; 3, possible downward leakage from Upper Carboniferous Coal Measures; 4, possible leakage of very old ^{4}He-bearing groundwater from Devonian sandstone and Lower Palaeozoics; 5, storage and chemical equilibration within the Carboniferous Limestone at 64–96°C; 6, rapid ascent, probably along Variscan thrust faults re-activated by Mesozoic extension; 7, lateral spread of thermal water into Permo-Triassic strata at Bath; 8, discharge of the thermal springs at Bath (46.5°C).

OD, compared with normal spring pool levels at *c.* 20 m OD. Burgess reasoned that the source area must be in a Carboniferous Limestone outcrop where groundwater levels were sufficiently in excess of the hot springs' head to drive recharge along a permeable pathway that took it to depths great enough to heat it, and from which it could rise up suitable structures to the springs themselves. The only areas that meet these twin criteria, according to Burgess *et al.* (1980, p.12), are Carboniferous Limestone outcrops in the Mendip Hills, 15–20 km S and SW of Bath, and Broadfield Down 20 km to the west. These areas lie respectively south and west of a structural basin containing the North Somerset coalfield (Figs 1a and 3a), beneath which the Carboniferous Limestone lies at depths exceeding 2700 m at the basin's centre, sufficient for groundwater to acquire the temperature indicated by its silica content. These considerations make the Mendips the most probable recharge area for the thermal water. This view was summarized by Andrews *et al.* (1982) as the 'Mendips Model' for the origin of the Bath springs (Fig. 2).

The Hotwells Spring was also studied by Burgess *et al.* (1980) and Andrews *et al.* (1982). It is cooler and contains similar tritium levels to recent groundwaters in the region, indicating a large component of modern recharge. The major ion chemistry suggests that it is a mixture of 'thermal water' similar to the King's Spring, and cold, 'normal' groundwater from the Carboniferous Limestone, in proportions of 1 : 2.3. Re-examination of the data in Burgess *et al.* (1980) indicates that for individual conservative species such as Na, SO_4 and Cl, the ratio varies within the range 2.0–2.6. The silica content of the Hotwells water indicates that the 'thermal' component has attained a maximum temperature of between 49 and 72°C.

Although the Mendip Hills are the most likely recharge area for both groups of thermal springs, other sources have been proposed, as summarized below.

(1) Burgess *et al.* (1991) suggested that part of the flow may be derived by vertical down-

wards leakage from Coal Measures in the basin to the Carboniferous Limestone beneath, although Kellaway (1994) dissents strongly from this view. Careful comparison of Li content and the Br/Cl ratio between Bath water and coalfield brines (McMurtrie 1886; Edmunds 1975) suggests that this source is unlikely to account for more than 0.4–3% of the total flow. Therefore, this is essentially a minor modification of the 'Mendips Model' (Fig. 2).

(2) Adequate natural heads formerly existed in the small outcrops of Carboniferous Limestone and Triassic Dolomitic Conglomerate around Wick, 9 km NNW of Bath, although it is doubtful whether there is a route from these to the Hot Springs that is deep enough to heat the groundwater flow sufficiently. The natural heads have since been reduced by quarry dewatering (see below). This area might be the source of the fluctuating minor component of flow detected by Burgess *et al.* (1980). Evaporites including celestite ($SrSO_4$) occur in the Keuper Marl facies of the Triassic and the high Sr content of the thermal water and its $^{87}Sr/^{86}Sr$ ratio may derive from this source (Wood & Shaw 1976; Burgess *et al.* 1980; Edmunds & Miles 1991). However, they are also consistent with derivation of Sr from post-Triassic mineralization in the Carboniferous Limestone (Burgess *et al.* 1980). Other geochemical indications of a fluctuating component in the thermal waters are variations in the Eh and in redox-sensitive species such as Fe and U (Andrews 1991; Edmunds & Miles 1991). Burgess *et al.* (1980) discuss the uranium geochemistry in detail, showing that the fluctuating component would amount to between 2 and 4 % of the total flow of the thermal water, if groundwaters in either Triassic or Liassic strata are its source.

(3) Broadfield Down (Fig. 1a) possesses the necessary water table elevation and may be connected to Bath via the Carboniferous Limestone beneath the coalfield.

(4) Kellaway has argued in a series of articles that a part of the thermal waters may be derived from the north, flowing to the Bath and Hotwells area via deep-seated faults in the Lower Palaeozoic 'basement' (Kellaway 1991, p. 21; Kellaway 1991*d*, 1993, 1994). A different northern source was proposed by Wilcock & Lowe (1999), but is convincingly dismissed by Stanton (2000).

Methodology

Future protection of the Bath springs requires a more detailed understanding of the thermal aquifer system as a whole than is apparent from Figure 2. The contribution of this paper is structured in four parts. The first reviews the geological and geophysical evidence that is relevant to the 'Mendips Model', paying particular attention to uncertainty regarding geological structures at depth. In the second part a series of simple numerical models of the thermal aquifer are formulated. These form a sequence of hypotheses as to what the flow system would be like given different assumptions about the uncertain aspects of the geology. This leads to the third part which discriminates between the models, eliminating some and retaining others by comparing the actual chemistry of the Hotwells spring with the chemical make-up implied by each. The final section of the paper examines the implications of the 'allowable' models for protection of the Hot Springs from disturbance by future groundwater withdrawal.

Stratigraphy, aquifer units and geological structure

Stratigraphy and aquifer units

Figure 1a shows the simplified outcrop geology of the region. The main stratigraphic units are listed in Figure 1b and divided into aquifers, aquitards and aquicludes. The succession can be divided into two groups of strata, separated by a regional unconformity beneath which the Palaeozoic rocks are deformed by folds and thrust faults formed during the late Carboniferous Variscan orogeny. Immediately above the unconformity are Permo-Triassic conglomerates, breccias and mudrocks formed under an arid or semi-arid climate, and representing the alluvial fans, screes, ephemeral rivers, lake beds and playa flats which made up the post-Variscan landscape. This sub-aerial sequence was followed in the Lower Jurassic by a succession of marine clays and thin limestones which show progressive overstep onto higher ground as the area underwent subsidence and marine transgression. By the time of deposition of the Inferior Oolite, the highest parts of the area were submerged and limestones were being deposited over the whole region. Further widespread marine clays and limestones were deposited in the Middle Jurassic. Erosion during the Tertiary and Quaternary has stripped the Jurassic rocks off almost the whole of the western part of the area shown in Figure 1a, exposing the Permo-Triassic and older strata that were formerly concealed, except where isolated outliers remain. In the eastern part of Figure 1a, however, the cover of Jurassic strata remains continuous and forms the high ground of the hills around Bath and the Cotswold escarpment east of the Coal Pit Heath basin.

The pre-Variscan strata contain a major aquifer, the Carboniferous Limestone. The Old Red Sandstone and Pennant Sandstone also form aquifers, although they are less permeable than the limestone at outcrop, and possibly at depth also. These aquifer units are separated by the aquitards of the Lower Limestone Shales and Lower Coal Measures, with the Upper Coal Measures an aquitard at the top of the sequence. The Permo-Triassic strata immediately above the unconformity show great lateral variability in facies and consequently in hydrogeological character. The well-jointed, carbonate breccias and conglomerates, known as Dolomitic Conglomerate, are an aquifer that forms a continuous hydrogeological unit with the Carboniferous Limestone where the two are in contact. In contrast, the Keuper Marl is generally an aquiclude, although thin limestones or sandstones may allow groundwater circulation locally. Near the base of the Jurassic, the interbedded limestones and shales of the Blue Lias form a local aquifer containing thermal water beneath Bath. Above this is the aquiclude of the Lias Clay.

Geological structure

Variscan structures. The Palaeozoic strata underwent progressive, compressive deformation over a long period during the Variscan orogeny, the greatest rates of deformation occurring towards the end of the Carboniferous. The Bristol–Bath region lies at the outer edge of the arcuate Variscan fold belt (Ries & Shackleton 1976; Shackleton *et al.* 1982) with a tectonic style dominated by anticlinal folds, overturned to the north in many places, with tight synclines between them. The Mendip Hills (Fig. 1a) consist of a belt of such folds arranged en echelon, while others form Broadfield Down and the limestone ridges north and west of Bristol. The British Geological Survey's maps show frequent thrust faults with northward tectonic transport, but as mapped these appear subsidiary in importance to the folds. This view of the structure was summarized by cross-sections on the published maps and in the accompanying memoirs (Kellaway & Welch 1948; Green & Welch 1965; Whittaker & Green 1983), and also in a review by Kellaway & Hancock (1983). Subsequently, the structural style of the region has been re-interpreted by Williams & Chapman (1986) as a thin-skinned foreland thrust belt in which thrust sheets are stacked in a piggyback fashion. The principal thrusts are thought to be laterally continuous for tens of kilometres along strike. The important aspects of this revised interpretation are summarized below.

Figure 3a is a map of the geological structure at the level of the post-Variscan unconformity, showing the outcrop pattern as it is likely to appear if the Jurassic and Permo-Triassic cover was stripped off. The interpretation of the tectonic features is aided by two balanced cross-sections (Fig. 3b), redrawn from Williams & Chapman (1986). Six major thrusts (or groups of closely related thrusts) have cut the strata and are labelled T1 to T6. The earliest,T1, lies in the south. When T2 later developed beneath and to the north of it, T1 became folded in its hanging wall. Development of further, younger thrusts proceeded northwards in the same successive piggyback fashion, with the earlier thrusts sometimes acting as roof-thrusts to the latter. The Mendip folds are variously interpreted as tip-line folds of blind minor thrusts or as ramp folds. The Kingswood Anticline, which runs west–east between Bristol and Wick (Fig. 1a) is a tip-line fold of T4 through which the thrust has later propagated northwards. North of T4, thrusts T5 and T6 strike obliquely south-west to north-east and probably have a significant sinistral strike-slip component.

The thin-skinned thrust-tectonic model elegantly synthesizes the structures of the area and explains the previously problematic occurrence of several klippen north of the Mendips in which overturned Carboniferous Limestone rests on Coal Measures. The total crustal shortening is 38% for both of the balanced cross-sections of Figure 3b, but because of the widening of the thrust belt the absolute displacements increase eastwards, from 19 km in the western cross-section to 27 km in the eastern one (Williams & Chapman 1986).

The Kingswood Anticline and thrust T4 form the northern boundary of a broad structural basin, the Bristol–Bath basin, which is picked out by the roughly circular sub-crop of the Pennant Series (Fig. 3a). The southern boundary of the basin is the Mendip Hills. Its western edge is formed by the back-thrust anticlinal structure of Broadfield Down. In the east the margin is not well known as the Variscan structures are largely buried beneath the Mesozoic (Fig. 1a) and there is very little subsurface information from mining. The reconstruction of sub-crop in

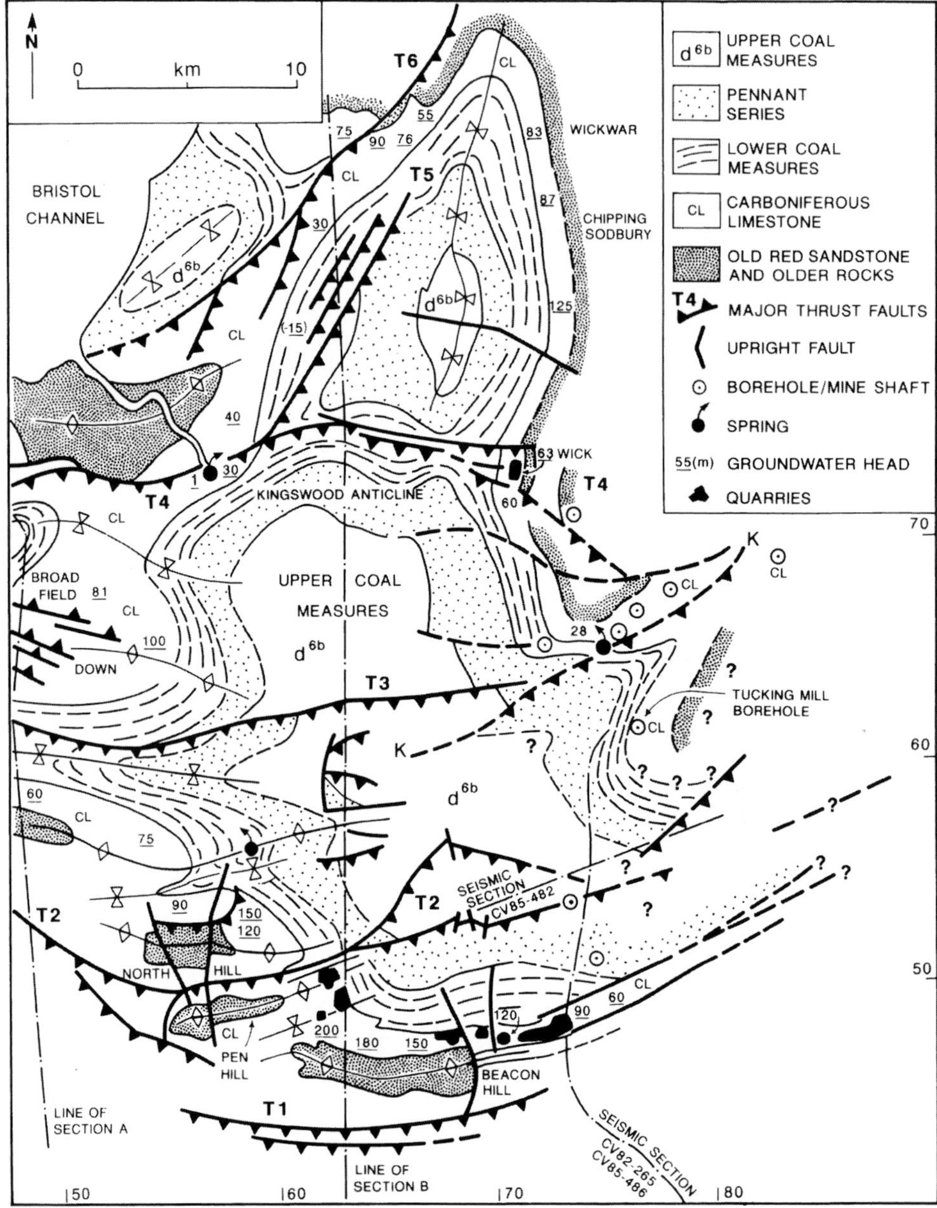

Fig. 3. (a) Geological structure of the Bristol and Bath region at the level of the post-Variscan unconformity, showing major thrust structures T1–T6. Also shown are the locations of boreholes into Palaeozoic strata, thermal springs, and generalized groundwater heads. (After Burgess *et al.* 1980; Andrews *et al.* 1982; Williams & Chapman 1986).

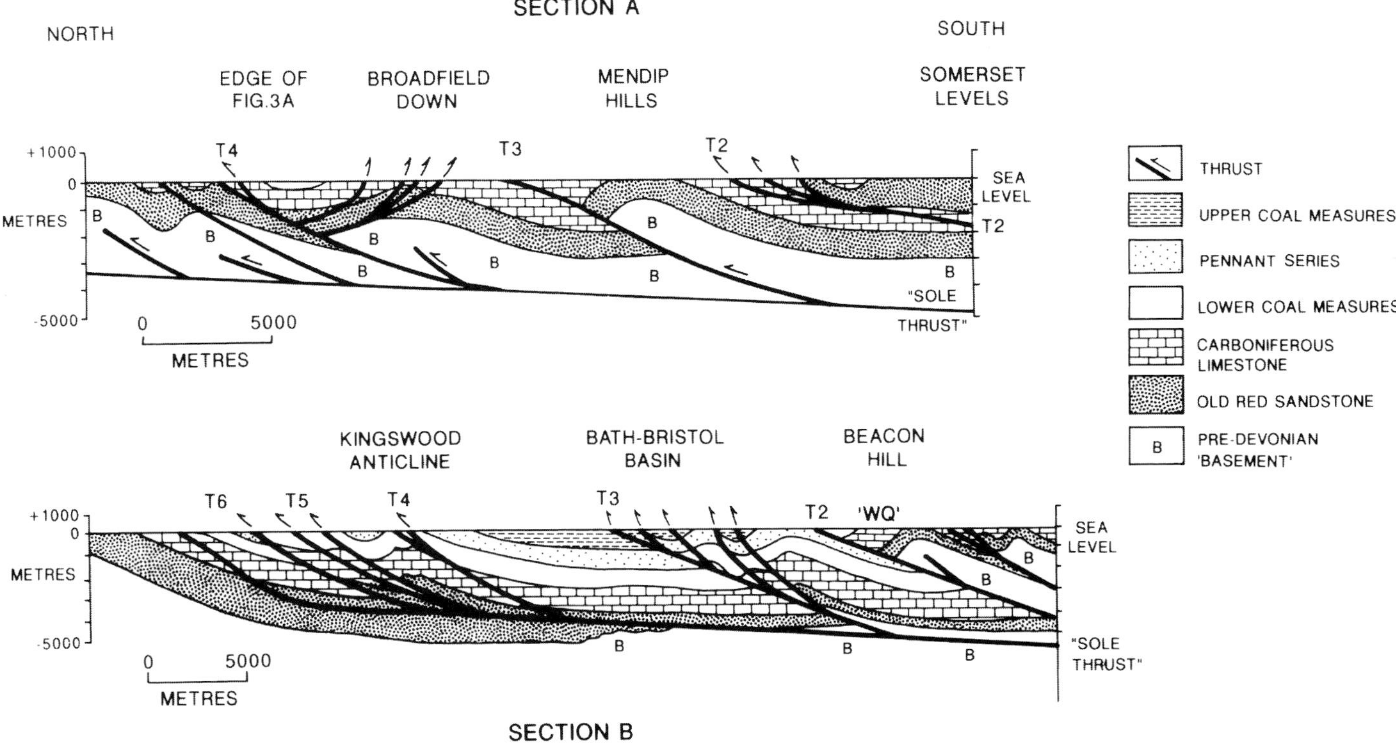

Fig. 3. (b) Cross-sections along lines A and B in Figure 3a, redrawn from balanced tectonic cross sections in Williams & Chapman (1986).

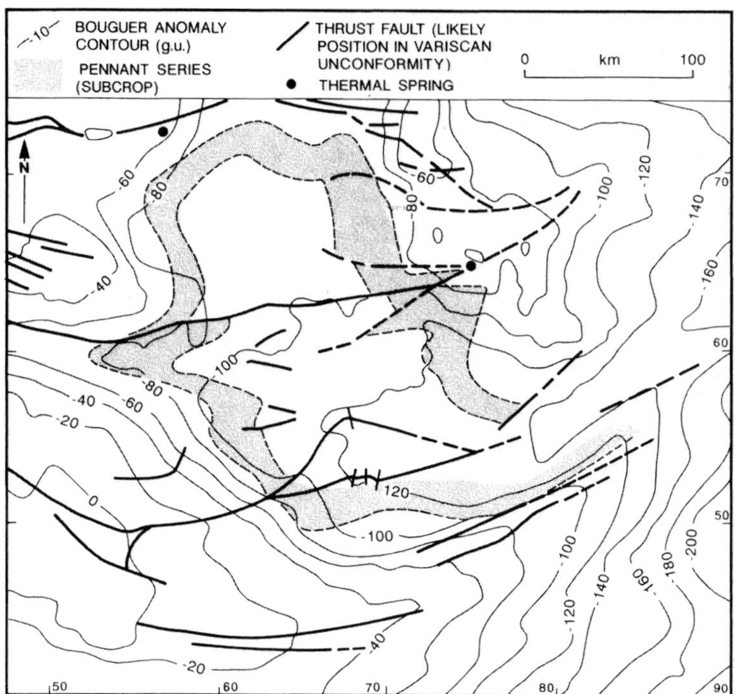

Fig. 4. Comparison of gravity anomaly contours (after Ates & Kearey 1993) with geological structures in the Bristol–Bath basin. Tick marks show UK National Grid.

this portion of Figure 3a follows Kellaway (1993) who relies primarily on sparse information from the few boreholes that penetrate through the post-Variscan unconformity.

The general form of the Bristol–Bath basin is reflected in gravitational and aeromagnetic data presented by Ates & Kearey (1993). The basin appears as an area of moderately negative magnetic anomaly and more pronounced negative Bouguer anomaly (Fig. 4). Ates & Kearey (1993) present an east–west seismic reflection profile (CV85-482) whose trace is indicated in Figure 3a. The top of the profile coincides with the tip of thrust T2 and the interpreted section is therefore through the footwall of this thrust. It shows an open synclinal structure with the base of the Coal Measures (identified from their position at outcrop in the section) at a depth of about 2.5 km between National Grid easting 370 and 375. Below this depth, however, the seismic reflection patterns are vague and indistinct, and there is no seismic indication of the position of the T3 thrust. In the eastern part of the profile, rather clear reflectors indicate the Palaeozoic strata rising eastwards, indicating that the top of the Carboniferous Limestone meets the sub-Permo-Triassic unconformity at a position a little east of gridline 380, where it forms the concealed rim of the coalfield basin.

Figure 5 shows structure contours drawn on the top of the Carboniferous Limestone in the western part of the basin. These are based on the conventional geological cross-sections drawn by the British Geological Survey on published maps (Sheets 263, 264, 280, 281 and Bristol District Special Sheet), plus six similar cross-sections drawn by the authors. The eastern of the two balanced sections produced by Williams & Chapman (1986) (Fig. 3b) crosses the contoured area. Though not used in constructing the contours, it shows excellent agreement in the horizontal position of thrust T3 at the level at which the top of the limestone in the footwall is truncated. Agreement with the contoured depth to the limestone in the centre of the basin is reasonable, the contour construction suggesting a maximum depth of more than 2000 m below OD in the line of section, whereas Williams & Chapman (1986) indicated the top of the limestone at 2500 m in the flat-lying part of the

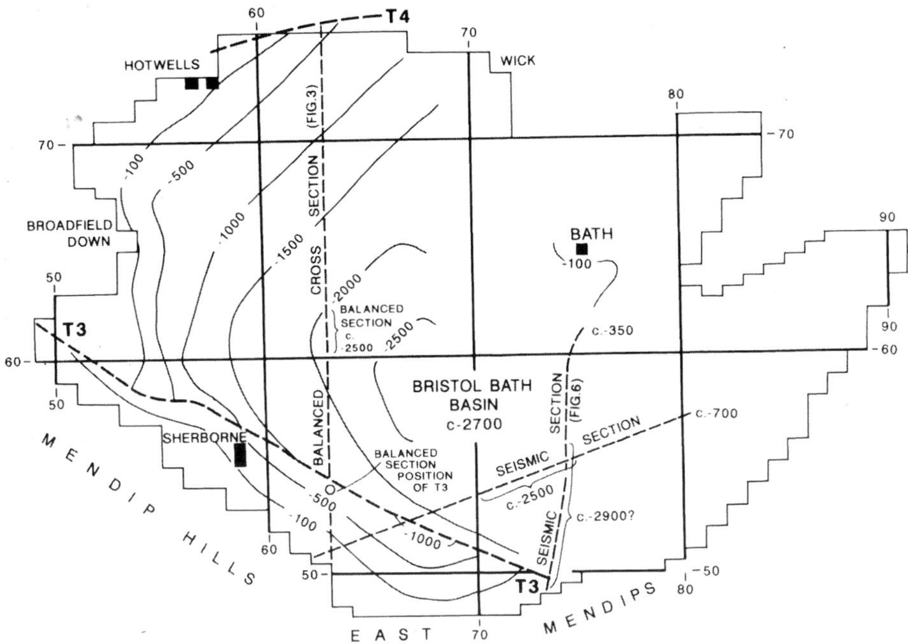

Fig. 5. Structure contours on the top surface of the Carboniferous Limestone superimposed on the UK National Grid and the outline of the groundwater model described in the text. Contours (in m below OD) are based on cross-sections drawn from Sheets 263, 264, 280, 281 and the Bristol District Special Sheet, all published by the British Geological Survey. Also shown are the lines of seismic sections discussed in the text (see Fig. 6) and the balanced tectonic section B (see Fig. 3b). The estimated position shown for thrust T3 is where it meets the top of the limestone in the footwall.

structural slice between thrusts T3 and T4. Bearing in mind that both approaches represent constrained guesses about the true situation, and that each embodies different starting assumptions, this degree of concord suggests that the depths to limestone indicated in Figure 5 are broadly correct. There is also broad agreement between the pattern of the structure contours and that of gravity contours (Fig. 4), and good agreement with the depths to limestone inferred from seismic sections (Ates & Kearey 1993).

The basin is traversed by two of the major thrust faults shown in Figure 3a. T3 crosses the middle of the basin and corresponds to the feature known from coal mining as the Farmborough Fault Zone or Compression Zone. Coal in this zone was impossible to win due to the shattered and discontinuous nature of the strata. Kellaway (1993) considered the Farmborough zone to run very close to Bath Hot Springs, but its precise position is necessarily speculative (Fig. 3a). Chadwick *et al.* (1983, p. 899) described a north-south seismic reflection profile some 20 km E of Bath, in which they recognized folded Palaeozoic strata riding on a major thrust which they tentatively identified as the along-strike continuation of T3. Williams & Brooks (1985) reinterpreted this section using volume-balancing techniques, but did not alter the significance of T3 which forms the frontal thrust of the Variscan orogen in the area 20 km E of Bath. In the area considered in the present paper, however, other thrusts, T4–T6, lie to the north of T3.

Thrust T2 can be traced in outcrop from Cheddar Gorge around the southern edge of the North Hill pericline and across the basin, where it has been mapped in coal mines (British Geological Survey Sheets 280 & 281). It divides into two faults, named the Radstock Slide Fault and the Southern Overthrust, which converge eastwards and probably reunite beneath the Mesozoic cover.

Post-Variscan structures. The geological map (Fig. 1a) shows two trends among the numerous faults which affect post-Carboniferous strata. Many trend almost E–W in the basin but swing to an E–NE direction in the eastern part of the area. These are almost all normal faults, and reflect crustal extension at some time since the Jurassic. The second trend is N–S. Faults with this trend often offset members of the E–W group, but not invariably. Both trends mirror the patterns of faults found in the Variscan structures at outcrop, the E–W group lying parallel to the major thrusts and the N–S group having their Variscan counterpart in upright faults which cut the thrust planes in the Mendips and in the Coal Pit Heath basin. Chadwick *et al.* (1983) demonstrate that the Vale of Pewsey and Mere Faults, members of the E–W group located some 20 km E of Bath, formed as growth faults during Triassic and Jurassic times. These normal faults appear to be listric to southwards-dipping Variscan thrust zones, reactivating them in an extensional regime. Similar extensional reactivation of Variscan thrusts during the Mesozoic was demonstrated by Brooks *et al.* (1988) in the Bristol Channel.

Uncertainties regarding geological structure and their relevance to the hot springs

The Mendips Model for the origin of the Bath and Hotwells thermal springs relies upon three conditions being fulfilled. One is that a recharge area exists where the heads are high enough to drive flow towards the springs; the second is that the geological structure provides a continuously permeable pathway that conducts the recharge waters to depths great enough for them to be heated; and the third is that a pathway exists along which they may rise to the surface rapidly enough to avoid cooling to ambient temperatures. The general structure of the basin and the outcrops of Carboniferous Limestone around its rim appear to fulfil these conditions, but unfortunately our knowledge is not detailed enough to be completely sure on all three points. The ambiguities will now be described by taking in turn the recharge area, pathway at depth, and rising pathways to exit at the springs. Figure 3a includes generalized values of head in the Carboniferous Limestone of the Mendip Hills (Stanton 1977; Harrison *et al.* 1992) and where known at other locations in Palaeozoic rocks. The highest heads are in the central Mendips but Broadfield Down and the Coal Measures of the Kingswood Anticline also possess values in excess of that at Bath. All three areas lie around the rim of the basin. It is unlikely that the Kingswood Anticline is the major source of the thermal springs, since the waters would have to move downwards through the thick aquitard of the Lower Coal Series. However, the Mendip Hills and Broadfield Down could both potentially be connected to Bath and Hotwells via the deeply buried Carboniferous Limestone, which is also the local source aquifer for the hot springs. The contours on the topmost beds of limestone (Fig. 5) indicate that the deepest pathways lie more than 2700 m below sea level. Although the geothermal gradient in the region is poorly known (see Burgess *et al.* 1980; Kellaway 1991*b* for discussion), an average figure of 20°C km^{-1} combined with a recharge temperature of 10°C would imply that circulation to this depth would produce maximum temperatures of at least 64°C, some 20°C hotter than the Bath springs themselves.

With recharge and temperature thus accounted for, the viability of the Mendips Model in general and the corollary question of whether the Mendips or Broadfield Down is more important as a particular source of recharge both rest on whether or not there is a continuous permeable pathway in the Carboniferous Limestone beneath the coalfield. Unfortunately, *direct* observations of the geological structures extend only to *c.* –500 m OD in the deepest mines. Deeper structures must be inferred from circumstantial evidence, as in Figures 3b & 5. The crucial issue is whether the major thrust faults which are such a prominent part of the thin-skinned tectonic model of the Variscan structures might cut off the continuity of the limestone at depth. It is almost impossible to decide this unequivocally using the geological evidence presently available. In the two balanced N–S cross-sections (Fig. 3b) there are limestone-to-limestone contacts across T3, but the eastern section shows the limestone as completely severed by T2. It is cut off north of Beacon Hill in the hanging wall (Fig. 3b), but in the footwall it is shown as extending as a ‘flat’ some 7–8 km further south. In the thrust plane of T2, the gap between lowest limestone beds in the hanging wall and highest in the footwall exceeds 4500 m. It is worth noting that there is no direct evidence for the ‘flat’ in the footwall. It is inferred to exist, being implied by the requirement of conservation of volume in constructing a balanced cross-section. However, a two-dimensional section must be drawn in the true direction of tectonic transport to be accurately balanced, and of course this is not known in detail. The eastern section of Williams & Chapman (1986) (Fig. 3b) suggests, but does not prove, a cut-off of the limestone north of Beacon Hill. Since the

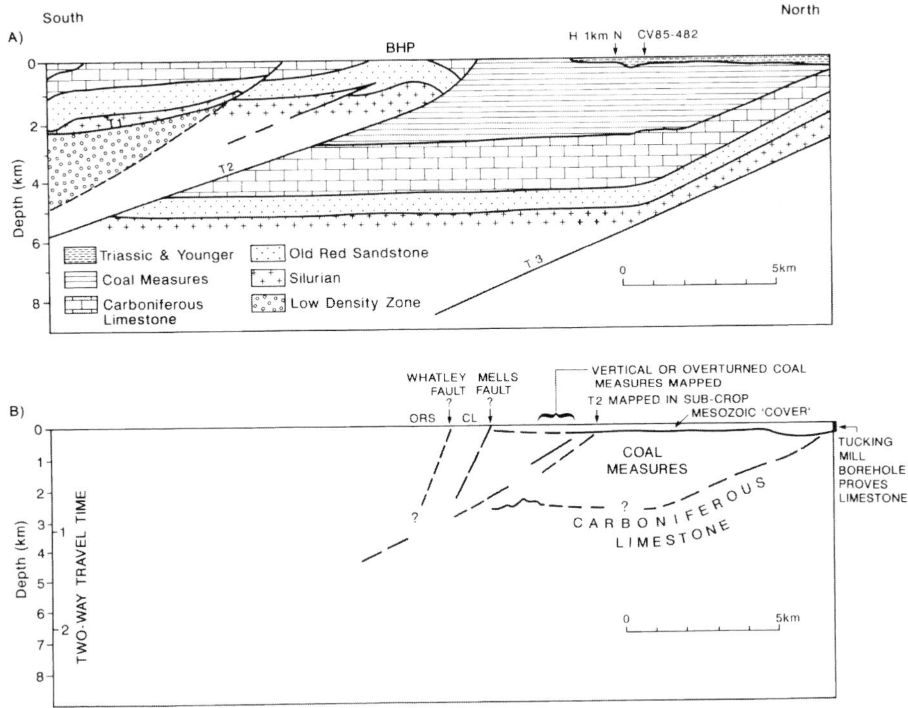

Fig. 6. Interpretations of a south–north seismic section through the eastern Mendips (based on seismic lines CV 82-265 and CV 85-486. Original black and white seismic image is figured by Ates & Kearey (1993). **(a)** Interpretation redrawn from Ates & Kearey (1993). **(b)** Minimalist re-interpretation of the northern part of the section – for discussion see text.

outcrop trace of T2 crosses the summit of the Mendips between the North and Pen Hill periclines (fig. 3a), only the hydraulic continuity of the eastern Mendips with the rest of the basin is likely to be affected.

A cut-off of the limestone by thrust T3 would also have an important effect on flow paths, since this thrust zone crosses the centre of the basin. Williams & Chapman (1986) identify T3 with the Farmborough Fault Zone, and Kellaway (1993) considered the eastwards extension of this feature to pass close to the Bath Springs. However, this location is somewhat speculative. All the cross-sections produced by Williams & Chapman (1986, including the E–W section in their fig. 8) indicate that limestone-to-limestone contact exists across T3, but as with the suggestion of a cut-off across T2, this is not completely proven.

Returning to T2 and the eastern Mendips, Ates & Kearey (1993) have presented an interpretation of seismic reflection lines CV82-265 and CV85-486, which are aligned N–S and perpendicular to the strike of T2. Their interpreted section is reproduced here as Figure 6a and shows a thrust labelled T2 cutting off the Carboniferous Limestone in a manner similar to the Williams & Chapman section (compare Figs 3b & 6a), with the base of the Coal Measures lying at *c.* 3 km depth in the footwall. This interpretation presents problems of compatibility with geological evidence. It does not agree well with the structure of the upper few hundred metres which is known moderately well from coal mines, and is shown by cross- sections and sub-crop patterns on maps published by the British Geological Survey (Sheets 280 & 281). These clearly show the Southern Overthrust, i.e. the mapped name for T2, at a position much further north than the T2 of Ates & Kearey's (1993) interpretation. We have examined the original large-scale seismic profiles and, on the basis of these, consider that the structure in the region of the Beacon Hill pericline is less intelligible and perhaps more complex than Ates & Kearey's (1993) interpreted section suggests. Figure 6b is a minimalist alternative interpretation of the part of the section north of the main Mendip ridge (Figs 1a, 3a and 4), based on recognition of bundles of strong reflectors, and on identifying discontinuities in their pattern at relatively large scales. Figure 6b shows the same sub-Mesozoic unconformity as Figure 6a but continues it somewhat further south where it meets a more steeply dipping, less obvious discontinuity which may be cautiously identified with the Mells Fault (Fig. 1a). A few hundred metres from the north end of the section lies the Tucking Mill borehole (Fig. 3a; not available to Ates & Kearey 1993), in which strata near the top of the Carboniferous Limestone were encountered below Dolomitic Conglomerate and Jurassic (Kellaway 1993), allowing tentative identification of the top of the limestone in the seismic section. Above it lies a bundle of strong parallel reflectors which probably represent the Lower Coal Measures and which can be traced southwards for 8 km where they lie between 0.6 and 1 s two-way time. Further south between the same depths is an area of irregular, grainy reflectors which is separated from the sub-horizontal reflectors of the 'Lower Coal Measures' by a slanting discontinuity that can be traced both upwards and downwards. If projected to the near-surface unconformity this discontinuity is in approximately the mapped position of T2, i.e. the Southern Overthrust. Tentatively, the discontinuity is identified as T2, dipping southwards at *c.* 30°.

North of our 'T2' Figures 6a & 6b are in substantial agreement; south of it, they differ substantially. A second steeply dipping discontinuity is identified in Figure 6b, parallel to but less obvious than the 'Mells Fault' and about 1.3 km south of it. This feature may correspond to a deep-seated westward extension of a mapped fault, named the 'Whatley Fault' in Figure 1a, which near the surface displays a post-Jurassic normal downthrow to the south. The Carboniferous Limestone crops out between these two 'fault' features of the seismic profile. About half of its stratigraphic thickness is exposed in Whatley Quarry, where it dips steeply north and is disturbed by minor thrust faults which dip southwards. It is tempting to regard the 'Mells and Whatley faults' identified in Figure 5c as steeply dipping splays from the T2 thrust which have been re-activated as normal faults during post-Mesozoic extension. Together these features bound a wedge-shaped portion of the seismic section in which the reflectors show grainy patterns that are impossible to interpret reliably in terms of geological units.

It is not clear from the minimalist interpretation of Figure 6b whether the Carboniferous Limestone is continuous across T2 or not, and we suggest that the matter is open to conjecture. At the time of writing, data from all the seismic lines in the Bath and Mendips area are undergoing reinterpretation using modern techniques of filtering and seismic image production (McCann *et al.* 2002). Future work along these lines may help to resolve the issue of whether T2 breaks the continuity of the limestone aquifer between the eastern Mendips and Bath.

New and older seismic evidence regarding structures beneath the City of Bath and its immediate environs is described and interpreted by McCann *et al.* (2002), who find that the uppermost Carboniferous Limestone appears to lie within 300 m or less below OD in all directions from the hot springs except south-west. In this direction the depth to limestone increases rapidly to 1.35 km below OD at a distance of 2.1 km from the springs, an apparent dip of 30°. These findings are broadly in line with the evidence of structures at the regional scale, discussed above. It seems likely that the thermal water ascends from depth to the Bath springs from the south-west direction. At Hotwells, in contrast, the thermal waters emerge from the lowest outcrops of limestone in the Avon Gorge. Hawkins & Kellaway (1991) describe the role that thrusts and upright fractures of Variscan age may play in channelling the hot waters and promoting various degrees of mixing with cold, local groundwaters.

To summarize, the thin-skinned tectonic model makes it likely that the basin is not a simple structure, but is affected by major thrust displacements. On the other hand, much of the considerable crustal shortening is accommodated on groups of splayed thrusts, rather than on single thrust faults, and by hanging wall folds. These features may ensure enough limestone-to-limestone contact across individual thrusts for there to be effective hydraulic continuity in the limestone aquifer across the whole basin. Large displacements on a single thrust seem most likely in the east Mendips, where there is the possibility of complete cutting of the limestone by T2. The major ambiguity in the Mendips Model lies in this question of the limestone's continuity across thrusts. This issue is examined by a novel approach using hydraulic models in the next two sections of this paper.

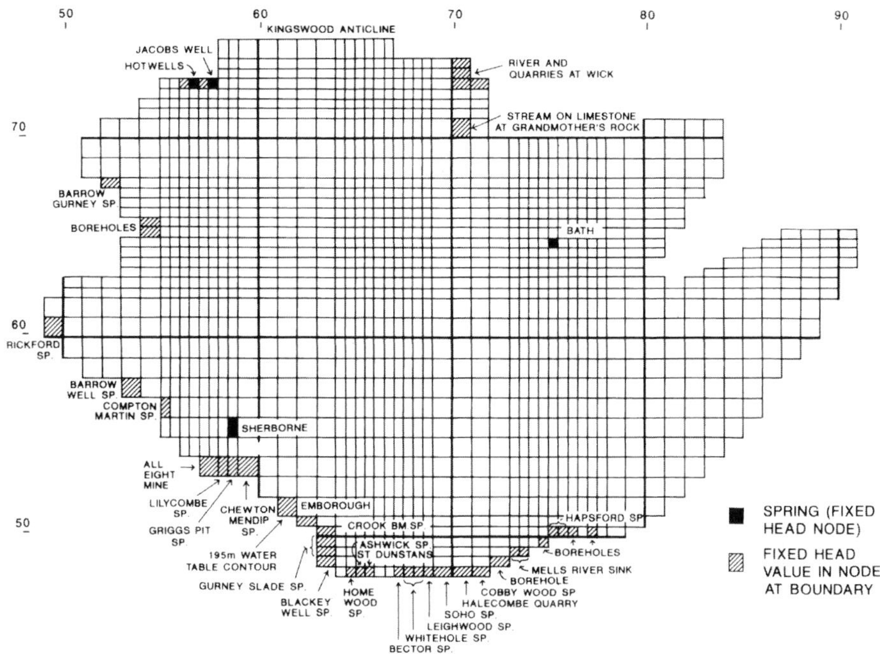

Fig. 7. The model grid showing springs, and fixed-head nodes.

Simple groundwater modelling of the thermal aquifer

Model construction and boundary conditions

The evidence reviewed above provides a framework in which the 2-D vertical section presented by Andrews *et al.* (1982) (Fig. 2) can be reconsidered in plan view and used as the basis of a numerical model. Using the simple groundwater modelling program ASM ('Aquifer Simulation Model'; Kinzelbach 1986; Kinzelbach & Rausch 1995) in confined aquifer mode we have developed a numerical model of groundwater flow in the basin in 2-D plan view. Two-dimensional modelling is justified by the vertical aspect ratio of the basin, which exceeds 5 : 1, and by the need to keep the model as simple as possible. Only steady-state groundwater flow was investigated as there is little information about transient effects or storativity in the real system. The first stage of model construction was therefore to define a 'base case' embodying the known aspects of geological structure and extrapolating them in the simplest way possible into the unknown parts of the model. The base case was later modified to explore the hydrogeological effects of other possible geological structures.

Figure 7 shows the grid of the base case model. The Carboniferous Limestone was assumed to possess uniform transmissivity and to be continuous at depth, i.e. unaffected by thrust faults. The limestone's permeability is due to fractures but these are not represented explicitly because the model's grid spacings of 500 and 1000 m are far larger than probable fracture spacings of *c.* 1–10 m. The boundaries of the grid coincide with the basin-ward edge of limestone outcrops along the Mendip Hills in the south and Broadfield Down in the west (Figs 1a & 5). The northern boundary was positioned along the Kingswood Anticline between Bristol and Wick. The eastern edge of the model is the least well constrained by data and represents the intersection of the Carboniferous Limestone aquifer with the post-Variscan unconformity, following Figure 1a. All external boundaries were made impermeable.

Table 1. *Comparison of modelled values with reported flows at Bath, Sherborne, Hotwells and Jacobs Well springs*

Spring	Modelled flow in base case ($l\ s^{-1}$)	Reported flow ($l\ s^{-1}$)	Reference
Bath	15	15	Stanton (1991)
Sherborne (Mendips)	52	57	Barrington & Stanton (1977)
Hotwells (Bristol)	9.4	*c.* 2.7	Hawkins & Kellaway (1991)
Jacobs Well (Bristol)	0.2	*c.* 2.7	Hawkins & Kellaway (1991)
Total, Bristol thermal springs	9.4	*c.* 5.4	

Given the above construction, the heads and steady-state flow-lines within the model will effectively be a numerical solution of Laplace's equation, subject to any boundary conditions that are chosen to represent the known parts of the head distribution. Known heads occur along the model's western and southern boundaries, where aquifer conditions change from unconfined to confined at the edge of the limestone outcrop. A large amount of groundwater circulates in the unconfined parts of the aquifer but almost all of this overflows as springs where the limestone dips beneath younger strata. The annual flow of Bath and Hotwells springs is less than 1.5% of total recharge to the unconfined limestone aquifer in the Mendip Hills (Atkinson 1977). Figure 7 shows grid elements in which heads were fixed at known values which were derived from the altitudes of springs plus one perennial stream where it crosses the edge of the limestone outcrop, and from average groundwater levels in six boreholes (Richardson 1928; Barrington & Stanton 1977). This procedure of fixing heads only where groundwater levels are actually known gives rise to model boundaries in which fixed-head nodes alternate with no-flow nodes. Although unconventional, such mixed conditions are mathematically allowable since they permit solution of Laplace's equation for flow under steady-state conditions. Experiment showed that there was no significant difference in results if these conditions for the western and southern boundaries were replaced by a continuous series of fixed heads taken from water table contour maps based on the same primary data (Stanton 1977; Harrison *et al.* 1992).

Four nodes shown in black in Figure 7 represent springs issuing from the confined aquifer – the Hotwells group, the Bath Hot Springs, and a cold water spring which rises through Keuper Marl at Sherborne a few kilometres north of the Mendips (Fig. 1a). The Hotwells and Sherborne springs were represented by fixed-head nodes, whereas the Bath Hot Springs were modelled as a leaky-aquifer node with a fixed head at the surface of 20 m OD, corresponding to the present level of the spring pools. The three Bath springs are point features but they are collectively represented in the model by upward leakage over a 500 m × 500 m grid element. An appropriate leakage factor, $6.10^{-4}\ d^{-1}$, was determined from data on yield-drawdown tests on all three springs (Stanton 1991).

With these boundary conditions the ASM code was used to compute the pattern of flow in the aquifer, results being presented as head contours, flow-lines and simulated spring discharges for Hotwells, Sherborne and Bath.

Model adjustment

Conventional groundwater modelling would continue from this point by adjusting the model's parameters until it successfully predicted a body of independent data that had not been used in model construction. The successful model would then have been calibrated against the independent data. The logic of this approach demands that the number of data items available for calibration be substantially larger than the number of adjustable parameters. However, the base case initially had only one adjustable parameter, the uniform transmissivity, and there are only two items of reliable independent data – the spring discharges at Bath and Sherborne – since the total discharge of the Hotwells group is poorly known (Table 1). The transmissivity was adjusted to give an exact fit between modelled and recorded discharges at Bath. This produced a serious underestimation of the flow at Sherborne spring which water tracing tests indicate is connected to a Carboniferous Limestone catchment with karstic flow (Atkinson 1977; Stanton 1977). To represent this, the model transmissivity within the Sherborne catchment was adjusted upwards, obtaining a

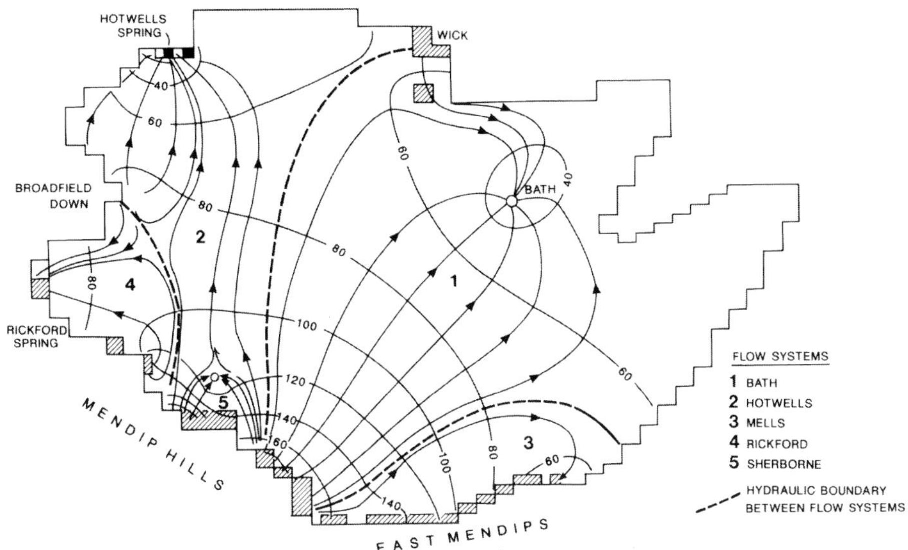

Fig. 8. The base case simulation showing calculated contours of groundwater head in the Carboniferous Limestone, flow-lines, and major groundwater flow systems. Note that the arrows indicate flow direction only, not time-of-travel.

satisfactory simulated discharge with very little effect on the pattern of flow in the whole aquifer. No further calibration of the base case model could be made due to the lack of other independent data. However, it should be noted that the total modelled discharge of the Hotwells group of springs is of the same magnitude as the sum of their reported flows (Table 1; Burgess *et al.* 1980; Hawkins & Kellaway 1991), although 1.8-times larger. The reported values may underestimate the true thermal water discharge as several of the springs emerge through inter-tidal mud in the channel of the River Avon and are difficult to locate and estimate. Other thermal water discharges may have gone undetected altogether or may form minor components of the several springs of cold water which lie in the Avon Gorge (Stanton 2000).

The base case simulation

The simulated flow pattern for the base case (Fig. 8) is independent of transmissivity and divides the aquifer into five flow systems. The Bath system is recharged in the central Mendips where boundary heads are highest. A distinct Mells system feeds water from the eastern Mendips into the confined aquifer and returns it to the southern boundary. In the real world, such a flow would supply the River Mells which prior to quarry dewatering was fed by several limestone springs that have much larger discharges than the small flows predicted by the model (Barrington & Stanton 1977). However, a direct comparison cannot be made because most of the spring waters are derived from the unconfined aquifer at outcrop, which is not represented by the model.

The base case provides an explanation for the geochemically mixed character of the Hotwells spring. The simulated flow is partly derived from the Mendips and partly from Broadfield Down (Fig. 8). Comparison with the geological structure (Figs 3a & 5) shows the flow-lines from Broadfield Down to be relatively short and shallow. It seems likely that water following these pathways will remain cold and retain the geochemical characteristics of 'normal' limestone groundwater. Only the flow-lines from the Mendips would pass through parts of the basin that are deep enough for water to acquire the elevated temperature and chemically evolved composition that characterizes the thermal component.

About 3% of the simulated flow at Bath is derived from the part of the model representing the area near Wick (Fig. 8). This agrees well with the geochemically based estimate of a 2–4%

contribution from a fluctuating component derived from the Permo-Triassic, possibly in the Wick area.

The base case simulation is broadly compatible with measurements of hydraulic conductivity in the Carboniferous Limestone. The model's regional transmissivity is 18 m^2 d^{-1} which implies an average hydraulic conductivity of 0.024 m d^{-1} for an aquifer thickness of 750 m. The statistical distribution of conductivity results from slug tests at outcrop has been considered by Smart *et al.* (1992) who concluded that there are three populations, each with a log-normal distribution. The most permeable corresponds to rock in which joints have been solutionally widened. The intermediate group was the largest, representing situations in which permeability is due to connected networks of fractures, whereas the small low-permeability group represents poorly connected fractures. Reworking the same data but retaining only those values that are derived from the limestone outcrop adjacent to the model gives a mean for the intermediate population of 0.035 m d^{-1}, just 1.5-times larger than the base case value. The difference lies at the upper 95% confidence limit of sampling error among the outcrop values (standard error = 0.092 in $\log_{10}$ units, $n = 45$) and is in the direction expected for deeply buried limestone in which joints will be fewer and have smaller apertures.

The simulated transmissivity for the Sherborne flow system is 170 m^2 d^{-1}, which is comparable with values determined for the karstic parts of the aquifer at outcrop (Atkinson 1977; Hobbs 1988).

The general features of the base case are compatible with all known aspects of the thermal system, and the simulation provides insight into the geochemical differences between Hotwells and Bath springs, explaining them in terms of hydraulically determined flow patterns. To this extent, the base case simulation supports the 'Mendips Model' for the origin of the thermal waters (Andrews *et al.* 1982).

Models simulating the effects of alternative geological structures

The sensitivity of flow patterns to the assumed geological structure was explored through almost 40 simulations incorporating features such as thrust faults and different shapes for the model's eastern boundary, these being the matters of greatest uncertainty during model construction. Even drastic adjustments of the eastern boundary proved to have very little effect on flow patterns, so the base case boundaries were adopted for investigating the effects of thrusts.

The incorporation of thrusts into the model raises two questions, the first being the location of the faults in the model grid, and the second the transmissivity of the grid elements designated as 'thrust zones'. Figure 9 shows four different configurations. In Figure 9a, thrust T3 is in the position where it intersects the post-Variscan unconformity (Fig. 3a), which is probably too far north relative to its position in the limestone at depth (Figure 3b). Figure 9b represents T3's position at depth, based on cross-sections drawn from maps in the west and centre of the basin, and on extrapolation of the trend of T3 into the basin's eastern part. This may place T3 too far south in the east, although it agrees with the eastern section in Figure 3b (Williams & Chapman 1986). Figure 9c illustrates locations for T3 and T2 together, but with both thrusts shown in their positions at the unconformity. Finally, Figure 9d shows a model in which T2 is the only thrust to affect the limestone, again in its unconformity position.

Five different simulations were made for each configuration. In the first, the thrusts completely cut the limestone reducing the transmissivity to zero. In the remaining four cases, partial cut-off merely tends to diminish the transmissivity through a local reduction in aquifer thickness while also creating a zone of intense fracturing. If the fractures are open they may enhance the transmissivity, but if filled with secondary minerals or fault gouge they may decrease it. These four cases were modelled by assigning transmissivities to the 'thrust zones' of one thousandth, one hundredth, one tenth, and ten times the regional value.

Flow patterns for completely impermeable thrust zones are shown in Figure 9. Completely impermeable thrusts give the greatest alteration from the base case, with west-to-east flow from Broadfield Down playing a prominent part in supplying the Bath Spring in three of the four cases (Figs 9a,b & c). Only the configuration in which T2 alone cuts the limestone somewhat resembles the base case (Fig. 9d). Increasing transmissivities of the thrust zones to one hundredth of the regional value (Fig. 10) provides a greater resemblance to the base case, but the low transmissivity zones still clearly affect the pattern by refracting the flow-lines. Raising the thrust zone transmissivity further, to one tenth or ten times the regional value, produces results that are closer to the base case but still distinguishable from it. Space precludes their illustration here.

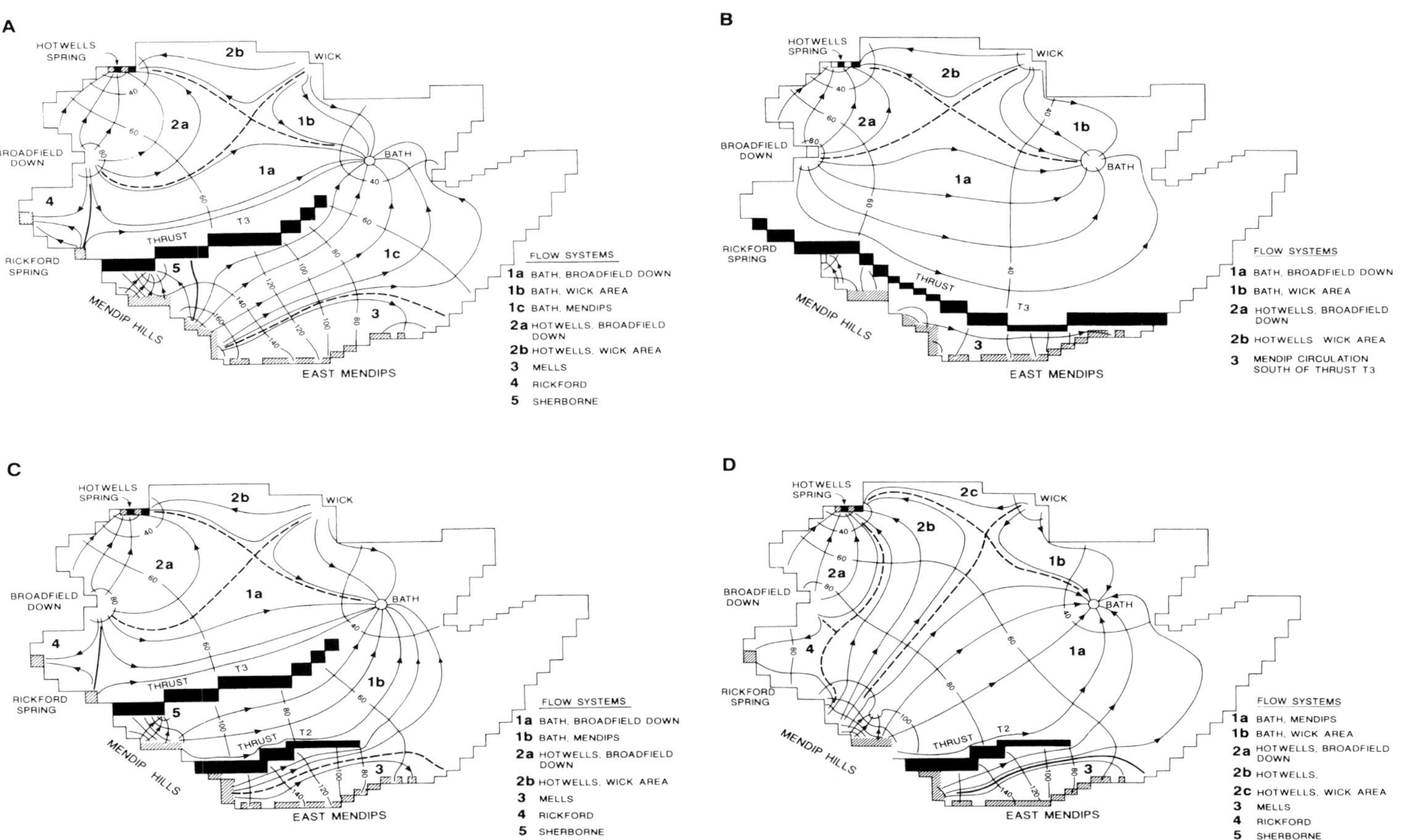

Fig. 9. Model simulations of the effects of various thrust configurations with zero transmissivity in the thrust zones. **(a)** Thrust T3 in its position at the level of the post-Variscan unconformity (cf. Fig. 3a). **(b)** Thrust T3 at its likely position at the level of cut-off of Carboniferous Limestone in the footwall (cf. Figs. 3b & 5). **(c)** Thrusts T2 and T3 at the level of the post-Variscan unconformity. **(d)** Thrust T2 acting to partially isolate the eastern Mendips from the rest of the thermal aquifer.

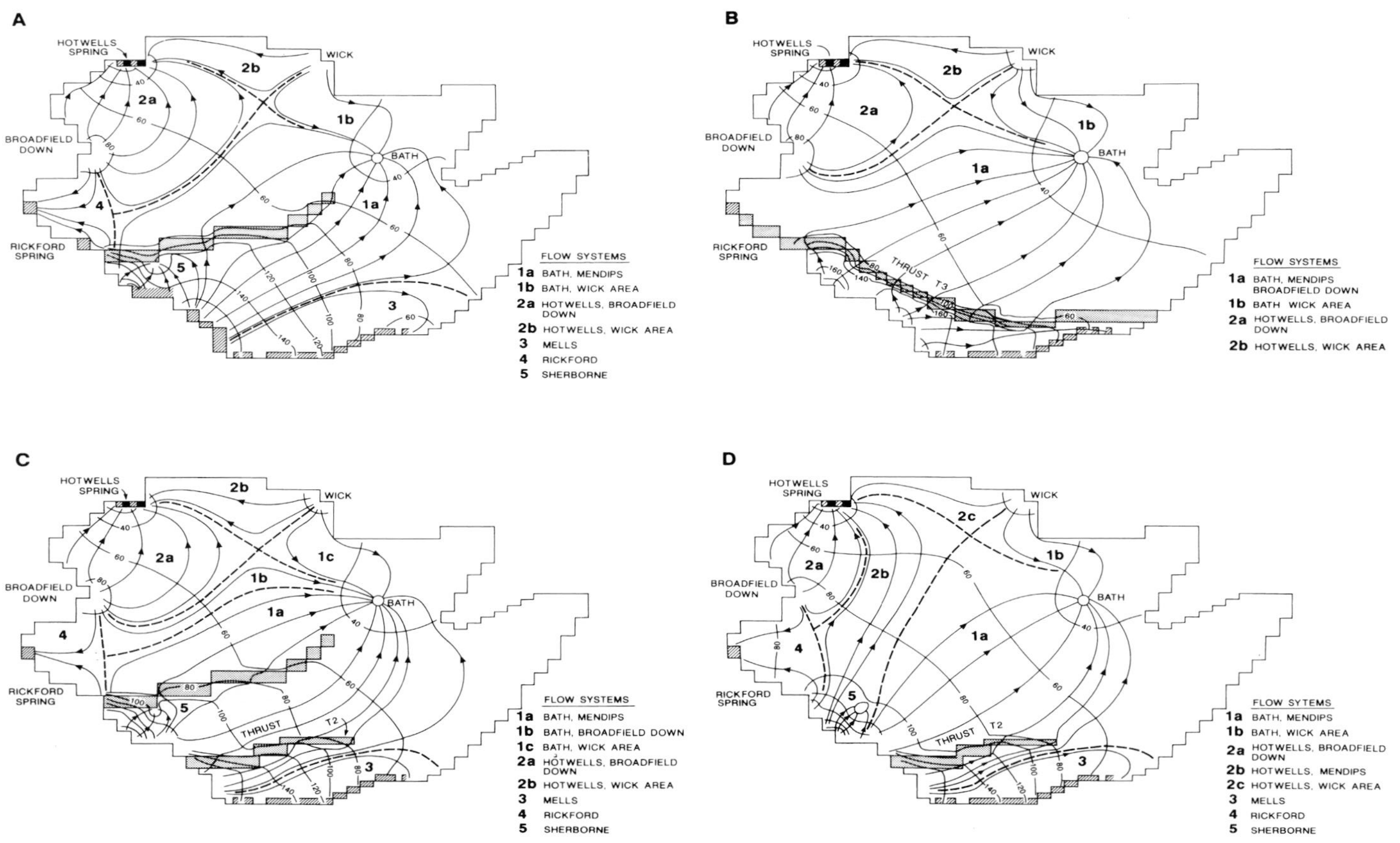

Fig. 10. Model simulations of the effects of various thrust configurations with transmissivity in the thrust zones one hundred times lower than the regional transmissivity. Dispositions of thrusts as for Fig. 9a, b, c, d.

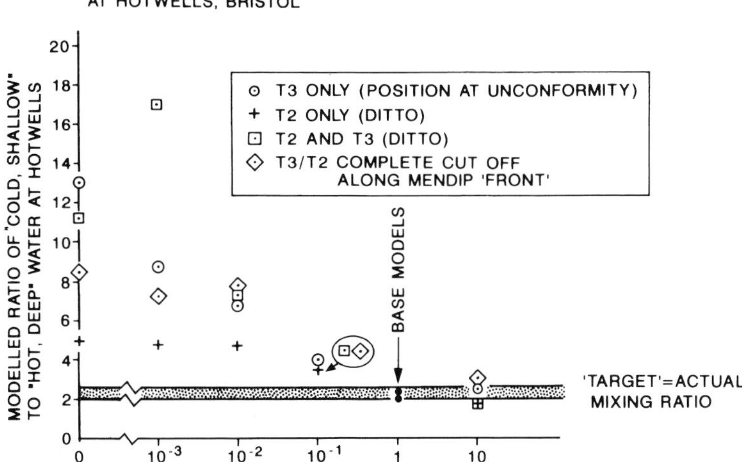

Fig. 11. Model predictions of the effects of the various thrust configurations upon the mixing proportions of 'deep thermal' and 'shallow cold' waters at Hotwells springs, as a function of the ratio between regional transmissivity and the transmissivity of thrust zones. The only models to predict the correct mixing proportions are those in which thrust transmissivities are either equal to or greater than the regional value.

Discrimination between alternative models

Simulations of the effects of thrusts imply that if the limestone is really cut off within the basin, the actual flow pattern would be very different from the base case, with west-to-east flow playing a prominent part. It is important to have a criterion for discriminating between the various possibilities, and if possible to eliminate some as being incompatible with known facts. Those that survive such testing can then be regarded as 'allowable' models. If only one model survives testing, then it is likely that it is more realistic than any of the others. This is the classical procedure followed by scientific logic in which no scientific hypothesis is ever regarded as completely proven, but in which good hypotheses survive all feasible attempts to disprove them. Some criterion of disprovability is needed which can be applied in the present situation of very sparse information.

The only available data that have not been used in model construction are the chemical characteristics of the thermal waters. For the Bath waters there is the inference that the water has circulated to a depth between 2.7 and 4.3 km. By careful comparison of the flow patterns with the structure contours in Figure 5, we can determine whether this is the case for each simulation. Dual criteria are available for the Hotwells spring in that the the major ion chemistry (Burgess *et al.* 1980) indicates that the spring waters comprise a 1 : 2.3 mixture of thermal : non-thermal water, and the thermal component has attained temperatures of at least 49°C, corresponding to circulation through parts of the basin where the top of the limestone lies deeper than 1.5 km. The ASM output for each realization was compared with Figure 5 to determine the proportions of simulated flow at Hotwells that passed through regions where the limestone is indicated to be deeper than this. The modelled ratio of 'deep, thermal' to 'shallow, cold' waters can then be found for each realisation, and compared with the actual ratio of 1 : 2.3.

The results of these comparisons are shown in Figure 11. Only a few simulations satisfy the target deep:shallow ratio at Hotwells. All the models with impermeable and low-permeability thrust zones can be eliminated. The only survivors are the base case and models in which the thrust zones possess *higher* transmissivity than the regional value. The flow patterns in these closely resemble the base case. It may be concluded that while the thin-skinned thrust-sheet model is very persuasive *tectonically*, the thrusts probably do not greatly reduce the transmissivity of the Carboniferous Limestone, nor cut off the hydraulic connection between the thermal springs and the Mendip Hills.

Sustainability and protection of Bath Hot Springs and the thermal aquifer system

The conclusion from comparing simple models with structure and chemistry is that the thermal aquifer most likely operates as a unified whole. There is therefore a possibility that permanent or long-term reductions in groundwater levels on a sufficient scale anywhere in the Carboniferous Limestone might affect the thermal springs.

In past centuries the largest lowering of groundwater levels anywhere in the basin was probably produced by dewatering of coal mines, the last of which closed in the early 1970s. Not all mines required dewatering, and many of those that did were affected by fresh water from shallow depths (Kellaway 1994). Nevertheless, confined waters (saline and fresh) were recorded down to 440 m below sea level in the Radstock coalfield (McMurtrie 1886) and the removal of these from the workings must have lowered groundwater heads in the vicinity. It is unlikely that this dewatering had any effect on the Bath springs, as it took place in the Upper Coal Measures which are separated from the Carboniferous Limestone by a large thickness of strata which comprise the Lower Coal Measures aquitard lying beneath the Pennant Sandstone aquifer. The Pennant Sandstone was not dewatered in the basin centre and heads in it would have been maintained despite the dewatering of mines above it.

The greatest potential threat to groundwater levels within the thermal aquifer today comes from quarrying on the Carboniferous Limestone outcrop at the rim of the basin. Recent decades have seen a trend in quarrying practice towards larger, deep workings below the water table. Figures 1a & 3a show the location of existing quarries and adjacent land with planning permission for quarrying. Workings below the natural water table are presently concentrated at Wick and in the eastern Mendip Hills, where some of the concerns involved have announced their intention to quarry to sea level. It is reasonable to use the ASM model to inquire whether reductions in the water table in these areas might constitute a threat to the thermal springs at Bath.

Potential impacts of quarry dewatering and further research needs

Groundwater modelling is commonly used for assessing the likely impact of developments on aquifers. Such assessments are *predictions* in the scientific sense in that a model constitutes a hypothesis about the nature of the aquifer. The modelled impact is a testable prediction about the aquifer's behaviour under the new conditions of the proposed development. To carry such a strictly scientific approach to its logical conclusion we would require the development actually to be carried out before we could use the results to test the validity of the original model. Such an attitude is useless for planning or conservation purposes, which require the model to act as a 'forecast' of what *might* happen, in order that a decision can be made about whether to follow certain courses of action. In effect, one must take a scientific prediction about the outcome, and on that basis make a non-scientific judgement about whether to proceed or not. This requires adequate confidence in the model for its predictions to be trusted. In the present case, the base model can be used to make an assessment of the possible impacts of quarry dewatering upon Bath springs, but the means of testing the model are very sparse and it is doubtful that the necessary confidence in it has been built up. Under these circumstances, the principal value of modelling is to highlight potential consequences of dewatering by quarries, and to identify critical topics for research in order to improve confidence in future models.

The potential impact of lowering groundwater heads in the east Mendips was examined using the base case to model a scenario in which the groundwater level at one quarry site (shown as WQ in Fig. 1) was reduced permanently to 6 m OD. The boundary conditions of the base case model were revised to take account of the local effects of this dewatering on heads in the unconfined aquifer, using figures from an environmental impact study conducted for the quarry's operators (Hydrotechnica 1992). The results are shown in Figure 12, expressed as contours of the reductions in head produced by the dewatering model compared with the base case. The head reduction at Bath Hot Springs corresponds to a 20% decline in flow, which is likely to be an underestimate because of the way the spring has been modelled as a node with upwards leakage towards a fixed head at the surface. This is clearly shown by the deflection of the 5 m contour in Figure 12. Without this influence of model construction a drawdown of

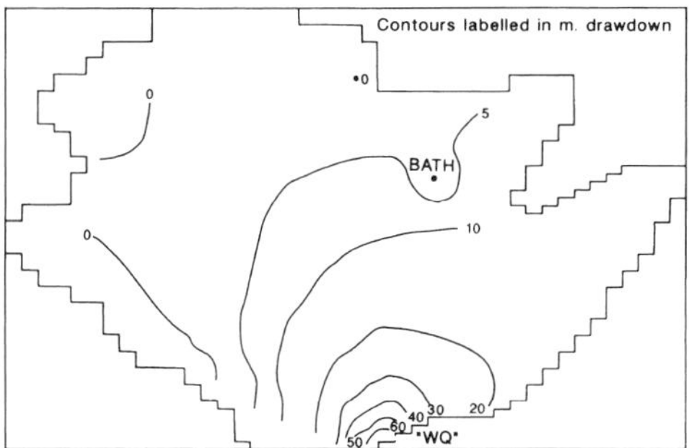

Fig. 12. Model simulation of the effects of prolonged dewatering of a deep quarry in the eastern Mendips. Contours show the steady-state reduction in modelled groundwater levels relative to the base case (see Fig. 8), produced by lowering of the water table within the quarry to 6 m OD.

around 7 m seems possible, which would result in a 70 or 80% reduction in flow.

This result must immediately be qualified, lest it be taken too literally. The model indicates only the steady-state pattern that would be likely to result if the lowering of water tables was allowed to persist indefinitely, which is an unrealistic assumption. In reality, quarry dewatering is likely to proceed in stages over several decades and may not persist for more than a few years before groundwater levels are allowed to recover partially. Thus, steady-state conditions may never become established before an upward movement of heads in the east Mendips starts to reverse the change in the pattern of flow. This issue cannot be clarified without better knowledge of the aquifer's storativity as without this parameter it is impossible to estimate the time required for the effects of dewatering in the Mendips to be seen at Bath. It should be a priority of future research to obtain reliable estimates of storativity in the confined parts of the thermal aquifer.

Another important area of uncertainty is the hydrogeological effect of the T2 thrust structure in the eastern Mendips. Kellaway (1994) points out that dewatering of east Mendip quarries began in 1978 and has yet to produce an observable effect at Bath, implying that the existence of a permeable pathway is therefore unproven. This inference could be correct and, if so, might imply that T2 presents an impermeable barrier between the east Mendip quarries and Bath. It is equally possible, however, that the lowering of water tables has not yet been great enough to affect the flow at Bath, or that the effects have not yet had sufficient time to spread across the aquifer. Once again, knowledge of storativity is a requirement before this issue can be investigated properly. In addition, research is needed on the sub-surface structure of the east Mendips. A deep borehole could test whether the east Mendip quarries lie in a klippe of limestone in the hanging wall of thrust T2. Figure 3b implies that a borehole at point 'WQ' should penetrate T2 at about 1 km depth, where Carboniferous Limestone should rest with tectonic contact upon Pennant Series, and that Lower Coal Measures should be reached at about 2 km. On the other hand, the presence of Carboniferous Limestone continuously to 2 km, or of Carboniferous overlying Devonian Old Red Sandstone, would suggest a structure in which cut-off by T2 was absent or limited.

The impact of the historical lowering of the water table at Wick quarries from 63 m OD to 40 m OD was examined in a separate simulation, resulting in a 3% reduction in flow at Bath; such a reduction is too small to be noticeable with the equipment used for monitoring the springs. However, a feature of the modelled impact was the elimination of any flow to the springs from the Permo-Triassic rocks around Wick. This might be expected to produce a change in the chemical composition of constituents thought to be derived from the Trias, such as uranium and strontium isotope compositions. These have not been monitored since prior to the lowering of water levels at Wick and there is a need for a monitoring scheme which might detect early signs of impact.

Protection of Bath Hot Springs

Two existing statutes, the City of Bath Act 1925 and County of Avon Act 1982, provide protection against disturbance to the Bath Hot Springs arising from excavations and boreholes, but only if these lie in the immediate vicinity. The only legal regulation of impacts that may arise from disturbances to groundwater levels further afield comes from national legislation on water abstraction and planning controls on mineral working. Various legislation since 1963 empowers the Environment Agency to regulate all forms of water abstraction for use, via a system of licensing, and to take environmental impacts of abstraction into account. This legislation exempts abstractions made for the purpose of keeping workings dry during the winning of minerals. In some *ad hoc* cases, these dewatering abstractions may be subject to prior control via planning conditions and agreements required as a condition for planning permission to be granted, but apart from this there is no legislative framework within which they can currently be regulated. Though it is beyond the scope of this article to analyse the legal situation in detail, planning conditions may in practice be a relatively weak form of restraint as they are often triggered only once the mineral operation has caused demonstrable harm. This places upon plaintiffs the burden of proving the cause of harm after it has already occurred, in contrast to the criterion of risk of future harm which applies in licensing water abstractions for all other purposes. The difficulties of proving and apportioning responsibility for harm increase in situations where several abstractors are involved. In effect, the existing planning framework for mineral workings does not deal effectively with situations in which it is desirable to safeguard the natural state of a whole aquifer, such as the thermal system considered here. A high level of scientific understanding, backed by good data from monitoring and scientific research, is likely to be at a premium if disputes arise in such cases.

Conclusions

The modelling approach summarized in Figure 11 demonstrates that the version of the Mendips Model which provides the best explanation for the origin of the Bath and Hotwells thermal springs is the base case. Comparison of very simple modelling with the chemical character of Hotwells spring has proved quite powerful in discriminating between different possible scenarios in a situation where reliable geological detail is extremely sparse. This reinforces Krabbenhoft & Anderson (1986), who describe the use of a numerical model for hypothesis testing, and Hiscock & Lloyd (1992) who point out the value of comparisons between models and hydrochemical data. An implication of the base case's superiority is that the Variscan thrust zones probably do not greatly reduce the continuity or transmissivity of the limestone in the thermal aquifer. Though it is possible that thrust T2 may partially isolate the eastern Mendips, the balance of modelling evidence favours the opposite view, that the eastern Mendip limestone outcrop has hydraulic continuity with Bath across all or part of T2.

This conclusion has important implications for the future protection of the hot springs from the effects of abstracting groundwater and lowering heads by many tens of metres in the areas of Carboniferous Limestone outcrop. Several quarries now have legal permission to dewater on a large scale and have commenced to do so. In effect this provides a full-scale experiment which may in time test the scientific hypotheses embodied in the base case model. In the writers' opinion, it would be unwise for such an experiment to continue for long without further research to establish the knowledge needed to predict its progress. This research should centre around the dual aims of (1) establishing the storativity and other data needed to predict the magnitude and timescale of the aquifer's response to progressive lowering of groundwater heads, and (2) determining the detailed geological structure of thrust zone T2 and testing the continuity of the Carboniferous Limestone at depth beneath the eastern Mendips by means of a borehole. Monitoring of the flow and temperature at Bath is already in progress and should continue on the most accurate and precise basis achievable. Regular chemical and radio-chemical monitoring of the spring should begin in order to establish the natural fluctuations and any longer term trends which may presage anthropogenic impacts on the thermal water.

We thank referees Peter Smart and William Stanton for their careful scientific comments, and Melanie Walters and Rhys Davies for comments on a draft version of this paper. The records department of British Petroleum kindly provided copies of seismic sections. TCA thanks the Royal Society for support via a Leverhulme Senior Research Fellowship.

References

Andrews, J. N., Burgess, W. G., Edmunds, W. M., Kay, R. L. F. & Lee, D. J. 1982. The thermal springs of Bath. *Nature*, **298**, 339–343.

Andrews, J. N. 1991. Radioactivity and dissolved gases in the thermal waters of Bath. *In:* Kellaway, G.A. (ed.) *Hot Springs of Bath*. Bath City Council, Bath, 157–170.

Ates, A. & Kearey, P. 1993. Deep structure of the East Mendip Hills from gravity, aeromagnetic and seismic reflection data. *Journal of the Geological Society, London*, **150**, 1055–1063.

Atkinson, T. C. 1977. Diffuse flow and conduit flow in limestone terrain in the Mendip Hills, Somerset (Great Britain). *Journal of Hydrology*, **35**, 93–110.

Barker, J. A., Downing, R. A., Gray, D. A., Findlay, J., Kellaway, G. A., Parker, R. H. & Rollin, K. E. 2000. Hydrogeothermal studies in the United Kingdom. *Quarterly Journal of Engineering Geology and Hydrogeology*, **33**, 41–58.

Barrington, N. & Stanton, W. I. 1977. *Mendip: The Complete Caves and a View of the Hills*. Cheddar Valley Press, Cheddar.

Brooks, M., Trayner, P. M. & Trimble, T. J. 1988. Mesozoic reactivation of Variscan thrusting in the Bristol Channel area, UK. *Journal of the Geological*

Society, London, **145**, 439–444.

BURGESS, W. G., BLACK, J. H. & COOK, A. J. 1991. Regional hydrodynamic influences on the Bath–Bristol springs. *In:* KELLAWAY, G. A. (ed.) *Hot Springs of Bath.* Bath City Council, Bath, 171–177.

BURGESS, W. G., EDMUNDS, W. M., ANDREWS, J. N. KAY, R. L. F. & LEE, D. J. 1980. *Investigation of the geothermal potential of the UK. The hydrogeology and hydrochemistry of the thermal water in the Bath–Bristol Basin.* Institute of Geological Sciences, London.

CHADWICK, R. A., KENOLTY, N. & WHITTAKER, A. 1983. Crustal structure beneath southern England from deep seismic reflection profiles. *Journal of the Geological Society, London,* **140**, 893–911.

EDMUNDS, W. M. 1975. Geochemistry of brines in the Coal Measures of Northeast England. *Transactions of the Institution of Mining and Metallurgy,* **Sect. B**, **84**, 39–52.

EDMUNDS, W. M. & MILES, D. L. 1991. The geochemistry of the Bath thermal waters. *In:* KELLAWAY, G. A. (ed.) *Hot Springs of Bath.* Bath City Council, Bath, 143–156.

GREEN, G. W. & WELCH, F. B. A. 1965. *Geology of the country around Wells and Cheddar.* Memoirs of the Geological Survey of Great Britain. H.M.S.O., London.

HARRISON, D. J., BUCKLEY, D. K. & MARKS, R. J. 1992. *Limestone resources and hydrogeology of the Mendip Hills.* British Geological Survey, Keyworth, Nottingham, Technical Report **WA/92/19**.

HAWKINS, A. B. & KELLAWAY, G. A. 1991. The hot springs of the Avon Gorge, Bristol. *In:* KELLAWAY, G. A. (ed.) *Hot Springs of Bath.* Bath City Council, Bath, 179–203.

HISCOCK, K. M. & LLOYD, J. W. 1992. Palaeohydrogeological reconstructions of the North Lincolnshire Chalk, UK, for the last 140 000 years. *Journal of Hydrology,* **133**, 313–342.

HOBBS, S. L. 1988. *Recharge, flow and storage in the unsaturated zone of the Mendip Limestone aquifer.* PhD thesis, University of Bristol.

Hydrotechnica. 1992. *Proof of evidence by Dr M. Walters, Director of Hydrotechnica Ltd., on behalf of ARC Ltd., at Public Inquiry into the Application by ARC Ltd. to Somerset County Council for a proposed Extension to Whatley Quarry.* 2 vols. Hydrotechnica Ltd., Abbey Foregate, Shrewsbury, Shropshire, SY2 6AL.

KELLAWAY, G. A. (ed.) 1991*a*. *Hot Springs of Bath: Investigations of the Thermal Waters of the Avon Valley.* Bath City Council, Bath.

Kellaway, G. A. 1991*b*. Preface. *In:* Kellaway, G.A. (ed.) *Hot Springs of Bath.* Bath City Council, Bath, 13–22.

KELLAWAY, G. A. 1991*c*. Investigation of the Bath Hot Springs (1977–1987). *In:* KELLAWAY, G. A. (ed.) *Hot Springs of Bath.* Bath City Council, Bath, 97–125.

KELLAWAY, G. A. 1991*d*. Structural and glacial control of thermal water emission in the Avon Basin at Bath. *In:* KELLAWAY, G. A. (ed.) *Hot Springs of Bath.* Bath City Council, Bath, 205–241.

KELLAWAY, G. A. 1993. The hot springs of Bristol and Bath. *Proceedings of the Ussher Society,* **8**, 83–88.

KELLAWAY, G. A. 1994. Environmental factors and the development of Bath Spa, England. *Environmental Geology,* **24**, 99–111.

KELLAWAY, G. A. & HANCOCK, P. L. 1983. Structure of the Bristol District, the Forest of Dean and the Malvern Fault Zone. *In:* HANCOCK, P. L. (ed.) *The Variscan Fold Belt in the British Isles.* Adam Hilger, Bristol.

KELLAWAY, G. A. & WELCH, F. B. A. 1948. *British regional geology: Bristol and Gloucester district (2nd ed.).* HMSO, London.

KELLAWAY, G. A. & WELCH, F. B. A. 1993. *Geology of the Bristol District.* Memoirs of the Geological Survey of Great Britain. HMSO, London.

KINZELBACH, W. 1986. Groundwater modelling – an introduction with sample programs in BASIC. *Developments in Water Science,* **25**, 1–333.

KINZELBACH, W. & RAUSCH, R. 1995. *ASM - Aquifer Simulation Model; Manual and Code.* International Groundwater Modeling Center, Colorado School of Mines, Golden, CO 80401, USA.

KRABBENHOFT, D. P. & ANDERSON, M. P. 1986. Use of a numerical groundwater flow model for hypothesis testing. *Ground Water,* **24**, 49–55.

MCCANN, C., MANN, A. C., MCCANN, D. M. & KELLAWAY, G. A. 2002. Geophysical investigations of the thermal springs of Bath, England. *In:* HISCOCK, K. M., RIVETT, M. O. & DAVISON, R. M. (eds) *Sustainable Groundwater Development.* Geological Society, London, Special Publication, **193**, 41–52.

MCMURTRIE, J. 1886. Notes on the occurrence of salt springs in the coal measures at Radstock. *Proceedings of the Bath Natural History and Antiquarian Field Club,* **6**, 84–94.

RICHARDSON, L. 1928. *Wells and Springs of Somerset.* Memoirs of the Geological Survey of Great Britain. H.M.S.O., London.

RICHARDSON, L. 1930. *Wells and Springs of Gloucestershire.* Memoirs of the Geological Survey of Great Britain. H.M.S.O., London.

RIES, A. C. & SHACKLETON, R. M. 1976. Patterns of strain variation in arcuate fold belts. *Philosophical Transactions of the Royal Society, London,* **Ser.A 283**, 281–288.

SHACKLETON, R. M., RIES, A. C. & COWARD, M. P. 1982. An interpretation of the Variscan structures in S.W. England. *Journal of the Geological Society, London,* **139**, 533–541.

SMART, P. L., EDWARDS, A. J. & HOBBS, S. L. 1992. Heterogeneity in carbonate aquifers; effects of scale, fissuration, lithology and karstification. *In:* BOWLING, A. B. & GREEN, G. H. (eds) *Proceedings of the Third Conference on Hydrology, Ecology, Monitoring and Management of Groundwater in Karst Terrains*, Kentucky, 1992, 475–492.

STANTON, W.I., 1977. Mendip Water. *In:* BARRINGTON, N. & STANTON, W. I. (eds) *Mendip: The Complete Caves and a View of the Hills.* Cheddar Valley Press, Cheddar, 202–213.

STANTON, W. I. 1991. Hydrogeology of the hot springs of Bath. *In:* KELLAWAY, G. A. (ed.) *Hot Springs of Bath.* Bath City Council, Bath, 127–142.

STANTON, W. I. 2000. Comments on the paper 'On the

origin of the thermal waters at Bath, United Kingdom: A sub-Severn hypothesis'. *In:* WILCOCK, J. & LOWE, D. J. (eds) *Cave and Karst Science,* **27**, 33–34.

TODD, D. K. 1976. *Groundwater Hydrology (2nd edn).* Wiley, Chichester, 363–369.

WHITTAKER, A. & GREEN, G. W. 1983. *Geology of the country around Weston-super-Mare.* Memoirs of the Geological Survey of Great Britain. H.M.S.O., London.

WILCOCK, J. & LOWE, D. J. 1999. On the origin of the thermal waters at Bath, United Kingdom: A sub-Severn hypothesis. *Cave and Karst Science,* **26**, 69–80.

WILLIAMS, G. D. & BROOKS, M. 1985. A reinterpretation of the concealed Variscan structure beneath southern England by section balancing. *Journal of the Geological Society, London,* **142**, 689–695.

WILLIAMS, G. D. & CHAPMAN, T. J. 1986. The Bristol-Mendip foreland thrust belt. *Journal of the Geological Society, London,* **143**, 63–73.

WOOD, M. W. & SHAW, H. F. 1976. The geochemistry of celestites from the Yate area near Bristol (U.K.). *Chemical Geology,* **17**, 179–193.

Geophysical investigations of the thermal springs of Bath, England

C. McCANN[1], A. C. MANN[2], D. M. McCANN[1] & G. A. KELLAWAY[3]

[1]*Postgraduate Research Institute for Sedimentology, The University of Reading, Whiteknights, Reading RG6 6AB, UK (e-mail: c.mccann@reading.ac.uk)*

[2]*IMC Geophysics Ltd., PO Box 18, Common Road, Sutton-in-Ashfield, Nottinghamshire NG17 2NS, UK*

[3]*Rowley Lodge, Hill Rise Road, Lyme Regis, Dorset DT7 3LN, UK*

Abstract: The origin of the thermal springs of Bath (England) remains unknown. As part of a programme of research into the structure of the thermal aquifer, the Carboniferous Limestone, an urban reflection seismic survey has been carried out to explore the deep geology of the Bath area. Existing gravity data have been used to provisionally identify the seismic reflectors and to map the depth of the (interpreted) Carboniferous Limestone in the area around the springs. The new seismic data show that at a distance of 2.1 km south-west of the springs, the depth of the (interpreted) Carboniferous Limestone surface increases from 0.4 km below Ordnance Datum (OD) to 1.35 km below OD within a distance of 1.8 km, an average apparent dip of nearly 30 degrees. In all other directions from the springs, the Carboniferous Limestone surface is at a depth of 300 m or less below OD. The work described in this paper is part of a continuing research programme.

Since Roman times the City of Bath in the west of England has been famous for its hot springs, which have been used for pleasure and medicinal purposes. The springs have been under the care of the Civic Authority (now the Bath and North-East Somerset Council) since 1590. The Bath springs are situated within a meander of the River Avon (Fig. 1). The water emerges from the Carboniferous Limestone aquifer at a temperature of about 47°C with a yield of 1.25 Ml d^{-1}. To achieve this temperature the water must have reached nearly 2 km beneath the ground surface. Geochemical data show that the water has been in the Carboniferous Limestone for at least several hundred years (Andrews 1991).

Understanding the origin and sustainability of the thermal springs requires a detailed knowledge of the deep geology of the Bath area. Boreholes near the springs (Fig. 1) show that the Carboniferous Limestone is at a depth below the ground surface of less than 200 m and that it is overlain unconformably by Mesozoic rocks. To the southwest the Mesozoic rocks overlie Palaeozoic Coal Measures, conformably resting on the Carboniferous Limestone which crops out 20 km S of Bath to form the Mendip Hills. To the north, in the Cotswold Hills, Palaeozoic rocks are hidden under hundreds of metres of younger Mesozoic rocks. The presence of complex landslips and slumping of the near surface rocks in the Bath area have made it difficult to establish the deep geology from surface mapping.

The origin of the hot springs remains uncertain. Three hypotheses have been advanced:

(1) A long-held view is that the source of the water is via the outcrop of the Carboniferous Limestone in the Mendip Hills, 20 km SW of Bath. Burgess *et al.* (1991) postulated an origin for the springs whereby cold water flows down from the Mendip Hills into the Carboniferous Limestone basin and up the northern limb of the syncline to reach the surface in the centre of Bath. The water flows through the limestone to depths of at

From: Hiscock, K. M., Rivett, M. O. & Davison, R. M. (eds) *Sustainable Groundwater Development*. Geological Society, London, Special Publications, **193**, 41–52. 0305-8719/02/$15.00

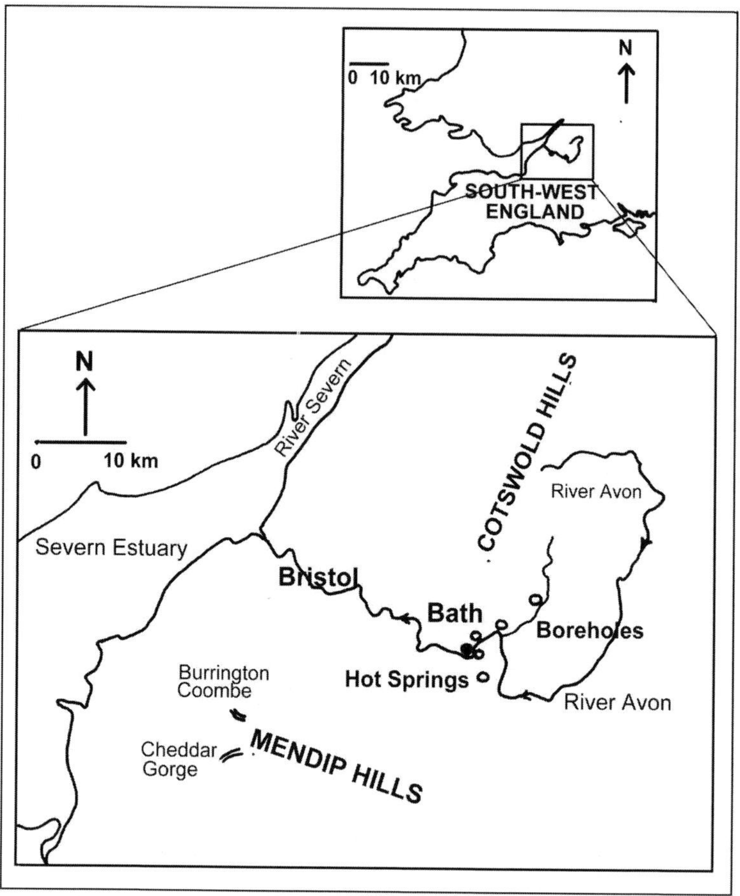

Fig. 1. Location map of the City of Bath, England.

least 2.5 km below the Upper Carboniferous before rising up fractures in the limestone and the overlying Trias and Lias in the Avon valley. Ates & Kearey (1993) analysed an existing seismic line CV82–482 that runs in an east–west direction at a distance of about 12 km S of Bath. These authors deduced that the Carboniferous Limestone reaches a depth of about 2.5 km below Ordnance Datum (OD) along this line.

(2) An alternative view is that the springs of both Bath and Hotwells (Bristol) lie on a NW–SE fracture zone that may be part of a much more extensive Avon–Solent Fracture Zone (Kellaway 1996). In this hypothesis the source is postulated to be near to Bath with the water descending to great depths within the Avon–Solent Fracture Zone. The water then rises into the springs as a result of the density reduction as it is heated. This implies a local thermal convection cell. This alternative model requires only *local* flow of meteoric water from the elevated catchment area around Bath.

(3) It has been suggested that the hot water may be associated with a granite batholith, postulated to extend to a depth of about 2.5 km below OD. As an example of such a granite-driven thermal system, water from a borehole in the Carmenellis Granite (Cornwall) reaches a temperature of 79°C at a depth of 2 km (Lee 1986). In this hypothesis the water would be heated locally in the granite batholith and would rise to the surface through fractures in the overlying Palaeozoic and Mesozoic rocks.

Without further information on the geological structure of the area it has not been possible to distinguish between these hypotheses. The proposal to create a new spa in Bath has made understanding the origin of the springs vital to predict their future behaviour.

Geological information on the depth and nature of the Carboniferous Limestone in the area is restricted to a small number of boreholes that have been drilled to investigate the hot springs themselves. In addition, information from shafts and workings associated with the extraction of coal to the south and west of Bath provides additional information on the near surface geological structure to a depth of approximately 400 m.

There is some existing geophysical information for the Bath area. The British Geological Survey has published maps of the regional Bouguer gravity (British Geological Survey 1986) and geomagnetic fields (British Geological Survey 1980) of the Bath area. These data are available as individual data points in digital format. Reflection seismic survey data, acquired as part of exploration programmes by oil companies in the period 1970 to 1987, are available from the UK Onshore Seismic Library for areas about 5 km to the south and east of Bath, but not for the area immediately within and around the city.

The primary aims of the research described in this paper were the acquisition of new seismic reflection data and the analysis of existing gravity data in the urban area of Bath, centred on the hot springs. One of the objectives of the work was to determine the structure of the Carboniferous Limestone aquifer within the area. As the Carboniferous Limestone has a significantly higher seismic velocity (6000–6500 m s^{-1}) and density (2720 kg m^{-3}) than either the Palaeozoic Coal Measure rocks or the Mesozoic rocks, it was expected that the Carboniferous Limestone interface would give rise to large amplitude seismic reflections and significant variations in the Bouguer gravity field.

The Bath Spa seismic surveys

Principles of the seismic method

The objective of a seismic reflection survey is to measure the time taken by a pulse of sound (seismic energy) to make the two-way journey between the Earth's surface and buried geological structures. The repetition of this procedure at numerous points along a line enables a detailed image of sub-surface geological structures to be created, to depths in excess of 4 km. In practice, the source of seismic energy is a large vibrating plate, forced into intimate contact with the ground surface. The reflected energy is received by geophones pushed into the ground at distances up to 2.0 km from the energy source. The electrical outputs from more than 2000 geophones are recorded on magnetic tape or disc and organized so that the reflections from individual areas of the sub-surface are collected together (Common Depth Point surveying, CDP). The change in the two-way travel time (TWTT) as a function of shot-to-geophone distance enables the average sound velocity of the rocks above each reflector to be determined and a process of adding together all the separate CDP data (Horizontal Stacking) significantly improves the signal-to-noise ratio and thus the quality of the final image. Computer processing of the data (migration) enables an accurate and stable image of the sub-surface geology to be obtained.

The method has a number of limitations. Firstly, the seismic reflection method cannot be used routinely to image the first 50 m depth of the sub-surface because of the interference of surface-seismic waves with the reflected energy. Secondly, sub-surface sound velocities cannot be determined very accurately because the TWTT of the reflected waves changes only slightly with source-receiver distance. This limits the accuracy with which the TWTT can be converted into depth below OD, and deep boreholes are normally required to convert TWTT accurately into depths.

Design and acquisition

The planning of the reflection seismic survey was carried out in the expectation that the interfaces, formed by either Palaeozoic Coal Measure rocks conformably overlying the Carboniferous Limestone (i.e. in its appropriate stratigraphic position) or Mesozoic rocks uncomformably overlying an eroded surface of Carboniferous Limestone, would both give rise to large amplitude (i.e. clear) reflectors on the seismic images.

IMC Geophysics Ltd. was awarded the contract to acquire six seismic lines, varying in length from 1.3 to 7 km (99-SPA-01, 99-SPA-02, 99-SPA-03, 99-SPA-06, 99-SPA-08, 99-SPA-09), using a Vibroseis energy source through the streets of Bath (Fig. 2). The locations of these lines (Fig. 3) were planned by the technical team using the following criteria, but taking cognizance of the logistical problems of using seismic sources within the city area: investigation of the

Fig. 2. Vibroseis units in front of the Royal Crescent, Bath.

shallow and deep geology of the area in and around the springs; defining the Palaeozoic equivalents of the known Mesozoic faults; determining the morphology of the Carboniferous Limestone; tying together both the seismic lines and existing and proposed boreholes; and tying the new seismic lines to existing seismic lines which terminated to the south and east of Bath.

Figure 3 also shows the CDP positions alongside the seismic lines and the locations of three of the boreholes proving Carboniferous Limestone below Mesozoic rocks in the urban area of Bath.

An initial test line, 99-SPA-01, was recorded in Victoria Park in the centre of Bath. A range of Vibroseis sweep parameters was tested. In addition, short portions of lines 99-SPA-02 and 99-SPA-06 were recorded during this period of testing. Processing was carried out on-site during the tests to provide an indication of the effects of changes in acquisition parameters. Although data were recorded with a 5 m group interval on line 99-SPA-01, group intervals of 10 m and 15 m were simulated in the on-site processing tests.

On the basis of results of tests on lines 99-SPA-01, 99-SPA-02 and 99-SPA-06, the seismic survey was continued and six seismic lines were recorded. These lines totalled 20.5 km of data. With the exception of line 99-SPA-01 for which a 5 m group interval was used, all remaining lines were recorded using a 10 m group interval.

A fleet of four Input/Output Type 362 Vibroseis units was used as the energy source on this survey although no more than two were operational at any one time on a line. Each unit was capable of producing a peak force of 8600 newtons.

Monitoring of peak particle velocity was carried out by Spectrum Acoustic Consultants Ltd. on all seismic lines in the proximity of buildings. The monitoring regulated the vibrator unit drive levels and the number of units used in the urban environment and in the proximity of sensitive structures. A linear sweep length of 12 s was selected, over 20 to 120 Hz. In urban areas up to 12 sweeps per vibroseis point (VP) were carried out especially where drive levels were low. These were not summed in the field. Elsewhere, however, generally four sweeps per VP were recorded. Vibroseis sweeps were carried out at all available stations.

Recording was carried out on an ARAM 24 recording system generally using 200 live channels with two strings of six 10 Hz Sensor SM4 geophones laid linearly about a peg. Data were recorded at a 2 ms sample rate with a record

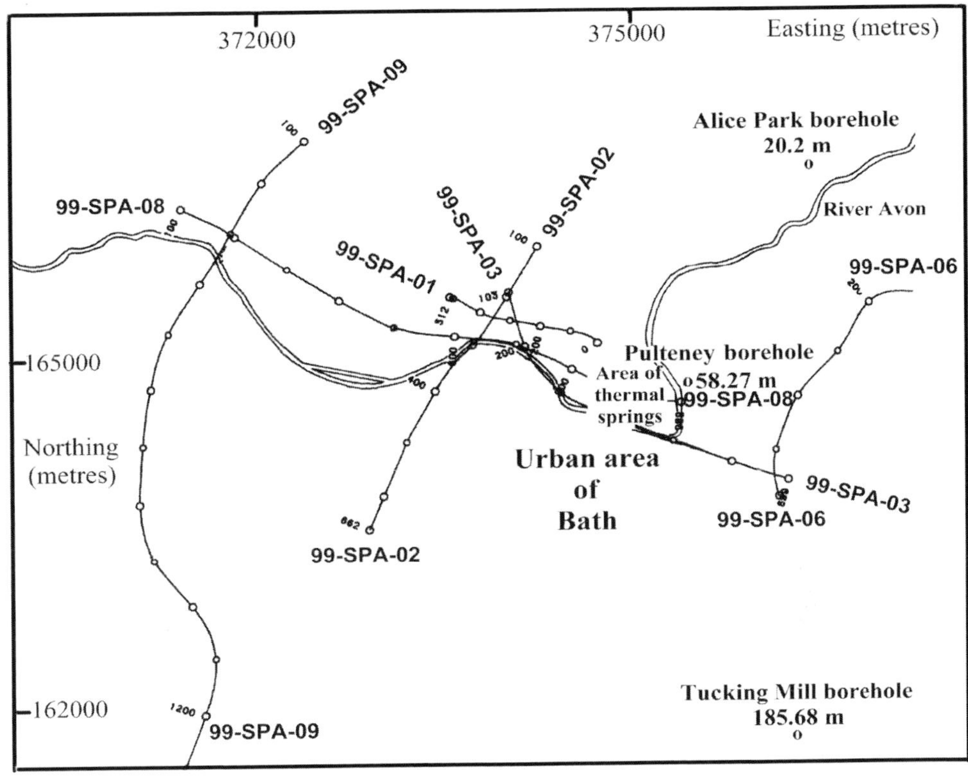

Fig. 3. Location map of the seismic lines. Common Depth Point (CDP) numbers are shown alongside each line.

length of 4 s for all lines apart from line 99-SPA-01 where only 3 s of data were recorded.

Processing

Seismic data were processed and interpreted by IMC Geophysics Ltd. to produce seismic time sections, two-way travel time contour maps of the key reflection horizons, depth contour maps of the key reflection horizons and shallow refraction velocity data. The quality of the data was generally good, but was affected locally by the near-surface geology and the urban environment, particularly traffic noise and reductions in permitted drive levels. Coherent reflectors are visible on line 99-SPA-09 to greater than 2 s two-way travel time and on other lines to greater than 1 s two-way travel time.

Seismic lines were processed to an individually selected flat datum using elevation statics.

Much processing effort was put into editing random noise from the urban seismic lines prior to summing and stacking the records in an attempt to improve the overall signal-to-noise ratio.

Data were processed to up to 4 s but only displayed to 2.5 s. Seismic sections were produced at a horizontal scale of 10 traces cm^{-1} and a vertical scale of 10 cm s^{-1}.

The following sections were available for interpretation: Filtered Stacked sections; Migrated Stacked sections; and Colour Enhancement of Seismic Attributes (CESA).

To illustrate the quality of the seismic data, Figures 4–6 show parts of the final processed seismic sections for lines 99-SPA-09, 99-SPA-02 and 99-SPA-06 respectively. In the section from line 99-SPA-09, coherent reflectors can be traced to TWTT greater than 1.8 s (about 3 km).

Initial interpretation of the seismic data

A strong reflector (Reflector A in Figs 4 & 5) was traced in the seismic sections to the west, north-west and south of the central Bath area. It was unambiguous with good agreement at the seismic ties. A similar strong reflector (Reflector B in Fig. 6) was traced in seismic sections to the

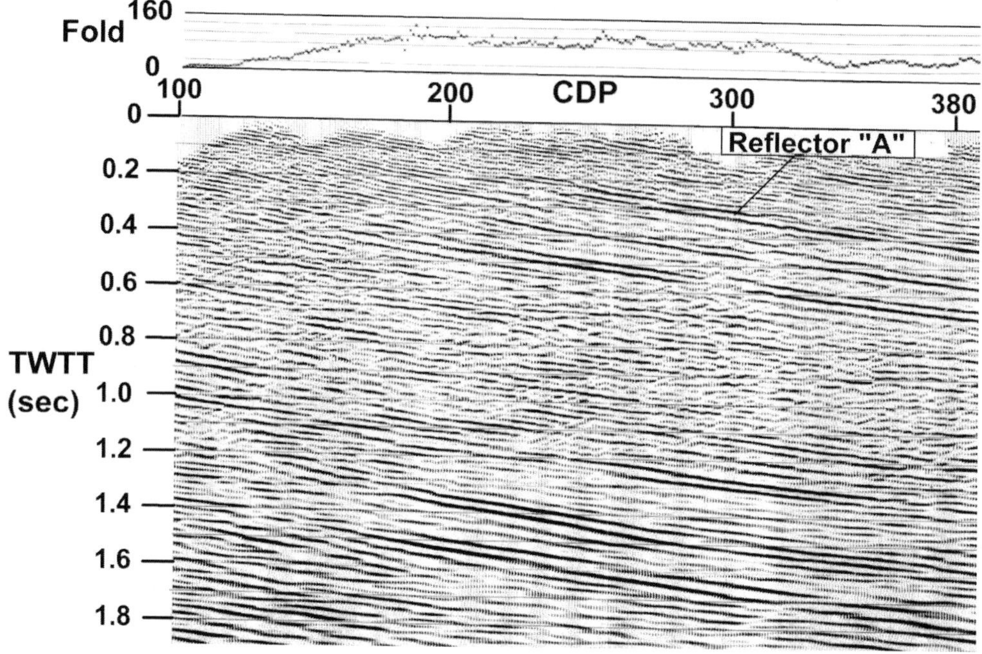

Fig. 4. Part of the final seismic section and the fold of cover along line 99-SPA-09.

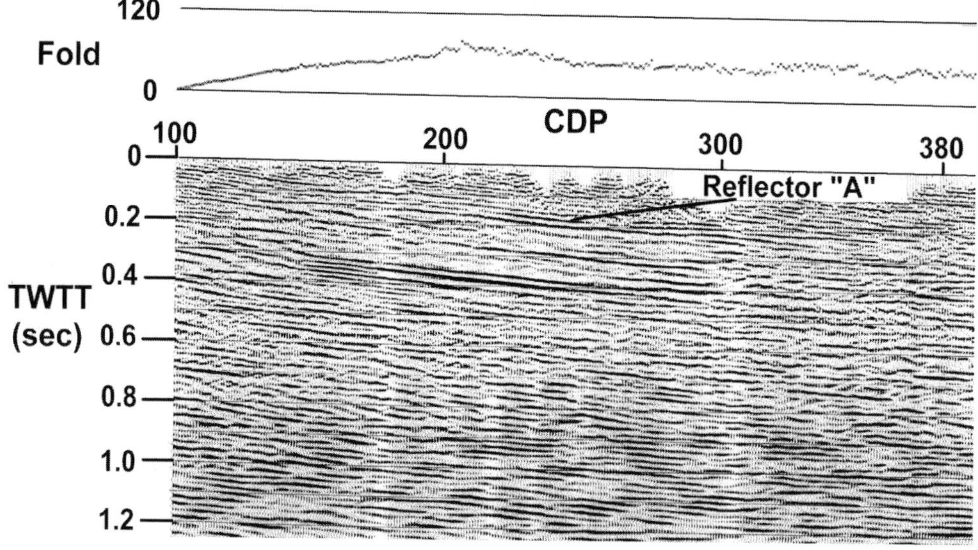

Fig. 5. Part of the final seismic section and the fold of cover along line 99-SPA-02.

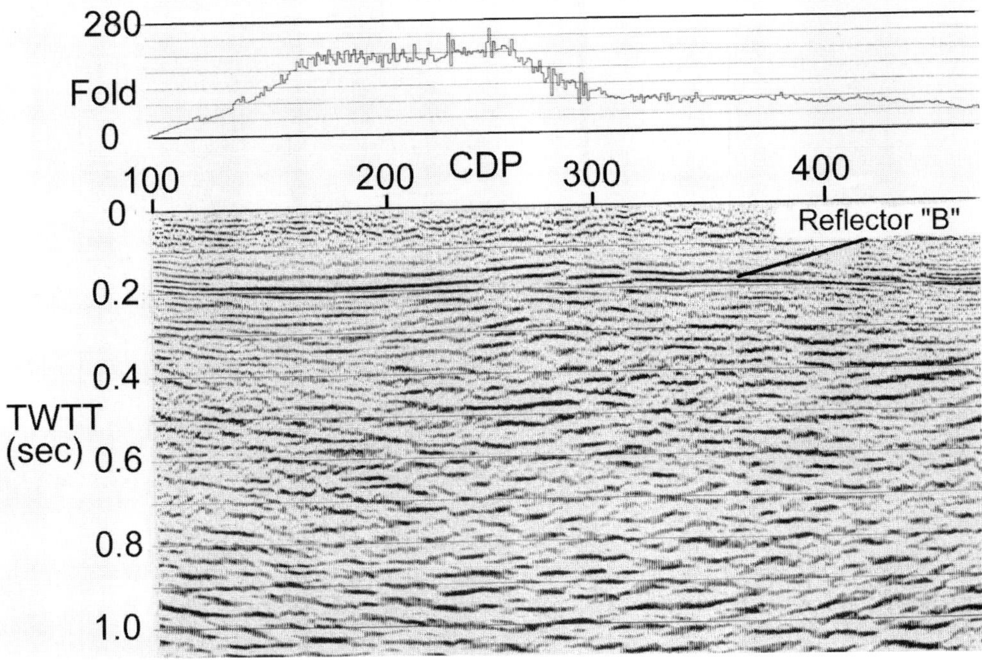

Fig. 6. Part of the final seismic section and the fold of cover along line 99-SPA-06.

east and south-east of Bath. Due to faulting, these two reflectors could not be directly tied together nor could they be tied to the boreholes. For this reason gravity data were used to correlate and identify the reflectors.

The Bath Spa Bouguer gravity map

Principles of the gravity method

The gravity method involves the measurement of variations of the gravity field of the Earth to detect differences of total mass (density times volume) of the sub-surface rocks. The basis of the method is the measurement of the acceleration due to gravity, g, which varies from 9.76 to 9.83 m s^{-2} over the surface of the Earth. Modern instruments (gravimeters) are capable of detecting (relative) spatial changes in the gravity field to an accuracy of 10^{-7} m s^{-2}.

The SI unit of acceleration is m s^{-2}. However, the spatial change in the gravity field caused by sub-surface variations in rock mass is typically only 10^{-4} m s^{-2}, so that smaller units are used to describe the gravity value, namely the gravity unit (g.u.), 10^{-6} m s^{-2}, and the milliGal, 10^{-5} m s^{-2}.

At a specific location, the value of g is influenced not only by the variation in the mass of the sub-surface rocks but also by elevation, latitude, local topography and tidal deformation. Thus, the precise location of the site and time of each observation must be recorded. In Britain, measurements are made at sites where both the elevation above sea level and the latitude/longitude are accurately known, such as at Ordnance Survey bench marks, spot heights and at local intersections of the height contours and identifiable topographic features. To achieve an accuracy of 0.01 milliGal in the final gravity map, the height and lateral position of each measurement station must be known to an accuracy of 4 cm and 10 m respectively. When the raw data have been corrected for these effects and are plotted at their correct locations, the final contoured map is known as the 'Bouguer gravity anomaly map'; this shows the variation of the gravity field, due solely to changes in the mass of the sub-surface rocks. The resulting field has not been 'downward-continued' to Ordnance Datum and still represents the anomaly measured along the topographic surface.

Gravity surveys carried out to investigate the sub-surface geology normally only measure the relative changes across the area of interest. To link these surveys together on a national and an

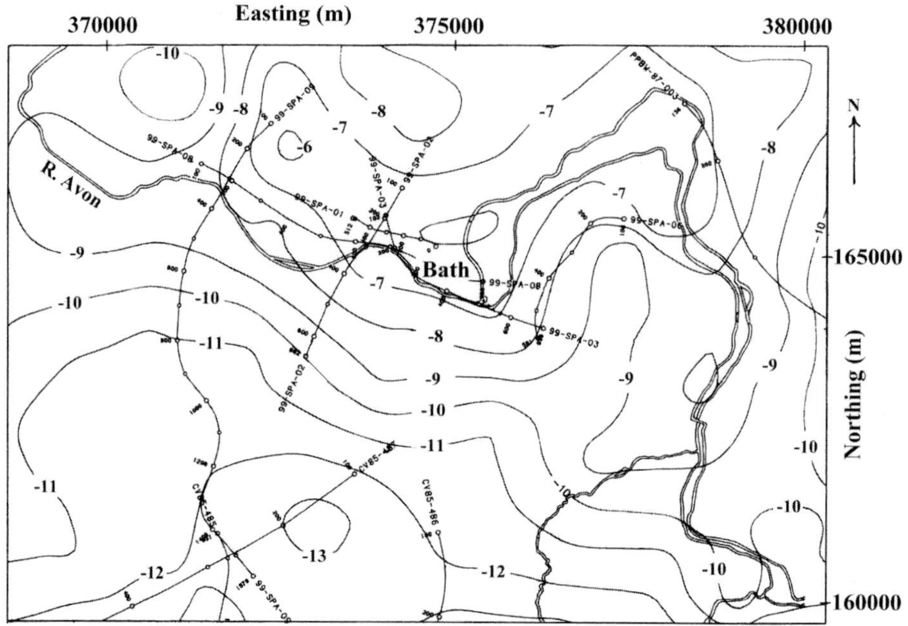

Fig. 7. Contoured map of the Bouguer gravity field in the area around Bath. Contour interval 1 milliGal. Locations of the seismic lines are shown. Based upon contoured digital gravity data supplied by, and with permission of, the British Geological Survey, IPR/18-2C.

international basis, a network of stations has been established at which the absolute value of gravity has been determined. This network is known as the International Gravity Standardisation Net (IGSN 71).

Production of the gravity map

The Bouguer anomaly gravity map of the Bath area was constructed by the British Geological Survey from data acquired as part of their national UK gravity survey over a period of 40 years. The local sub-surveys which form the national grid were interconnected using the National Gravity Reference Net (1973) which in turn was linked to the IGSN 71 described above. The density used to reduce the Bath area field gravity data to Ordnance Datum was 2650 kg m^{-3} (2.65 g cm^{-3}).

The Bath Spa gravity map (Fig. 7), was created by computer plotting and contouring all available gravity data from the British Geological Survey digital database on a 30 km by 30 km grid with the city at its centre. The map is based on 729 data values, approximately one per square kilometre. Careful analysis of the relationships between the contours and the individual data points indicates that the computer software has achieved accurate interpolation, and that the resulting map is stable and reliable.

To understand the interpretation of the gravity data two important points must be borne in mind from above. First, in deriving the Bouguer anomaly an attempt has been made to remove the gravity effects of the rocks above Ordnance Datum. As pointed out in the previous section, the Bouguer anomaly is still measured along the topographic surface. Any density variations that are different from the applied Bouguer density above the Ordnance Datum will show on the Bouguer anomaly. Secondly, and nevertheless, it is expected that the major variation of the Bouguer gravity values will arise from lateral changes in the mass of the sub Ordnance Datum rocks; for example, an increase in the thickness of a slab of low density rock overlying a higher density rock will cause a decrease in the Bouguer anomaly.

It was assumed that the regional variation of the field was not significant to the interpretation for the limited area over which the Bouguer gravity field has been analysed.

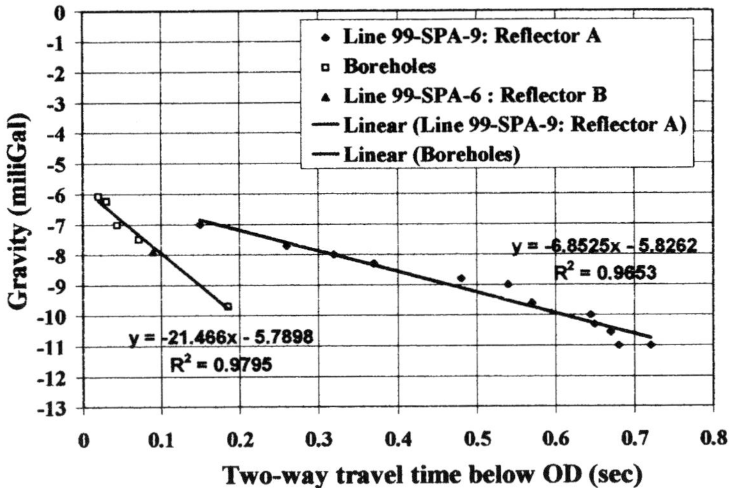

Fig. 8. Bouguer gravity value versus two way travel-time for reflector 'A' on line 99-SPA-09, reflector 'B' on line 99-SPA-06 and the Carboniferous Limestone interface from the boreholes.

Provisional correlation of Reflectors A and B by use of the gravity data

Falcon & Tarrant (1951) correlated increasing thickness of Mesozoic rocks unconformably overlying Palaeozoic basement with decreasing Bouguer gravity values, with a density contrast of 400 to 700 kg m^{-3}. They showed a similar correlation within the Palaeozoic between thickening Coal Measures conformably overlying Carboniferous Limestone and decreasing Bouguer gravity, with a density contrast of 150 kg m^{-3}. It can be assumed that the two-way travel time to a horizon on a seismic section is proportional to the thickness of the rock above that horizon. The correlation between TWTT and Bouguer gravity anomaly can therefore be used to identify density contrasts between different rock types. It will also be demonstrated in the next section that rock types can be identified by comparing the Bouguer gravity anomalies when their surfaces are at Ordnance Datum, that is, when the Bouguer effect of overlying rocks, with a significant density contrast, has decreased to zero.

Bouguer gravity values and the equivalent TWTT to Reflector A were picked along line 99-SPA-09 (CDP 340–740). Along line 99-SPA-06 both the Bouguer gravity value and the TWTT to Reflector B were almost constant and a single value was picked. For the local boreholes the equivalent two-way travel times to the tops of the Carboniferous Limestone (or the Dolomitic Conglomerate) were calculated for an average velocity of 2.0 km s^{-1} and the Bouguer gravity values were picked from the map. Three of the boreholes are shown in Figure 3; of the remaining three, two are outside the area of the map and one is in the 'area of thermal springs'.

Gravity and TWTT data are plotted in Figure 8. Regression lines were calculated for the data. The graphs show:

(1) that the Bouguer gravity values along line 99-SPA-09 are correlated with the TWTT to Reflector A with an intercept of –5.8 milliGal (R^2 = 0.97);
(2) that the borehole gravity data show a good correlation with the calculated TWTT to the Carboniferous Limestone, with a significantly greater gradient but with an intercept of –5.8 milliGal, identical to the line 99-SPA-09 intercept (R^2 = 0.98); and
(3) that the line 99-SPA-06 Reflector B gravity value clusters with the borehole data.

The density contrast for the borehole data, calculated from the gradient of graph 'Boreholes', is 520 kg m^{-3}, typical of that for Mesozoic overlying Palaeozoic rocks. The density contrast for Reflector A on line 99-SPA-09 (CDP 340–740), calculated from the gradient of

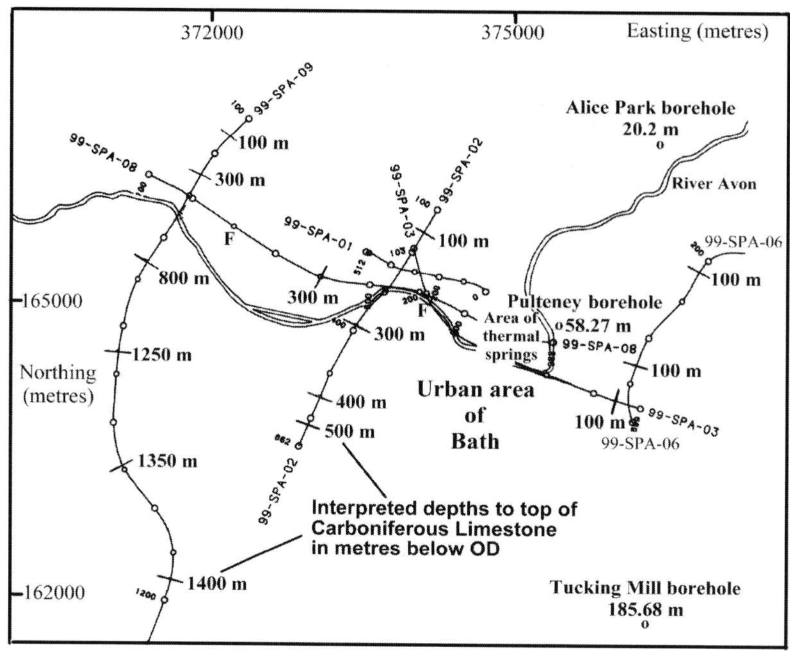

Fig. 9. Interpreted depths to the top of the Carboniferous Limestone in the Bath area.

the graph and assuming a surface-to-reflector velocity of 4 km s^{-1}, is 80 kg m^{-3}, typical of that for Coal Measures thickening over Carboniferous Limestone. The observed Bouguer gravity will also depend on the density contrast at the lower interface of the Carboniferous Limestone with the Upper Old Red Sandstone. However, the smaller density contrast between these two rocks means that the gravity effect of the Coal Measures/Carboniferous Limestone density contrast is likely to dominate.

The similar intercepts of the seismic/gravity regression lines for the 'Borehole' Carboniferous Limestone data and for Reflector A data from line 99-SPA-09 indicate that identical gravity values would be obtained when the two rocks are at Ordnance Datum. This is good evidence that Reflector A is close to the surface of the conformable Carboniferous Limestone beneath Coal Measures. The good fit between the gravity value for Reflector B of line 99-SPA-06 with the borehole data is good evidence that this is close to the unconformable Carboniferous Limestone beneath Mesozoic rocks. Of course it is not possible to draw the conclusion from the combined gravity and seismic data that Reflector A and Reflector B exactly represent the Carboniferous Limestone surface, without confirmation from a deep borehole. For example, on lines 99-SPA-09 and 99-SPA-02, the Carboniferous Limestone may be closely overlain by a high impedance rock layer within the Coal Measures. The conclusions drawn below should be read with this caveat in mind. Nevertheless, our provisional conclusion from the combined gravity and seismic data, subject to confirmation by a borehole, is that both Reflector A and Reflector B can be mapped as being stratigraphic surfaces 'somewhere near' to the Carboniferous Limestone interface. At the least, the seismic reflectors that have been identified can be used to map variations in depth of the Carboniferous Limestone surface. In the central area of Bath, the unconformable surface of the Carboniferous Limestone below the Mesozoic can be mapped from the borehole data, or is less than 100 m below OD, in the parts of the seismic sections where it cannot be clearly seen.

Results

Figure 9 shows interpreted spot depths to the Carboniferous Limestone below Ordnance Datum. These have been calculated from the TWTT via the stacking velocities. It has not been possible to calibrate these data because of the absence of deep boreholes in the critical areas. Consequently, there is a possible error of 20% in the interpreted depths.

Along line 99-SPA-09 the Carboniferous Limestone interface has a steep southerly apparent dip. At the north end of the line it is interpreted as being about 50 m below OD (about 120 m below the ground surface), probably its closest approach to the ground surface. Major faults in the Carboniferous Limestone, with throws from 15 to 90 m, have been tentatively identified on line 99-SPA-09.

Along line 99-SPA-02, which lies parallel to line 99-SPA-09, the Carboniferous Limestone interface has a much shallower apparent dip to the south and reaches an interpreted depth of 500 m below OD, while again reaching about 50 m below OD (about 150 m below the ground surface) at its northern end. There is a difference in depth between the two lines (99-SPA-09 and 99-SPA-02) at northing 163000 of some 800 m over a distance of some 3 km. Along line 99-SPA-02 the Carboniferous Limestone interface has a shallower apparent southerly dip and again reaches about 50 m below OD (about 150 m below the ground surface) at its northern end.

Along line 99-SPA-08 two major faults have been provisionally identified (marked F in Fig. 9) with throws of 85 m (upthrow to the east) and 90 m (upthrow to the east) respectively, bringing the Carboniferous Limestone interface to about 50 m below OD.

East of the springs on line 99-SPA-06, the minimum elevation of the ground surface is 60 m above OD. The ground surface rises rapidly to the east and south. The Carboniferous Limestone interface is at an interpreted constant depth of about 100 m below OD, about 160 m below the ground surface.

Discussion and conclusions

The proposal to create a new spa in the City of Bath has made understanding the origin of the thermal springs vital to predict their future behaviour. This paper describes the design, acquisition and provisional interpretation of a reflection seismic survey, undertaken in and around the urban area of Bath as part of an ongoing programme of geological and geophysical research into the mechanisms of the springs. By combining seismic data with existing gravity data from the British Geological Survey, it has been possible to identify stratigraphic horizons 'close to' the surface of the thermal water aquifer, the Carboniferous Limestone.

The data show that the Carboniferous Limestone surface is at rather shallow depths (less than 300 m) in and around Bath, except in the sector to the southwest of the city. In this sector the Carboniferous Limestone dips rapidly to great depth. For example, at a distance of 2.1 km SW of the springs, the depth of the (interpreted) Carboniferous Limestone surface increases from 0.4 km below OD to 1.35 km below OD within a distance of 1.8 km, an average apparent dip of nearly 30 degrees. This clearly demonstrates that towards the south-west of the city, and within 4 km of the King's Spring, the Limestone sequence approaches the depth at which the temperature is sufficient to heat the spa water.

It can be speculated that the steeply dipping Limestone could provide a path for the rapid transport of hot water into the region of the springs, if it has the appropriate permeability. This may indicate that the source of the water is more local to Bath than is implied by the Mendip Theory.

In this paper we have described the techniques being used to investigate the geological structure beneath the urban area of Bath, where the hot springs rise to the ground surface, and in the surrounding area. We have presented a provisional interpretation of the seismic survey and some speculations arising from the data. Whilst new information has been added to our knowledge of this structure, it is not yet possible to identify conclusively the source, pathways and heating mechanism of the water. Further work is clearly necessary and the research programme is ongoing. We are undertaking a detailed analysis of the shallow and deep geology of the Bath area using existing borehole and mineshaft data, together with the results of the new seismic survey. We are re-processing and re-interpreting existing seismic lines and gravity data in the areas to the south and east of Bath to determine the structure of the Palaeozoic floor. The work will be presented in future papers and reports to enable geologists, hydrogeologists and geochemists to incorporate the findings into existing and new models of the thermal springs of Bath.

This research was funded by The Millennium Commission and by Bath and North East Somerset Council. We thank the staff of the Bath Spa Project, particularly P. Simons, R. Samuel and R. Davies, for their continuing interest and support.

References

ANDREWS, J. N. 1991. Radioactivity and dissolved gases in the thermal waters of Bath. *In*: KELLAWAY, G. A. (ed.) *Hot Springs of Bath*. Bath City Council, Bath.

ATES, A. & KEAREY, P. 1993. Deep structure of the East Mendip Hills from gravity, aeromagnetic, and seismic reflection data. *Journal of the Geological Society of London*, **150**, 1055–1063.

BRITISH GEOLOGICAL SURVEY. 1980. *Aeromagnetic anomaly map. Bristol Channel Sheet, 51N-04W*. Natural Environment Research Council, Swindon.

BRITISH GEOLOGICAL SURVEY. 1986. *Bouguer gravity anomaly map. Bristol Channel Sheet, 51N-04W*. Natural Environment Research Council, Swindon.

BURGESS, W., BLACK, J. & COOK, J. C. 1991. Regional hydrodynamic influences on the Bath-Bristol springs. *In*: KELLAWAY, G. A. (ed.) *Hot Springs of Bath*. Bath City Council, Bath.

FALCON, N. L. & TARRANT, L. H. 1951. The gravitational and magnetic exploration of the Mesozoic covered areas of South-Central England. *Quarterly Journal of the Geological Society*, **106**, 141–170.

KELLAWAY, G. A. 1996. Discovery of the Avon-Solent Fracture Zone and its relationship to the Bath Hot Springs. *Environmental Geology*, **28**, 34–40.

LEE, M. K. 1986. Hot dry rock. *In*: DOWNING, R. A. & GRAY, D. A. (eds) *Geothermal Energy – The Potential in the United Kingdom*. HMSO, London.

Groundwater resource management in eastern England: a quest for environmentally sustainable development

D. B. BURGESS

Environment Agency, Orton Goldhay, Peterborough PE2 5ZR, UK
(e-mail:david.burgess@environment-agency.gov.uk)

Abstract: The management of groundwater resources in England and Wales was initially based only on measures of the renewable resource. This has been extended to include the need to preserve the springs, river flows and surface water levels dependent on groundwater discharges as a key objective of a sustainable management of groundwater resources. The impacts of all new groundwater abstraction proposals on the surface water environment are now evaluated using a number of techniques, some of which are under further development. The sustainable management of groundwater catchments also includes the control of pumping from existing groundwater sources to meet agreed environmental targets. This approach is illustrated by examples of groundwater catchments managed to augment surface flows, to prevent saline intrusion and to preserve the integrity of wetland conservation sites.

Groundwater pumping provides about of a third of the public water supplies of England and Wales. The total annual amount of groundwater abstracted amounts to over 2300 million m^3 per year. In recent years there has been increasing concern over the impact of groundwater pumping on rivers, springs and wetlands. To accommodate these concerns, the concept of the sustainable management of groundwater resources has evolved as giving equal consideration to the needs of groundwater users together with the needs of the water environment.

The Environment Agency of England and Wales controls the abstraction of groundwater by issuing abstraction licences under the Water Resources Act 1991. This, together with the Water Resources Act 1963, requires the licensing authority to assess the impact of all new pumping proposals on both existing water rights and river flows. This concern for river flows has now been extended to include all aspects of the water environment such as springs, seepages and wetlands. The Environment Act 1995 not only set up the Environment Agency but also made it responsible for ensuring the sustainable development of land, air and water.

This paper reviews how the concept of environmentally sustainable development is being applied to the management of groundwater resources of eastern England. It examines how groundwater resource limits are estimated, the methods used for assessing the impact of new groundwater abstraction proposals on surface waters and the management of groundwater pumping to meet local environmental targets.

Defining groundwater resource availability

The rational planning and management of a renewable resource, such as groundwater, requires measures that indicate both the current and the potential resource availability status of individual aquifer units or groundwater catchments. Within England and Wales, such measures have usually been developed for two purposes. Firstly, as a means of limiting the issue of new groundwater abstraction licences to levels that prevent environmental damage or

From: HISCOCK, K. M., RIVETT, M. O. & DAVISON, R. M. (eds) *Sustainable Groundwater Development*. Geological Society, London, Special Publications, **193**, 53–62. 0305-8719/02/$15.00

derogation of existing water rights. Secondly, to produce a measure of groundwater resource availability for water resource planning purposes that is analogous to the yield calculations used for surface reservoirs.

Ideally, groundwater resource estimates are best based on the results of regional numerical models constructed to simulate groundwater flow for the whole catchment or aquifer unit (e.g. Rushton *et al.* 1989). Such models are necessary to provide an accurate assessment of groundwater yield and in forecasting how different pumping patterns are likely to affect groundwater levels and groundwater to surface discharges. The Environment Agency has a 10-year programme for deriving such models for all the major aquifer units of England and Wales. In the interim, there remains the need to evaluate groundwater resources at the catchment scale, employing simple, albeit less accurate, estimates of resource availability for those units where groundwater models are not available.

Simple estimates of the renewable resource of a groundwater catchment are usually based on the assumption of steady-state or dynamic equilibrium so that in the long term, recharge to groundwater equals groundwater discharge. In the UK, the first estimates of groundwater resources were derived from the infiltration over a notional recharge area, the infiltration being derived from long-term records of rainfall minus actual evaporation with some adjustment for direct surface run-off (e.g. Day 1964). Estimates based on groundwater discharge employed baseflow separation of hydrographs from gauging stations with records exceeding 10 years (e.g. Ineson & Downing 1965).

Both these methods have limitations. For example, estimates of recharge are reliant on soil moisture accounting techniques, which are limited by assumptions about the extent and initial status of the soil moisture store. These assumptions can only be tested at a small number of sites within England and Wales where regular measurements of soil moisture have been made (Ragab *et al.* 1997). Equally, the estimation of groundwater discharge is dependent on the arbitrary methods of baseflow separation from river hydrographs (Gustard *et al.* 1992). The method also requires a flow record free of upstream artificial influences such as sewage treatment works or industrial effluents as well as pumping from surface and groundwater. This restriction severely limits the successful use of baseflow separation techniques in many of the highly developed groundwater catchments of southern England.

Both methods require some knowledge of the groundwater table in order to define the groundwater catchment that is receiving recharge or contributing to groundwater discharge. The seasonal fluctuation of water tables can make the definition of steady-state recharge areas and groundwater catchment boundaries difficult. For example, some Chalk groundwater catchment areas are known to fluctuate seasonally by over 30% (e.g. Gypsy Race, East Yorkshire; Foster & Milton 1976).

Although such methods help define the renewable resource, this should not be taken as equating to the environmentally sustainable groundwater resource. Allowing an aquifer to be pumped up to the renewable rate will result in widespread lowering of groundwater levels and reductions in spring and river flows as groundwater storage is seasonally depleted. This would result in the derogation of water rights and environmental damage associated with a depletion of springs and flows to wetland conservation sites (Sophocleous 1997). It would also result in a reduction of downstream river flows with consequent negative effects on fisheries, river ecology and the yield of water supply intakes. In coastal aquifers, pumping up to the renewable resource limit can reverse groundwater table gradients, which would result in widespread water quality changes.

In order for estimates of groundwater resources to be considered sustainable, estimates of the renewable resource must be adjusted downwards by amounts that protect both the existing groundwater users and the environmental features dependent on groundwater. An example of this approach is that taken in planning the groundwater resources for the relatively dry, East Anglian region of England (National Rivers Authority 1994). Over this region, and for each groundwater unit, the following simple accounting procedure was adopted to estimate the sustainable resource:

(1) The renewable resource was calculated from long-term, annual average estimates of either infiltration to groundwater, baseflow from river hydrographs or, where available, from groundwater models. The renewable resource was then reduced to allow for the depletion of groundwater storage that occurs during droughts that last longer than 1 year, such as that experienced between 1989 and 1991. The figure used to reduce the renewable resource of the Chalk aquifer units of East Anglia was an arbitrary 20%. The derivation of a more objective method for calculating this reduction figure is the

subject of ongoing research and development (Environment Agency 1997).

(2) The amount of groundwater discharge required to maintain the quality, the ecological status and the reliability of any downstream river intakes was then subtracted from the renewable resource calculated above. This was again arbitrarily taken as the 95 percentile flow from a flow duration curve calculated at a river gauging station nearest to the outfall of the catchment. Where a gauging station did not exist or where records were of insufficient length, the 95 percentile was estimated from regression equations based on catchment characteristics (Gustard *et al.* 1992). For aquifers bounding coastal areas, a groundwater flow of $0.5 \mathrm{l\,s^{-1}}$ per kilometre wide groundwater flow front was used to estimate the discharge required to prevent saline incursion.

(3) The renewable resource was further adjusted to include the net effect of pumping under current abstraction licences to give the sustainable resource as available for further abstraction. The net effect is calculated from a series of rules on how pumping from current groundwater sources depletes river flows and how this is offset by the return of effluents discharged to the rivers in the groundwater catchment in question.

The above procedure has provided a simple and effective method of indicating resource availability. It has also provided an important tool for the control of groundwater abstraction licences or permits. Tables and maps have been drawn up that indicate those groundwater units where existing authorized abstraction amounts exceed the available resource and where groundwater is no longer considered to be available for further abstraction. They also show those groundwater units where resources are still available and where applications for new groundwater licences may be considered (Fig. 1).

Methods for assessing the local impact of groundwater pumping

It is the policy of the Environment Agency that all new proposals for significant groundwater pumping must be accompanied by a report detailing the likely effect on surrounding groundwater users, springs, rivers and wetlands. Detailed guidance on this procedure and the test pumping of new boreholes is given in the Agency's water resources licensing manual (Environment Agency 1998).

Evaluating and forecasting the likely impact of groundwater pumping on springs and river flows is a requirement for most proposals received for unconfined groundwater units. The methods used range from analytical solutions for the initial evaluation of a pumping proposal, through to the use of calibrated numerical models when abstraction proposals are thought to impact on contentious rivers or wetlands.

The main analytical technique used is based on the Theis (1941) solution as adapted by Jenkins (1970). In common with other analytical techniques, these equations include a number of restrictive assumptions, the most notable being that the river and borehole fully penetrate the aquifer, that the river is the only source of recharge and that the aquifer is isotropic and homogenous. Nevertheless, the techniques provide quick and useful 'first pass' estimates of stream depletion used for designing test pumping and as a basis for deciding the need for further investigations. The techniques also provide a useful insight into both the time lag between the start of pumping and depletion of river flow as well as the persistence of stream depletion after groundwater pumping has stopped. This can be particularly useful in evaluating the effect of seasonal pumping cycles on river flows associated with spray irrigation licence applications (Wallace *et al.* 1990).

Such analytical techniques can be readily adapted for easy use within spreadsheet computer programmes for use by licensing hydrogeologists (Environment Agency 1999). Even so, the oversimplification and idealization of such simple analytical models limits their use to the preliminary screening of groundwater proposals and to the design for test pumping of new boreholes. For most of the aquifers of south-east England, where the rivers run on extensive floodplain deposits, the assumptions of a fully penetrating river and isotropic conditions within the analytical solutions could be considered an oversimplification and unrealistic.

In such situations stream depletion may be better estimated by reference to a series of numerical models, calibrated for a number of simplified, idealized, aquifer-valley cross-sections. Such methods have been used to explore the sensitivity of stream depletion to uncertainties in aquifer parameters (Sophocleous *et al.* 1995). The Agency has commissioned the development of similar techniques to develop methods to provide estimates of stream depletion resulting from single boreholes with simple pumping patterns for typical aquifer-valley cross-sections. However, estimates of stream depletion for the more complex groundwater

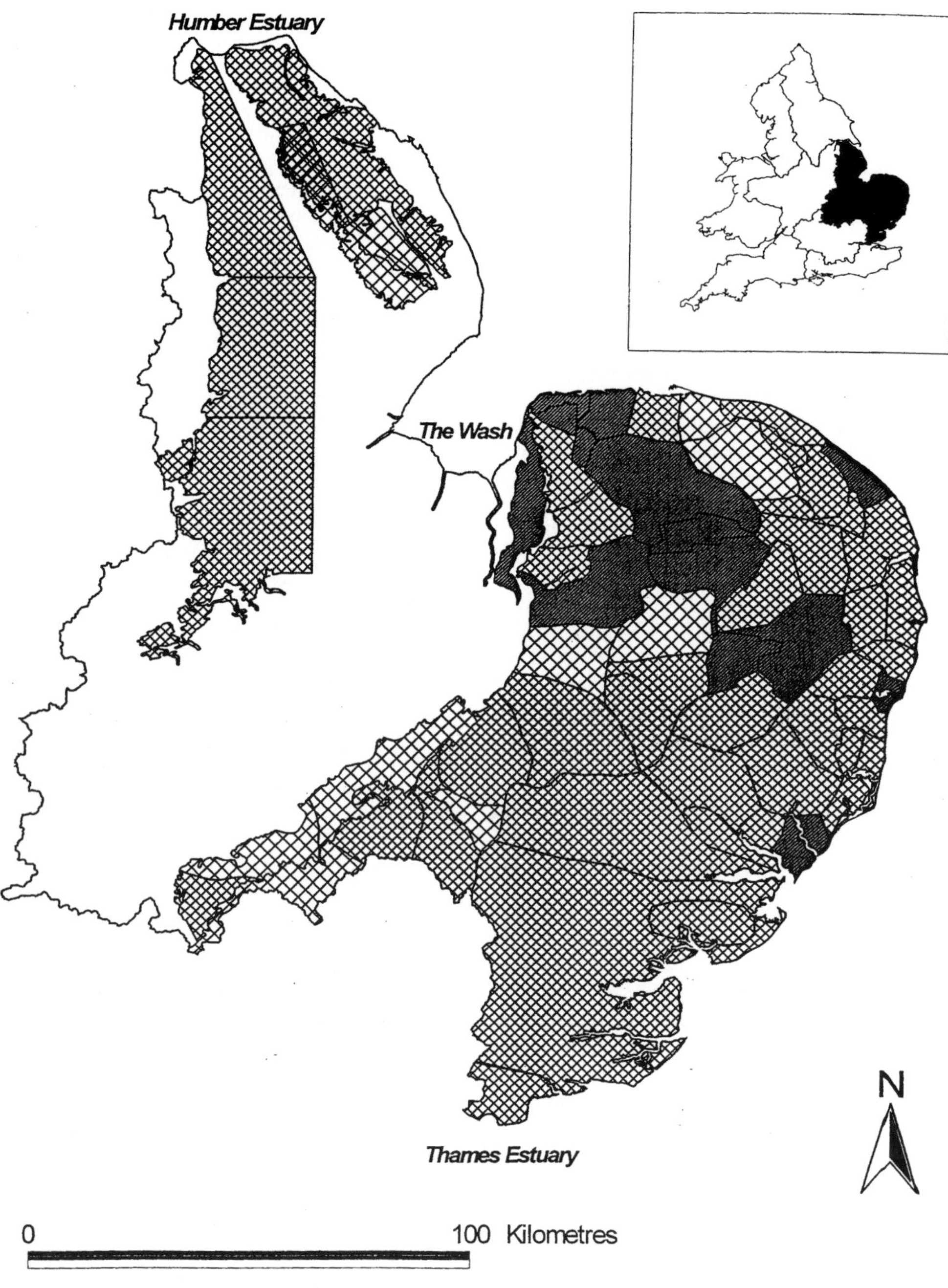

Fig. 1. Groundwater resource availability map: Environment Agency, Anglian Region, 1998.

conditions, the large number of boreholes and the variety of pumping patterns experienced on the major aquifers of the UK, will continue to rely on the results from calibrated regional numerical groundwater models.

Managing groundwater pumping to meet environmental targets

While the local impact of individual proposals to pump groundwater has tended to be evaluated on a site by site basis, catchment-wide development of groundwater resources has required a more integrated and considered approach. The management of groundwater pumping to meet environmental targets such as surface water flows or to prevent deterioration in quality remains a key objective for catchment-wide schemes for the sustainable development of groundwater resources.

Seasonal groundwater pumping to augment river flows

The seasonal pumping of groundwater into rivers to augment lower flows is now an established pattern of development for the major aquifers in England and Wales. The effectiveness and efficiency of this type of pumping depends on the lag and attenuation between the start of pumping and depletion of flows already mentioned above. A full account of the theory and methods used for this type of development are given in Downing (1986).

The conventional use of seasonal groundwater pumping to rivers is to increase the yield and reliability of downstream river water supply intakes. A good example of this type of scheme is that constructed within the Thet catchment, 12 km NE of Thetford in the county of Norfolk (Fig. 2). Here groundwater is seasonally pumped from the underlying Chalk aquifer into tributaries of the river Thet to augment low, summer river flows at an intake at Denver, 45 km downstream. The scheme consists of 18 production boreholes constructed to depths of between 79 and 122 m below ground level. Groundwater pumped from each of these is discharged via pipelines to the nearest watercourse. The total length of the pipeline is over 12 km. The gross output of the boreholes is 78 $Ml\,d^{-1}$ declining to 61 $Ml\,d^{-1}$ at the end of a season of pumping. The net gain to flows at the Denver intake is estimated to be between 45 and 30 $Ml\,d^{-1}$ (Backshall *et al.* 1972).

Increasingly, seasonal groundwater pumping is used to augment flows on rivers where traditional direct borehole supplies have locally depleted river flows to environmentally unacceptable levels. A good example of this type of scheme is that within the Lodes–Granta catchment to the north-east of Cambridge, England (Fig. 2). This is an area of some 608 km^2 largely underlain by the mainly unconfined Chalk aquifer. The catchment lies in one of the driest areas of the UK, with a mean annual effective rainfall of 140 mm and frequent summer droughts. As a result, local springs and rivers are heavily dependent on groundwater discharges to maintain summer flows into a number of internationally important wetland and spring conservation sites.

The Lodes–Granta Chalk aquifer is also a major source of public water supply. Increased demand of up to 55 $Ml\,d^{-1}$ from 14 sources throughout the 1980s resulted in concern for the integrity of the wetlands and summer flows. The options to resolve this potential conflict were explored using a numerical model (Rushton & Fawthrop 1991) to select a pattern of pumping that met all the local environmental targets for maintaining spring flows and wetlands. This resulted in a scheme, finished in 1994, that included six new boreholes to seasonally pump water to 12 springheads via nearly 40 km of pipeline. The maximum output of the scheme is 20 $Ml\,d^{-1}$ but has resulted in the groundwater pumping allowed for public supply to be raised to 80 $Ml\,d^{-1}$.

Managing groundwater pumping to control saline intrusion.

Saline intrusion has occurred at a limited number of locations in England where Chalk and Triassic sandstone aquifers are over-pumped near coastlines and estuaries (Downing 1986). For these aquifers, setting the sustainable rate of abstraction requires knowledge of the rate of recharge, amounts of groundwater pumped, and the groundwater flow within the aquifer required to prevent or limit the extent of saline intrusion.

For the north Lincolnshire Chalk aquifer unit (Fig. 2), a more dynamic approach has been adopted. Historically, this aquifer has provided a source of relatively cheap water to meet the public water supply and industrial demands of South Humberside. As a result of over-pumping in the 1950s a saline intrusion of over 2.5 km occurred inland around Grimsby (Gray 1964). The licences to pump groundwater predate the rational control and assessment required by the 1963 and 1991 Water Resources Acts. Conse-

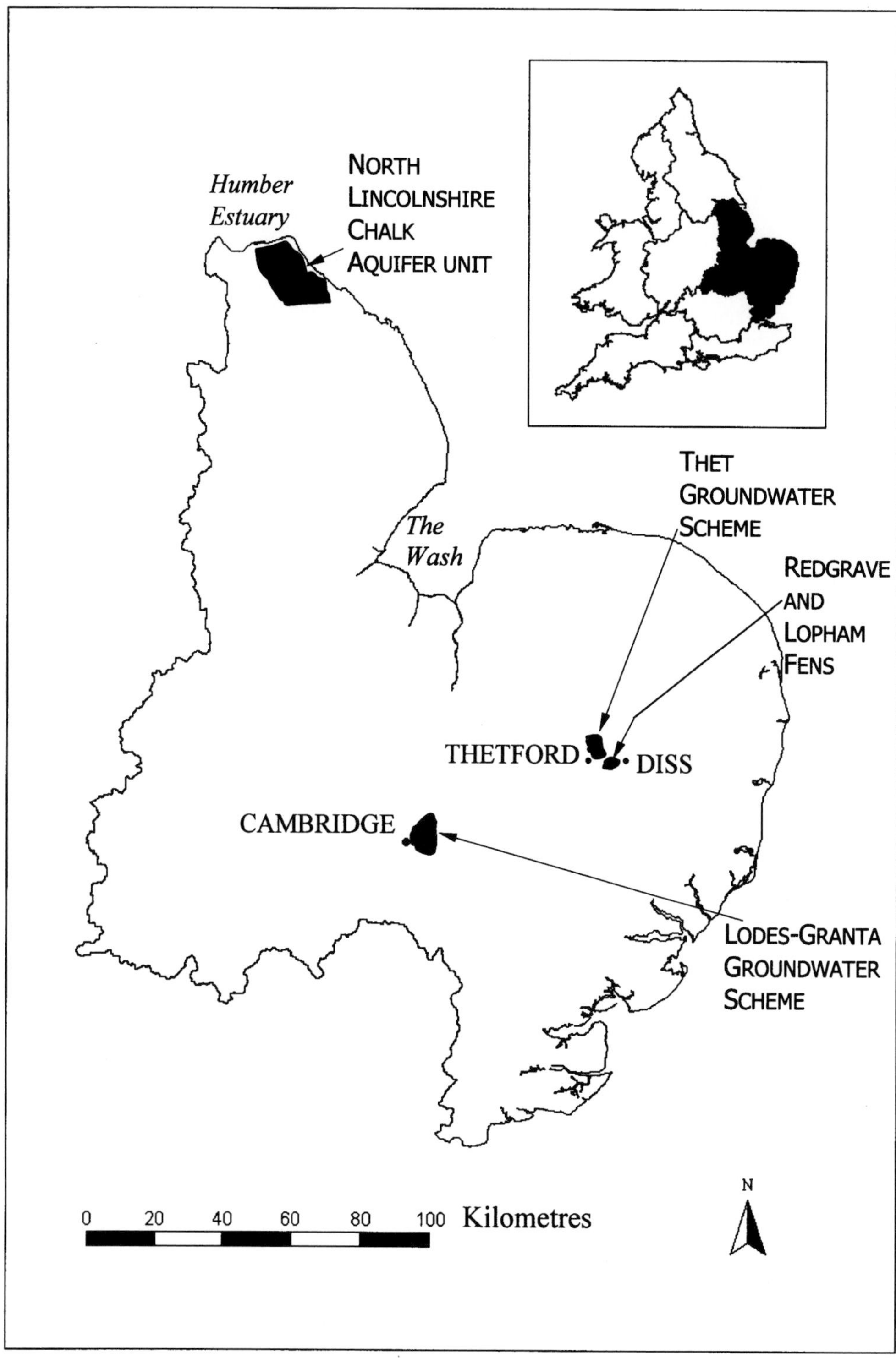

Fig. 2. Eastern England: location of groundwater development schemes and groundwater-fed wetlands mentioned in the text.

quently, the permitted rate of groundwater abstraction still remains at 193 Ml d^{-1}, which is over 85% of the estimated mean annual groundwater recharge.

Groundwater investigations over the last 20 years were aimed at the development and calibration of a reliable numerical model for the aquifer unit. This model was originally developed to assess the strategic options for managing the aquifer (Spink & Wilson 1989). However, following the significant groundwater droughts of 1989/92 and 1995/96 its use was extended to include the operational control of groundwater pumping in conjunction with surface water storage. To achieve this, the recharge and groundwater pumping are regularly updated within the model on a 3-monthly basis. The model is then used to forecast the movement of saline water for different combinations of likely recharge and different pumping patterns. The results from these runs are then discussed with groundwater users to decide the groundwater pumping rates that can be achieved without increased and unacceptable salt concentrations. In this way any shortfall in groundwater source outputs is forecast, agreed and the deficit made up from planned pumping of surface water transfers and storage. Consequently, during the drought of 1989 to 1992, groundwater pumping from the Chalk aquifer was kept at between 110 and 135 Ml d^{-1} but with no significant rise in chloride concentrations as measured at key monitoring boreholes (Spink & Watling 1995).

Relocating groundwater pumping away from wetland conservation sites.

The discharge of groundwater to the surface is characterized by poor land drainage, perennial ponds, areas of fen, springs and seepage to rivers. Even in the densely populated, agricultural landscapes that cover the aquifers of England and Wales, some of these features have survived in an isolated, semi-natural state. The ecological value of such wetland sites has been increasingly recognized with many being conserved and protected under National (Wild Life and Countryside Act 1981) and European legislation (EC Habitats Directive (92/43/EC); The Conservation (Natural Habitats, etc.) Regulations 1994).

The dispersed nature of these features, matched against the widespread demand in borehole sources for both rural water supplies and spray irrigation, has resulted in a number of sites where groundwater pumping is perceived or has been shown to reduce flows or water levels to neighbouring wetlands. For new groundwater pumping proposals likely to affect wetland conservation sites, it is common for the proposed sites of boreholes to be either relocated away from the wetland, or for conditions to limit the effects of pumping to be included within the abstraction licence, or for the application for a licence to be refused.

Early recognition of the role of groundwater in maintaining flows and water levels in wetlands can help in assessing their susceptibility to being adversely affected by nearby groundwater pumping proposals. Such an approach was adopted in a study of 60 different wetland sites in East Anglia. This study resulted in the wetlands being classified according to susceptibility to nearby groundwater pumping, together with guidance on the investigations necessary to forecast the impact of groundwater pumping on nearby wetlands (Lloyd *et al.* 1993). Water balance and analytical equations have also been developed to provide a preliminary and rapid estimate of the impact of different groundwater pumping proposals on groundwater-fed wetlands (Williams *et al.* 1995).

Evaluating the effects of existing groundwater abstraction licences on neighbouring wetlands requires both a hydrogeological investigation to quantify changes in flows and water levels together with ecological surveys to evaluate the conservation significance of the changes. These investigations often involve detailed field work both in terms of vegetation surveys and accurate piezometry. Where an adverse effect of groundwater pumping is evident, the Agency has taken steps either to reduce the pumping rate or to relocate the borehole source away from the affected wetland (National Rivers Authority 1993).

A good example of this approach is provided at the Redgrave and Lopham Fen in the headwaters of the River Waveney in East Anglia (Fig. 2). The fen is an internationally recognized conservation site, covers nearly 125 ha and is the largest fen of its type in lowland England. In the late 1950s, two public water supply boreholes were drilled into the underlying Chalk aquifer immediately adjacent to the fen. These supplied up to 3.6 Ml d^{-1} to surrounding towns and villages. The close proximity of the pumping resulted in abnormal changes to the pattern of groundwater flow to the fen. The normal condition of perennial, high water levels with Chalk groundwater discharging through the fen was replaced by a seasonal downward movement of surface water (Fig. 3a). During summer months the fen dried out more frequently with groundwater heads reduced to a metre below the

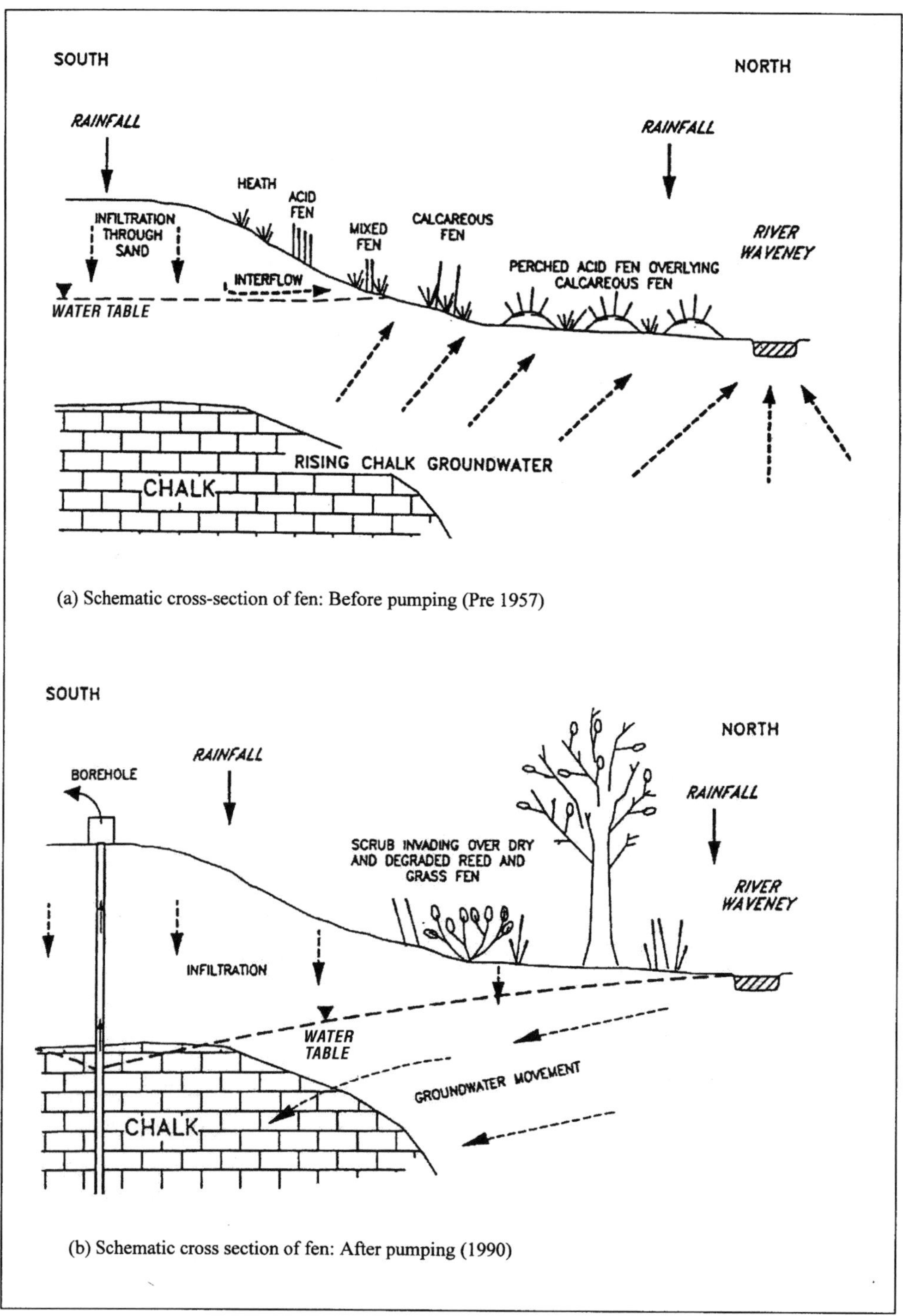

(a) Schematic cross-section of fen: Before pumping (Pre 1957)

(b) Schematic cross section of fen: After pumping (1990)

Fig. 3. Schematic cross–sections through Redgrave and Lopham Fen, eastern England, illustrating groundwater and ecological conditions (**a**) before groundwater pumping and (**b**) after several years of groundwater pumping from a neighbouring public water supply borehole.

fen surface (Fig. 3b). Test pumping and radial flow modelling suggested that some 26% of the pumped groundwater was at the expense of seepage and spring flow into the fen. These hydrogeological changes were matched to a deterioration of the flora and fauna at the site (Harding 1993). Following discussions with the water company and English Nature, the decision was taken to relocate the groundwater pumping to a borehole 3.5 km E of the site and downstream of the fen; this replacement borehole became fully operational in 1999. The original borehole near the fen is no longer in use and the recovery of water levels and of fen habitats is currently being evaluated. The total cost of the replacement supply is in the order of £3.3 million, which includes the cost of the investigations, source works, pipeline and restoration work on the fen.

Conclusions

The wide variety of examples of control of groundwater pumping at both the site and the catchment scale illustrates well how the concept of environmentally sustainable development is being applied to the management of groundwater resources in eastern England.

Initial, far-sighted legal obligations to consider the effects of groundwater pumping on downstream river flows have now been replaced by a widespread public concern to maintain water levels and flows to preserve the health and ecology of the whole of the water environment. The outcome is that groundwater and surface water are increasingly being managed as a combined resource.

In the past, groundwater resource development relied on single measures such as estimates of 'safe yield' to describe the limits of resource development. Consideration of the effects of future climate change has added uncertainty to such measures. However, of far greater importance is the change in the value that society puts on the different components of groundwater discharge and their importance to preserving the water environment. Thirty years ago, features such as groundwater-fed springs and wetlands were given little or no protection by groundwater managers and were often seen as a challenge to the land drainage engineer. Such features are now valued for their biological diversity. As such, the maintenance of groundwater levels within wetlands and groundwater seepage to springs and rivers are seen as important considerations for inclusion within all environmentally sustainable, catchment management plans.

The methods and results described in this paper owe a lot to my colleagues too numerous to mention. However, the opinions and views given are those of the author only and are not those of the Environment Agency.

References

Backshall, W. F., Downing, R. A. & Law, F. M. 1972. Great Ouse groundwater study. *Water and Water Engineering*, **76**, 215–223.

Day, J. B. W. 1964. *Infiltration into a groundwater catchment and the derivation of evaporation*. Geological Survey of Gt. Britain, Research Report **2**.

Downing, R. A. 1986. Development of groundwater. *In*: Brandon, T. W. (ed.) *Groundwater: Occurrence, Development and Protection*. Institution of Water Engineers & Scientists, London, 485–542.

Environment Agency. 1997. *Groundwater resource reliable yield*. Environment Agency Research & Development Technical Report **W9**.

Environment Agency. 1998. *Water Resources: Manual of Licensing*. Environment Agency, Bristol.

Environment Agency. 1999. *Impact of groundwater abstractions on river flows*. National Groundwater & Contaminated Land Centre Report **NC/06/28**.

Foster, S. S. D. & Milton, V. A. 1976. *Hydrological basis for large scale development of groundwater storage capacity in the East Yorkshire Chalk*. Institute of Geological Sciences, London, Technical Report **76/3**.

Gray, D. A. 1964. *Groundwater conditions in the Chalk of the Grimsby area, Lincolnshire*. Water Supply Papers of the Geological Survey of Gt. Britain, Research Report **1**.

Gustard, A., Bullock, A. & Dixon, J. M. 1992. *Low flow estimation in the United Kingdom*. Institute of Hydrology Report **108**.

Harding, M. 1993. Redgrave and Lopham Fen, East Anglia, England: A case study of change in flora and fauna due to groundwater abstraction. *Biological Conservation*, **66**, 35–45.

Ineson, J. & Downing, R. A. 1965. Some hydrological factors in permeable catchment studies. *Journal of the Institution of Water Engineers*, **19**, 59–80.

Jenkins, C. T. 1970. *Computation of the rate and volume of stream depletion by wells*. Techniques of Water–Resources Investigations of the United States Geological Survey, Book 4, Chapter **D1**.

Lloyd, J. W., Tellam, J. H., Rukin, N. & Lerner, D. N. 1993. Wetland vulnerability in East Anglia; a possible conceptual framework and generalised approach. *Journal of Environmental Management*, **37**, 87–102.

National Rivers Authority. 1993. *Low Flows and Water Resources*. Bristol.

National Rivers Authority. 1994. *Water Resources in Anglia*. Peterborough.

Ragab, R., Finch, J. W. & Harding, R. J. 1997. Estimation of groundwater recharge to chalk and sandstone aquifers using simple soil models. *Journal of Hydrology*, **190**, 19–41.

RUSHTON, K. R., CONNORTON, B. J. & TOMLINSON, L. M. 1989. Estimation of groundwater resources of the Berkshire Downs supported by mathematical modelling. *Quarterly Journal of Engineering Geology*, **22**, 329–341.

RUSHTON, K. R. & FAWTHROP, N. P. 1991. Groundwater support of stream flows in the Cambridge area, UK. *International Association of Hydrological Science Publication*, **202**, 367–376.

SOPHOCLEOUS, M. 1997. Managing water resource systems: why 'safe yield' is not sustainable. *Ground Water*, **35**, 561.

SOPHOCLEOUS, M., KOUSIS, A., MARTIN, J. L. & PERKINS, S. P. 1995. Evaluation of simple stream–aquifer depletion models for water rights administration. *Ground Water*, **33**, 579–588.

SPINK, A. E. & WATLING, D. J. 1995. Use of regional groundwater models in an operational role. *In:* SKINNER, A. (ed.) *Regional Groundwater Resource Modelling*. Institution of Water & Environmental Management, Reading, 131–142.

SPINK, A. E. & WILSON, E. M. 1989. Groundwater management in coastal aquifers. *International Association of Hydrological Science Publication*, **173**, 101–110.

THEIS, C. V. 1941. The effect of a well on the flow of a nearby stream. *American Geophysical Union Transactions*, **22**, 734–738.

WALLACE, R. B., DARAMA, Y. & ANNABLE, M. D. 1990. Stream depletion by cyclic pumping of wells. *Water Resources Research*, **26**, 1263–1270.

WILLIAMS, A., GILMAN, K. & BARKER, J. 1995. *Methods for the prediction of the impact of groundwater abstraction on East Anglian wetlands*. British Geological Survey Report **WD/95/SR**.

Institutional development for sustainable ground-water management – an Arabian perspective

M. E. YOUNG

Water Management Consultants Chile Ltda., Isidora Goyenechea 3162, Las Condes, Santiago, Chile (e-mail: myoung@watermc.com)

Abstract: In the Arabian Peninsula, hydrogeological and engineering solutions have failed to solve the severe and worsening problem of unsustainable groundwater abstraction, which threatens rural environments and livelihoods. Conventional western fiscal and regulatory measures to reduce abstractions seem to be impracticable in the present institutional and social contexts. In the region, groundwater rights without volume limitations are distributed mostly among numerous private well owners, and individual interests predominate over a communal imperative for aquifer sustainability. The solution may lie more in modifying the institutional context than in attempting to introduce official controls. This would involve the decentralization of water resources management to a basin or aquifer level and the development of local users associations. Water users associations could improve users' understanding of local hydrological limitations, promote conservation among irrigators, and cooperatively develop sustainable strategies and rules, which might ultimately include tradable rights and quotas. Government subsidies and incentives are necessary. Essential components of this participatory approach are strong leadership at national and local levels, the active engagement and leadership of Islamic institutions, and the use of modern communication methods.

Of all countries, the states of the Arabian Peninsula (Fig. 1) are among the most reliant on groundwater (Fig. 2), and among the most threatened by its depletion (Fig. 3). Aquifer stress in the region is frequently reported, reflecting the effects of high population growth, industrial development, and national policies that encouraged the growth of irrigated agriculture; 60–95% of water consumption in the Arabian Peninsula states is for irrigation (Fig. 4).

The possible social and economic consequences of aquifer depletion are serious, and include the demise of rural communities, urban migration, and the disruption of the social fabric. Inappropriate allocation of water to sectors of national economies with poor returns is a constraint on economic growth (Schiffler 1998; World Bank 1999). Internal social conflict over scarce groundwater resources has already occurred and the prospects of international conflict over shared resources are often discussed.

Fig. 1. Map of the Arabian Peninsula.

The groundwater crisis has been widely debated in the region. Various technical, fiscal

From: Hiscock, K. M., Rivett, M. O. & Davison, R. M. (eds) *Sustainable Groundwater Development*. Geological Society, London, Special Publications, **193**, 63–74. 0305-8719/02/$15.00

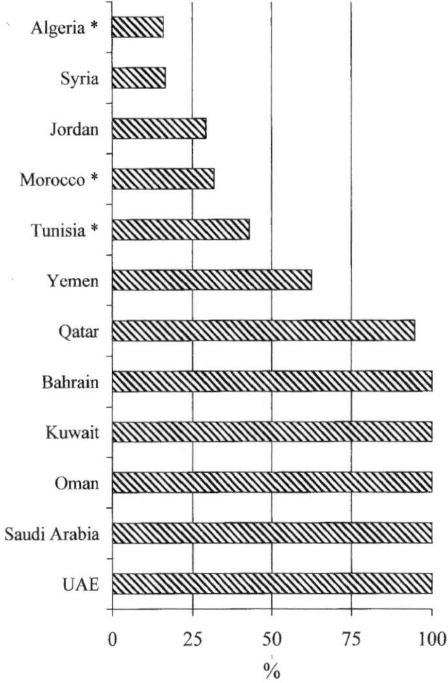

Fig. 2. Groundwater as a percentage of total renewable water resources (Lajaunie 1999). Asterisks indicate countries outside the region; shown for comparison.

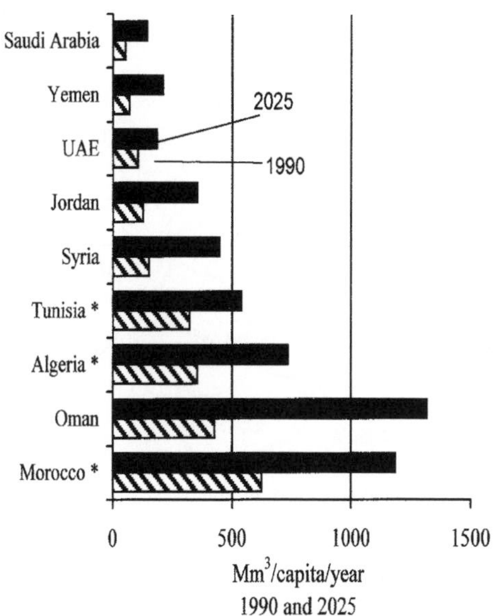

Fig. 3. Renewable water per capita per year in 1990 and estimated for 2025 (INWRDAM 1999). Asterisks indicate countries outside the region; shown for comparison.

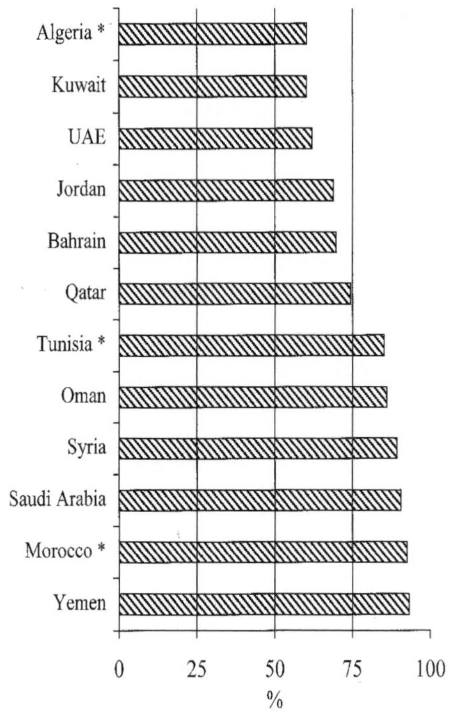

Fig. 4. Share of water allocated to agriculture (1995) (Lajaunie 1999). Asterisks indicate countries outside the region; shown for comparison.

and institutional proposals for reducing abstraction have been made, but there is still no clear consensus about how, in practice, these can be implemented successfully. Governments have generally chosen the politically easier solution of supply augmentation through aquifer exploitation, rather than the more difficult alternative of reducing demand. Concerted initiatives to reduce demand, particularly in the agricultural sector, have yet to begin in most countries.

Throughout the Arabian Peninsula, most boreholes and wells used for irrigation are private, rather than falling under the control of government or an irrigation agency. Restrictions on abstraction are either limited, absent or not enforced. Consequently, as land holdings and cropped areas have expanded and the use of mechanical pumps has increased, groundwater abstraction has grown beyond sustainable levels. The extent to which groundwater use exceeds recharge in the Arabian Peninsula, and, for comparison, some of the other countries of the Middle East and North Africa (MENA) region, is shown in Figures 5 and 6. This has led to depletion of fossil aquifers, lowering of water

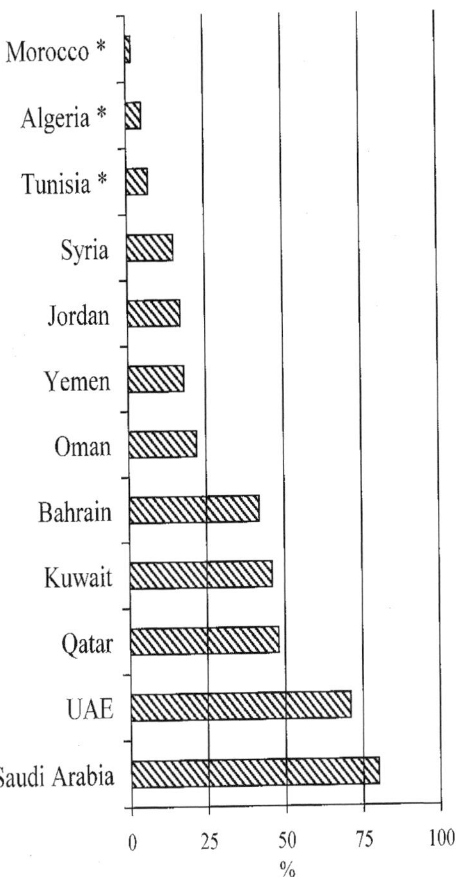

Fig. 5. Groundwater overdraft (withdrawal less renewal) as a percentage of total supply (Lajaunie 1999). Asterisks indicate countries outside the region; shown for comparison.

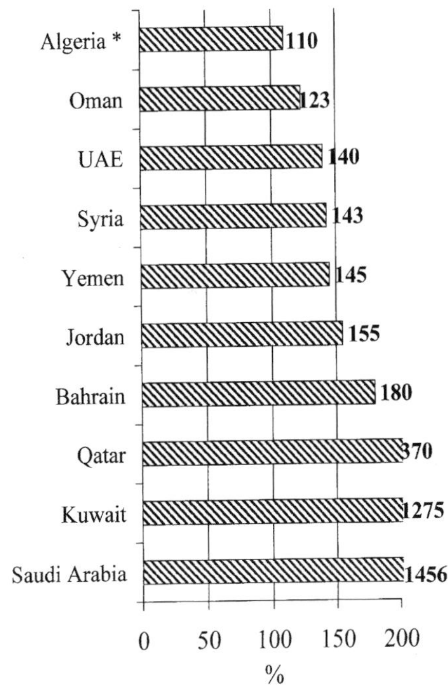

Fig. 6. Groundwater withdrawal as a percentage of renewal (Lajaunie 1999). Asterisks indicate countries outside the region; shown for comparison.

levels and increased salinization of irrigation water. In coastal districts, the resulting sea water intrusion into coastal freshwater aquifers has caused farms to be abandoned and the loss of farm communities.

In the 1980s and 1990s, most states undertook water exploration and development programmes. Agricultural expansion was encouraged and irrigation was subsidized under national policies to pursue food self-sufficiency. Supply augmentation projects have included: artificial recharge; the construction of retention dams in upland areas; experimentation with re-injection of treated waste water; cloud seeding; and condensation of fog. Water importation has been proposed. Many cities of the Gulf States rely on desalination of seawater for municipal supply. In comparison, expenditure on demand management programmes has been low.

How is this situation to be corrected? The consensus of international agencies and observers is that to manage water resources sustainably requires a mix of technical, institutional and financial approaches. This paper reviews the background and origins of this thesis, and the recurring recommendations of international agencies that decentralization of water management and stakeholder participation in managing water supply are principal components of the solution.

Why are institutional issues relevant to hydrogeologists? Simply because, in general, the many hydrological and hydrogeological investigations and technical demand management initiatives made in the region in recent years have failed, ultimately, to solve the problem. If technical recommendations cannot be implemented because of the local political, institutional, or economic contexts, then they are not, in practice, useful recommendations.

The development agency approach to water resources conservation

Non-engineering interventions to promote conservation of groundwater fall into four classes (Rosegrant 1997):

(a) legal: permits, quotas, water rights, restrictions;
(b) fiscal: tariffs, fuel taxes, subsidies, trading of water rights;
(c) political: by government taking over and managing supplies; and
(d) institutional: through expert consultancy, social pressures, public awareness, outreach and participation, including water users associations.

The combination of these approaches that is appropriate for promoting groundwater conservation depends upon the extent and effectiveness of national and local institutions. Legal measures are most applicable where water law is clearly defined, and where clear economic and financial frameworks facilitate the operation of market-based systems and use of fiscal regulation. This is not yet the case in much of the developing world. Political interventions, whereby governments operate at least primary systems of water-lifting and distribution, are found where substantial public investment is required in the absence of private capital. Such is the case, for example, with large groundwater schemes in Libya and Bangladesh. Institutional approaches are appropriate where organizations and networks of civil society are amenable to development through training, leadership and incentives.

Most discourse on the integrated water resources management starts with reference to the 'Dublin Principles' formulated at the International Conference on Water and the Environment in 1992 (World Meteorological Organization 1992):

(a) fresh water is a finite and vulnerable resource, essential to sustain life, development and the environment;
(b) water development and management should be based on a participatory approach, involving users, planners and policy-makers at all levels, and making decisions at the lowest appropriate level;
(c) women play a central part in the provision, management and safeguarding of water; and
(d) water has an economic value in all its competing uses and should be recognized as an economic good.

After Dublin, the World Bank (1993) and the Food and Agriculture Organization (1995) set out guidelines for the development of these neoliberal economic and institutional approaches to groundwater management in developing countries. These principles have been reinforced at several subsequent plenary meetings, the most recent being the Second World Water Forum (SWWF) held in March 2000, in The Hague. For the development agencies, the participatory approach and the principles of full economic valuation of water are central to successful implementation of integrated water resources management. This approach stresses the importance of developing an enabling environment with appropriate policy and legal frameworks, institutional strengthening and human resources development.

Under the agency model, the components for promoting sustainable water management are:

(a) a centralized national policy and regulatory body;
(b) decentralization of supply and distribution management to the lowest appropriate level organized on a river basin, catchment or aquifer scale;
(c) a well established and respected water law;
(d) clearly defined property rights to water and a market or mechanism for the sale, lease or exchange of water rights or for the sale of water; and
(e) participation of stakeholders, where appropriate, in policy formulation, operations, and maintenance functions.

Are these concepts and methods, specified by western agencies, appropriate to the socio-political, economic and institutional contexts of the Arabian Peninsula? The next section examines these components in more detail.

Components for developing sustainable groundwater management

National policy and administration

In countries of the Arabian Peninsula, development plans have often been formulated without considering the sustainability of water resources. In the absence of coordination by a central water resources authority, development policies of individual ministries may conflict, and taken together may not be sustainable. In the 1980s and 1990s, many governments pursued policies of food self-sufficiency, apparently unaware that the water requirements for such policies exceeded the available renewable resource. The

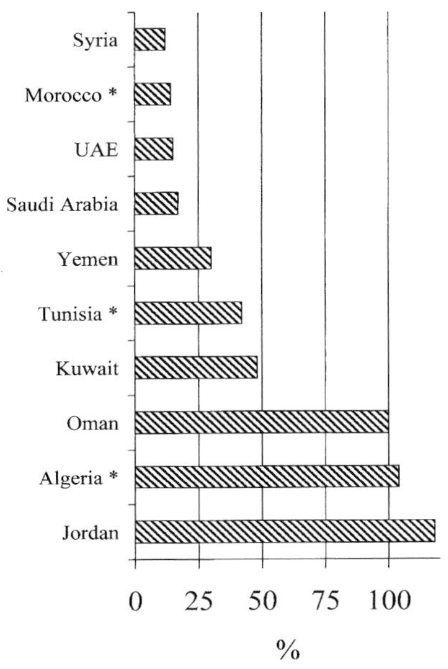

Fig. 7. Water embedded in net food imports as a percentage of total supply, 1990 (Lajaunie 1999). Asterisks indicate countries outside the region; shown for comparison.

concept of virtual water, the volume of water effectively imported in the form of food, was not widely understood as an essential part of the national water budget (Fig. 7).

A unitary or central authority, independent of user interests and responsible for setting policy, assessing the resource, monitoring consumption, and licensing abstraction is now widely regarded as essential for coordinating water policy, regulation and management on a national scale. Most countries of the region have established separate bodies for this purpose, but their power and authority, by comparison with longer established ministries, may not yet be fully developed.

Decentralization

In principle, decentralization of water supply administration should improve local management efficiency and accountability. Local accountability should improve the standard of local water resources management. However, there are obstacles to decentralization, including the central authority's resistance to regional devolution of power; dominance by elites; reduced management capability on a local scale; coordination; and aquifer management across political boundaries (Carney 1995). These problems are particularly likely to apply in countries with authoritarian and centralized political systems, such as the Arabian Peninsula.

Water law, rights and markets

Clearly understood and nationally accepted water laws are central to promoting change in water resources management. In the Peninsula, landowners have traditional rights to abstract groundwater from wells they have constructed or inherited. Construction of new wells and deepening of existing wells is now widely restricted in water-stressed areas but the application of groundwater abstraction quotas for irrigation is ubiquitously rare. Consequently, there is little incentive to save groundwater through more efficient irrigation practice. Without regulation, groundwater is exposed to the 'tragedy of the commons' whereby an open access, or common-pool, resource is over-used by self-interested parties, perhaps ultimately to destruction.

Allocating volumetric rights to groundwater use on a basin or aquifer scale is a rational method of distributing the resource sustainably and effectively, having regard to users' needs and groundwater availability. However, in the Arabian Peninsula, such rights are not defined or registered, and there are no established systems of allocating or trading water rights within an aquifer. To restrict groundwater use through application of volumetric quotas or the withdrawal of well and abstraction licences would require major changes to or clarification of law, which would only be feasible after extensive public awareness and incentive campaigns to counter resistance from farmers.

Sharia, Islamic law, applies throughout the Peninsula. The law is the result of interpretation by scholars and divines and the interpretations of Sunni and Shi'ite sects differ on some issues, such as the transferability of irrigation water rights (Caponera 2001). Although charges may be made for purification and delivery, water in its natural state and location is common property; consequently, the introduction of tradable water rights in an aquifer may be controversial. However, an overriding Islamic principle is that irrigation is only the third priority for water use, after supplying human and animal needs. Governments, as the modern custodians of the common property of water, have the right to reallocate the resource for the wider public good and in the interests of sustainability (Faruqui 2001).

A cooperative non-profit trust that holds rights in the public interest, but distinct from government, may be a fairer alternative to individual water rights that would better protect public interests (Moench 2000). Over the long term, such a model could provide enhanced security and reduced cost of water more effectively than a conventional water rights model. Owners of water rights might be willing to relinquish their existing, but deteriorating, water security in return for longer-term security supported by collective investment in aquifer protection and sustainability (Loehman *et al.* 1999).

Stakeholder participation in irrigation management

'Participation' and 'participatory approaches' are terms from the lexicon of international development which describe community-based processes that help alleviate poverty, improve working skills, and empower disadvantaged groups. In this paper these terms are used in the sense of promoting community management of a common-pool groundwater resource on an aquifer scale.

Most case experience of participatory irrigation management (PIM) comes from South Asia, where communal water management schemes have been widely developed. Most have involved the management of surface water flows, which are more easily divided between users than is groundwater in an aquifer. The hydrological characteristics and limits are more tangible and visible than is the case with groundwater.

Stakeholder participation is central to the agency approach to improving the management of water resources. This two-way process should:

(a) ensure that everyone's interests are considered;
(b) assist data collection and monitoring;
(c) improve transparency and accountability;
(d) educate and accustom stakeholders to difficult choices;
(e) build commitment through realistic discussion of benefits, risks and costs of options; and
(f) increase the likelihood of success of chosen options.

Studies in South Asia show that the success of participation efforts in irrigation depends on how well projects mobilize local support and build effective organizations. Vital aspects include incentives, training, successful pilot projects and well-trained and accessible local community organizers. Transparency and accountability, financial viability, facilitating legislation and strong commitment by government are seen as essential institutional factors (Meinzen-Dick *et al.* 1997; World Bank 1997; Stacey 1999). Even then, in communities with rigid hierarchical social structures, it may be unrealistic to expect democratic local organizations to develop easily. Practical pitfalls in implementation include stakeholder complexity and competition, vested interests, duplication of management, and under-design of the social structures (Rhoades 1998).

Promoting cooperative user participation in groundwater management is likely to be more difficult than that of surface-(gravity) fed schemes of South Asia, Turkey and elsewhere. In the Arabian Peninsula, the situation is complicated because:

(a) sources of irrigation water lie on numerous small individual plots rather than at a central irrigation source such as a river, canal or central wellfield; and
(b) subsistence farmers are accustomed to, and have rights to, unrestrained free access to the water from their own wells.

Development of policy perspectives for the Arabian Peninsula

Prior to the 1992 Dublin conference, water resources policies in the region were oriented towards securing municipal supplies and satisfying the irrigation requirements for agricultural policies of pursuing full food self-sufficiency. Supply augmentation remained paramount throughout the 1990s, although external agencies and advisers continued to reiterate the Dublin recommendations with little, if any, deviation.

Rogers (1994) showed that water for basic human needs is abundant in the region but 90% is used for agriculture, which is barely, if at all economic, and which only partly supplies the food needs of the region. Reallocation of part of this water to more profitable uses would benefit the economy and reduce the water balance deficit. The solution, for economists, lies in pricing water at values that reflect the opportunity costs of not applying it for alternative municipal and industrial uses. These sectors either go short in the poorer countries, or in the richer countries are supplied with desalinated water at high cost. Rogers (1994) cites the need for institutional development as the process for implementing the necessary changes, including the establishment of unitary water ministries, the development of water user associations, and increased promotion of the private sector as a provider of services.

Allan (1994) predicted that the introduction of realistic water pricing would be inevitable in the region, together with new engineering and institutional instruments. To date, rulers have met the water gap by importing food and have thereby avoided addressing these difficult challenges. Allan (1995) identified a complex of stakeholder interests, including users with entrenched rights, wealthy interest groups, professional groups, government bureaucrats, and contractors.

The World Bank (1998) continues to promote the importance of national water policy coordination; private sector involvement; wider public awareness campaigns; centralized policy but decentralized management; and improved irrigation efficiency. Public participation and the development of water users associations (WUAs) are seen as vital components of the process in developing groundwater sustainability. Economic growth in the region has been adversely affected by the failure to allocate water resources to the most effective economic sectors (World Bank 1999).

Western prescriptions have tended to ignore social priorities in rural Arabia and the practicalities of implementing such changes in the contexts of Islamic law and existing political systems. By the late 1990s there was a growing awareness that these agency prescriptions had not been widely assimilated in the Arabian Peninsula states. Government representatives in the region appear sceptical about the practicality of implementing legal and fiscal components of the agency agenda, particularly the allocation and trading of water rights, due to the weakness of existing organizations (World Bank 1999).

Resistance to western agency prescriptions was articulated at the Arab Region Session of the SWWF in March 2000 (Shady 2000). While delegates supported the supply-side engineering and institutional approaches to the regional crisis, they declared that water was a common resource and not a commodity. The acceptance of the concept of economic valuation and the development of western systems of water rights in the region in the near future therefore seem unlikely.

The potential influence of religious leaders and institutions

It is widely but incorrectly thought that Islamic teaching and law oppose the provision of water as anything but a free good, and thus prevent the development of practices such as water trading. In fact, although Islamic law as applied to water is complex and subject to interpretation, it is essentially pragmatic, adaptable to changing needs, and offers scope for interpretation in the context of the groundwater crisis. Recent examples include a *fatwa* declared in Saudi Arabia to allow the use of wastewater in irrigation, and demand management measures implemented in accordance with Islamic guidelines (Abderrahman 2001).

The holy writings from which Islamic law is derived contain many prescriptions relevant to water management, and concur with the spirit of the Dublin principles regarding community participation (*shura*). Water is seen as a social good that must be managed sustainably. Concepts of full-price costing for delivery, use of wastewater, and privatization of service delivery are now widely seen to be permissible. There is also an increasing feeling in the Islamic world that it is appropriate and correct that Islamic institutions should play a proactive role in promoting and driving water demand management (Faruqui 2001).

Islamic institutions play a prominent, central and universally respected role in the region. The status and practical power of Islam, an institution distinct from – but working closely with – both state and people, are great. In the Arab region people look to the mosques for leadership; harnessing this could be a powerful force for change.

Oman – progress at an appropriate pace

Although most political systems in Arabia are autocratic, effective change generally occurs only after a process of consultation and careful consideration of effects on local interests. Oman serves as an example where, after a long period of uncontrolled groundwater use in the 1970s and 1980s, appropriate policies and strategies have been developed. An independent ministry was established in 1989 to coordinate policy, optimize groundwater and surface water development, and regulate use. In the 1990s, all existing wells were inventoried and licensed; wellfield protection zones established; new well construction regulated; and abuses effectively policed. Some aquifers are now closed to new

Table 1. *Institutional requirements for user participation in irrigation water management in the Arabian Peninsula*

Institutional prerequisites:	*Exists in region?*
At the national level	
Government policy on water use and food security which is coherent and consistent, requiring excellent cooperation between ministries	Not yet
Strong government commitment to demand management	Not yet
Centralized resources assessment and policy formulation by a well informed agency independent of other government agencies with interests in water use	Well developed in some countries (e.g., Oman)
Good understanding of the sustainability and extent of the water resources on a basin/aquifer level	Variable
Decentralization of regulation and monitoring to a regional and then to an aquifer level	In progress in some countries
Secure, legalized, registered water rights, to facilitate trading of rights and water	No
At the local level	
Groundwater crisis	Yes, in many places
Tradition of cooperation at the user level	Well developed in places
Respected potential local leadership	Yes
Absence of elite interest groups	Generally present
Homogeneous user groups	Yes
Clearly comprehensible benefits	Not yet understood
Farmer incentives	Subsidies already common
Transparency and accountability	Satisfactory in places
Financial viability	Unlikely

well construction. Regulatory functions are already partly decentralized to regional offices, which also coordinate technical and financial support for the existing local user cooperatives that manage and maintain the *aflaj* systems (Al-Shaqsi 1999).

Nevertheless, groundwater abstraction (estimated in 1998) averaged approximately 120% of recharge nationally, and significantly more in some regions. Eighty-seven percent of water is used for agriculture, even though financial returns to groundwater from irrigated agriculture average less than 10 US cents m^{-3}, while the production cost of desalinated water for the capital exceeds 1US $. Virtual water imports as food are approximately equal to the renewable water resources (Al-Shaqsi 1999). Only in the late 1990s was the policy to pursue full food self-sufficiency abandoned. Government now actively encourages removal of fodder cultivation to areas of brackish water; provides subsidies for modernizing irrigation systems; and has removed customs barriers to the importation of food.

Although increasingly few people work in the agricultural sector, the rural villages and *aflaj*-supplied oases lie at the heart of Omani culture and tradition. It is essential, socially and politically, to support and maintain this heritage, a contingency cost typically ignored in estimating the economic value of water. The system of participatory management of communal *aflaj* (sing. *falaj*) is an ancient tradition in Oman. This is a naturally adjusting sustainable system by which common-property flows are channelled and distributed to irrigation plots on a time-based system, under the management of a local committee. In the mountains, these flows are from springs, while on the foreland they are from near-horizontal tunnels intersecting the water table. The system provides for trading of water allocations between plot owners (Dutton 1995; Rahman & Omezzine 1996).

The *aflaj* system provides an existing social model for introducing participatory management of aquifers tapped at central wellfields. An area where the approach could be attempted is the Batinah coastal plain west of the capital, Muscat, where the density of private wells is highest. Here, falling groundwater levels and saline intrusion have destroyed coastal farms in what is the country's most important agricultural zone. One solution would be to exchange the existing private rights of hundreds of wells in which water quality is deteriorating for a secure supply from a central wellfield in the same aquifer, up-gradient and further inland. The allocation of quotas and management of the supply would be the responsibility of a water users association in which was vested the collective water rights to the aquifer, and which would be representative of all stakeholders.

Analysis

Institutional options for water conservation

It is now widely agreed that promoting sustainable groundwater management requires a combination of legal, fiscal, technical and institutional measures, appropriate to each country. In the Arabian Peninsula, however, difficulties in implementing legal and fiscal restraints on private groundwater users have been identified.

In this region democratic systems are not well developed. In some countries state systems are fragmented, water policies may be inconsistent and technical capabilities are still growing. Market frameworks and conditions are improving but private capital is not yet well dispersed. Civil society institutions are emerging gradually, but society is generally family and tribally focused.

Successful policy formulation, inter-ministry coordination, and the development of the fiscal and institutional measures to facilitate water conservation require a strong, confident and informed central water resource authority. The relationships between government departments and user groups will vary, but the role played in each country by the Ministry of Agriculture in influencing irrigation consumers is particularly significant. Unless this ministry can coordinate its policy with the central water resources authority, there is little chance of developing a coherent or consistent national policy.

Institutional strengthening through promoting participatory involvement at a local level is regarded as desirable and necessary by agency and government observers. Table 1 summarizes the necessary characteristics that have been identified. The difficulties lie as much at the government level as at the user level.

Multiple stakeholders with many different interests may be identified, at the government, regional and local levels. Conflicts and tensions between stakeholders at and between these levels in managing change are likely to be substantial, and it seems unlikely that existing institutions would be able to resolve these.

Managing change through coordination

It is helpful to analyse institutional relationships in terms of competition, coordination and cooperation, to reveal possible avenues for introducing participatory interventions. Water allocation and management naturally involve competition, negotiation and brokering, between demand sectors, communities and individuals. Coordination between central government and local administrations, advisers and water users is essential. Cooperation is obviously central to successful participatory management.

Robinson (2000) identified some of the sources of competition and problems in facilitating public/private sector coordination. Many of these are recognizable in this context:

(a) multiple agendas and conflicting policies of different government ministries, and departments within ministries;
(b) lack of technical skills, poor internal communications;
(c) government systems lack flexibility to tackle complex social and institutional relationships; and
(d) government departments are project- and task-centred and not comfortable with a protracted process approach necessary for introducing participatory interventions.

In this context where high levels of competition exist between primary stakeholders there is an essential role to be played by an external coordinating agency in balancing these perspectives. Certain specialized development organizations, independent of the lending agencies, have this necessary technical and institutional expertise.

Most importantly, Islamic institutions could play a prominent leadership role in initiating and driving the necessary legal and social changes. These institutions are highly revered and respected, locally and nationally, and in many countries are more potent influences than national governments. Islamic law applies throughout the Arabian Peninsula and the Holy Qur'an provides many insights and directives specifically regarding water that may be used to guide institutional development in this context.

A practical approach

Some issues, such as policy coordination, water law reform and organizational development of central and regional water resources authorities can only be addressed at senior government level. These developments will take time, even in the most enlightened states.

Greater progress at the user level is possible, through appropriate participatory interventions on a pilot scale, using one or more of the now well tried methods of social intervention known as 'action research' (World Bank 1996; Borrini-Feyerabend *et al.* 2000). These methods are well known to non-governmental organizations and other development workers but are, as yet, less practised in the Arabian Peninsula.

An intervention would involve the implementation of a pilot participation scheme in a selected area, designed by a specialist external agency with the active collaboration of the regulatory agency (i.e. water ministry or authority) and Islamic institutions. The intervention would have two components:

(a) to educate government stakeholders in the need for water conservation, the available options, the consequences of different options, and the process of participatory interventions. Relevant approaches at this level are described by Thompson (1998);
(b) to plan and implement a pilot participatory intervention in a pilot catchment.

The pilot scheme should have these characteristics:

(a) be in, or of, a catchment or aquifer that is obviously in crisis, with chronically falling water levels and deteriorating water quality;
(b) have a high proportion of educated individuals likely to be responsive to an innovative approach;
(c) include respected national figures who could provide a focus and example;
(d) employ an action research methodology appropriate to the situation;
(e) include an intensive, modern public awareness campaign; and
(f) incorporate incentives and subsidies to reduce consumption and modernize irrigation.

The objectives of the pilot study would be to develop a communal, participatory approach to the problem of water resources degradation and overabstraction in the hydrological unit, by:

(a) sharing legal, religious, social, economic and technical perspectives on the groundwater situation;
(b) removing errors of perception and understanding of the groundwater situation, by a well designed programme of public awareness using modern methods and on-the-ground contact between specifically trained facilitators and water users;
(c) agreeing a programme of reduction in abstractions, linked to subsidized programmes of agricultural extension, cropped area reduction, crop modification, produce marketing, and training;
(d) involving water users in a communal monitoring programme with regular reviews of progress and sharing of information; and
(e) establishing a water users association of primary stakeholders to continue the monitoring and extension programmes.

By this means, the community of water users within a hydrological boundary would work out their own set of long-term solutions to the problem of groundwater overuse. The package of solutions may require voluntary or subsidized withdrawal or limitation of irrigated agriculture, modification of practices, and improvement in irrigation efficiency. The secondary effects would be valuable, particularly in terms of other environmental benefits, improved civil society institutions, and increased profitability. A successful pilot study would serve as a model for other areas of the country.

Elsewhere in the MENA region Tunisia has already begun a process with many of the above features, and may serve as a useful model to other Arab states. Legislation is being revised, a national consultation process is underway, and regional management bodies will be strengthened. The existing model of surface water users associations will be modified for groundwater management, beginning with pilot-scale participatory schemes for shallow aquifers (Findikakis 1998).

Conclusions

The practicality of conserving groundwater by introducing legal or fiscal policies or constraints, according to the western development agency paradigm, is not high in the Arabian Peninsula. The power of agricultural interests, the absence of market processes and of defined water rights, and the nature of Islamic law would seem to inhibit this approach. The concept of restricting abstractions is undeveloped, and where attempted has met with opposition.

Institutional strengthening, combined with improvements in agricultural efficiency, is an alternative route. Processes for implementing water management by decentralization and user participation in irrigation have been widely developed elsewhere. However, persuading landowners to cooperate in restraining their use of the open-access, common-pool groundwater resource will require, in the Arabian Peninsula, strong and impartial leadership, coherent sectoral policies, clearer water law, technical inputs, incentives and subsidies, and better focused public awareness campaigns. Strong communal participation, usually through water users associations, is essential before water policy reforms can be successfully implemented in practice.

Although user-participation has been successful where water is obviously common property, the process is less easy to apply in complex situations of multiple private groundwater rights where the political power of elites prevails. Substantial financial support from government will be necessary, but the contingent value of agriculture to the environment and to social structures is more important in this context than its small contribution to gross domestic product.

A participatory intervention at the aquifer level offers scope for introducing reform gradually, through education and engagement of all stakeholders. Such an intervention should be led by an independent specialist agency working closely with Islamic community leaders, as well as government agencies. A two-pronged strategy is needed, aimed to manage change at government level and at the user level, initially through pilot schemes.

These pilot schemes and the water users associations that would emerge from them could presage gradual changes in national water policy. These changes might include the establishment of groundwater rights or cooperative trusts and the decentralization of regulatory functions, such as the allocation of quotas, licensing of wells and regulation of water markets. Successful pilot studies could be scaled up for wider application in the same area and would provide a model for adaptation to other aquifers.

This strategy requires much more than the establishment of user committees and a general public awareness campaign. Seriously educating and engaging all stakeholders requires real commitment from national and local government, forceful leadership, financial resources, focused and detailed public awareness efforts using modern communication techniques, and a deep experience of modern approaches to managing change through participation, in an Islamic context.

I am grateful to Saif bin Rashid Al-Shaqsi and Zahir bin Khalid Al-Suleimani, of the Ministry of Water Resources in Oman, who asked me to review Oman's water resources policy; to Andrew Macoun and Ashok Subramanian for their invitation to participate in the World Bank's 1999 Regional Conference on Water Policy Reform; and to Ruth Meinzen-Dick, Marcus Moench and Bob Rout who kindly suggested many useful information sources.

References

ABDERRAHMAN, W. A. 2001. Water demand management in Saudi Arabia. *In:* FARUQUI, N. I., BISWAS, A. K. & BINO, M. J. (eds) *Water Management in Islam*. United Nations University Press, Tokyo, Japan.

ALLAN, J. A. 1994. Overall perspectives on countries and regions. *In:* ROGERS, P. & LYDON, P. (eds) *Water in the Arab World*. Harvard University Press, Harvard, USA.

ALLAN, J. A. 1995. Water deficits and management options for arid regions with special reference to the MENA region. *In: Proceedings of the International Conference on Water Resources Management in Arid Countries*. Ministry of Water Resources, Muscat, Oman.

AL-SHAQSI, S. 1999. Case study: water policy reforms in Oman. *In: Summary Report, Second Regional Seminar of the MENA/MED Initiative on Water Policy Reform*, Amman, Jordan, May 1999. World Bank, Washington, USA.

BORRINI-FEYERABEND, G., FARVAR, M. T., NGUINGUIRI, J. C. & NDANGANG, V. A. 2000. *Co-management of Natural Resources: Organising, Negotiating and Learning-by-Doing*. GTZ and IUCN, Kasparek Verlag, Heidelberg, Germany.

CAPONERA, D. A. 2001. Ownership and transfer of water and land in Islam. *In:* FARUQUI, N. I., BISWAS, A. K. & BINO, M. J. (eds) *Water Management in Islam*. United Nations University Press, Tokyo, Japan.

CARNEY, D. 1995. *Management and supply in agriculture and resources: is decentralization the answer?* Overseas Development Institute, London, Natural Resources Perspectives, No. 4.

DUTTON, R. W. 1995. Towards a secure future for the aflaj in Oman. *In: Proceedings of the International Conference on Water Resources Management in Arid Countries, March 1995*, **3**, 25-30. Ministry of Water Resources, Muscat, Oman.

FARUQUI, N. I. 2001. Islam and water management: overview and principles. *In:* FARUQUI, N. I., BISWAS, A. K. & BINO, M. J. (eds) *Water Management in Islam*. United Nations University Press, Tokyo, Japan.

FAO 1995. *Water Sector Policy Review and Strategy Formulation, a General Framework*. , FAO, Rome, Food and Agriculture Organisation Land and Water Bulletin No. **3**.

FINDIKAKIS, A. N. 1998. Elements of a groundwater management strategy. *In: Proceedings of the 8th Stockholm Water Symposium*. Stockholm International Water Institute.

INWRDAM. 1999. *Table of water scarcity in OIC states*. Inter-Islamic Network on Water Resources Development and Management, Amman, Jordan. World Wide Web address: http://www.nic.gov.jo/inwrdam/

LAJAUNIE, M-L. 1999. MENA/MED regional water data report. *In: Proceedings of the Second Regional*

Seminar of the MENA/MED Initiative on Water Policy Reform, Amman, Jordan, May 1999. World Bank, Washington, USA.

LOEHMAN, E., BECKER, N., WEG, E. & HENDRICK, M. 1999. Cooperation in a hydrogeologic commons: new institutions and pricing to achieve sustainability and security. *Workshop in Political Theory and Policy Analysis*, Indiana University, USA. World Wide Web address: http://www.indiana.edu/~workshop/papers/loehman.pdf

MEINZEN-DICK, R., MENDOZA, M., SADOULET, L., ABIAD-SHIELDS, G. & SUBRAMANIAN, A. 1997. Sustainable water users associations: lessons from a literature study. *In: User Organizations for Sustainable Water Services.* World Bank, Washington, USA.,World Bank Technical Paper No. **354**.

MOENCH, M. 2000. The underground frontier: managing groundwater development for environmental sustainability and social equity. *In: Conference on Water Security in the 21st Century*, The Hague, Netherlands, March 2000. World Wide Web address: http://www.worldwaterforum.org

RAHMAN, H. A. A. & OMEZZINE, A. 1996. Aflaj water resources management: tradable water rights to improve irrigation productivity in Oman. *Water International,* **21**, 70–75.

RHOADES, R. E. 1998. *Participatory Watershed Research and Management: Where the Shadow Falls.* International Institute for Environment and Development, London.

ROBINSON, D. 2000. Reforming the state: coordination, regulation or facilitation? *In:* ROBINSON, D., HEWITT, T. & HARRISS, J. (eds) *Managing Development: Understanding Inter-organizational Relationships.* Sage Publications, London, 143–165.

ROGERS, P. 1994. The agenda for the next thirty years. *In:* ROGERS, P. & LYDON, P. (eds) *Water in the Arab World.* Harvard University Press, Harvard, USA.

ROSEGRANT, M. W. 1997. *Water Resources in the Twenty-First Century: Challenges and Implications for Action.* International Food Policy Research Institute, Washington, USA, Discussion Paper **20**.

SCHIFFLER, M. 1998. *The Economics of Groundwater Management in Arid Countries.* Frank Cass, London.

SHADY, A. 2000. Summary: Arab countries session. *In: Conference on Water Security in the 21st Century.* The Hague, Netherlands, March 2000. World Wide Web address: http://www.worldwaterforum.org

STACEY, D. 1999. Water users organizations. *Agricultural Water Management,* **40**, 83–87.

THOMPSON, J. 1998. Participatory approaches in government bureaucracies: facilitating institutional change. *In:* BLACKBURN, J. & HOLLAND, J. (eds) *Who Changes? Institutionalizing Participation in Development.* Intermediate Technology Publications, London, 108-117.

WORLD BANK. 1993. *Water Resources Management a Policy Paper.* World Bank, Washington, USA.

WORLD BANK. 1996. *Participation Source Book.* World Bank, Washington, USA.

WORLD BANK. 1997. *User organizations for sustainable water services.* World Bank, Washington, USA, World Bank Technical Paper No. **354**.

WORLD BANK. 1998. *From scarcity to security – averting a water crisis in the Middle East*, World Bank, Washington, USA.

WORLD BANK. 1999. *Summary Report, Second Regional Seminar of the MENA/MED Initiative on Water Policy Reform*, Amman, Jordan, May 1999. World Bank, Washington, USA.

WORLD METEOROLOGICAL ORGANIZATION (WMO). 1992. *The Dublin statement on water and sustainable development.* International Conference on Water and Environment, January 1992. WMO, Geneva.

Who needs sustainability?

M. PRICE

Centre for Earth and Atmospheric Sciences, The University of Reading, Whiteknights, PO Box 227, Reading, Berkshire RG6 6AB, UK (e-mail: m.price@reading.ac.uk)

Abstract: Three inescapable factors face us at the turn of the century. First, the climate, and with it the supply of water, is becoming more variable. Regardless of whether this variability is the result of man-made global warming, it seems to be with us. Secondly, demand for water will increase; in many countries domestic water use is currently less than 25 l/person-day and needs to increase if living conditions are to improve. In Great Britain, *per capita* use of water is low compared with many other western countries – including those that pay for water on a metered basis – and it is certain to rise as living standards improve. Thirdly, concern for the environment is likely to increase. We can counter the first problem only by ensuring that we have adequate water storage; most water storage in Great Britain and the rest of the world is in the ground. We need to counter the second by recognizing that the vast majority of public water use is non-consumptive, but that very little waste water is re-used efficiently. We need to counter the third by welcoming the concern but ensuring that the response is sensible. We must be prepared to question some of the accepted wisdom of the environmental movement, and remember that many of the developments in human history were non-sustainable. Do they represent environmental disasters, or human progress?

There is a widespread assumption that sustainability is a 'good thing', and that anything non-sustainable is a 'bad thing'. I would challenge that assumption. Most human advances have been non-sustainable in the long term. For example, the first use of metals was of native metals such as copper and gold. These were obviously limited in quantity, and the use of metals could not have been sustained, even with recycling, if man had not learned to smelt ores. However, our ancestors worked and hammered native metal into practical items and trinkets and developed metal-working skills (Bronowski 1973; Andrews 1991). Once smelting began, charcoal was used to smelt iron. In Britain and Europe, this use was also non-sustainable – large areas of European forest were cleared to provide wood for charcoal (Andrews 1991). Our present use of fossil fuels clearly cannot be sustained indefinitely. The simple fact is that natural resources have always been exploited by the first people who needed them and had the technology to exploit them. Without that exploitation we should, in effect, still be living in the Stone Age.

In terms of groundwater, there are many examples of its supposed non-sustainable use. The use of groundwater from the Chalk of the central London Basin during the 19th and early 20th centuries was clearly not sustainable in the long term. However, as Downing (1986, 1993) points out, this supply of water enabled London to develop as a major centre of population and manufacturing. That use of water lowered the potentiometric surface in the Chalk and caused springs and streams to dry up. It would not be allowed today and was one of the factors that led to the introduction of controls on abstraction. Are we to conclude that Britain would have been better served if the abstraction had not taken place and London had not developed as it did?

Ironically, the reduction in use of groundwater since 1945 has also turned out to be non-sustainable, with pumping schemes having to be introduced to lower the potentiometric surface again to prevent damage to foundations and tunnels (Simpson *et al.* 1989; Anon 2000). This raises the point that when talking of sustainable use, it is necessary to stipulate the period over which the use is planned or implemented.

California is one of the most productive agricultural regions in the world. In this semi-arid environment such agricultural productivity

From: Hiscock, K. M., Rivett, M. O. & Davison, R. M. (eds) *Sustainable Groundwater Development*. Geological Society, London, Special Publications, **193**, 75–81. 0305-8719/02/$15.00

has been made possible only by widespread use of irrigation – much of it using groundwater from the deposits of the Central Valley. This abstraction has led to significant falls in water levels and to the largest volume of land subsidence in the world; the ground surface has fallen by as much as 9 m (Farrar & Bertoldi 1988). Groundwater still supplies 41 000 Ml d^{-1} for irrigation in California (USGS 2001). In the long term, the abstraction will probably be self-limiting, as fresh water is limited to about the upper 300 m of the aquifer, and compaction reduces the storage coefficient (Farrar & Bertoldi 1988). Groundwater abstraction began to reduce in the 1970s but increased again during the drought of 1977–1978 to take advantage of the groundwater storage. Again, we have to ask whether the non-sustainable use of groundwater irrigation that led to the development of California as one of the most dynamic economic regions in the world should have been prevented.

Libya's Great Man-Made River project involves the abstraction of fossil groundwater from beneath the Libyan Desert and its transfer to coastal cities and towns. This groundwater has had no significant recharge during the last 15 000 years. It has been argued by some workers that because the water is not being replenished, it should not be abstracted, or that it should be left for the benefit of 'future generations'. Such arguments overlook the point that left to itself, much of the water will naturally flow northwards and discharge at coastal sabkhas, benefiting no one. Clearly the water cannot be a permanent, renewable supply unless the climate changes and significant rain again falls on the Sahara. However, for a period of at least decades, and possibly centuries, this fossil water can supply most of the population of Libya with an essential resource. This abstraction therefore seems to meet the 'ethical requirements' of groundwater mining (Llamas & Priscoli 2000), namely that: (1) evidence is available that pumping can be maintained for a long period; (2) the negative impacts of development are smaller than the benefits; and (3) the users and decision-makers are aware that the resource will eventually be depleted.

In these circumstances it seems eminently sensible for Libyans to use this water now, when they have the need and the technological ability to do so, rather than leave it in the ground to discharge or deteriorate in quality. Therefore, non-sustainable use is not in itself undesirable. However, the way that some of the water will be used is a cause for concern. In California, there is some recharge to groundwater, albeit at a rate lower than abstraction; in Libya, there is effectively none. Using non-renewable groundwater for inefficient agriculture is not a sensible long-term plan.

The current challenges

Three facts seem to face mankind, and especially mankind in Britain, as we start a new century. The first is that the supply of water is becoming less predictable. Since 1975, Great Britain has suffered four droughts, each of which is estimated to have a return period of more than 200 years (Price 1998). At the time of writing, parts of England are experiencing some of the worst, if not the worst, instances of flooding on record. In part, these probably result from bad planning decisions in the past as much as from unprecedented weather. Nevertheless, the rainfall in autumn 2000 was the highest since records began in 1766 (CEH/BGS 2000). Current global climate models indicate that the rise in mean global temperature in the 20th century can be explained only if man-made emissions of greenhouse gases are taken into account; they predict that global temperature will rise by 4°C by 2100. The modellers are less confident about predicting possible changes in precipitation, but the best estimates are still that summer rainfall will decrease and winter precipitation will increase – largely as the result of more intense precipitation events (CCIRG 1991, 1996; Price 1998). In other words, we can expect to see more droughts and more periods of heavy rainfall accompanied by increased occurrences of flooding. Marsh (2001) shows evidence of general stability in long-term rainfall and evaporation, but with significant short-term variation.

The second fact is that the global demand for fresh water will increase, as population increases and the standard of living improves. In Britain and other 'developed' countries, that improved standard of living may involve increased ownership of dishwashers, power showers and swimming pools; in the Third World it may simply involve having a supply of piped water available to more dwellings, perhaps with adequate sewerage. Figures 1 and 2 show some estimates of water use in the developed and developing world. The figures for the developed world are based on data for 1995 (Anon. 1998), with the value for the UK towards the top of the range given by the Environment Agency for England and Wales (Anon. 2001); those for the developing world are from Gleick (2000) for various years. There is some uncertainty in these figures, particularly those from Gleick, as he readily acknowledges. There is, however, little

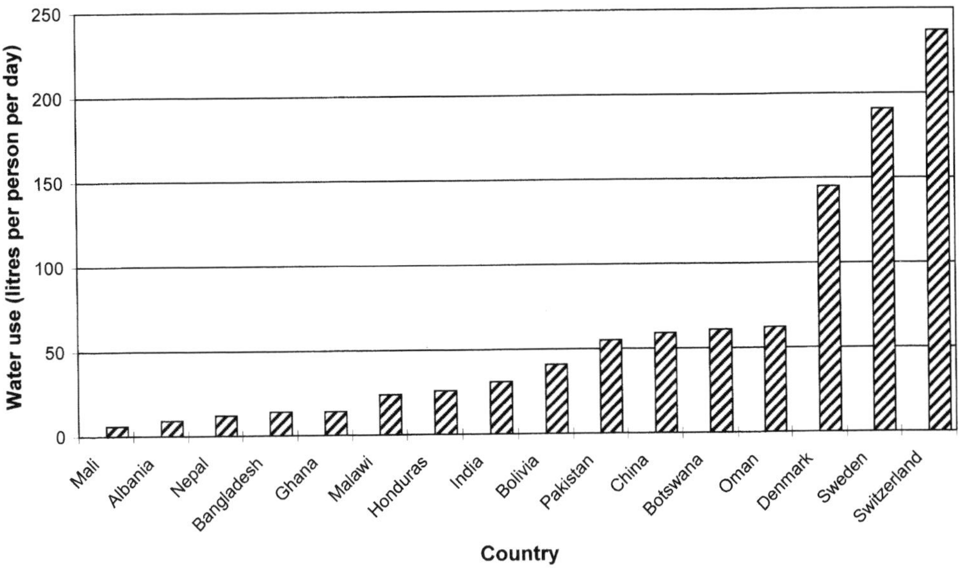

Fig. 1. Domestic water use in representative countries (data from Anon 1998; Gleick 2000).

doubt about the disparity they display. I assume that few people would deny the right of those in other parts of the world to aspire towards the standards of living that the more fortunate take for granted.

In Britain the demand managers will argue that we can counter this increased demand by metering water and making it more expensive. I believe they are wrong. As Figure 2 shows, consumption of water per head of population in Britain is below that of many other countries, despite the fact that in many of those countries domestic water is metered. In most parts of England and Wales, domestic water use is less than 160 l/person-day (Anon. 2001). It is true that Gleick (2001) was able to say that demand for water in some parts of the world has actually fallen over the last few years, but the falls are mainly in countries where *per capita* use is very high. The switch in the United States since 1992 from lavatories that used more than 20 l per flush to a new standard requiring less than 6 l per flush has probably played a part in this.

The stark truth underlying Figures 1 and 2 is that demand for water is almost directly related to prosperity: as disposable income increases, water use increases. A second stark truth is that the population will no longer sit by and let water managers dictate to them how much water they can use. This was shown by events in West Yorkshire in 1995, when attempts to install standpipes were halted by public opposition and threats. In Britain, for better or worse and despite the aims and claims of the European Water Framework Directive (EU 2000), water is now a commodity, not a public service.

The third fact is that concern for the environment has increased and is likely to go on increasing. This too is a reflection of increased prosperity and – for at least some of the population – increased leisure time to spend enjoying the countryside and thinking about environmental issues.

Response to those challenges

We can counter the first problem i.e. possible long-term stability but significant short-term fluctuation in the supply of water, only by ensuring that we have adequate storage (Price 1998). Most fresh water storage in this country and the rest of the world is in the ground. It is inconceivable that we should not make full use of that storage.

We need to counter the second problem, that of increased demand, by recognizing that the vast majority of public water use is non-consumptive. Apart from water used for garden watering, almost all water used in the home is returned to sewers, with a small proportion

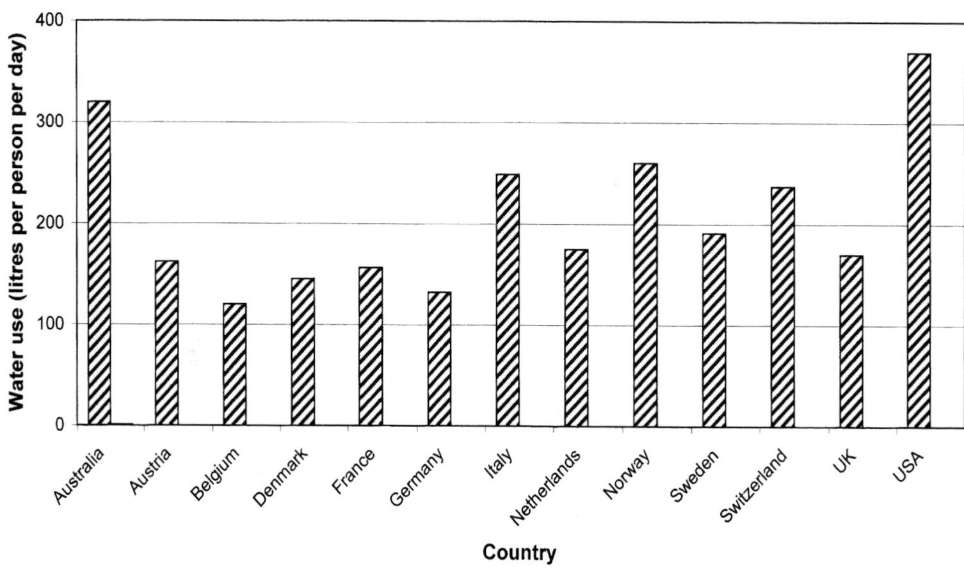

Fig. 2. Domestic water use in the developed world (data from Anon 1998; Anon 2001).

going to the ground via septic tanks. In some parts of Britain, sewage is returned to rivers and the water is effectively re-used. The water returned to the Thames from Oxford and Reading, for example, is abstracted for use in London. People who are unhappy about this should remember that all the water that we use is part of the hydrological cycle, and that every glass of water we drink contains molecules that have been drunk by someone else in the past. Unfortunately, the Thames Valley is something of an exception and in most parts of Britain water is used only once after abstraction. In many places, sewage is discharged to coastal waters or estuaries where it cannot be re-used.

The many issues of sustainability

To deal with the third challenge, that of increased need for environmental protection, we must aim for some form of sustainability – but we have to decide what it is that we are trying to sustain.

Is the target for sustainability the groundwater as a source of public supply? If that were the case, we could go back to the idea of safe yield. We would have to ensure only that average abstraction is less than average recharge. The only difficulties would be, as in the case of London, deciding over what period to take that average, and in remembering that what was average recharge during the twentieth century might not be the same as average recharge during the present one.

Is it the groundwater for its own sake? This seems unlikely. There are some natural phenomena that are so uplifting to the human spirit that they are worth preserving for that reason alone. In the case of water, a mountain lake or a waterfall would fit into that category. Groundwater, by strict definition, does not fit into that group. Groundwater is, almost by definition, invisible unless we enter caves that intersect the water table. Springs could just about be regarded as groundwater, though strictly by the time the water emerges it is surface water. Springs certainly have the property of uplifting the spirit. Even then, it is the effects produced by the spring – vegetation in a desert, or the wildlife attracted to the water with its constant quality and constant temperature – that are usually the real phenomena of interest and attraction to the general public.

I suspect that to most people the target for groundwater sustainability centres on the environmental benefits of natural groundwater discharge – the 'sublime phenomena' of springs, wetlands and Chalk streams, and the habitats, flora and fauna that they support. This seems quite acceptable, but we then have to do three things: (a) be clear about exactly what we are trying to preserve and why; (b) be sensible about the way we achieve it; and (c) realize that most things in life are a compromise, and that few people set out to damage these phenomena – the problem usually is a conflict of interests, and each interest alone may be perfectly reasonable.

So under (a) we should be clear that we are usually trying to preserve the environment for the benefit of mankind. The animals and plants on Earth got along very well without us before we evolved and will continue to do so when we have become extinct. Of course, our extinction will not guarantee their preservation – species will continue to die out just as they did before Man arrived on the scene.

We should also be clear that the aim is to preserve the environment for general enjoyment, not for individuals or small groups to enjoy or profit from the environment at the expense of others. When action is taken to reduce abstraction from an aquifer to protect a spring or wetland, for example, it usually involves expense in replacing that source of water with another one. That cost is passed on to society, in the form of increased water charges if it is met by a water undertaker or in some other way if it is met by the regulator. If the whole of society is able to enjoy the benefits of the spring or wetland then the increased cost may be worthwhile. When abstraction is reduced to improve private amenity then it is usually not justified. Reducing abstraction to preserve an endangered habitat may be a valid activity for the regulator; reducing abstraction to increase property prices is not.

Under (b) we have to be careful not to be swayed too readily by emotional claims, especially those from single-interest pressure groups. For many decades, for example, sewage from many coastal towns in Britain was discharged to sea with little or no treatment. This was aesthetically disgusting and a risk to public health. Latterly, it was also in conflict with legislation arising from the EC Bathing Waters Directive (Directive 76/160/EEC). Pressure groups rightly campaigned for the situation to be improved but did not take account of the wider requirements.

In response to those concerns, water companies around Britain have invested hundreds of millions of pounds to treat sewage to very high standards and then pump it several kilometres out to sea, where it will not affect surfers or bathers. It would surely have been more sensible, having treated the effluent to high standards, to return it inland for re-use, perhaps by recharging it to aquifers. For decades sewage with very little treatment has been disposed of directly to land at sites such as the unconfined Chalk at Winchester, with no evidence of detriment to groundwater quality (Beard & Giles 1990).

Now, instead of this approach to sewage disposal, there is pressure to re-use 'grey' water – water from baths and showers – in the home. Such suggestions overlook the fact that we long ago abandoned domestic-scale systems of energy conversion, water supply and waste treatment in favour of safer, more efficient and convenient central systems. The cholera epidemics of the 19th century and the smogs of the 20th are among the compelling reasons for not going back.

The other important point is that matters have to be kept in perspective. The privatization of the water industry in England and Wales and the separation of the roles of water undertaker and regulator occurred in April 1989, coinciding with the period of one of the longest and most severe droughts on record. By 1992 stream networks in many parts of the country had shrunk dramatically. For several years, thinking in the National Rivers Authority (the predecessor to the Environment Agency) was dominated by low flows. In this, single-interest groups, notably fly-fishermen, were particularly vocal and influential with the NRA and the media. The public was constantly told that low flows and shrinking streams were caused by over-abstraction not lack of recharge, and that water tables all over the country were in decline. The general impression was that groundwater was over-exploited and should not be considered for future development. In the NRA water resources development strategy, groundwater was scarcely mentioned and concepts such as artificial recharge and wastewater re-use were dismissed alongside schemes like water transfer by sea in large polythene bags (National Rivers Authority 1994).

At the beginning of 2001 we have had record recharge to aquifers, and stream networks have expanded to historical maxima. Given that several towns have suffered repeated flooding, that groundwater levels in many areas are at record highs and groundwater flooding is now a serious issue, it seems pertinent to ask whether the priorities of the agency would have been the same had it been formed in April 2001 instead of in April 1989. Have lessons of climatic change, or at least of climatic fluctuation, been learned or not?

The balance of evidence suggests that they

have not. The near-hysteria over low flows has simply been replaced by panic over high flows. The lesson that climatic variability is causing more extremes in the water environment is given lip service but does not seem to have been generally accepted.

As far as (c) is concerned, there is still a tendency to look for villains instead of accepting the conflicts of interest inherent in modern living. Those currently portrayed as villains include the people who build, or allow building, on flood plains. The simple fact is that although buildings on flood plains are more vulnerable to flooding, building almost anywhere increases the quickflow component of river discharge and hence the likelihood of flooding. Householders want their properties drained, so that their houses are dry and their paths and gardens free of standing water. Rainfall that would previously have been allowed to infiltrate the soil or to occupy depression storage and which, in many cases, would eventually have increased groundwater reserves, is now channelled swiftly to surface drainage systems where it merely increases the flood risk.

Similarly, roads serving these properties need rapid drainage for the convenience and safety of road users. This rapid drainage conflicts with the need to prevent pollution of local waters. Desire to prevent polluted water entering aquifers means that there is now a general presumption against allowing road drainage to soakaways, where it would increase recharge to aquifers and add to groundwater resources (Price 1994): instead it too is sent to add to peak flows in streams, increasing the risk of flooding.

Bad planning – such as poor siting of abstractions, and poor planning for the disposal of urban drainage – has had a role to play in all these problems, but it is difficult to see how the problems can be totally avoided given the pressures of 60 million people living on a small island, most of whom constantly seek more living space and an improved road network. Many of the problems, in short, are caused by simple conflicts of interest.

Agriculture is one of the areas where conflicts of interest abound. Irrigated agriculture is a minor user of water in Britain but by its nature tends to take water in the very places and at the very times when water is most scarce (Anon. 2001, p. 25). Irrigation is used in Britain not necessarily because there is insufficient rainfall to grow crops, but because uniform crop growth requires a uniform supply of soil moisture and because attractive-looking fruit and vegetables result from uniform growth (Anon. 2001, p. 26). Supermarkets, which claim to reflect the requirements of today's more discerning customer, insist on uniformly-attractive produce. Thus, the alleged desire for nicely-shaped potatoes can have an indirect impact on groundwater discharge to wetlands.

World-wide, the situation is no better. Roughly two-thirds of the fresh water used in the world is used for agriculture and 40% of the world's food is grown on irrigated lands (Gleick 1993). The need for irrigation may be unavoidable but at present we have some absurd situations. In some countries with arid and semi-arid climates, inefficient irrigation techniques and possibly non-renewable groundwater are being used to produce cash crops for export. Unfortunately, it is not easy to see how the situation can be improved under current economic regimes.

Conclusions

The following conclusions can be drawn:

(a) Sustainability should not be seen as an end in itself. Most human advances have not been sustainable in the long term; rather, mankind has advanced by a series of unsustainable developments.
(b) It follows that when we talk of sustainability we should define the time period over which we are measuring sustainability.
(c) Traditionally resources have been exploited by the first people to need them and to have the technology to use them.
(d) Although bad planning decisions play a part, environmental problems are usually the result of a conflict of legitimate interests rather than deliberate lack of concern for environmental well-being.
(e) Selfish desire to preserve individual benefits should not be allowed by regulatory bodies to masquerade as environmental concern.
(f) The best ways to ensure sustainable use of water are to re-use water after abstraction and to reduce irrigated agriculture wherever possible in favour of rain-fed agriculture. The first will require a change in public attitude; the second will require a change in economic approach that is outside the control of hydrogeologists.
(g) Finally, and perhaps most importantly, we should remember that a water environment that was sustainable under one climatic regime may become unsustainable if (or when?) the climate changes. Water tables will rise and fall and stream networks expand and shrink depending on the amounts of recharge. If the changes are

short-term fluctuations, groundwater storage will be important in helping to ameliorate the effects on both the environment and water supplies. If the climatic changes are long-term trends, then no amount of planning or legislation will prevent the resulting changes to the environment.

References

ANDREWS, M. A. 1991. *The Birth of Europe*. BBC Books, London.

ANON. 1998. *Waterfacts '98*. Water UK, London.

ANON. 2000. *Rising groundwater levels in the Chalk-Basal Sands Aquifer of the central London Basin*. Environment Agency, Reading.

ANON. 2001. *Water resources for the future: a strategy for England and Wales*. Environment Agency, Bristol.

BEARD, M. J. & GILES, D. M. 1990. Effects of discharging sewage effluents to the chalk aquifer in Hampshire. *In:* BURLAND, J. B., MORTIMORE, R. N., ROBERTS, L. D., JONES, D. L. & CORBETT, B. O. (eds) *Chalk*. Thomas Telford, London, 597–604.

BRONOWSKI, J. 1973. *The Ascent of Man*. British Broadcasting Corporation, London.

CCIRG. 1991. *The potential effects of climate change in the United Kingdom*. UK Climate Change Impacts Review Group, HMSO, London.

CCIRG. 1996. *Review of the potential effects of climate change in the United Kingdom*. UK Climate Change Impacts Review Group, HMSO, London.

CEH/BGS 2000. *Hydrological summary for Great Britain, November 2000*. Centre for Ecology and Hydrology/British Geological Survey, Wallingford.

DOWNING, R. A. 1986. Development of groundwater. *In:* BRANDON, T. W. (ed.) *Groundwater Occurrence, Development and Protection*. Institution of Water Engineers and Scientists, London, 485–542.

DOWNING, R. A. 1993. Groundwater resources, their development and management in the UK: an historical perspective. *Quarterly Journal of Engineering Geology*, **26**, 335–358.

EU. 2000. *The EU Water Framework Directive*. Official Journal of the European Union, OJ L327, 22 December 2000.

FARRAR, C. D. & BERTOLDI, G. L. 1988. Region 4, Central Valley and Pacific Coast Ranges. *In:* BACK, W., ROSENSHEIN, J. S. & SEABER, P. R. (eds) *Hydrogeology: The Geology of North America, Vol. O-2*. Geological Society of America, Boulder, Colorado.

GLEICK, P. H. 1993. An introduction to global fresh water issues. *In:* GLEICK, P. (ed.) *Water in Crisis: a Guide to the World's Fresh Water Resources*. Oxford University Press, New York and London.

GLEICK, P. H. 2000. *The World's Water 2000–2001: The Biennial Report on Freshwater Resources*. Island Press, Washington DC.

GLEICK, P. H. 2001. Safeguarding our water: Making every drop count. *Scientific American*, **284** (2), 29–33.

LLAMAS, M. R. & PRISCOLI, J. D. 2000. Report of the UNESCO Working Group on the ethics of freshwater use. *Papeles del Proyecto Aguas Subterráneas*. Fundación Marcelino Botín, Madrid.

MARSH, T. J. 2001. Climate change and hydrological stability – a look at long-term trends in south-eastern Britain. *Weather*, **56** (10), 319–326.

NATIONAL RIVERS AUTHORITY. 1994. *Water: Nature's precious resource*. HMSO, London.

PRICE, M. 1994. Drainage from roads and airfields to soakaways: groundwater pollutant or valuable recharge? *Journal of the Institution of Water and Environmental Management*, **8**, 468–479.

PRICE, M. 1998. Water storage and climate change in Great Britain – the role of groundwater. *Proceedings of the Institution of Civil Engineers, Water, Maritime and Energy Journal*, **130**, 42–50.

SIMPSON, B., BLOWER, T., CRAIG, R. N. & WILKINSON, W. B. 1989. *The engineering implications of rising groundwater levels in the deep aquifer beneath London*. Construction Industry Research and Information Association, London, Special Publication, **89**.

USGS. 2001. United States Geological Survey, Table 8 (Ground-water withdrawals by water-use category and State, 1995). World-Wide Web Address: http://water.usgs.gov/watuse/.

Incorporation of groundwater modelling in the sustainable management of groundwater resources

P. HULME[1], S. FLETCHER[1] & L. BROWN[2]

[1]*Environment Agency, National Groundwater and Contaminated Land Centre, Olton Court, 10 Warwick Road, Solihull, West Midlands B92 7HX, UK*
(e-mail paul.hulme@environment-agency.gov.uk)
[2]*Environmental Simulations International, Priory House, Priory Road, Shrewsbury SY1 1RU, UK*

Abstract: Numerical groundwater models have been used by the Environment Agency and its predecessors for over 30 years to help understand the behaviour of aquifer systems and as one of the tools to help manage groundwater resources effectively. The Agency has recently reviewed the past use of distributed numerical models with a view to improving their utilization for groundwater resource management. This is particularly important due to the changes in groundwater management strategies that will occur as a consequence of the new European Union Water Framework Directive. The review has highlighted a number of areas where changes are desirable. Three in particular are the role and importance of the conceptual model, the requirement for a nationally coordinated programme of modelling and the need to develop further the Agency's in-house modelling expertise. The Agency is currently developing proposals in these and other areas as part of a Research and Development project.

The term 'groundwater modelling' is often taken to be the process resulting in the development of a distributed numerical groundwater model. However, in a broader sense, it can be applied to the process that produces a conceptual understanding of a groundwater flow system, which is then tested quantitatively using an appropriate distributed numerical model, lumped water balance model or analytical model. A distributed numerical model may not always be required, nor be the most appropriate tool in all cases.

Distributed numerical models used for resource estimation and groundwater management are generally built on a catchment scale or larger to represent the spatial and temporal variation of the flows and groundwater heads in the real system. In England and Wales, this type of model has been used since the mid-1970s as a tool to support decisions required for the effective management of groundwater resources. The models have generally been designed to answer specific questions such as: how much water can we pump from a particular location; what is the likely influence on nearby rivers, wetlands or other abstractors; and can we abstract from elsewhere with less impact? Questions of this type arise as part of the responsibility of the Environment Agency in England and Wales (the Agency) to manage water resources effectively.

Whether a distributed numerical model or one of the simpler tools is employed as part of the resource estimation, it must be used within the context of the resource management strategies operating at the time. Recently, the Agency's new national water resources strategy (Environment Agency 2001*a*) and accompanying regional strategies were published. These strategies encompass other recent changes such as the development of Catchment Abstraction Management Strategies (Environment Agency 2001*b*)

From: HISCOCK, K. M., RIVETT, M. O. & DAVISON, R. M. (eds) *Sustainable Groundwater Development*. Geological Society, London, Special Publications, **193**, 83–90. 0305-8719/02/$15.00

and the new European Union Water Framework Directive (European Parliament Directive 2000).

Framework for groundwater resources management

The most significant recent change to the way water resources will be managed in the future is the introduction of the Water Framework Directive (European Parliament Directive 2000), which came into force on 22 December 2000. This has introduced a system of water resources management based on the River Basin District as the main management unit. For the management of groundwater resources, the significant impact is that within this River Basin District surface water and groundwater management will be integrated into a single system. This link is explicitly stated in the Directive's definition of 'available groundwater resources'. This defines available groundwater resources as 'the long-term annual average rate of overall recharge of the body of groundwater less the long-term annual average rate of flow required to achieve the ecological quality objectives for associated surface waters specified under Article 4, to avoid any significant diminution in the ecological status of such waters and to avoid any significant damage to associated terrestrial ecosystems'. This definition, which classifies groundwater on the basis of its effects on surface water ecology, links the two more strongly than hitherto has been the case. The Directive also introduces the concept of a 'groundwater body', defined as 'a distinct volume of groundwater within an aquifer, or aquifers'.

As far as groundwater resources management in England and Wales is concerned, the Water Framework Directive will bring about three main changes: (1) groundwater and surface water will be considered together and available groundwater resources will be defined with respect to ecological quality objectives that have been set for its associated surface waters; (2) the River Basin District will be introduced as the water resource management unit and will be defined by a surface water catchment, with groundwater bodies defined by their association with that surface water system; and (3) systematic monitoring of surface water and groundwater for both water quality and quantity will be required, with obligations to report this to the EU every 6 years.

The Agency has already gone some way to implementing the Directive's requirements with the introduction of Catchment Abstraction Management Strategies (CAMS), which will provide information required to formulate the River Basin Management Plans. Each CAMS will require a resource assessment leading to a statement of the resource availability status of the unit. The Agency is currently developing a resource assessment framework within which the variety of tools available can be used in a consistent manner (Environment Agency Research & Development (R&D) project W6-066: A unified framework for abstraction licensing and reporting water resource assessments for catchment abstraction management strategies).

Tools for groundwater resources assessment

The changes to groundwater resources management have prompted the Agency to review some of the tools it uses to quantify groundwater resources and there are current and planned Research and Development projects in a number of areas.

The first consideration under the Directive is the requirement to define groundwater bodies. The current management of aquifers in England and Wales is based on so-called 'groundwater units'. The need for an integrated management of surface and groundwater will require an examination of such terms. There will need to be a programme within the Agency to facilitate the move from the current management units where these conflict with the new groundwater bodies.

The Directive defines the available resource as the recharge minus the flow necessary to maintain the ecological health of the associated surface water. To be truly effective, the recharge term must include all inflows to the groundwater body, such as direct recharge, induced recharge, drift leakage, etc. This implies a much greater understanding of the water balance for any groundwater body than has been the case until now. Development is currently being undertaken within the Agency to improve methods of estimating recharge input to distributed numerical groundwater models.

A critical input to the assessment of the available resource is the definition of the ecologically acceptable flows and levels in the associated surface water. The Agency has looked at a number of techniques but an acceptable solution has not yet been found and there is likely to be considerable R&D effort in this direction in the future.

In order to be able to select, and apply, the correct tool and evaluate the results, we need a good conceptual understanding of the hydro-

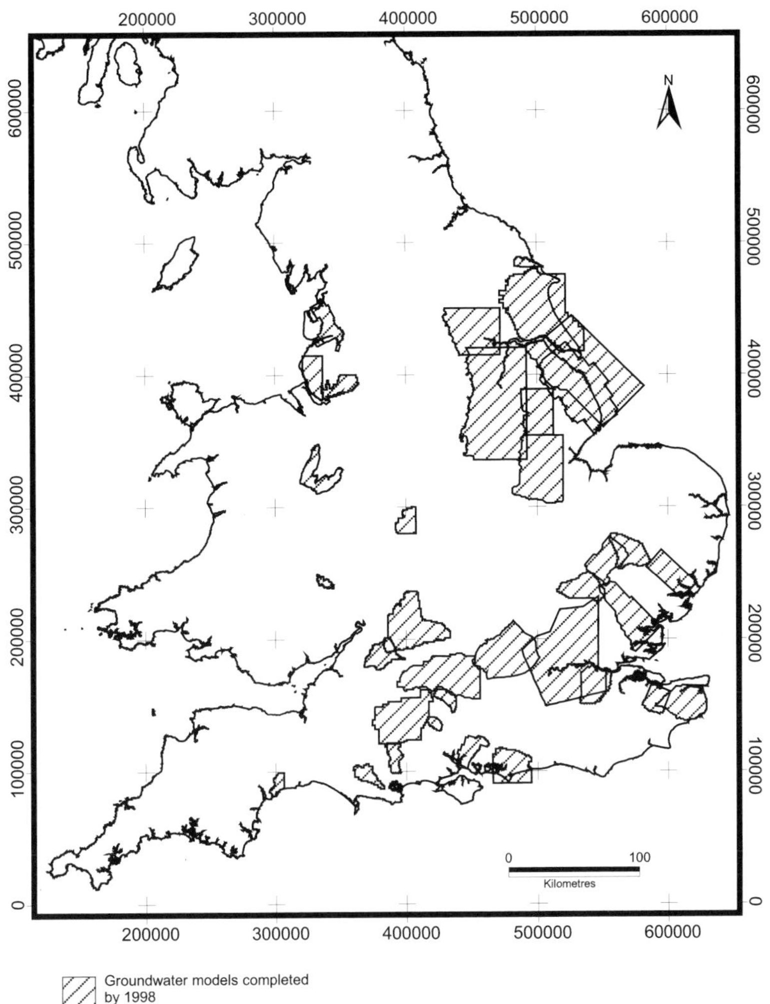

Fig. 1. Location of regional groundwater models developed for the Environment Agency and its predecessors (1975–1998).

logical system being evaluated. For example, to be able to estimate the likely effect of an abstraction on a nearby river flow, we need a conceptual model of the interaction between the river and the aquifer. If this is lacking we might select a tool which assumes that a river is always hydraulically connected to the aquifer, when in fact it is perched along some reaches, which significantly reduces the flow to the aquifer. The importance of the conceptual model has been emphasized during an Agency R&D project to develop and improve the use of distributed numerical groundwater models within the Agency (Environment Agency 2001*c*). The main objective of this project is to 'To promote a nationally consistent framework for the use of groundwater modelling as a tool for groundwater resources management'. The project has allowed the Agency to review past groundwater modelling projects in the light of the new changes to groundwater resource management and to propose changes in the way the Agency approaches modelling in the future.

Regional numerical groundwater modelling in the Agency (1970–1998)

The use of regional numerical groundwater models by the Agency, and its predecessors, for managing groundwater resources dates back to the early 1970s. Most of the earliest models were developed as part of research projects by universities, mainly in the School of Civil Engineering at the University of Birmingham, UK (for example, University of Birmingham 1981). From the mid-1980s, as modelling expertise developed and model codes became more widely available, model development was increasingly undertaken by consultants. By 1998, the Agency had developed, and inherited, 34 regional models covering approximately 30% of the major aquifers. The location of these models is shown in Figure 1.

A review of past modelling projects was undertaken in 1998 as part of a Research and Development project (Environment Agency 2001*c*). One of the objectives of the review was to establish how past experience could be used to improve future projects. All the models had been developed to answer specific questions about how better to manage groundwater resources, for example, to investigate the options for alleviating low river flows and declining groundwater levels, the management of rising groundwater levels or of coastal saline intrusion.

In almost all of the 34 models reviewed, these questions were successfully addressed. In addition, the results from many of them have subsequently been used to underpin management decisions, for example to formulate aquifer management rules, to support the redistribution of abstractions or to redefine groundwater management units. A few models have also been used for other purposes, for example to produce Source Protection Zones, or to assess the potential effects of climate change on groundwater resources. However, at the time of the survey, only five models were being used actively to support on-going management decisions, and most of the models reviewed had not been updated beyond the original project input data. The following reasons were given for this: firstly, some of the early models were never intended to be used beyond solving an immediate problem, as was the case for some of the early University of Birmingham models; secondly, there was a perception by Agency staff that some models did not adequately represent the processes within the aquifer, thus suggesting that there was dissatisfaction with the behaviour of the numerical model and, by implication, the conceptual understanding upon which the model was based; and thirdly, there was a lack of modelling capability within the Agency and its predecessors which prevented models being updated, sometimes as a result of a lack of specialist modelling expertise within an Agency Region, but more often because operational priorities took precedence.

Future of regional groundwater modelling within the Agency

As a result of further work undertaken during the R&D project (Environment Agency 2001*c*), and in co-operation with the Agency's Regional Offices in England and with the Environment Agency in Wales, a number of areas were recognized which required improvement. Three areas considered to be of particular importance are the conceptual model, the modelling programme, and the development of modelling expertise within the Agency.

The conceptual model

The process of developing a conceptual understanding requires us to simplify reality. In doing so, the key processes that most influence how the real system behaves must be identified . If this is not done properly, the most sophisticated numerical model is likely to be unsatisfactory. A conceptual model has value only to the extent that we have confidence in its representation of reality. Therefore, it needs to be tested and a numerical model is one means of doing this. Both conceptual and numerical modelling can, therefore, advance our understanding of the behaviour of the real system. Furthermore, it is only when we have confidence in both our conceptual understanding and our quantitative representation of the real groundwater system that we can begin making useful predictions.

The importance of the conceptual model as a tool in its own right has also been recognized within the Agency (Environment Agency 2000*c*). It provides a summary of all the available data and an explanation of how the hydrological system is thought to behave. A comprehensive conceptual model report can provide useful resources, such as potentiometric surface maps, hydrogeological cross-sections showing water levels and preliminary lumped water balances.

Modelling programme

The Agency has begun developing a groundwater modelling programme stretching over the next 10 to 15 years via the development of modelling strategies for each Agency Region. The overall aim of the programme is to develop a network of conceptual models covering all the major, and regionally important, aquifers in England and Wales. Distributed numerical models will only be developed where they are

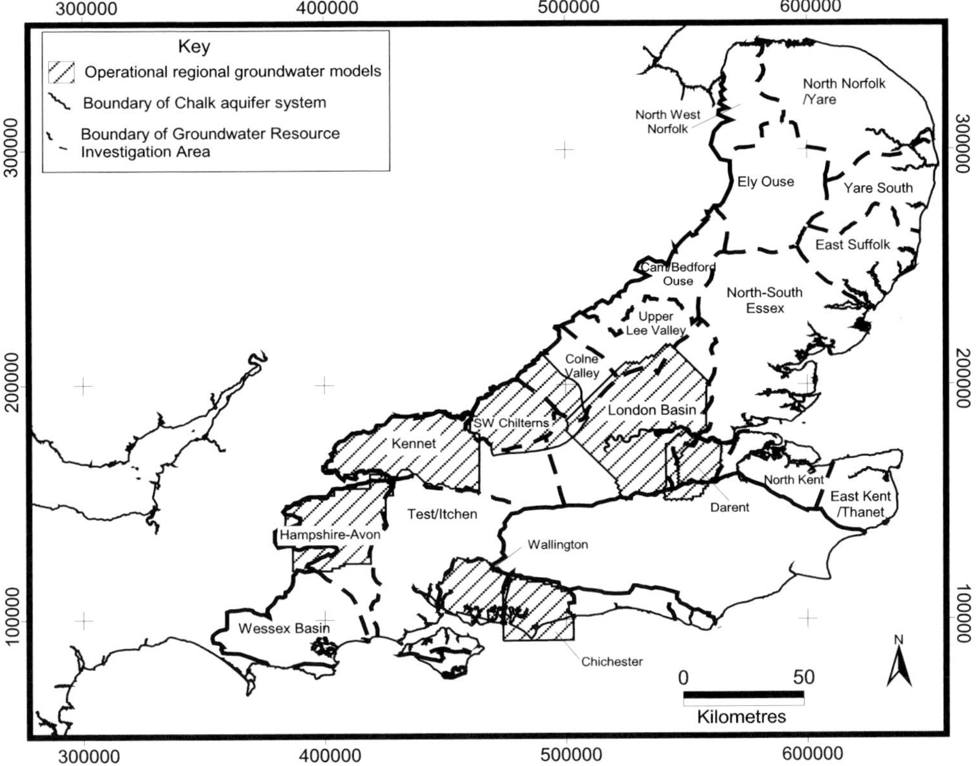

Fig. 2. Location of Groundwater Resource Investigation Areas on the Chalk in south-east England.

the most appropriate tool either for testing our understanding or for predicting the behaviour of the aquifer system in response to possible changes in, for example, recharge or abstraction patterns. If a simpler means can be used to give a satisfactory degree of confidence in the results, it will be preferred.

To provide a basis for this planning, the major aquifers have been divided into Groundwater Resource Investigation Areas. These investigation areas, identified by the Regional Offices of the Environment Agency in England and in Wales, define areas of aquifer where water resource issues have been identified and need to be addressed. Two examples are illustrated, the first for the Chalk in south-east England (Fig. 2) and the second for the Magnesian Limestone and the Sherwood Sandstone in eastern England (Fig. 3).

It is important to note that the boundaries shown are intended to act only as an outline planning tool. The boundaries do not represent the extent of a conceptual model as this will inevitably need to consider the area beyond these boundaries. Nor are they meant to signify the boundaries of any future numerical model as these will only be decided during the development of the conceptual model. For example, although the Magnesian Limestone and Sherwood Sandstone have been identified as separate investigation areas (Fig. 3), conceptual modelling may reveal that they need to be considered as a single connected unit in some areas.

To aid planning, provisional timetables for developing this network of quantitative conceptual models are being developed. An example of this is shown in Figure 4 for the Sherwood Sandstone and Magnesian Limestone in eastern England. At this stage, these projects are only provisional and their implementation will depend on other priorities within the Agency. However, this initial planning provides an indication of the potential workload over the next 10 years. It is intended that all future models, both numerical and conceptual, will be updated on a regular basis. A provisional 3-

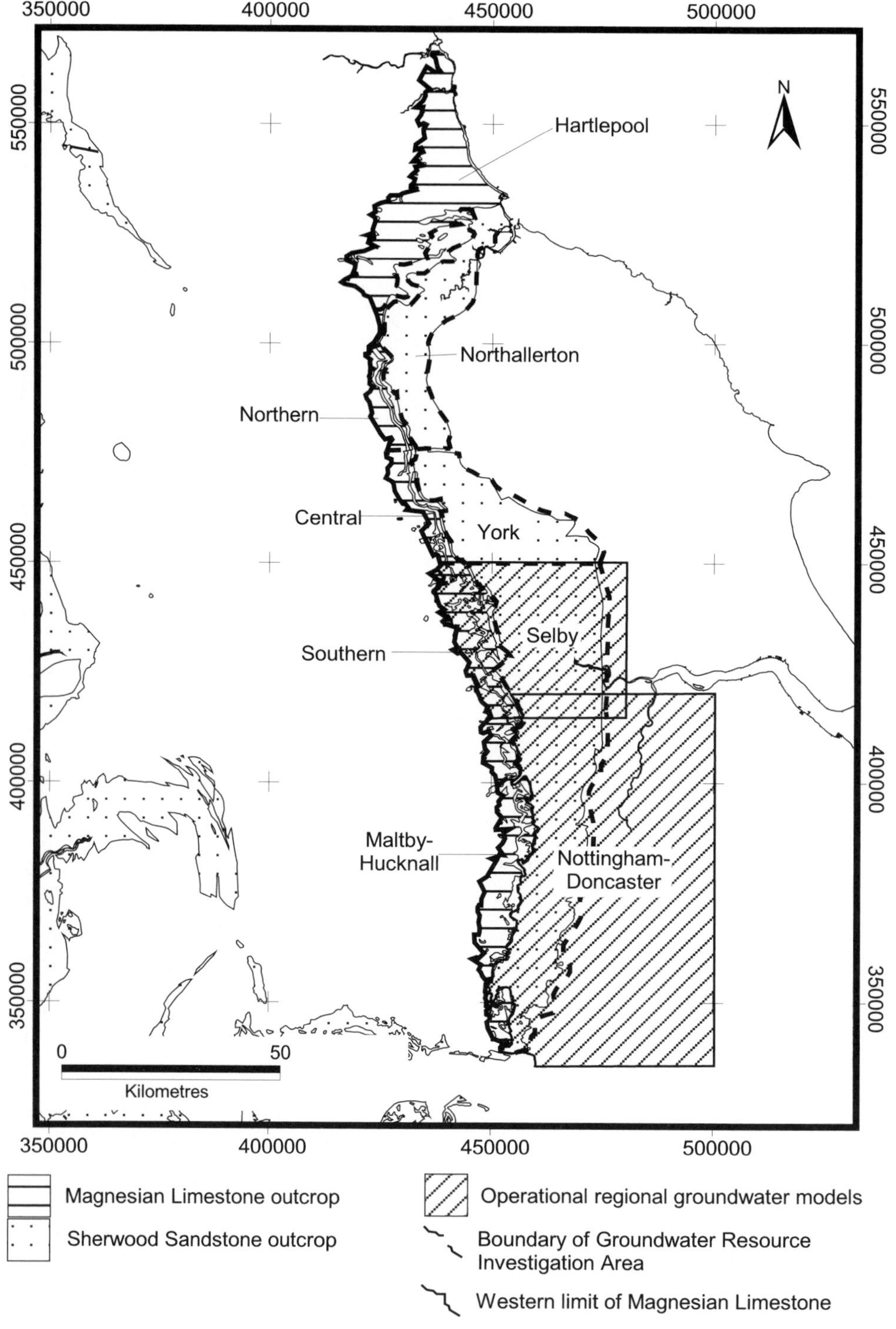

Fig. 3. Location of Groundwater Resource Investigation Areas in the Magnesian Limestone and Sherwood Sandstone of eastern England.

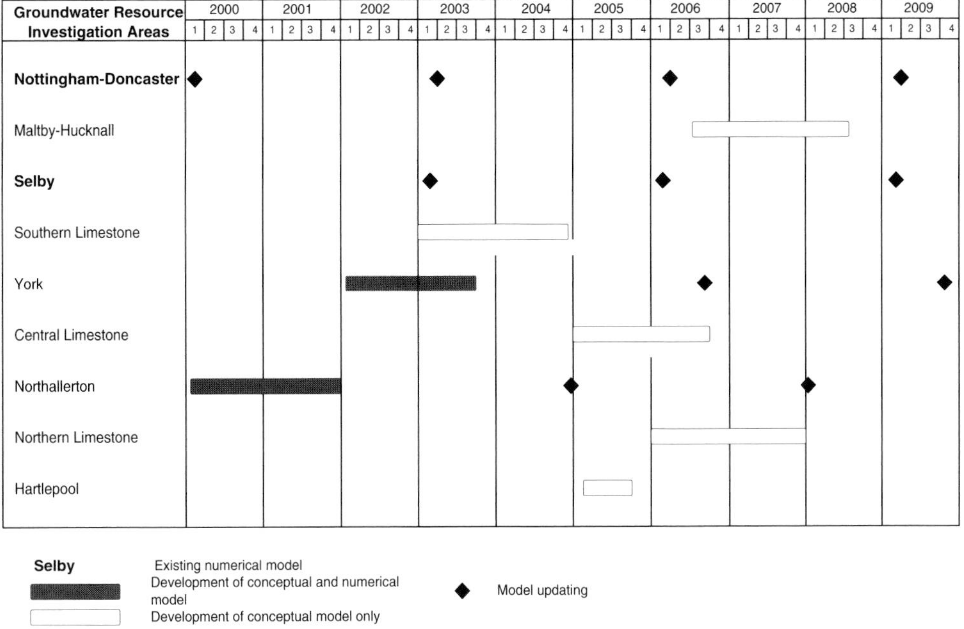

Fig. 4. An example of possible project timescales for groundwater investigation areas in eastern England

yearly interval for numerical models is indicated in Figure 4. Conceptual models will be updated on a 6-yearly cycle, in line with CAMS.

In-house modelling capability

Finally, it has been recognized that the Agency should further develop its own groundwater modelling capability and make more use of its staff for modelling projects. Many staff have developed considerable experience of particular aquifers over 10 or 20 years and they should be included in conceptual modelling teams.

It is not considered effective for the Agency to rely on consultants for all, or even the majority, of its modelling work. The experience and knowledge gained from conceptual and numerical modelling is too valuable to be vested purely in Agency contractors. The discipline of the modelling process leads us to struggle to make sense of the data, to unravel why the real system behaves as it does, and to challenge our assumptions. The Agency's own staff should reap these benefits so that they can carry them into their operational work of managing the hydrological system. In addition, the Agency needs its own knowledge base and modelling capability to be able to update its groundwater models and to run individual predictive scenarios as new questions arise.

The level of this staff development will depend on the modelling workload within each Region. Some Agency Regions make more use of numerical groundwater models as operational or predictive tools than others. However, it is likely that all Regions will be using numerical modelling in some way, both as a means of exploring and testing their conceptual models and for making predictions.

To further develop the Agency's groundwater modelling capability in line with these needs the following options are being considered: combining Agency and consultancy staff on technical teams which are developing conceptual and numerical models; encouraging more inter-regional co-operation; updating models in-house, rather than relying on external organizations; and utilizing the specialist modellers from the Agency's National Groundwater and Contaminated Land Centre as a 'floating resource' to be made available to Regions at key stages in a modelling project.

To establish the expertise currently available and likely to be needed, the Agency Regions have been asked to review their in-house modelling capabilities based on their 10-year modelling programmes.

Summary

The Environment Agency is reviewing and developing tools to aid groundwater resource management in anticipation of the introduction of the European Union Water Framework Directive into the legislative framework of water resource management in England and Wales. As part of this effort, the Agency has undertaken a review of groundwater modelling. This review highlighted a number of areas where changes were thought to be required, as follows: increased investment in well-documented and adequately tested conceptual models; the need for a planned national programme of conceptual modelling and, where appropriate, regional numerical modelling projects; and further development of the Agency's in-house capability in groundwater modelling so that the benefits of the modelling process are more effectively vested in the Agency's own staff.

This investigation is being carried out through an Agency Research and Development project (Environment Agency 2001*c*). A provisional nationwide modelling programme has been developed and is currently under review by Regional teams. Other improvements in regard to modelling methodology and the Agency's in-house modelling capability have also been proposed.

The views expressed in this paper are those of the authors and not necessarily those of the Environment Agency.

References

DETR. 1999. *Taking Water Responsibly: Government decisions following consultation on changes to the water abstraction licensing system in England and Wales.* Department of the Environment, Transport and the Regions, London.

Environment Agency. 2001*a*. *Water Resources for the Future. A Strategy for England and Wales.* Environment Agency, Bristol.

Environment Agency. 2001*b*. *Managing Water Abstraction. The Catchment Abstraction Management Strategy process.* Environment Agency, Bristol.

Environment Agency. 2001*c*. Environment Agency Framework for Groundwater Resources Conceptual and Numerical Modelling. R&D Technical Report W214. Environment Agency, Bristol (in press).

European Parliament Directive. 2000. Directive 2000/60/EC of the European Parliament and of the Council of 23 October 2000 establishing a framework for Community action in the field of water policy. *Official Journal*, **L327**.

University of Birmingham. 1981. *Saline Groundwater Investigations Phase 1: Lower Mersey Basin.* Final report to North West Water Authority.

Groundwater sustainability and water resources planning for the East Midlands Resource Zone

M. HUDSON

Water Strategies, Severn Trent Water Ltd, 2297 Coventry Road, Birmingham B26 3PU, UK (e-mail: matthew.hudson@severntrent.co.uk)

Abstract: The East Midlands Resource Zone is one of three water resource zones identified by Severn Trent Water in its recent Water Resources Plan, submitted to the Environment Agency. The public water supply to this resource zone is derived from a complex mix of surface and ground water sources, and supplies nearly 3 million people. The most significant aquifer in the East Midlands is the Nottinghamshire Sherwood Sandstone which supplies approximately one quarter of the deployable output for the resource zone (some 200 Ml d^{-1}). This aquifer is over-licensed and historically abstraction has exceeded the long-term average recharge. Severn Trent Water has agreed reductions in groundwater licences from this aquifer; however more reductions are targeted by the Environment Agency following its recent reassessment of resources and the introduction of the 'environmental fraction'. In order to meet the existing resource shortfall and the continuing pressure to reduce abstraction licences, Severn Trent Water propose to support its new surface water abstraction on the River Trent by developing groundwater resources beneath the City of Birmingham. Significant further work is required by water industry regulators and Severn Trent to resolve the projected long-term supply deficit.

Severn Trent Water (STW) is the second largest water and sewerage company in the UK, both in terms of area and population served. Broadly speaking, the area served by the company stretches from the Severn estuary in the south to the Humber estuary in the north, and from mid-Wales to the Lincolnshire border. The company supplies water and sewerage services to over seven million people, including the major cities of Birmingham, Stoke, Derby, Coventry, Nottingham and Leicester.

In order to satisfy the Environment Agency's requirement for a Water Resources Plan and as part of the 1999 Periodic Review Process the company was divided into three resource zones (RZs): RZ1 – Stoke, Stafford and Telford Resource Zone; RZ2 – East Midlands Resource Zone; and RZ3 – Severn Resource Zone (see Fig. 1).

Dry-year water demands and the water resources available in drought years were calculated for each resource zone and projected over the next 25 years, using national water resource planning guidelines (Environment Agency 1998). Dry-year demands were calculated using the base demands at the time (1997), but incorporating an allowance for a weather-related summer peak (from 1995). The available water resources were calculated following a reassessment of source reliable yield (deployable output) with an allowance for supply interruptions (outage), resulting in a Water Available For Use (WAFU).

The East Midlands Resource Zone covers most of Derbyshire, Leicestershire and Nottinghamshire, comprising around 40% of customers, with a dry-year demand of approximately 750 Ml d^{-1}. The resource zone (see Fig. 2) is supplied by a complex mix of impounding reservoirs (e.g. Derwent Valley), pumped storage reservoirs (e.g. Carsington), river abstractions (e.g. the river Derwent and the river Dove) and groundwater sources on the Nottinghamshire Triassic sandstone.

The East Midlands Resource Zone is the most critical zone in the Severn Trent area, with

From: HISCOCK, K. M., RIVETT, M. O. & DAVISON, R. M. (eds) *Sustainable Groundwater Development*. Geological Society, London, Special Publications, **193**, 91–98. 0305-8719/02/$15.00

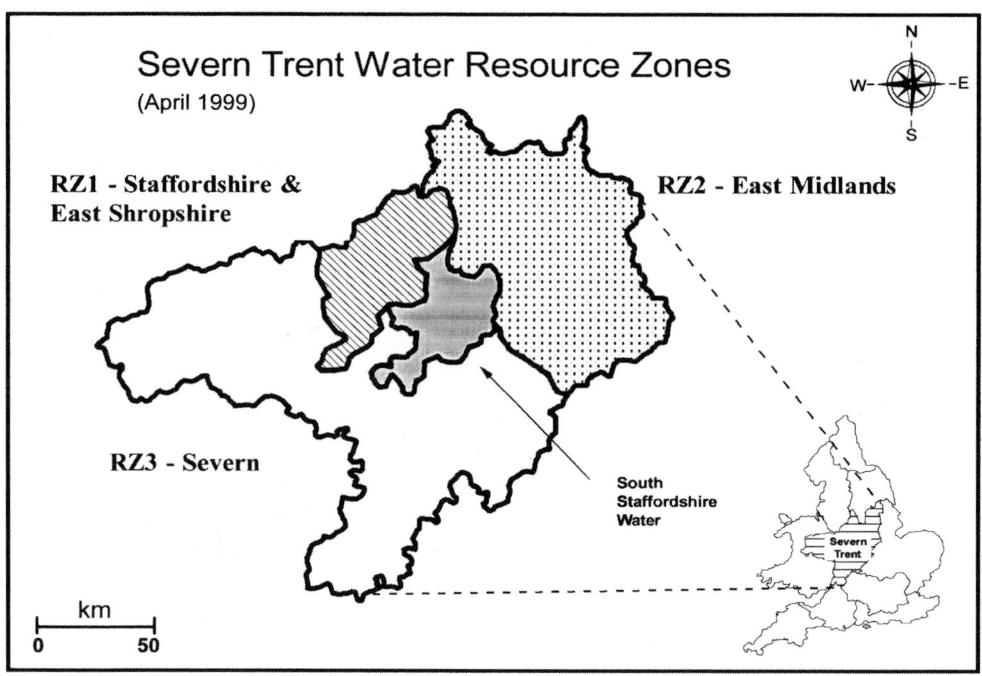

Fig. 1. Severn Trent Water Resource Zones.

available resources (WAFU) already below the acceptable planning margin (headroom). Headroom is the allowance made for the uncertainty in the planning process and was expressed as a demand profile approximately 6% higher than the calculated dry-year demand.

With only limited potential for further reductions in demand, the supply-demand position is projected to worsen over the next 25 years due to the increasing downward pressure on the resource base. There are two main drivers for the falling resource base: (1) groundwater licence reductions targeted on the Nottinghamshire Triassic sandstone aquifer by the Environment Agency; and (2) the calculated average effects of climate change on the surface water sources (Crookall & Bradford 2000).

The Nottinghamshire aquifer resource position

The outcrop of the Nottinghamshire Triassic sandstone stretches from the City of Nottingham north to Doncaster (and beyond), and has been exploited for groundwater supply for over 100 years. The sandstone dips eastwards beneath the Mercia Mudstone and this confined part of the aquifer is also exploited, as far east as the River Trent. Severn Trent Water currently has some 30 groundwater sources abstracting from the aquifer, providing a drought deployable output of around 200 $Ml\,d^{-1}$. The potential yield of the aquifer is significantly greater than this figure, however the deployable outputs are constrained by a series of group licences. The majority of STW's licences for the aquifer are licences of right granted to the Trent River Authority as part of the Water Resources Act (1963).

The current resource assessment of the Nottinghamshire Triassic sandstone aquifer by the Environment Agency is based on the results of a groundwater model developed by the University of Birmingham (Bishop & Rushton 1993) for the former National Rivers Authority. The model predicts an annual average recharge to the sandstone in the Severn Trent area of 354 $Ml\,d^{-1}$ (excluding the Wollaton unit).

A proportion of the recharge figure for the aquifer has also been identified as an 'environmental fraction' by the Environment Agency,

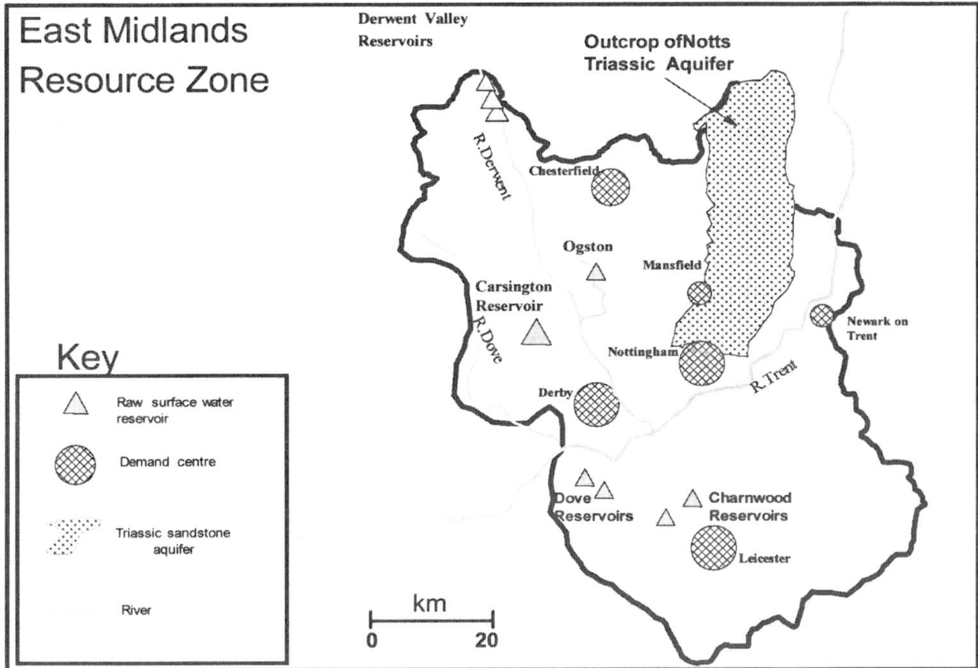

Fig. 2. East Midlands Resource Zone.

with the aim of protecting baseflows and other surface water features (Table 1). This fraction is set at 35% of the available recharge (124 Ml d^{-1}). The resulting annual average resource available for licensing purposes is therefore calculated to be 230 Ml d^{-1}. This is significantly lower than the total current licensed abstraction of 454 Ml d^{-1} and the total actual abstraction of approximately 400 Ml d^{-1}. As a result, the Environment Agency has set a preliminary target figure of reducing total abstraction licences to 90% of the assessed long-term average recharge. This translates to a reduction in Severn Trent Water abstraction licences of around 65 Ml d^{-1}.

Table 1. *Resource balance for whole Nottinghamshire Triassic sandstone aquifer*

Environment Agency Resource Assessment	Ml d^{-1}
Total licences	454
Actual abstraction	*c.* 400
Modelled recharge*	354
Environmental fraction	124
Available resource	230

(*excluding the Wollaton groundwater unit)

The Environment Agency has also identified two specific 'low flow' sites on the outcrop of the sandstone which they consider require additional water: the Dover Beck and Rainworth Water. Specific compensation discharges are being sought from STW, either from existing groundwater abstractions or from new boreholes to alleviate low flows.

Determining the effects of abstraction

Environment Agency groundwater level hydrographs for the Nottinghamshire Triassic sandstone aquifer show that groundwater levels have declined across some parts of the aquifer, with particularly low groundwater levels between 1975–1977, 1990–1993 and 1997–1999. Groundwater level fluctuations over the last 25 years on the unconfined aquifer are typically between 1 and 3 m due to the high storage characteristics of the sandstone. It is also clear that annual average rainfall over this period has decreased and there have been a number of significant droughts which precede the lowest groundwater levels.

In order to demonstrate a link between abstraction and low groundwater levels or low flows it is necessary to separate out the effects of abstraction from any changes resulting from periods of low recharge. One of the few methods of testing the effects of abstraction and climate is by the use of groundwater modelling.

The Midlands Region of the Environment Agency is currently examining the Nottinghamshire Triassic sandstone aquifer using the existing groundwater model (Bishop & Rushton 1993). A number of scenarios have been run against a baseline to examine the impacts of abstraction, and Severn Trent Water (along with the other relevant water companies) has proposed a number of the modelling scenarios. It is hoped that the modelling work will provide answers to: (1) the impact of abstraction on groundwater levels and surface water flows; (2) the effect of the already agreed reductions; (3) the effect further reductions will have on groundwater levels and surface water flows, and the timing of any benefits; (4) the effects of transferring more abstraction from the unconfined aquifer to the confined aquifer; and (5) the effect of further conjunctive use of the resource.

Only once these answers are available will it be possible to calculate meaningful costs and benefits of the different options for the future sustainable management of the Nottinghamshire Triassic sandstone aquifer.

Preliminary results of the modelling work suggest that the model is not suitable for examining detailed local scale impacts of abstraction in several places, because the model grid refinement is too coarse and the model calibration is poor in some areas. The model is still a useful tool when considering broad abstraction scenarios and water balance components.

It is interesting to note that the preliminary results of the modelling work suggest that in some areas, such as the southern part of the aquifer, even significant reductions in abstractions produce limited benefits to surface water flows. This is considered to be the result of a number of factors: reducing the groundwater abstraction in turn reduces the leakage input from the overlying transition beds (the Colwick Formation); some groundwater is still returned to storage even after the 25-year model run; some watercourses remain influent; and the model is poorly calibrated in some areas.

The effects of groundwater abstraction on the unconfined outcrop of the sandstone could be divided into two main categories: (1) local effects restricted to the cone of depression around an individual source, where a source is located close to a watercourse or water feature, and abstraction has reduced groundwater heads so that groundwater flow to the surface water feature is reduced; and (2) regional dewatering of groundwater levels due to widespread abstraction at a rate greater than the annual average recharge.

Where it is clear that abstraction is affecting surface water features it will become important to identify whether the effects are local (and might be attributable to a particular source) and where the effects are a result of a regional lowering of groundwater levels (from a number of sources). Local-scale solutions are likely to be considerably cheaper and much more likely to succeed. It is possible therefore that for some low-flow sites the proposed short-term engineering solution (compensation flow discharges) will actually be more successful and will have a more favourable long-term cost-benefit than large-scale reductions in groundwater abstraction with significant replacement water costs.

Climate change impacts

As part of an ongoing programme to investigate the potential impact of climate change on future water resources, STW has commissioned modelling work using the Environment Agency's groundwater model for the Nottinghamshire Triassic sandstone aquifer (Entec UK Ltd 1999). The most recent UK Climate Impacts Programme (UKCIP) climate change scenario data (Hulme & Jenkins 1998) were incorporated into the model and comparisons were made against a baseline model run. The results of the modelling showed that annual average recharge (and groundwater levels and baseflows conditions) are predicted to increase over the next 50 years for all the climate change scenarios (see Fig. 3). Although the warmer drier summers increase soil moisture deficits during the summer and autumn, this is more than compensated for by the increase in recharge over the winter period (when soil moisture deficits are zero). The results of this modelling work suggest climate change is an issue that needs to be considered and that any future strategy for the Nottinghamshire Triassic sandstone aquifer should take account of the latest climate change predictions.

The Nottinghamshire Triassic sandstone aquifer is an example of how techniques such as groundwater modelling can help to examine the impacts of abstraction on the environment. Unfortunately, there are many groundwater units in the Severn Trent area where groundwater models are not yet available.

STW believes that the Environment Agency needs to look more closely at the costs and benefits of reducing abstraction, particularly the quantity and timing of benefits to surface water

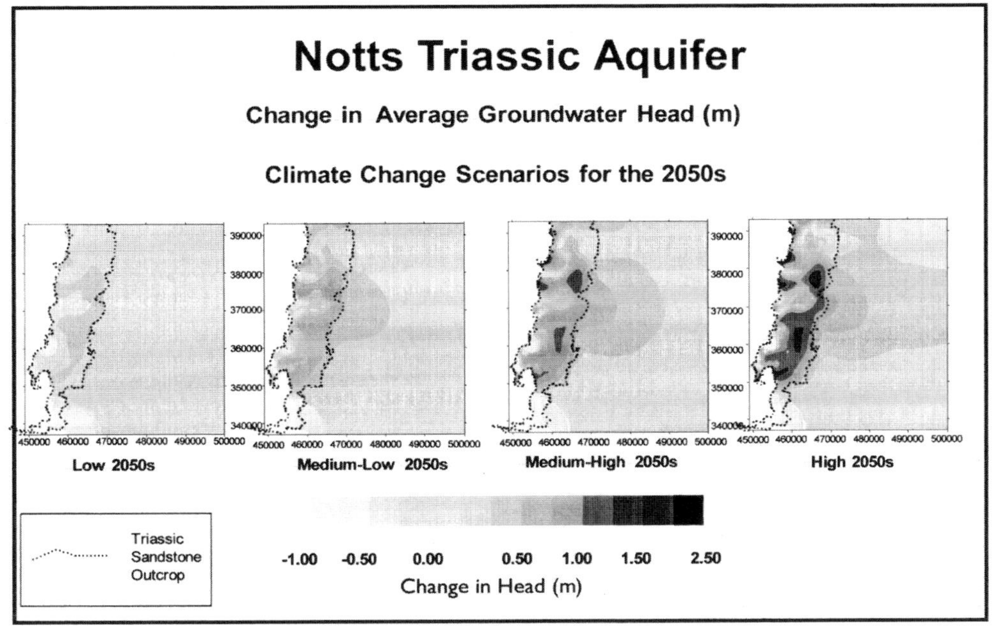

Fig. 3. Nottinghamshire Triassic sandstone aquifer climate change modelling results.

features, and at the same time, be more rigorous in the calculation of resource availability and the environmental fraction.

Severn Trent Water's position

Severn Trent Water has already made significant voluntary licence reductions from the Nottinghamshire Triassic sandstone aquifer, and is committed to further licence reductions of 25 $Ml\,d^{-1}$ over the next five years. These licence reductions were agreed prior to the recent resource reassessment and will increase the current dry-year supply-demand imbalance in the East Midlands to *c.* 55 $Ml\,d^{-1}$ by the end of 2005.

Water company investment plans (Strategic Business Plans) are submitted to the water industry regulator (the Office of Water Services – OFWAT) every five years. OFWAT reviews these plans and sets the tariff levels that water companies can charge their customers (the Price Determination). This is the mechanism by which all investment in the UK water industry is funded. For the recent company Strategic Business Plan, STW decided to plan on the basis that the abstraction licence reductions identified by the Environment Agency would be implemented, as long as the funding of replacement water resources was made available through the Price Determination.

In order to make up for the loss of groundwater licences a number of new water resource developments were proposed. In the short to medium term the schemes proposed for the East Midlands include: the development of the Birmingham Groundwater scheme to support the River Trent abstraction licence; improvements to the treatment capacity at Church Wilne treatment works; improvements to the strategic water network; and the further conjunctive use of groundwater and surface water resources in the East Midlands. The cost of these developments was calculated at approximately £40 million in 1999.

Funding of replacement water resources

STW's Strategic Business Plan identified a requirement for water resources funding of £80 million (for all three resource zones). However, only two items have been funded through the recent Price Determination by OFWAT: £10

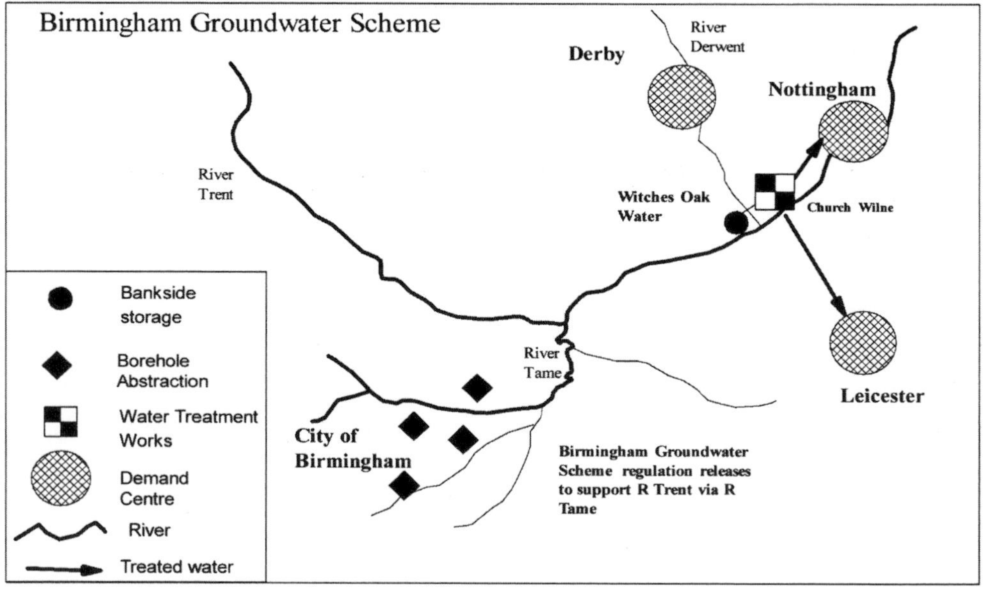

Fig. 4. The Birmingham Groundwater Scheme.

million to resolve the current supply-demand imbalance in the East Midlands (*c.* 30 Ml d^{-1}), and a small amount (£2 million) to carry out the engineering work required for 11 low flow alleviation schemes across Severn Trent Water. Replacement water resources have not been funded for licence reductions or for the low flow schemes. OFWAT has stated that the funding of replacement water should be a matter for the Environment Agency under the current abstraction licence compensation mechanism (Water Resources Act 1991). However, this compensation mechanism appears inappropriate as it is really intended only for small-scale licence reductions. Due to the scale of the proposed licence reductions the Environment Agency has refused to accept the compensation mechanism as an appropriate method of funding replacement water. One of the few ways the Environment Agency could raise the amount of money required for large-scale compensation payments would be to increase abstraction licence charges, but this is likely to be politically unacceptable as many smaller abstractors would also be affected.

The current situation is therefore that the majority of Severn Trent Water's water resources plans remain unfunded with the exception of a limited amount of funding for the East Midlands.

Proposed water resource developments for the East Midlands

The key water resources scheme to be developed over the next 2–3 years to reduce the dry-year supply shortage in the East Midlands is the Birmingham Groundwater Scheme (see Fig. 4). During drought periods, groundwater will be abstracted from the Sherwood Sandstone of the Birmingham groundwater unit and discharged to surface water courses which feed the River Trent. Current plans are to abstract a total of 50 Ml d^{-1} for 150–200 d per year, during dry years. The first phase of the project, which has been funded, will deliver 30 Ml d^{-1}. This water is required to support the existing surface water abstraction licence on the River Trent at Shardlow, near Derby (see Fig. 4). The current licence at Shardlow is linked to a prescribed flow condition on the river and cannot be used when flows at North Muskam are less than 2650 Ml d^{-1} (see Fig. 4).

Initial scoping calculations and, more recently, pilot drilling suggest that 50 Ml d^{-1} could be obtained from eight new sites producing 5 Ml d^{-1}, with a further 10 Ml d^{-1} from an existing groundwater source in Birmingham (Short Heath well). A test pumping and monitoring programme has been completed at Short Heath pumping station and the first pilot borehole has been drilled and tested in the Newtown area of Birmingham (Fig. 4).

Organic contamination of the Triassic sandstone aquifer under Birmingham is widespread (Rivett *et al.* 1990); however it is anticipated that groundwater away from industrial areas and at depth within the aquifer will be of a sufficiently high quality to discharge to the local watercourses. The test pumping carried out to date has confirmed this, but it is recognized that further abstraction and long-term monitoring will be required.

The scheme is likely to produce a number of potential benefits to Severn Trent Water customers, to the environment and to Birmingham City Council (which is responsible for many of the surface water courses in the city). The scheme is a low-cost solution whereby relatively clean water will be added to urban watercourses and surface sewers during drought periods, improving the quality of the surface water system. There is also the potential (albeit limited) to alleviate shallow water table problems in selected areas of the city by reducing groundwater levels which have risen following the decline of industrial abstraction.

Support for the scheme has been received from the Environment Agency, and the former Department of Transport, Energy and the Regions has stated that the scheme is the preferred solution to help meet the deficit in dry-year supplies in the East Midlands.

Some alternative solutions

If further groundwater abstraction licence reductions are required from the Nottinghamshire Sherwood Sandstone aquifer then STW and the Environment Agency must be innovative in finding the most environmentally friendly and cost-effective solutions to meet the resource deficit, whilst ensuring that any reductions are backed up by the most rigorous supporting arguments.

There are a number of alternative resource options which could be considered, including:

(1) Further conjunctive use of surface water and groundwater sources. Abstraction from the Nottinghamshire Triassic sandstone aquifer could be reduced during years with above-average rainfall and surface water taken from the River Derwent system could be used to replace this groundwater (Fig. 2). During dry years, surface water abstraction would be cut back to reserve resources as long as possible and groundwater abstractions could be used at a higher rate to make up the shortfall. This concept appears straight-forward; however, there are significant issues to be resolved. Surface water is more expensive to abstract, treat and distribute and there may be distribution constraints and unacceptable changes in water quality to customers.

(2) Redistributing abstraction. The pattern of abstraction could be shifted further from the unconfined part of the aquifer to the confined part. This could reduce the impact on surface water features.

(3) Artificial recharge. Recharge of the aquifer with surplus surface water during wet years would increase the available groundwater resources. This could remove the need for further reductions in groundwater abstractions or even increase groundwater abstraction from the aquifer. However, the nearest supplies of treated surface water are not close to the aquifer (Fig. 2).

(4) Low flow alleviation schemes. One alternative could be to provide compensation discharges to surface water courses as a permanent solution in some areas. If the proposed reduction in abstraction is large and the timing and nature of the benefit to surface water is unclear or not cost-effective then this solution could be the best option. This could remove the need for some of the identified groundwater licence reductions.

(5) Targeted reductions. Any reductions should be targeted at specifically sensitive areas, i.e. where local effects within a specific cone of depression have been proved. This should provide the maximum benefit for the environment.

Severn Trent Water identified a number of feasibility schemes in the recent Water Resources and Strategic Business Plan. These schemes were intended to examine the feasibility of the alternative water supply options such as artificial recharge and further conjunctive use. No funding was forthcoming in the Price Determination.

Conclusions

Within the East Midlands Resource Zone, dry-year demand exceeds the available water supplies (Water Available For Use). Future reductions in groundwater abstraction will increase the dry-year supply deficit.

Abstraction from the Nottinghamshire Triassic sandstone aquifer contributes a significant proportion of the water resources supply to the East Midlands. Recent recalculation of the available resources and the identification of an environmental fraction by the Environment Agency have increased pressure for further reductions in groundwater abstraction from this aquifer.

The Birmingham Groundwater Scheme should provide a cost-effective and environmentally beneficial solution to the current dry-year supply deficit.

There are, however, a number of key technical and regulatory conflicts which still need to be resolved in order to reconcile the future environmental and economic pressures on the Nottinghamshire Triassic sandstone aquifer, and on other aquifers in the Severn Trent area.

Funding of new water resources schemes through OFWAT, and in particular replacement water supplies, has so far been very limited. The issues of funding replacement water supplies have yet to be resolved.

STW would like to see a more rigorous cost-benefit assessment of increasing river flows at the expense of groundwater abstraction, and of developing replacement water supplies.

A more technically rigorous assessment of groundwater resources for all aquifers is required, particularly the quantification of the environmental fraction.

Alternative strategies for abstraction management such as conjunctive use and abstraction redistribution should be considered alongside proposals to close groundwater sources.

Compensation discharges to low flow rivers and wetlands could provide more successful and better value solutions when compared to large-scale reductions in groundwater abstraction.

Uncertainties associated with climate change need to be included in the 25-year planning horizon.

The work using the Environment Agency's Nottinghamshire-Doncaster Sherwood Sandstone groundwater model was performed with permission from the Environment Agency, Midlands Region. However, the Environment Agency does not necessarily agree either with the changes made to the model or any conclusions drawn from the results.

References

Bishop, I. & Rushton, K. 1993. *Water Resource Study of the Nottinghamshire Sherwood Sandstone Aquifer System of Eastern England, Mathematical Model of the Sherwood Sandstone Aquifer*. Department of Civil Engineering, The University of Birmingham.

Crookall, D. & Bradford, W. 2000. Impact of climate change on water resources planning. *Proceedings of the Institution of Civil Engineers*, **138**, 44–48.

Entec UK Ltd, 1999. *Notts Trias model: climate change scenarios*. Report to Severn Trent Water Ltd.

Environment Agency. 1998. *Water resources planning guideline*. Environment Agency, Bristol.

Hulme, M. & Jenkins, G. J. 1998. *Climate change scenarios for the United Kingdom: scientific report*. Climatic Research Unit, Norwich, UK Climate Impacts Programme Technical Report No. **1**.

Rivett, M. O., Lerner, D. N., Lloyd, J. W. & Clark, L. 1990. Organic contamination of the Birmingham aquifer. *Journal of Hydrology*, **113**, 307–323.

The High Plains Aquifer, USA: groundwater development and sustainability

K. F. DENNEHY, D. W. LITKE & P. B. McMAHON

U.S. Geological Survey, Denver, Colorado, 80225, USA

Abstract: The High Plains Aquifer, located in the United States, is one of the largest freshwater aquifers in the world and is threatened by continued decline in water levels and deteriorating water quality. Understanding the physical and cultural features of this area is essential to assessing the factors that affect this groundwater resource. About 27% of the irrigated land in the United States overlies this aquifer, which yields about 30% of the nation's groundwater used for irrigation of crops including wheat, corn, sorghum, cotton and alfalfa. In addition, the aquifer provides drinking water to 82% of the 2.3 million people who live within the aquifer boundary. The High Plains Aquifer has been significantly impacted by human activities. Groundwater withdrawals from the aquifer exceed recharge in many areas, resulting in substantial declines in groundwater level. Residents once believed that the aquifer was an unlimited resource of high-quality water, but they now face the prospect that much of the water may be gone in the near future. Also, agricultural chemicals are affecting the groundwater quality. Increasing concentrations of nitrate and salinity can first impair the use of the water for public supply and then affect its suitability for irrigation. A variety of technical and institutional measures are currently being planned and implemented across the aquifer area in an attempt to sustain this groundwater resource for future generations. However, because groundwater withdrawals remain high and water quality impairments are becoming more commonplace, the sustainability of the High Plains Aquifer is uncertain.

Use of the High Plains Aquifer as a source of irrigation water has transformed the mid-section of the United States into one of the major agricultural regions of the world. During 1995, water use in the High Plains Aquifer was estimated to be 7.5×10^7 m^3 d^{-1} (U.S. Geological Survey National Water Information System data base). About 96% of the water pumped from the aquifer is used to irrigate crops on about 27% of the irrigated land in the United States and withdrawals amount to about 30% of the nation's groundwater used for irrigation (Dennehy 2000).

The High Plains Aquifer underlies 450 660 km^2 in parts of eight States (Fig. 1). For discussion purposes in this paper, the High Plains Aquifer has been geographically subdivided into northern, central and southern regions. Principal crops are cotton, alfalfa and grains, especially wheat, sorghum and corn. Grains provide feed for the 15 million cattle and the 4.25 million swine (1997) that are raised over the aquifer. In addition, the aquifer provides drinking water to 82% of the people who live within the aquifer boundaries.

Substantial pumping of the High Plains Aquifer for irrigation since about the 1940s to 1980 had resulted in water-level declines in some parts of the aquifer of more than 30 m (Luckey *et al.* 1981). In 1984, concern about these declines led the U.S. Congress to mandate a water-level monitoring programme for the aquifer. The water issue of most concern to people living in the area is the declining water levels, meaning that their way of life cannot be sustained into the future. Although the rate at which water levels are declining has slowed, the downward trend continues in many areas across the High Plains Aquifer (McGuire & Fischer 1999).

Groundwater quality is a more recent concern. State and local governmental entities

From: Hiscock, K. M., Rivett, M. O. & Davison, R. M. (eds) *Sustainable Groundwater Development*. Geological Society, London, Special Publications, **193**, 99–119. 0305-8719/02/$15.00

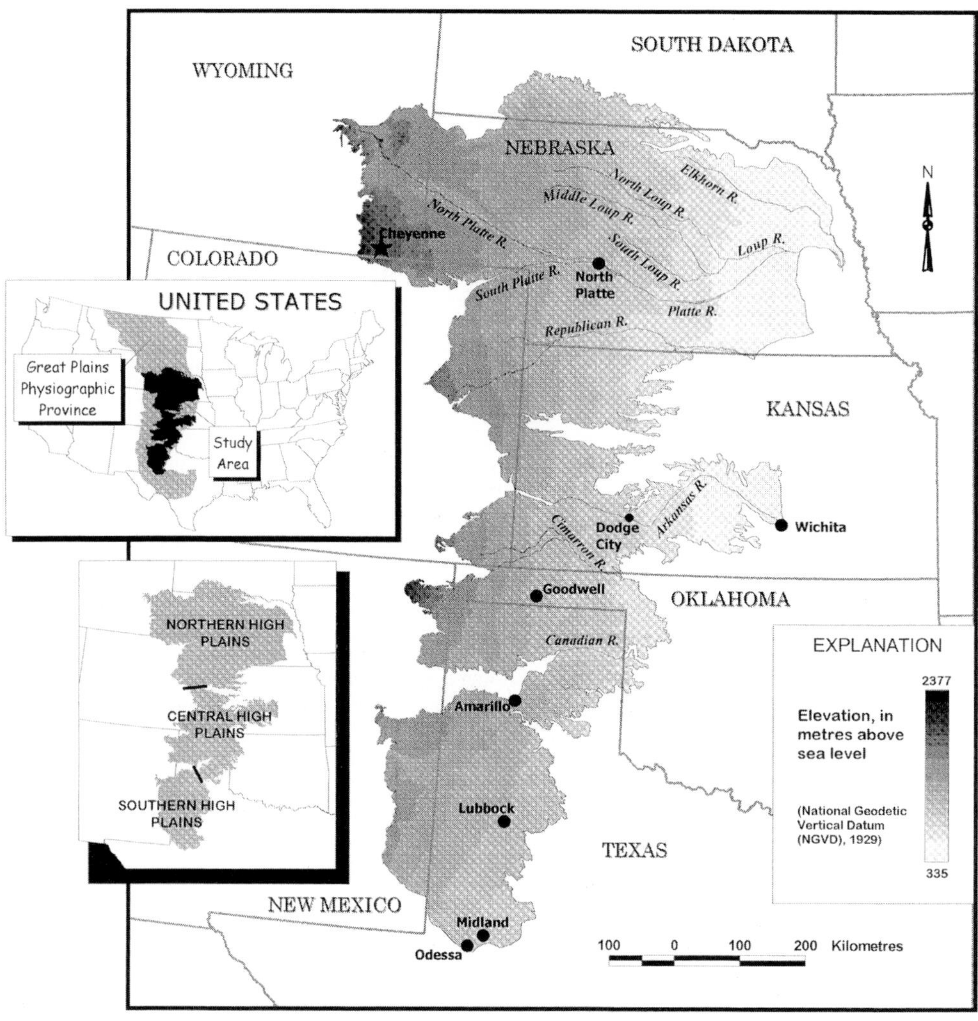

Fig. 1. Location map of the High Plains Aquifer.

(groundwater management districts, municipalities and universities) have carried out local studies of groundwater quality, but no comprehensive aquifer-wide assessments have been made. The U.S. Geological Survey's (USGS) National Water-Quality Assessment (NAWQA) Program is an ongoing effort to document water-quality conditions across the United States, and the High Plains Regional Ground-Water Study Area (hereafter referred to as the High Plains study area) was created as part of NAWQA in 1998. In that same year, the USGS solicited the help of water professionals in the eight-State study area and produced the following list of their high priority, regional-scale water-quality issues of concern:

(a) Nutrient contamination of groundwater from the operations of confined animal feeding operations;
(b) Effects of saline groundwater from bedrock aquifers discharging into the High Plains Aquifer and the potential impairment of drinking and irrigation supplies;
(c) Effects of agricultural and urban land-use practices on groundwater quality and the potential for degradation of drinking water;
(d) Deterioration of groundwater quality within the shallow alluvial aquifers as a result of induced infiltration of degraded surface

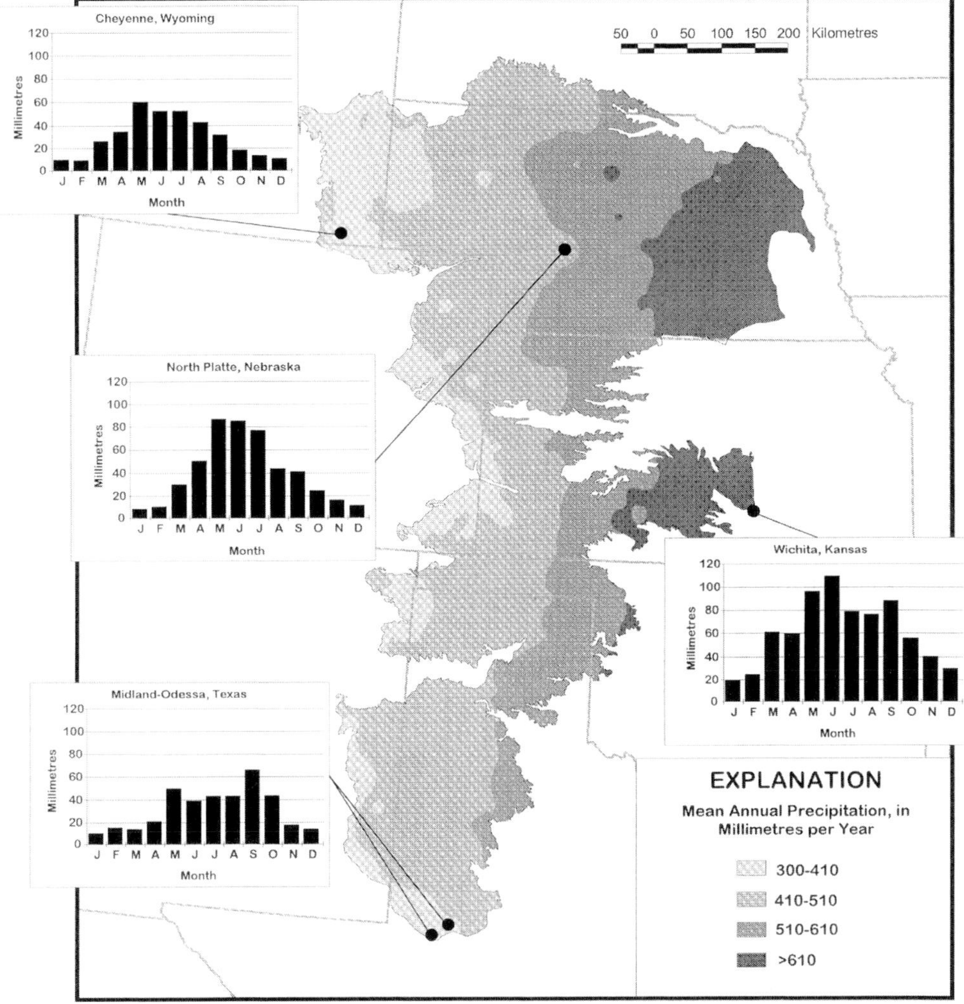

Fig. 2. Average annual precipitation in the High Plains study area and monthly precipitation at selected sites (source of data: Daly & Taylor 1998; National Climate Data Center 2000).

water; and

(e) Effects of focused recharge through playas on local groundwater quality.

This paper describes the physical and cultural characteristics of the High Plains Aquifer and explains how natural and human activities have affected the quantity and quality of the groundwater resource and thus its sustainability.

Physical and cultural characteristics

The High Plains study area occupies the higher elevation parts of the much larger Great Plains Physiographic Province, which lies between the Rocky Mountains in the west and the Central Lowlands in the east. Elevations range from about 2377 m on the western boundary to 355 m on the eastern boundary. This area is a remnant of a vast plain formed by sediments that were deposited by streams flowing eastward from the ancestral Rocky Mountains. Gently sloping, smooth plains characterize the High Plains study area, which makes it ideal for agricultural use.

Climate

The High Plains study area has a middle-latitude dry continental climate with abundant sunshine, moderate precipitation, frequent winds, low humidity and a high rate of evaporation. This combination of characteristics leads to extreme weather patterns, including droughts, floods, blizzards, tornadoes and hail. Mean annual temperature ranges from about 6°C in the north to 17°C in the south, and mean annual precipitation ranges from 406 mm in the western part of the study area to about 711 mm in the east (Fig. 2). Rainfall patterns are perhaps most notable for their temporal and geographic variability. Winter, spring and summer precipitation totals often vary by more than 100% from one year to the next (Doesken 1998). About 75% of the annual rainfall occurs between April and September (Fig. 2), usually in the form of localized thunderstorms; even weather stations that are relatively close often show noticeable differences (Doesken 1998). Droughts have occurred on the High Plains during 1887–1897, 1910–1913, 1932–1938 and 1952–1957, causing severe disruptions to the area's agriculture because water sources for irrigation had not been developed. Since the 1950s, however, there has not been a period of prolonged, widespread drought.

Evaporation rates measured from free water surfaces in the High Plains range from 1520 mm in the north to 2670 mm in the south (Gutentag *et al.* 1984). These rates are among the highest in the United States because of high summer temperatures and persistent winds. The mean annual number of days during which the daily temperature reaches 32°C or above ranges from about 40 d in the north to about 100 d in the south (High Plains Associates 1982). The average, annual sustained-peak daily wind speed for most of the study area exceeds 19 km h^{-1}, and reaches 26–29 km h^{-1} in the Texas–Oklahoma Panhandles (Thornthwaite 1936). Because evaporation rates are so high relative to precipitation, there is little water available to recharge the aquifer, and in some areas recharge events may be years or possibly decades apart (Luckey & Becker 1999). Potential annual recharge under non-irrigated conditions (1951–1980) in the High Plains study area has been estimated to range from less than 6 mm along the western boundary of the aquifer to as much as 127 mm in the north-eastern part of the High Plains Aquifer (Dugan & Zelt 2000).

Soils

Soils of the High Plains study area are fine-textured and sandy, silty, clayey or loamy. Soils are generally developed on windblown sand or loess, and the loess soils would be classified as prime soils but for the lack of available precipitation. Calcium carbonate is a common soil constituent, especially in the southern High Plains where a thick caliche horizon has developed (High Plains Associates 1982). Over much of the High Plains, slopes are flat to gentle and soil texture is sufficiently coarse to allow moderate to high infiltration rates through soils. As a result, run-off is small, ranging from less than 13 mm along the western boundary of the High Plains study area to 102 mm in the north-eastern part of the study area (Dugan & Zelt 2000).

Soil is a fragile resource in the High Plains. Marginal farming lands brought under cultivation during wet years can become susceptible to wind erosion during dry years. During the drought of 1910–1913, dust storms occurred over about 26 305 ha of land in north-west Kansas (Helms 1981). The drought of 1932–1938 produced an area called the Dust Bowl in a ten-county area centred on Texas County, Oklahoma. At Goodwell, Oklahoma, 22 d of severe dirt blowing were recorded in 1934, followed by 53 d in 1935, 73 d in 1936 and 134 d in 1937. A dust storm on 14 April 1935 darkened skies from Colorado to the East Coast of the United States and layered dust on ships 480 km out in the Atlantic Ocean (Opie 1993). The drought of 1952–1957 brought even more severe conditions over a much larger area that encompassed the western High Plains from southern New Mexico to northern Colorado. During a storm in February 1952, 129 km h^{-1} winds forced a dust cloud to an altitude of 3658 m; Hamilton County, Kansas, lost 95% of its wheat crop; and sand dunes 11 m high were formed (Opie 1993). As a result of the 1950s drought, the U.S. Congress established the Great Plains Conservation Program in 1956, with the goals of removing marginal lands from production and of establishing conservation measures on remaining farmlands.

Population and land use/land cover

The High Plains were primarily the domain of native Americans and bison until the 1860s when the area was reached by the westward push of the transcontinental railways. For a brief 15-year period, cowboys rode through the unfenced prairie as they drove cattle from their winter ranges to railheads such as Dodge City, Kansas. During 1870–1890, farmers lured by the free land grants of the Homestead Act of 1862 began settling and fencing areas of the High Plains in Colorado, Kansas and Nebraska, along the

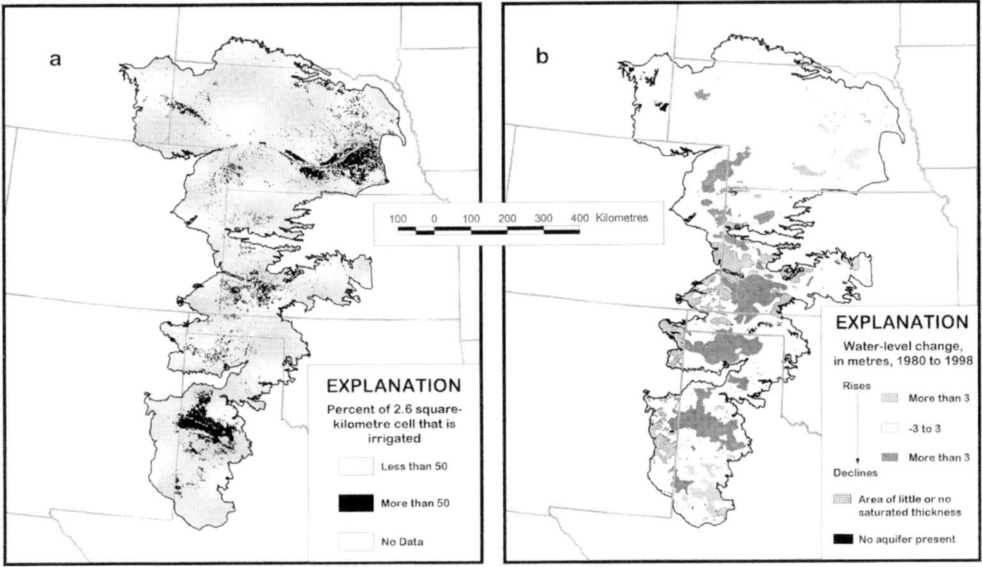

Fig. 3. Location of (**a**) irrigated lands in the High Plains study area, 1980 (source of data: Thelin & Heimes 1987) and (**b**) water-level changes in the High Plains Aquifer, 1980 to 1998 (source of data: McGuire & Fischer 2000).

newly established routes of the Santa Fe, Kansas Pacific and Union Pacific Railroads. Land-development corporations from as far away as England promoted the area as a land of golden opportunity for dryland farmers, and a climate anomaly of a series of wet years during 1878 to 1887 led to the saying that 'rain follows the plough', and to the first large population surge. The population across the High Plains rose and fell dramatically from 1887 to 1950 in response to the dry and wet weather cycles that followed; bankrupt farmers left the region during dry years and were replaced by new immigrants during wet years. The postwar economic boom of the 1950s, the development of irrigation technologies, and federally funded irrigation projects in the 1960s allowed farmers to work their land independent of dry and wet weather cycles.

About 2.3 million people (1990 census) now live within the boundaries of the High Plains study area. The population is evenly distributed between the northern and the southern High Plains, with the smallest population (25%) being located in the central High Plains. However, 38% of the people live in the ten largest cities, primarily on the fringes of the study area. Within the interior of the study area the population is largely rural, with more than half of the counties having populations less than 2500 people; the percentage of rural counties in the area is six-times the national average (Donofrio & Ojima 1998). However, even within the High Plains, recent population trends indicate that people are migrating from farms to regional trade centres, partially in response to an increase in the number of corporate farms relative to the number of family farms.

Within the High Plains study area, rangeland accounts for 55.6% of the land use, agriculture for 41%, and the remaining 3.4% of the land use is a combination of wetlands, forest, urban, water and barren lands (U.S. Geological Survey 1999–2000). The agricultural land use is comprised of 53% rowcrops (such as sorghum, corn and cotton), 33% small grains (mostly wheat), and 14% pasture, alfalfa and fallow land. The USGS mapped irrigated cropland in the High Plains study area (Thelin & Heimes 1987) using satellite imagery with a nominal date of 1980. The percentage of irrigated land in 2.6 km^2 cells across the study area was estimated; the total irrigated area amounted to 5.5×10^6 ha (Fig. 3a). At that time, agricultural land use was 57% dryland farming, 26% irrigated farming, and 17% pastureland. Within the dryland area, about half (6×10^6 ha) comprised harvested crop area (mostly winter wheat) and the other half was fallow.

Of the total crop production in the United States, the High Plains Aquifer area accounts for about 19% of the wheat, 19% of the cotton,

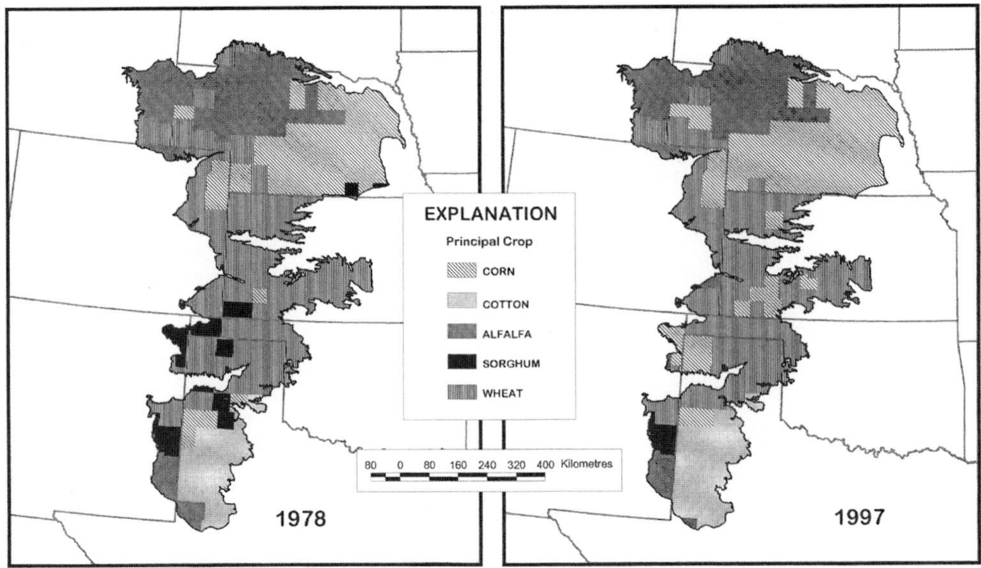

Fig. 4. Principal crop by county in the High Plains study area, 1978 and 1997 (source of data: U.S. Department of Commerce 1990; U.S. Department of Agriculture 1999).

15% of the corn and 3% of the sorghum (U.S. Department of Agriculture 1999). The geographic distribution of crops in 1978 and 1997 is shown in Figure 4. Counties with wheat as the principal crop occur primarily in the central and north-western part of the northern High Plains. Corn was the second most common crop in both 1978 and 1997, and has expanded to cover an area from about 3.2×10^6 ha in 1978, to about 4.5×10^6 ha in 1997. Corn is the predominant crop in the northern High Plains, with increases in corn production occurring in the central as well as in the northern High Plains. Alfalfa, sorghum and cotton are the next most abundant irrigated crops; alfalfa predominates in the northern High Plains, while sorghum and cotton predominate in the southern High Plains.

In addition to crops, the region accounts for nearly 18% of the total cattle production in the United States and is rapidly becoming a centre for swine production. The number of cattle on the High Plains remained fairly steady from 1975 to 1997 with modest increases from 1991 to 1997; however, the number of swine has increased steadily from 2.5 million in 1975 to 4.2 million animals in 1997. Approximately a five-fold increase in the number of swine has occurred in Oklahoma from 1993 to1997; there were roughly corresponding decreases in Nebraska (U.S. Department of Agriculture 1999, 2000).

Hydrologic characteristics

More than 80% of the water used in the area is derived from the High Plains Aquifer. With a total of 3.9×10^{12} m^3 of drainable water in storage, it is probably one of the largest aquifers in the world (Gutentag *et al.* 1984). The quantity of surface water entering the High Plains study area was determined from long-term mean annual data for its five major rivers (North Platte, South Platte, Arkansas, Cimarron and Canadian Rivers) to be about 2.5×10^9 m^3 per year. Although a smaller resource, surface water provides wildlife habitats; the valleys of the major rivers were the first areas to be developed for agriculture and are important for their population centres and transportation corridors.

Surface water

Surface water hydrology is important to High Plains Aquifer studies because the High Plains Aquifer is in hydraulic connection with the major river systems (Weeks *et al.* 1988). The principal river systems in the High Plains study area originate in the Rocky Mountains as snowmelt run-off streams and then traverse the High Plains study area from west to east (Fig. 1).

Streamflow in these rivers depends largely on the quantity of snowmelt run-off originating in the Rocky Mountains. Among High Plains

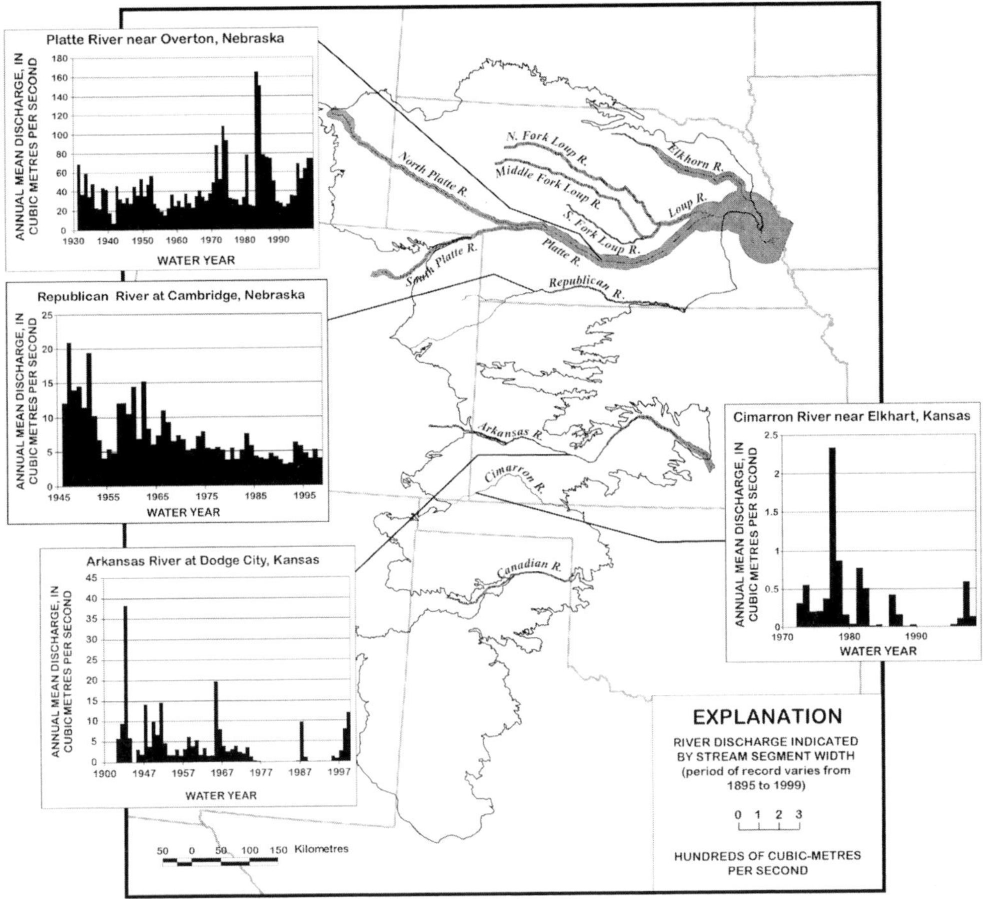

Fig. 5. Streamflow statistics at selected sites on major rivers in the High Plains study area (source of data: U.S. Geological Survey National Water Information System).

rivers, streamflow is largest in the Platte River system (Fig. 5). Long-term flow in the Platte River near Overton, Nebraska, was relatively constant from 1930 to 1970, but increased greatly during years of large snowmelt run-off in the 1970s, 1980s, and 1990s. Streamflow is relatively constant as the Platte River traverses central Nebraska, with significant streamflow gains seen in the eastern reaches where tributary inflows, agricultural return flows (surface and groundwater), and large rainfall run-off rates add water to the river (Fig. 5). Streamflow patterns are made more complex by widespread diversion of water from the river into canals for irrigation, and subsequent returns to the river from agricultural surface and groundwater inflows.

Streamflow in the Arkansas, Cimarron and Canadian Rivers is much smaller than in the Platte River because less mountain snowmelt run-off water reaches these rivers. Streamflow quantities generally decrease from north to south within the study area, and no major streams occur in the dry southern High Plains. Annual flow variability is large in these rivers as illustrated by annual mean flows of 0 to 40 $m^3 s^{-1}$ on the Arkansas River at Dodge City, Kansas, and annual mean flows of 0 to 2.3 $m^3 s^{-1}$ at the Cimarron River near Elkhart, Kansas (Fig. 5). The smaller streamflow in these rivers is also a direct response to increasing quantities of surface and groundwater withdrawals for irrigation.

Several tributary streams and one river system derive all their water from within the High Plains study area. Several tributaries to the

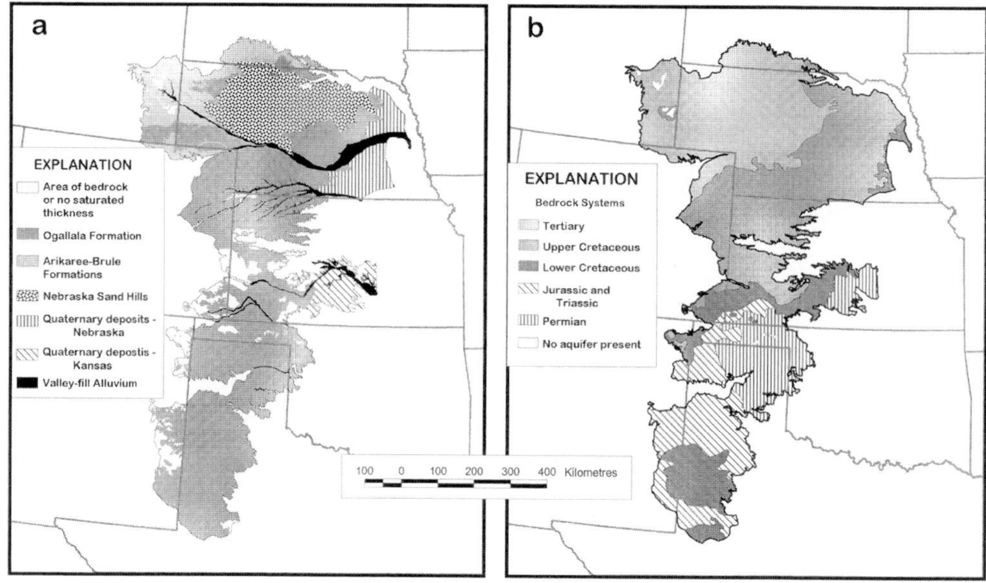

Fig. 6. Location of (**a**) hydrogeological units comprising the High Plains Aquifer and (**b**) bedrock units underlying the High Plains Aquifer (source of data: Weeks & Gutentag 1981).

Platte River in eastern Nebraska (Elkhorn and Loup Rivers) derive their flow primarily from groundwater discharge in the Sand Hills area where the land surface often intersects with the shallow water table. The Republican River originates within the study area (Fig. 5) and derives its flow in approximately equal proportions from rainfall run-off and aquifer discharge (M. Landon, USGS, pers. comm. 2000). Long-term flow patterns at the Republican River at Cambridge, Nebraska (Fig. 5), indicate that streamflows are decreasing. This decrease is due to increasing diversions and pumpage of water for irrigation, decreasing run-off due to land-use changes, and a general declining regional water table in the aquifer (U.S. Bureau of Reclamation 1985, 1996). The effect of a regional declining water table on streamflow is even more pronounced to the south in the North Canadian River part of the High Plains study area in the Oklahoma Panhandle. In this area, average streamflows have decreased by as much as 91% of historical averages (Wahl & Tortorelli 1997).

Groundwater

The High Plains Aquifer, as defined by the USGS's Regional Aquifer System Analysis (RASA) Program (Weeks *et al.* 1988), consists mainly of near-surface deposits of late Tertiary or Quaternary age and can be stratified into six hydraulically connected hydrogeologic units (Fig. 6a, Table 1), forming one water-table (unconfined) aquifer. The Ogallala Formation of Miocene age, which underlies 347 060 km^2, is the principal geological unit of the aquifer and consists of a heterogeneous sequence of clay, silt, sand and gravel. Within the Ogallala Formation, zones cemented with calcium carbonate are resistant to weathering and form the escarpment that typically marks the south-western boundary of the High Plains. Other hydrogeological units making up the High Plains Aquifer include the Arikaree and Brule Formations of Tertiary age in the northern part of the study area consisting of sandstones and siltstones; windblown sands of Quaternary age on top of or adjacent to the Ogallala Formation primarily in the Sand Hills of Nebraska; alluvial deposits in central Kansas and along present-day river valleys; and glacial deposits in eastern Nebraska. For a more detailed discussion of the hydrogeology of the High Plains Aquifer see Gutentag *et al.* (1984).

The most-used hydrogeological units in the aquifer, as measured by water use and percentage of irrigated area, are the Nebraska Quaternary deposits, the valley-fill alluvium, and the Ogallala Formation, where the intensity of use increases from north to south. Median depth to

Table 1. Selected characteristics of hydrogeological units within the High Plains study area

Hydrogeological Unit	Area (km^2)	Estimated water use in 1995 ($m^3 s^{-1}$)	Median saturated thickness (m)	Median depth to water (m)	Percentage of area irrigated
Ogallala Formation–Northern High Plains	97 645	118	34	32	10
Ogallala Formation–Central High Plains	90 910	180	36	48	15
Ogallala Formation–Southern High Plains	69 930	184	18	36	19
Arikaree-Brule Formations	50 250	21	66	19	5
Nebraska Sand Hills	58 275	31	164	6	5
Quaternary deposits–Nebraska	27 970	92	51	24	34
Quaternary deposits–Kansas	14 500	16	27	8	9
Valley-fill Alluvium	20 720	*44	*35	6	26

* Water use and saturated thickness in the Valley-fill Alluvium may include contributions from underlying High Plains aquifer units.

Table 2. *Ground-water budget information for the High Plains Aquifer based on U.S. Geological Survey's Regional Aquifer System Analysis ground-water model* [Luckey et al. 1986]*

Budget parameter	Northern High Plains	Central High Plains	Southern High Plains
Primary inflows (in cubic metres per year)			
Recharge from precipitation on rangeland and streams	[†]5.98×10^{8}	[†]4.66×10^{8}	1.97×10^{8}
Recharge from precipitation on agricultural land	2.89×10^{9}	—	1.43×10^{9}
Groundwater irrigation return (pumpage minus crop demand)	2.31×10^{9}	2.07×10^{9}	3.61×10^{9}
Recharge from other human activities (e.g. seepage from reservoirs and canals)	2.31×10^{9}	—	—
Recharge from other aquifers across subunit boundary	—	[§]1.88×10^{7}	—
Totals	$\mathbf{8.11 \times 10^{9}}$	$\mathbf{2.55 \times 10^{9}}$	$\mathbf{5.24 \times 10^{9}}$
Primary outflows (in cubic metres per year)			
Total pumpage	6.48×10^{9}	[‖]6.89×10^{9}	8.59×10^{9}
Discharge to streams and shallow water-table areas	2.87×10^{9}	4.15×10^{8}	—
Discharge along eastern boundary	—	6.97×10^{7}	1.05×10^{8}
Totals	$\mathbf{9.35 \times 10^{9}}$	$\mathbf{7.37 \times 10^{9}}$	$\mathbf{8.70 \times 10^{9}}$
Net residual	$\mathbf{-1.24 \times 10^{9}}$	$\mathbf{-4.82 \times 10^{9}}$	$\mathbf{-3.46 \times 10^{9}}$

* Assumptions: Inflow/Outflow values determined using 1960-1980 estimates; base of aquifer modelled as no-flow boundary; vertical flow in aquifer considered negligible on regional scale.

[†] Recharge distributed unevenly based on soil type.

[‡] Additional recharge from precipitation on agricultural land because of changes in soil character due to tillage.

[§] Flow only from northern and southern subunits to central subunit.

[‖] Municipal and industrial pumpage is 3.2% of this amount.

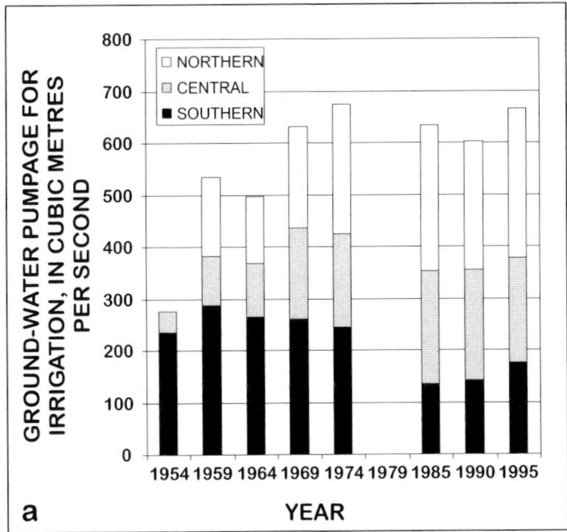

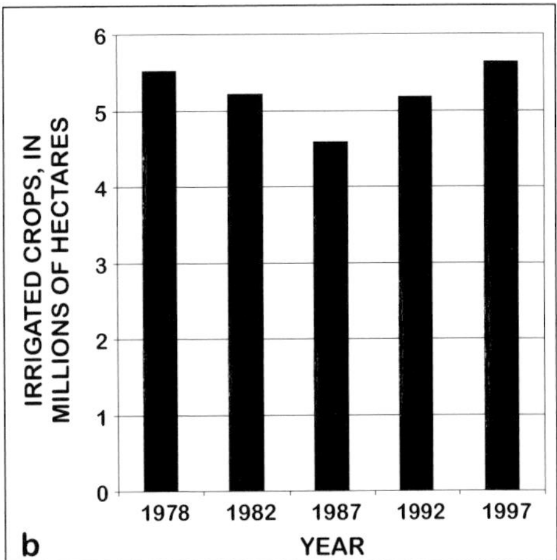

Fig. 7. Histogram of (**a**) historical groundwater pumpage (source of data Weeks *et al.* 1988 and U.S. Geological Survey National Water Information System data base) and (**b**) area of crops irrigated from the High Plains Aquifer (source of data: U.S. Department of Agriculture 1999).

water is smallest and saturated thickness is largest in the Nebraska Sand Hills, but water use in this unit is small because of rough topography and sandy soils. In the alluvial deposits water use is large and median depth to water is small, while water use is large and median depth to water is largest for the Ogallala Formation.

Saturated thickness of the High Plains Aquifer ranges from about 0 to more than 305 m and averages about 61 m. However, the saturated thickness of the aquifer in Nebraska averages about 104 m. Depth to water ranges from 0 to 152 m and has an average depth of about 30 m. Groundwater generally flows from west to east, discharges naturally to streams and springs and is subject to evapotranspiration in areas where the water table is near the land surface. Pumping from numerous irrigation wells across the study area is the primary mechanism of groundwater discharge. Precipitation is the principal source of recharge to the aquifer.

Regional variability of water-level changes in the High Plains Aquifer results from large regional differences in climate, soils, land use and groundwater withdrawals for irrigation (Fig. 3b). Withdrawals greatly exceed recharge in many areas, causing large declines in water levels. For example, water levels have declined from 30 to 43 m since the 1940s when irrigation began in parts of Kansas, New Mexico, Oklahoma and Texas. Due to water-level declines in some areas irrigation has become impossible or cost prohibitive. In general, declines have been largest in the south. From 1980 to 1997, the average area-weighted water level in the High Plains Aquifer declined 0.8 m (McGuire & Fischer 1999).

A water budget prepared by the High Plains RASA project provides estimates of the primary groundwater inflow and outflow values using 1960 to 1980 information over the study area (Table 2). It is apparent that each region of the High Plains Aquifer is in a deficit situation. Outflows are greater than inflows in all regions. In the southern and central regions, outflows are about twice the inflows.

Prior to irrigation development, water fluxes were in dynamic equilibrium. Precipitation recharged the aquifer at an average rate of about 15 mm a^{-1} (Weeks *et al.* 1988) and small quantities of water left the aquifer in the form of discharge to springs and streams. Subsequent to development, groundwater pumpage has removed an amount of water equivalent to about 0.6 m of water from the aquifer, or about 7% of the original total water volume. The water budget (Table 2) further indicates the decrease in storage would have been greater if 30–40% of the water pumped for irrigation had not percolated downward to the aquifer as irrigation return flows each year (Luckey *et al.* 1986). Agricultural practices can enhance infiltration and increase the rate of recharge to the aquifer; for example, water can more easily percolate downward below fallow fields and where crop growth is at an early stage compared to fields covered with native vegetation (Luckey & Becker 1999).

Factors affecting groundwater quantity

Irrigated agriculture is the primary driving force behind the economic development of the High Plains area. In 1901, a vast groundwater reservoir underlying the High Plains was described by Johnson (1901), but at that time the technology was lacking to bring this deep water to the surface. Beginning in the 1930s, advances in well technology made deeper wells feasible. Development of the aquifer started in Texas, where depths to water were smaller, and expanded northward; irrigated area in the High Plains increased from about 8.1×10^5 ha in 1949 to 5.3×10^6 ha in 1978, and groundwater pumpage increased from about 4.9×10^9 to 2.8×10^{10} m^3 (Gutentag *et al.* 1984).

Irrigation throughout the High Plains developed rapidly in the 1960s and 1970s, resulting in a 36% increase in groundwater pumpage from 1964 to 1974 (Fig. 7a). However, in the 1980s, groundwater pumpage for irrigation decreased along with irrigated area. The area of irrigated crops decreased by about 17% from 1978 to 1987 (Fig. 7b). During the 1990s, the High Plains experienced moderate growth in irrigated area accompanied by a corresponding increase in groundwater pumpage. The increase in groundwater pumpage was largest, percentage wise, in the southern High Plains (24%) followed by the northern High Plains (16%) with the central High Plains actually decreasing slightly. As water-use efficiency increased so too did the land area in crop production. Advances in agricultural science and technology, in addition to a recent 20-year wet period across much of the area (Doesken 1998), have stimulated an expansion in agricultural production. Additionally, the types of crops being grown have shifted to crops that require more irrigation. The shift from wheat, alfalfa and sorghum to corn across the High Plains from 1978 to 1997 is shown in Figure 4. Corn requires about 660 mm of irrigation water to produce higher yields compared to as little as 0 mm for dryland wheat (Waskom *et al.* 1994). This shift to corn production can be traced to the demand for corn as livestock (cattle and swine) feed, but even so corn production does not meet feed demand in many areas in the central and southern regions of the High Plains, resulting in importation of corn from outside the area (Guru & Horne 2000).

During 1995, about 83% of the water used in the High Plains Aquifer was pumped from aquifers, and about 17% was withdrawn from rivers and streams. Almost all the surface water withdrawn for use (about 85%) is from the Platte River in Nebraska. Outside of the Platte River Valley, 92% of water used in the High Plains is supplied by groundwater. Groundwater from the High Plains is used primarily to grow crops; irrigation accounts for 96% of the groundwater use. The second largest groundwater use, 1.6×10^6 m^3 d^{-1}, is for public supply. Almost 2 million people rely on the High Plains Aquifer for their drinking water. Surface water is used for drinking water primarily in the larger cities near the periphery of the High Plains Aquifer [Cheyenne (Wyoming), Lubbock, Odessa, and Amarillo (Texas)]. Other uses of groundwater include livestock (8.4×10^5 m^3 d^{-1}), mining (8.0×10^5 m^3 d^{-1}), and industry (5.9×10^5 m^3 d^{-1}).

The declining water table impacts local agriculture in several ways. Rapid water withdrawals have lowered the water table and therefore increased the vertical lift needed to get the water to the surface. The greater depths to groundwater result in increased energy consumption, which can be seen in higher electricity and natural gas costs. The resultant decrease in saturated thickness reduces well yields, thus requiring that the discharge from several wells be combined to produce a sufficient volume of water to operate irrigation equipment. The continued declines in water levels, in addition to the fact that groundwater pumpage remains high, threaten the future agricultural economy of the area.

Factors affecting groundwater quality

The six hydrogeological units that comprise the High Plains Aquifer differ from each other spatially, lithologically, and in terms of depths to water, water use, and land use (Fig. 6a, Table 1). Each of these factors can affect water quality and its role in the sustainable development of the High Plains Aquifer. A general characterization of water quality, by hydrogeological unit, in the High Plains Aquifer is given in Table 3. It is beyond the scope of this paper to examine the interplay between water quality and sustainable groundwater development in each of the hydrogeological units; therefore, because the Ogallala Formation is the most important hydrogeological unit in the High Plains Aquifer in terms of size and water use (Table 1), it will be the focus of the following discussion.

Table 3. *General characterization of water quality, by hydrogeological unit, in the High Plains Aquifer.*

Hydrogeological unit	Median value of pH (standard units)	Median concentration of alkalinity (mg l^{-1} as $CaCO_3$)	Median concentration of dissolved solids (mg l^{-1})	Water type, based on median concentrations of major ions	Median concentration of dissolved oxygen (mg l^{-1})	Median concentration of nitrite + nitrate (mg l^{-1} as N)	Number of pesticide detections
Ogallala Formation–Northern High Plains	**7.4** (691)	**151** (457)	**315** (457)	Ca-HCO_3	**5.4** (72)	**2.8** (1787)	**76** (13 988)
Ogallala Formation–Central High Plains	**7.6** (1327)	**200** (1331)	**320** (1264)	Ca-Na-HCO_3	***6.2** (74)	**2.1** (1468)	**56** (5910)
Ogallala Formation–Southern High Plains	**7.7** (2185)	**239** (2147)	**517** (2077)	Na-HCO_3	†—	**2.5** (2249)	**60** (18 184)
Arikaree-Brule Formations	**7.6** (454)	**180** (238)	**295** (277)	Ca-Na-HCO_3	**6.1** (270)	**3.6** (939)	**61** (4022)
Nebraska Sand Hills	**7.3** (287)	**65** (123)	**155** (112)	Ca-HCO_3	**1.8** (90)	**2.4** (840)	**86** (5306)
Quaternary deposits–Kansas	**7.4** (204)	**180** (214)	**290** (222)	Ca-HCO_3	**15.1** (20)	**5.4** (302)	**20** (2222)
Quaternary deposits–Nebraska	**7.1** (809)	**222** (133)	**382** (132)	Ca-HCO_3	**5.9** (284)	**4.3** (3325)	**309** (12 291)
Valley-fill Alluvium	**7.4** (1030)	**227** (727)	**603**(469)	Ca-Na-HCO_3	**1.7** (310)	**4.0** (2544)	**1750** (29 819)

Data from Litke (2001) unless otherwise noted.
Number of observations in parentheses.
*U.S. Geological Survey, National Water Information System data base.
†No data available.

Natural factors

Water in the Ogallala Formation is generally of high quality and is suitable for most agricultural, domestic and industrial uses without treatment. In areas unaffected by natural or anthropogenic sources of contamination, the water is primarily a calcium bicarbonate type, the pH ranges from 7 to 8, median concentrations of dissolved solids are less than 517 mg l^{-1}, and median concentrations of dissolved oxygen are greater than 5.4 mg l^{-1} (Table 3). Nitrate is the dominant form of dissolved nitrogen in the water, and median concentrations range from 2.1 to 2.8 mg l^{-1} as nitrogen (N). These concentrations are slightly above the estimated national background nitrate concentration in groundwater of 2 mg l^{-1} (Mueller & Helsel 1996; USGS National Water Information System data base).

Concentrations of nitrate plus ammonia in precipitation in the study area ranged from about 0.6 to 1.0 mg l^{-1} as N, based on measurements made at eight sites in the High Plains from 1994 to 1999 (National Atmospheric Deposition Program 2000). Thus, atmospheric deposition could account for about one-fifth to one-half of the dissolved nitrate in the Ogallala Formation, assuming that precipitation collected during the measurement period was representative of precipitation that recharged the formation during earlier time periods. Based on measurements of $\delta^2H[H_2O]$ and $\delta^{18}O[H_2O]$ (Nativ & Smith 1987; Dutton 1995), there is no evidence that water recharging the Ogallala Formation is extensively evaporated, a process that would increase the concentration of N derived from the atmosphere in recharge.

The primary natural process that leads to degradation of water quality in the Ogallala Formation is the upward leakage of mineralized water into the formation from underlying Permian to Tertiary System geological units (Fig. 6b). Several of these geological units contain mineralized water, which has migrated into the Ogallala Formation, causing an increase in the concentrations of dissolved-solids in groundwater in the formation. For example, the upward movement of highly mineralized water from rocks of the Lower Cretaceous System of marine origin to the Ogallala Formation in the southern High Plains of Texas has resulted in increased dissolved-solids concentrations in Ogallala groundwater from less than 400 mg l^{-1} to more than 1000 mg l^{-1} (Reeves & Miller 1978; Gutentag *et al.* 1984). Seepage from natural saline lakes and dissolution of volcanic ash deposits in the Ogallala Formation also produce increased concentrations of dissolved solids in the Ogallala Formation in Texas.

In some areas of the central High Plains that are underlain by Permian System evaporite deposits, dissolution of gypsum and halite by deeply circulating meteoric water and its subsequent upward discharge into the Ogallala Formation has produced increased concentrations of calcium, sodium, sulphate, and chloride in Ogallala groundwater (Whittemore 1984; Mehta *et al.* 2000). Reported concentrations of both dissolved sulphate and chloride in the Ogallala Formation in these areas range from about 50 to 600 mg l^{-1}. Flow of meteoric water through the Permian System deposits is largely topographically driven; therefore, the flow began prior to development of the Ogallala Formation as a water resource.

Salinization of water in the Ogallala Formation by the upward movement of mineralized water from underlying geological units decreases the value of the resource as a public- and irrigation-water supply. For example, the United States Environmental Protection Agency's (USEPA) (1996) primary drinking-water standard for fluoride (4 mg l^{-1}) and the secondary drinking-water standards for sulphate and chloride (both 250 mg l^{-1}) are commonly exceeded in some areas underlain by Lower Cretaceous and Permian System bedrock, respectively, in the southern and central High Plains. Higher concentrations of dissolved solids and sodium relative to calcium and magnesium in irrigation water can be deleterious to soils and crops to which it is applied. The U.S. Salinity Laboratory (1954) characterized sodium and salinity hazards for irrigation water. Although the discharge of mineralized water into the Ogallala Formation is largely a natural process, excessive pumping in the formation and the resulting declines in water levels can enhance the upward movement of mineralized water into the aquifer by increasing the upward hydraulic gradient (Whittemore 1993).

Anthropogenic factors

Agriculture is the dominant land use comprising irrigated and dryland crop production, rangeland grazing, and confined-animal feeding operations. The input of nitrogen to croplands from manure and fertilizer applications is often substantial when compared with other sources such as atmospheric deposition. In 1999, nitrate-plus-ammonia loading to the High Plains from atmospheric deposition ranged from 2.5 to 6 kg-N h^{-1} (National Atmospheric Deposition Program 2000). Of the major crops grown in the High Plains, corn has the largest nitrogen

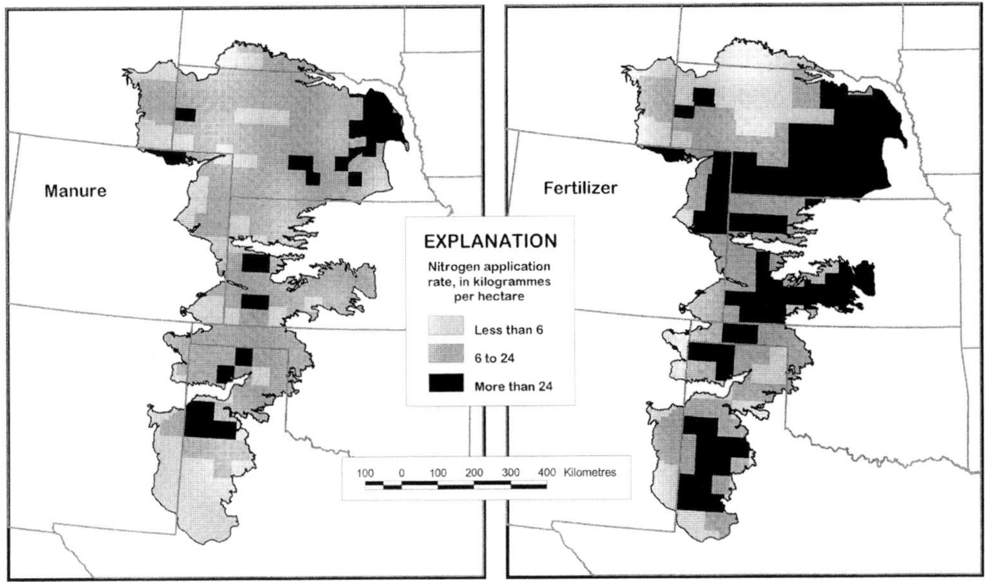

Fig. 8. Nitrogen application rate by county in the High Plains study area, 1992 (source of data: Puckett *et al.* 1998; D. Lorenz, U.S. Geological Survey, pers. comm. 1999).

requirement, at about 130 to 180 kg-N ha^{-1} per year; cotton and wheat require less than 100 kg-N ha^{-1} per year, dependent on soil texture; and alfalfa requires little or no nitrogen (U.S. Department of Agriculture 2000). Nitrogen application rates from manure are relatively uniform across the High Plains (Fig. 8). In contrast, nitrogen application rates from fertilizer show strong regional patterns (Fig. 8). Generally, the largest nitrogen application rates from fertilizer coincide with areas of predominantly corn and cotton production (Fig. 4).

In addition to crop production, confined-animal feeding operations for cattle and swine are a major agricultural endeavour in the High Plains. Proper disposal of the animal manure produced by these operations is a major water-quality concern (Miller 1971; McMahon 2000). For example, in the southern High Plains of Texas, cattle feedlots are frequently located near playa lakes that dot the landscape. Playa lakes are large, natural depressions that collect run-off and are considered the primary conduits for recharge to the southern High Plains Aquifer. However, increased environmental awareness on the part of the producers in the High Plains has resulted in improved methods to handle and dispose of animal waste, including composting, lining of waste retention ponds, and improved run-off collection systems.

In 1992, a large number of pesticides were applied to agricultural land in the High Plains; 162 different compounds were applied at a total rate of 2.2×10^7 kg of active ingredient per year (G. Thelin, USGS, pers. comm. 2000). The most commonly used herbicide was atrazine (more than 5×10^6 kg applied per year), which is used on corn, wheat and sorghum. Other common herbicides were metolachlor, 2,4-D, alachlor, cyanazine and trifluralin. Common insecticides were chlorpyrifos, terbufos, methyl parathion and carbofuran. Total pesticide application rates in 1992 (Fig. 9) indicate that pesticides were most heavily applied in areas where irrigation was most intense, in eastern Nebraska, south-western Kansas, and in the Panhandle of Texas. At least 30 pesticides were applied in most counties, and more than 60 pesticides were applied in counties in the Panhandle of Texas and in areas of western Nebraska where crop diversity is large.

The amount of land cultivated for corn in the High Plains has increased in the last 20 years (Fig. 4). This trend may have important consequences with respect to water quality in the High Plains Aquifer for several reasons. The land areas dominated by corn production are some of the areas that receive the largest inputs of nitrogen and pesticides (Figs 8 and 9), have the greatest irrigation intensity (Fig. 3a), and

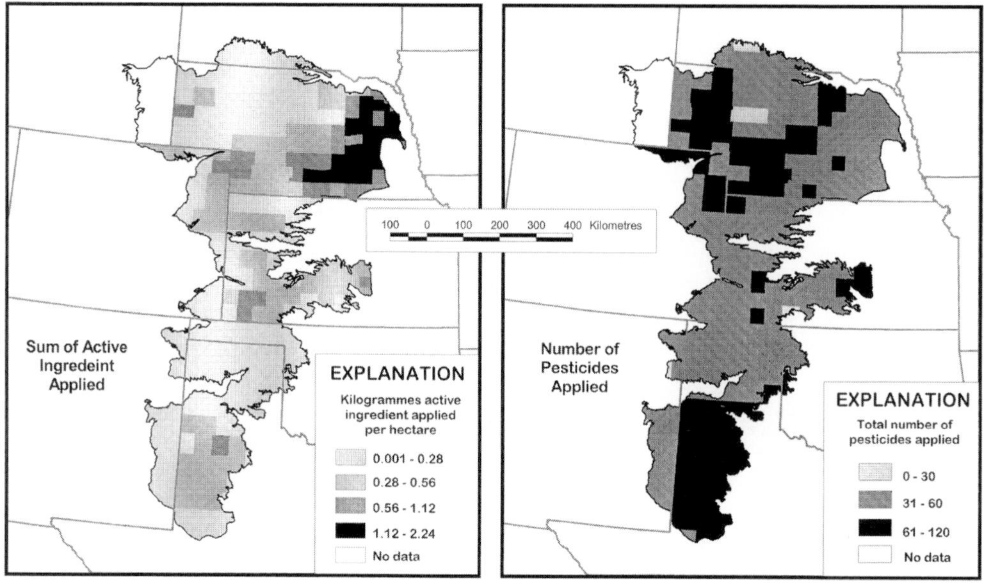

Fig. 9. Pesticides applied by county in the High Plains study area, 1992 (source of data: G. Thelin, U.S. Geological Survey, pers. comm. 2000).

have moderately- to well-drained soils. This combination of factors increases the potential for contamination of groundwater by nitrogen fertilizers and pesticides. A study demonstrated that recently recharged groundwater beneath irrigated cornfields in the Ogallala Formation, central High Plains, contained increased concentrations of nitrate (maximum concentration of 54 mg l^{-1} as N) and several of the most commonly used pesticides (alachlor, atrazine, metolachlor and simazine) in that region (McMahon 2000). The occurrence of increased nitrate concentrations in groundwater is not restricted to the Ogallala Formation; there are large areas of the High Plains Aquifer in which more than 10% of the reported nitrate analyses exceed the USEPA (1996) primary drinking-water standard of 10 mg l^{-1} as N (Fig. 10). These areas include the Quaternary System deposits in Kansas and Nebraska, where fertilizer application rates are high (Fig. 8) and the median depths to water are small to moderate (Table 1). Nitrate contamination already present in the High Plains Aquifer is not likely to be reduced substantially by natural processes such as denitrification because of the overall oxidizing conditions in the aquifer (Table 3). Recent improvements in farming practices, such as increased irrigation efficiencies and careful monitoring of crop nitrogen requirements to minimize over-fertilization, should decrease the amount of nitrogen leaching to groundwater in the future.

Salinization of groundwater and surface water occurs in the High Plains as a result of recycling water for multiple irrigation applications. This process is most evident along the major rivers of the High Plains – the Arkansas, Republican and Platte Rivers – where water is diverted from the rivers for irrigation, irrigation return flows discharge back to the rivers, and the water is diverted again for irrigation further downstream. Additional soluble salts and agricultural chemicals may be dissolved in the water with each application of the water to cropland. Salinization of groundwater by irrigation practices typically takes place in the alluvial aquifer adjacent to the river. However, along the Arkansas River in parts of south-western Kansas, where there has been excessive pumping in the underlying Ogallala Formation, mineralized water has moved downward from the alluvial aquifer to the Ogallala Formation. Concentrations of dissolved solids as large as 4900 mg l^{-1} have been measured in groundwater from the Ogallala Formation in these areas (USGS National Water Information System data base). Improvements in irrigation efficiencies in addition to reduced application of agricultural chemicals to cropland will probably

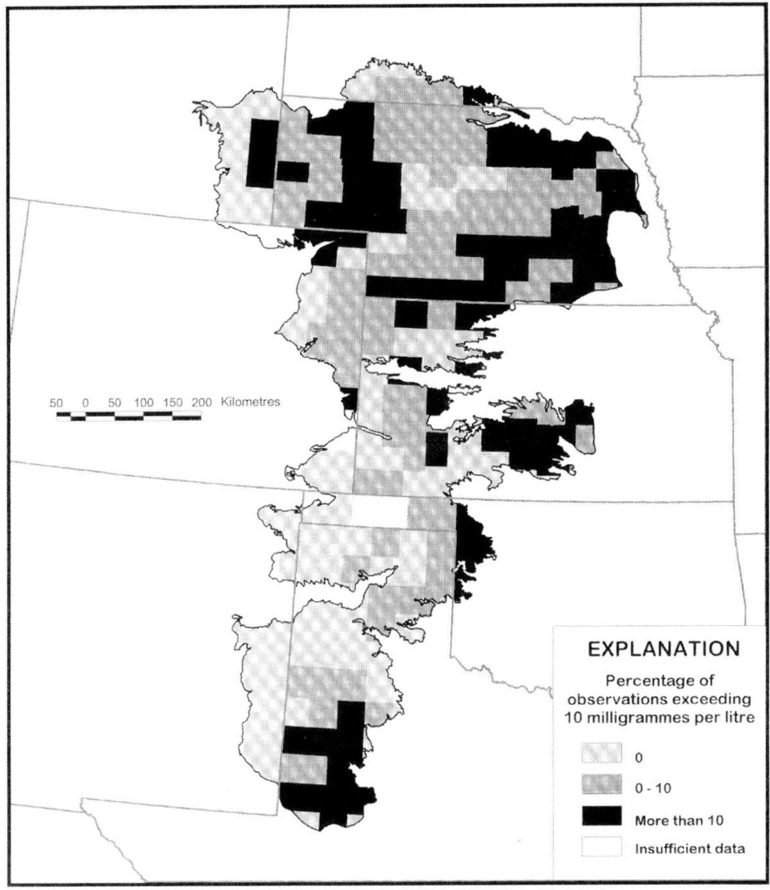

Fig. 10. Percentage of nitrate analyses exceeding the U.S. Environmental Protection Agency's (1996) drinking-water standard (10 mg l^{-1} as N), by county, in the High Plains study area.

help slow the salinization process associated with irrigation. However, owing to the relatively slow movement of groundwater, the time required to flush mineralized water from the aquifer can be in the order of thousands of years (P. McMahon, USGS, pers. comm. 2000).

Historical trends in groundwater quality

Historical groundwater-quality data were assembled from 43 local, state and federal agencies as part of the High Plains NAWQA study and used to evaluate trends in water quality. Minimum requirements were established for inclusion of data into a data base to ensure compatibility of data for analysis. Upon review of the data base, the determination was made that sufficient information was available to examine nitrate and dissolved-solids concentrations over time (Litke 2001).

Nitrate concentration data for the northern, central, and southern Ogallala Formation were plotted by sampling date and were smoothed using the LOWESS (LOcal WEighted Scatterplot Smoothing) method (Cleveland 1979) (Fig. 11a). To test for statistically significant trends, the Wilcoxon rank-sum test (Wilcoxon 1945) was performed at the 0.05 level of significance. The nitrate data indicate that statistically significant increases in nitrate concentrations have occurred throughout the Ogallala Formation, although increases were small in the central High Plains part of the Ogallala Formation where depth to water is greatest. Nitrate concentrations in the northern and southern Ogallala Formation are above the estimated

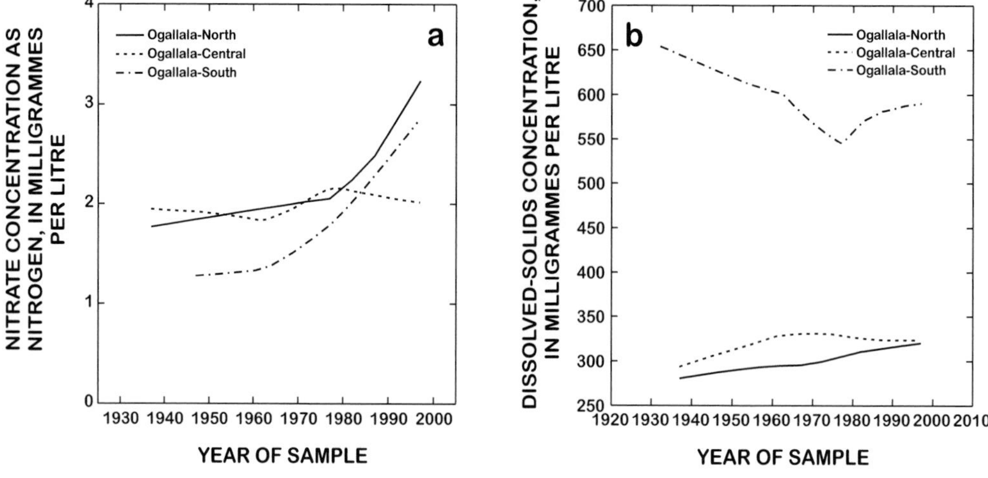

Fig. 11. Temporal variability of (**a**) nitrate concentrations and (**b**) dissolved solids in the Ogallala Formation (lines are LOWESS smooth curves, f = 0.7).

national background nitrate concentration in groundwater of 2 mg l^{-1} as N (Mueller & Helsel 1996). The increases seen in nitrate concentrations starting in the late 1970s and early 1980s indicate that the water quality in these areas is being impacted by human activities.

Historical trends in dissolved-solids concentrations were examined (Fig. 11b) using the same techniques as those used for nitrate. These data indicate that dissolved-solids concentrations have increased slightly in the northern and central Ogallala Formation, but have decreased in the southern Ogallala Formation. However, the concentrations of dissolved solids in the southern Ogallala Formation are still above the USEPA (1996) recommended limit for drinking water of 500 mg l^{-1}; the median dissolved-solids concentration for the southern Ogallala Formation is 517 mg l^{-1} (Table 3). Additionally, dissolved solids in irrigation water can be a concern, dependent on crop type, when concentrations exceed 500 mg l^{-1}. Differences in dissolved-solids concentrations were determined to be statistically significant (Wilcoxon 1945) when comparing data from 1930 to 1970 with data from 1980 to 1998.

Implications for the future

The history of water development in the High Plains Aquifer centres on irrigation and has transformed the area into one of the major agricultural regions of the world. This transformation from a dry prairie environment to what we see today occurred because agriculture took advantage of cheap energy and improved well-drilling and pumping systems. The rapid growth in irrigation over the last several decades has resulted in increases in population, employment and income. An additional consequence of this rapid expansion in irrigation is declining groundwater levels across the study area. Water in the High Plains Aquifer is being mined faster than it is being recharged. Water level declines are greatest in those parts of the aquifer where irrigation was developed first and where it was most intense. It is apparent that at today's rate of water use, the High Plains Aquifer cannot provide irrigation water very far into the future. In response to this threat to the irrigation economy of the High Plains, water users are implementing technical and institutional measures that would permit continued farming of the land. However, because groundwater withdrawals and the amount of irrigated land remains high, the future of the High Plains Aquifer is uncertain. What is certain is that water withdrawals in excess of water recharge are not sustainable.

The agricultural community realizes the need to improve farming methods throughout the High Plains to conserve water. As advances in irrigation technology become available farmers are implementing these improvements. Irrigation water-use efficiency data reported by the High Plains Underground Water Conservation District No. 1 (1997–1998) in the Panhandle of Texas service area indicate that average water-use efficiency improved from about 50% in mid-1970s to approximately 75% in 1990. Current state-of-the-art, low-pressure, full-dropline, centre-pivot systems are about 95% efficient; buried drip lines approach 100% efficiency. Other less obvious improvements include institutional and policy changes that focus on management of scarce water supplies. Evidence of these changes can be seen in the establishment of local/regional water management groups, who set limits on the spacing and number of wells, meter well discharge, and promote water conservation.

Chemical data indicate that agricultural activities are impacting the water quality of the High Plains Aquifer; thus, water quality should also be considered when evaluating the sustainability of the High Plains Aquifer. Farming methods can be improved through the use of best-management practices, but the response time of the aquifer to land-use changes may be in the order of thousands of years. It is the responsibility of all High Plains residents to work together so that the groundwater resource will be available to future generations.

If we look to the future of irrigated agriculture in the High Plains, several realities become apparent. Groundwater overdrafts cannot be sustained indefinitely, and some irrigated lands will ultimately disappear as a result. Where there is a decline in irrigated agriculture, the land may be converted to dryland farming or shifted to other uses such as grazing or conservation. The conversion period from irrigated agriculture to dryland farming may be delayed with the adoption of new technologies but probably cannot be avoided. These shifts in agriculture will have far-reaching implications for the regional economy. Historically, federal water and agricultural policies have been supportive of irrigated agriculture. Future policies may be less favourable, including a reduction or phasing out of agricultural support programs (CAST 1996). Environmental policies could also play a role in the conservation of the natural resource. Simultaneously, the economics of agriculture in this area will continue to feel the squeeze from markets outside the United States. Nevertheless, irrigation will play a part in the High Plains well into the future; however, the public's perceptions of agriculture and its effect on their lives (environment, culture and economy) will surely change.

Funding for this paper was provided by the U.S. Geological Survey's National Water-Quality Assessment (NAWQA) Program High Plains Regional Groundwater Study. The authors would like to recognize Sharon Qi for her assistance with the production of the article graphics. Reviews of earlier versions of this manuscript by Kathy Ogle, Virginia McGuire, and anonymous reviewers for the journal are gratefully acknowledged.

References

CAST. 1996. *Future of irrigated agriculture*. Council for Agricultural Science and Technology, Task Force Report No. **127**, 4420 West Lincoln Way, Ames, IA 50014-3447.

CLEVELAND, W. S. 1979. Robust locally weighted regression and smoothing scatterplots. *Journal of the American Statistical Association*, **74**, 829–836.

DALY, C. & TAYLOR, G. 1998. 1961–90 Mean monthly precipitation maps for the conterminous United States. World Wide Web Address: http://www.ocs.orst.edu/prism/state_products/cent_maps.html.

DENNEHY, K. F. 2000. *High Plains Regional Ground-Water Study*. United States Geological Survey Fact Sheet **091-00**.

DOESKEN, N. J. 1998. High Plains water and weather impacts, *In*: AL-KAISI, M. (ed.) *Ogallala Aquifer Symposium Proceedings*. Sterling, Colorado, 57–63.

DONOFRIO, C. & OJIMA, D. S. 1998. The Great Plains Today. *In*: OJIMA, D. S., EASTERLING, W. E. & DONOFRIO, C. (eds) *Climate Change Impacts on the Great Plains*. Ft. Collins, Colorado, 1–24.

DUGAN, J. T. & ZELT, R. B. 2000. *Simulation and analysis of soil-water conditions in the Great Plains and adjacent areas, central United States, 1951-80*. United States Geological Survey Water-Supply Paper **2427**.

DUTTON, A. R. 1995. Ground-water isotopic evidence for paleorecharge in U.S. High Plains aquifers. *Quaternary Research*, **43**, 221–231.

GURU, M. V. & HORNE, J. E. 2000. *The Ogallala Aquifer*. The Kerr Center, Poteau, Oklahoma.

GUTENTAG, E. D., HEIMES, F. J., KROTHE, N. C., LUCKEY, R. R. & WEEKS, J. B. 1984. *Regional aquifer-system analysis of the High Plains aquifer in parts of Colorado, Kansas, Nebraska, New Mexico, Oklahoma, South Dakota, Texas, and Wyoming—Geohydrology*. United States Geological Survey Professional Paper **1400-B**.

HELMS, D. 1981 The Great Plains Conservation Program, 1956–1981: A short administrative and legislative history. *In: Great Plains Conservation Program: 25 years of accomplishment*. SCS National Bulletin Number **300-2-7**.

HIGH PLAINS ASSOCIATES. 1982. *Six-State High Plains Ogallala aquifer regional resources study*. Report to the U.S. Department of Commerce and the High Plains Study Council. Austin, TX.

HIGH PLAINS UNDERGROUND WATER CONSERVATION DISTRICT NO. 1. 1997–1998. *The Ogallala Aquifer*. World Wide Web Address: http://www.hub.of the.net/hpwd/ogallala.html.

JOHNSON, W. D. 1901. *The High Plains and their utilization*. United States Geological Survey 21st Annual Report, **pt. 4-C**, 601–741.

LITKE, D. W. 2001. H*istorical water-quality data for the High Plains Regional Ground-Water Study area in*

Colorado, Kansas, Nebraska, New Mexico, Oklahoma, South Dakota, Texas, and Wyoming, 1930-98. United States Geological Survey Water-Resources Investigation Report **00-4254**.

LUCKEY, R. L. & BECKER, M. F. 1999. *Hydrogeology, water use and simulation of flow in the High Plains aquifer in northwestern Oklahoma, southeastern Colorado, southwestern Kansas, northeastern New Mexico, and northwestern Texas.* United States Geological Survey Water-Resources Investigations Report **99-4104**.

LUCKEY, R. R., GUTENTAG, E. D., HEIMES, F. J. & WEEKS, J. B. 1986. *Digital simulation of ground-water flow in the High Plains aquifer in parts of Colorado, Kansas, Nebraska, New Mexico, Oklahoma, South Dakota, Texas, and Wyoming.* United States Geological Survey Professional Paper **1400-D**.

LUCKEY, R. R., GUTENTAG, E. D. & WEEKS, J. B. 1981. *Water-level and saturated-thickness changes, predevelopment to 1980 on the High Plains aquifer in parts of Colorado, Kansas, Nebraska, New Mexico, Oklahoma, South Dakota, Texas, and Wyoming.* United States Geological Survey Hydrologic Investigations Atlas **HA-652**, scale 1 : 2,500,000, 2 sheets.

MCGUIRE, V. L. & FISCHER, B. C. 1999. *Water-level changes, 1980 to 1997, and saturated thickness, 1996-97, in the High Plains aquifer.* United States Geological Survey Fact Sheet **124-99**.

MCGUIRE, V. L. & FISCHER, B. C. 2000. Water-level changes in the High Plains aquifer—1980 to 1998 and 1997 to 1998. World Wide Web Address: http://www-ne.cr.usgs.gov/highplains/hp98_web_report/hp98fs.html.

MCMAHON, P. B. 2000. *A reconnaissance study of the effect of irrigated agriculture on water quality in the Ogallala Formation, central High Plains aquifer.* United States Geological Survey Fact Sheet **FS-009-00**.

MEHTA, S., FRYAR, A. E. & BANNER, J. L. 2000. Controls on the regional-scale salinization of the Ogallala aquifer, southern High Plains, Texas, USA. *Applied Geochemistry*, **15**, 849–864.

MILLER, W. D. 1971. *Infiltration rates and groundwater quality beneath cattle feedlots, Texas High Plains.* Report to the U.S. Environmental Protection Agency by Texas Tech University, USEPA Project #16060EGS, Contract #14-12-804.

MUELLER, D. K. & HELSEL, D. R. 1996. *Nutrients in the Nation's waters—Too much of a good thing?* United States Geological Survey Circular **1136**.

NATIONAL ATMOSPHERIC DEPOSITION PROGRAM. 2000. *Nitrogen in the Nation's rain.* National Atmospheric Deposition Program Brochure **2000-01a**.

NATIONAL CLIMATE DATA CENTER 2000. Comparative climatic data for the United States through 1999. World Wide Web Address: http://nndc.noaa.gov/?http://ols.nndc.noaa.gov/plolstore/plsql/olstore.-prodspecific? prodnum = C00095-PUB-A0001.

NATIV, R. & SMITH, A. 1987. Hydrogeology and geochemistry of the Ogallala aquifer, southern High Plains. *Journal of Hydrology*, **91**, 217–253.

OPIE, J. 1993. *Ogallala—Water for a Dry Land.* University of Nebraska Press, Nebraska.

PUCKETT, L. J., HITT, K. E. & ALEXANDER, R. B. 1998. *County-based estimates of nitrogen and phosphorus content of animal manure in the United States for 1982, 1987, and 1992.* World Wide Web Address: http://water.usgs.gov/lookup/getspatial?manure

REEVES, C. C. & MILLER, W. D. 1978. Nitrate, chloride, and dissolved solids, Ogallala aquifer, West Texas. *Ground Water*, **16**, 167–173.

THELIN, G. P. & HEIMES, F. J. 1987. *Mapping irrigated cropland from Landsat data for determination of water use from the High Plains aquifer in parts of Colorado, Kansas, Nebraska, New Mexico, Oklahoma, South Dakota, Texas, and Wyoming.* United States Geological Survey Professional Paper **1400-C**, C1-C38.

THORNTHWAITE, C. W. 1936. The Great Plains. *In*: GOODRICH *et al. Migration and Economic Opportunity.* University of Pennsylvania Press, Philadelphia, 202–250.

U.S. BUREAU OF RECLAMATION. 1985. *Republican River Basin Water Management Study, Colorado, Nebraska, Kansas.* Special Report.

U.S. BUREAU OF RECLAMATION. 1996. *Resource Management Assessment-Republican River Basin.* Water Service Contract Renewal, various pagination.

U.S. DEPARTMENT OF AGRICULTURE. 1999. *1997 Census of Agriculture Geographic Area Series.* National Agricultural Statistics Service CD-ROM **AC97-CD-VOL1-1B**.

U.S. DEPARTMENT OF AGRICULTURE. 2000. *National Agricultural Statistics Service On-Line Data Base.* World Wide Web Address: http://www.usda.gov/nass.

U.S. DEPARTMENT OF COMMERCE. 1990. *1987 Census of Agriculture County Data File.* Bureau of the Census CD-ROM **CDRM 327500**.

U.S. ENVIRONMENTAL PROTECTION AGENCY. 1996. *Drinking water regulations and health advisories.* U.S. Environmental Protection Agency Office of Water Report **822-B-96-002**.

U.S. GEOLOGICAL SURVEY. 1999–2000. *National Land Cover Data Base.* World Wide Web Address: http://edcwww.cr.usgs.gov/pub/edcuser/vogel/states/

U.S. SALINITY LABORATORY. 1954. *Diagnosis and improvement of saline and alkali soils.* U.S. Department of Agriculture Handbook **60**.

WAHL, K. L. & TORTORELLI, R. L. 1997. *Changes in flow in the Beaver-North Canadian River Basin upstream from Canton Lake, western Oklahoma.* United States Geological Survey Water-Resources Investigations Report **96-4304**.

WASKOM, R. M., CARDON, G. E. & CROOKSTON, M. A. 1994. *Best management practices for irrigated agriculture.* Colorado Water Resources Research Institute, Completion Report No. **184**, p. 45.

WEEKS, J. B. & GUTENTAG, E. D. 1981. *Bedrock geology, altitude of base, and 1980 saturated thickness of the High Plains aquifer in parts of Colorado, Kansas, Nebraska, New Mexico, Oklahoma, South Dakota, Texas, and Wyoming.* United States Geological Survey Hydrologic Investigations Atlas **HA-648**, scale 1 : 2,500,000, 2 sheets.

WEEKS, J. B., GUTENTAG, E. D., HEIMES, F. J. & LUCKEY, R. R. 1988. *Summary of the High Plains regional aquifer-system analysis in parts of Colorado,*

Kansas, Nebraska, New Mexico, Oklahoma, South Dakota, Texas, and Wyoming. United States Geological Survey Professional Paper **1400-A**.

WHITTEMORE, D. O. 1984. *Geochemical identification of the source of salinity in ground water of southeastern Seward County, Kansas*. Kansas Geological Survey Open-File Report **84-3**.

WHITTEMORE, D. O. 1993. *Ground-water geochemistry in the mineral instrusion area of Groundwater Management District No. 5, south-central Kansas*. Kansas Geological Survey Open-File Report **93–2**.

WILCOXON, F. 1945. Individual comparisons by ranking methods. *Biometrics*, **1**, 80–83.

Sustainable groundwater resources in a hard-rock island aquifer – the Channel Island of Guernsey

N. S. ROBINS, K. J. GRIFFITHS, P. D. MERRIN & W.G. DARLING

British Geological Survey, Wallingford, Oxfordshire OX10 8BB, UK (e-mail: N.Robins@bgs.ac.uk)

Abstract: The annual volume of water in the Guernsey public supply, which derives largely from surface storage, is approximately 5 Mm^3. Additional abstraction from private surface and groundwater sources amounts to a further 1.5 Mm^3. A shallow weathered zone in ancient crystalline metamorphic rocks forms the main aquifer, and this has a significant resource potential in maintaining baseflow to streams. The average annual water budget for the island is 831 mm rainfall, which supports 613 mm potential evapotranspiration, 90 mm surface runoff and 128 mm groundwater recharge. These figures contrast with a poor rainfall year in which infiltration may be zero; the annual variation in rainfall from the long-term mean is often considerable. Annual rainfall has also been declining on the island since the 1940s, and although Guernsey has survived droughts in the past it may be less able to do so in the future. Groundwater on the island is moderately mineralized, but over half of the 21 samples collected recently contained nitrate at concentrations greater than the EU maximum admissible concentration 11.3 mg-N l^{-1}. Some of the nitrate may derive from leaking cesspits, but past application of nitrogenous fertilizer to cultivated land accounts for the major component. Attempts at groundwater dating by analysis of chlorofluorocarbon species at a small number of sites was hindered by contamination, although much of the water sampled is apparently young, and recently recharged. The long-term sustainability of the shallow island aquifer and its associated surface waters requires careful husbanding to protect it from conflicting land use interests and water demands.

The groundwater body, its discharge as baseflow to surface waters, and runoff, have together sustained the island community of Guernsey throughout history. The sustainability of the resource is now under threat from a variety of directions. Perhaps the most alarming is an apparent decline in rainfall over the island of some 10% since the 1940s which, coupled with demographic growth, means that the stress during years of water shortage may become more prevalent in future years. However, the resource is also under threat from pollution by horticultural nutrients and leaking cesspits; the occurrence of nitrate in Guernsey waters, which exceed the EU maximum admissible concentration (MAC) for nitrate, is a major cause for concern.

Management of water resources requires great sensitivity on small but heavily populated islands. Guernsey has achieved a very high level of resource recovery and storage so that little, if any, economically recoverable water is allowed to discharge to the coast. So far, the island has kept pace with demand despite a reduction in overall rainfall. Good management of the resource will ensure that it is utilized to the best advantage of the whole island, although continuing decline in rainfall may force the island to look to desalination at some time in the future.

The public water supply in Guernsey derives largely from surface storage, and is provided by the States of Guernsey Water Board. In addition, private surface and groundwater abstraction satisfies some domestic and commercial demand amounting to between 10 and 30% of the total amount of water used; much of this usage is seasonal. Private consumption not only draws on groundwater and stream off-takes but also on rooftop collection.

Guernsey is the second largest of the Channel

From: Hiscock, K. M., Rivett, M. O. & Davison, R. M. (eds) *Sustainable Groundwater Development*. Geological Society, London, Special Publications, **193**, 121–132. 0305-8719/02/ $15.00

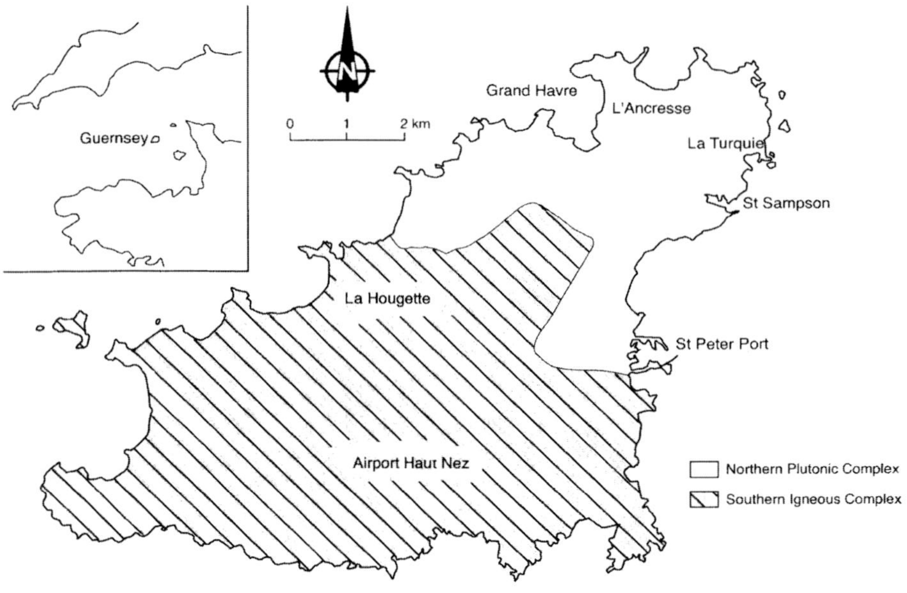

Fig. 1. Simplified geological map of Guernsey (after Roach *et al.* 1991).

Islands (Fig. 1). It has an area of 63 km^2 and is situated 130 km from the English south coast and 30 km from the French Cherbourg Peninsula. The population reported in the 1996 census was 58 700 of which 16 200 lived in St Peter Port. The greatest population density is concentrated around St Peter Port and St Sampson on the east coast of the island. The present population is about 62 000. Over 300 000 visitors come to the island each year, increasing the demand for water supply during summer months.

Dominant land-uses on Guernsey are horticultural (glasshouses), agricultural and urban. Native woodland and open countryside make up the remainder. Horticultural activity has diminished greatly since tomato cultivation declined in the 1970s; cut flowers, pot plants and plant production have now taken over as the main horticultural products. There has been a recent significant development of hydroponic culture in glasshouses.

The topography of the island divides into two distinct areas. The south is a cliff-bounded plateau rising to 107 m above sea level, with a number of steep-sided valleys, mostly trending north-west. There is a marked topographical change in the north of the island which is flat and low-lying.

The southern plateau is the main source for surface water drainage. Most of the streams drain to the north-west and to the north, but there is also drainage from the plateau area to the south. Ultimately, these streams should discharge to the sea on the north-west coast, but like most of the stream discharge from the island the water is largely intercepted for use in the public water supply system.

The island enjoys a temperate climate with over 2000 h of sunshine per annum. Prevailing wind directions are from the west and south-west, although dry easterly continental air prevails from time to time. Rainfall has been monitored at six stations on the island, although only three are maintained today (Table 1). Rainfall decreases from the south to north of the island: 838 and 837 mm per annum respectively at the Airport and Haut Nez in the south of the island to 792 and 766 mm at L'Ancresse and La Turquie (Fig. 1) in the extreme north-east of the island (based on 1961–1990 long-term averages). However, there is a general temporal

Table 1. *Long-term mean average rainfall data*

Name	Guernsey grid ref.	Altitude (m above sea level)	Data Range	Mean (mm)
L'Ancresse	338 830	3	1966– present	792
La Turquie	357 829	12	1935–1969	766
Les Quartiers	323 804	12	1934–1978	846
Haut Nez	311 762	107	1907–1974	837
Airport	292 759	104	1949– present	838
Kings Mills	292 787	14	1923– present	832

decline in rainfall over the last few decades and there has been concern about overuse of resources during drier periods (e.g. during the years 1989–1992 and 1996–1997). The 1990–1997 average for the Airport is only 789 mm (Guernsey Airport Meteorological Office 1998) and for L'Ancresse is 722 mm.

The mean annual air temperature at the airport is 10.8°C and varies from 8.4°C to 13.1°C. Mean wind speeds are consistently above 10 knots each month, with an average of over 12 knots. Average sea temperatures vary from approximately 8.5°C in February and March to approximately 16.5°C in early September. Potential evapotranspiration has not been measured, and appropriate meteorological data from which to derive it are not available.

Following a period of drought in 1976, a survey was commissioned by the States of Guernsey Water Board to investigate the potential surface and groundwater resources of small catchments on the south and west of the island (Hawksley 1977). The results of this study indicated that the potential role of groundwater in augmenting the public supply was minimal compared to the benefit of increasing the number of surface water collection reservoirs. The only subsequent work to be completed before the present study is that by Jehan (1993), who reported elevated nitrate concentrations in Guernsey groundwaters and first raised the issue of their sustainability.

The volume of water put into public supply, largely from surface storage, peaked in 1970 at 5.7 Mm^3 per annum. In recent years, water in public supply has fluctuated between 4.7 and 5.2 Mm^3 per annum, with increased domestic consumption offset by a declining demand from horticultural and commercial uses. Quantities of surface and groundwater abstracted for private consumption are not known, but it is thought that about 5000 properties have access to private sources, i.e. given the average UK consumption of 0.6 m^3 d^{-1} per household, this represents an annual total of 1.1 Mm^3. However, as many of these properties include supply to glasshouses and other uses, overall consumption may be as high as 1.5 Mm^3 per annum. Total water use on the island in a normal year, therefore, amounts to about 6.5 Mm^3.

There are few data available on groundwater and groundwater sources. Groundwater level monitoring at a single private well has been maintained since 1989 while the States Water Board has been monitoring water levels in its own wells only since 1997. Groundwater quality data are limited to occasional analyses for potability. Drilling records are maintained by the island drillers, but these record well dimensions and trial pumping rates only. The two main Guernsey drilling contractors report that between 12 and 15 new boreholes are being drilled each year. Some of these are replacement sources, but the majority are new, many of which are for garden use.

The States of Guernsey Water Board commissioned a desk and field investigation in 1999 to gather evidence to support the optimal exploitation of the surface and groundwater resource. The field campaign recorded the physical dimensions of a sample of boreholes and wells (Table 2). The sample sites were selected to provide representative coverage of the island reflecting different geological and topographical settings. Depth measurements were made using a plumb line, and static and dynamic water levels using an electric dipper. Pumping rates were estimated by timing the discharge of a known volume of water.

Geological setting

The island is formed by rocks of Precambrian age (Fig. 1). In the south there are metamorphic, sedimentary and igneous rocks with numerous intrusive dykes which together form the Southern Igneous Complex. The remainder of the

Table 2. *Summary characteristics of wells, boreholes and springs*

	Depth (m)	Depth to static water (m)	Average abstraction ($m^3 d^{-1}$)	Majority use
Boreholes ($n = 22$)	Mean 35 ($n = 17$) Range 6–100	Mean 6.2 ($n = 18$) Range 0.6–12.3	99 ($n = 10$)	Horticulture
Wells ($n = 11$)	Mean 18 ($n = 3$) Range 9–24	Mean 5.8 ($n = 6$) Range 2.9–8.4	14 ($n = 1$)	Various
Spring ($n = 1$)	–	–	–	Industry

Some sources have no data at all, some have only incomplete data.

island is made of intrusive igneous rocks that have been relatively unaltered and which form the Northern Plutonic Complex (Roach *et al.* 1991). There are also some intrusive dykes which are of late Palaeozoic age. Younger sedimentary rocks occur offshore and range in age from Cambrian to Upper Cretaceous and Palaeocene.

Marine erosion during the Quaternary has formed a series of raised erosion surfaces. Head, loess and blown sand deposits form extensive outcrops on inland areas and these are important for soil formation. There is a partial cover of loess, in places up to 5 m thick, over the plateau area in the south of the island. Head deposits up to 20 m thick occur in some coastal sections and comprise reworked loess and solifluction deposits. Inland, the head is generally less than 3 m thick.

Recent deposits are of both marine origin (marine alluvium and storm beach deposits), and freshwater origin (peat and alluvium). There are also deposits of blown sand. The area east of St Sampson was reclaimed from the sea in 1808 (Collonette 1916).

The groundwater system

The island-wide water budget is given by:

Rainfall (P) − actual evapotranspiration (AE) = runoff (Q) + net groundwater recharge (ΔG)

Rainfall is measured daily by the Guernsey Airport Meteorological Office. The long-term (30 year mean) annual rainfall is 831 mm, although it may be as much as 65 mm less in the north of the island. Variation from the long-term annual mean is significant. For example, in 1997 the monthly rainfall varied by 11% in September and 208% in June, whilst the annual rainfall for that year was only 86% of the 30-year average. There has been a steady decline in rainfall of about 10% since the 1940s.

Although potential evapotranspiration data are not currently available for Guernsey, data for the neighbouring island of Jersey are comprehensive. Potential evapotranspiration and groundwater recharge have recently been determined through intensive study of an instrumented catchment on Jersey (Robins & Smedley 1994, 1998). Comparison of long-term rainfall data for the respective island airport sites (Fig. 2) indicates little monthly variation (although specific rainfall events may be quite different between the islands), and the respective long-term means (1961–1990) are 831 mm for Guernsey and 836 mm for Jersey (Meteorological Department 1991). Rather than attempt crude and unjustified attempts at determining evapotranspiration for Guernsey, the existing 30-year mean data for Jersey have been transposed (*pro rata* against rainfall) for use at Guernsey as follows: $P = 831$ mm; $AE = 613$ mm; $Q = 90$ mm; and $\Delta G = 128$ mm.

The function ΔG also includes change in soil moisture content, although in the long term this is likely to be small on an island. For much of the year there is a significant soil moisture deficit, and in very dry winters (e.g. during the early 1990s) field moisture capacity was not attained and recharge did not take place.

A hydrograph for a private well at La Hougette (Fig. 3) indicates that the response of the groundwater level to specific rainfall events in this area of the island is short, usually only a matter of days. This suggests rapid movement of water through the thin fractured and weathered unsaturated zone to the water table. Furthermore, many springs increase in flow following periods of heavy rain, and some boreholes reportedly contain suspended sediment after intensive rainfall events. Seasonal water level fluctuations are typically about 2 m. As the mean annual recharge is about 130 mm, a rise in groundwater level of 2 m (plus a similar amount of recession caused by groundwater discharge from the aquifer) suggests an effective aquifer

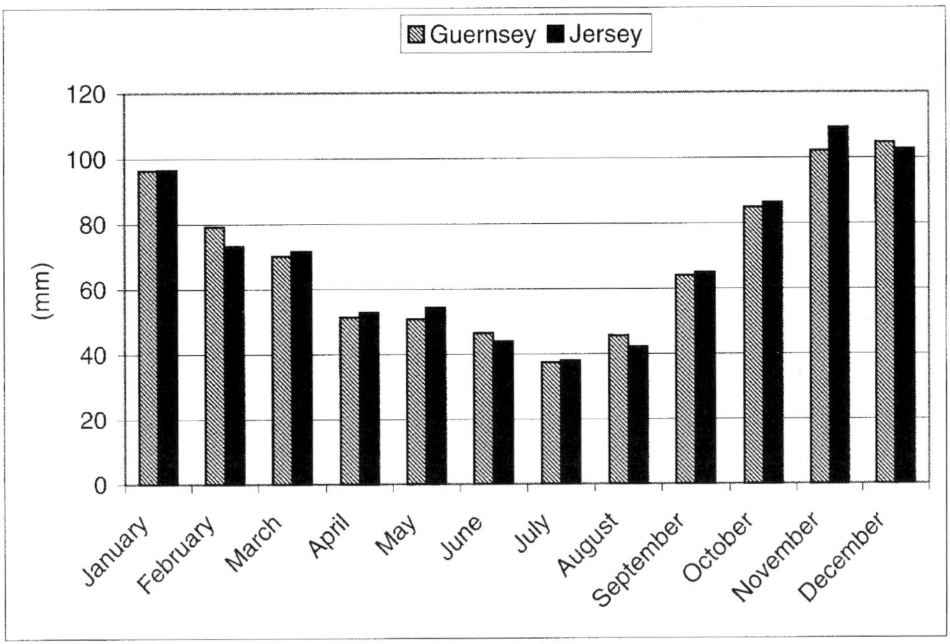

Fig. 2. Average monthly rainfall for the Channel Islands.

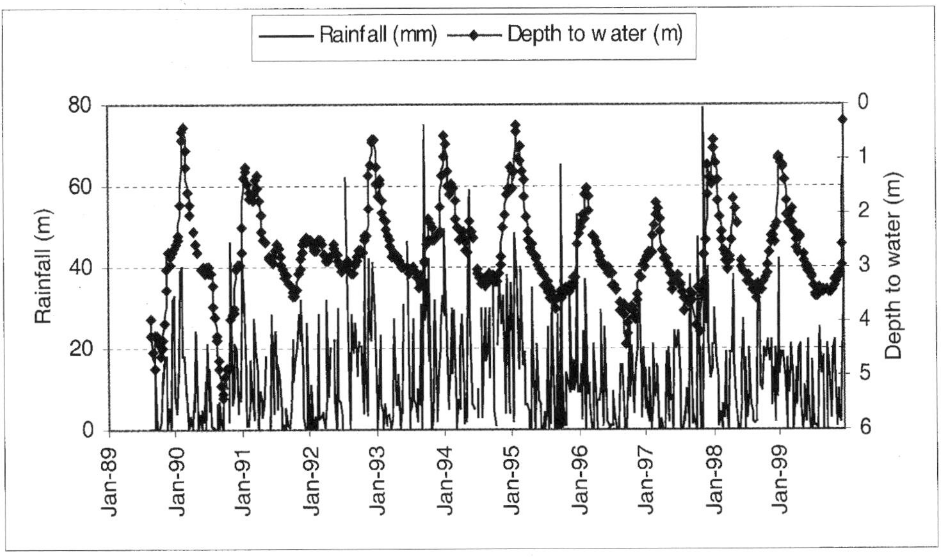

Fig. 3. Groundwater hydrograph for a private well at La Hougette in the centre of the island.

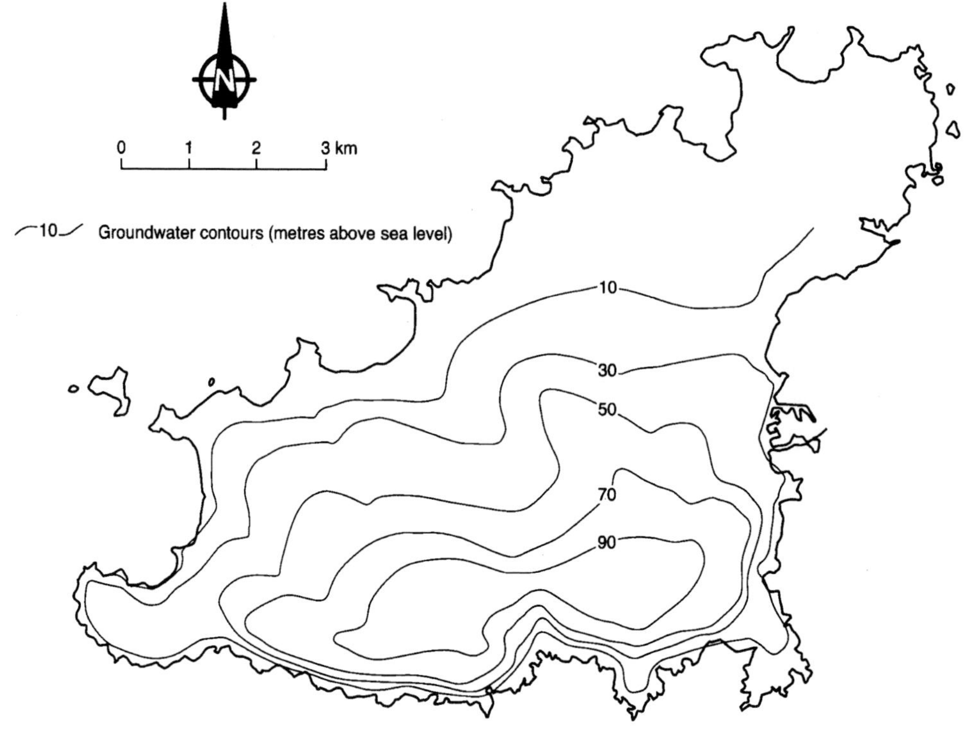

Fig. 4. Groundwater level contours (m above sea level).

specific yield (approximating to porosity) of about 130/4000, i.e. about 3%.

The potentiometric surface of the main aquifer (Fig. 4) has been constructed from water level data collected during the current survey, and that of Jehan (1993). The potentiometric surface is closely related to topography, with the highest water levels occurring in the elevated southern part of the island. The highest level measured during the field visits was approximately 97 m above mean sea level. There is a steep hydraulic gradient from the southern plateau to the south coast, reflecting the steep topographic gradient and low transmissivity of the aquifer, which is probably thin, in this area. North of the plateau the predominant groundwater flow direction is towards the north-west, except in the north where the general flow direction is to the north-east. Rest water levels are commonly less than 5 m above mean sea level along the north-west coast. In the northernmost part of the island, there is both a low hydraulic gradient and small groundwater flux with water levels generally about 9 or 10 m above mean sea level, i.e. within a metre or two of the ground surface.

The bedrock aquifer has a three-dimensional joint and fracture network which is independent of larger scale geological fault structures. The depth of the weathered zone may be about 30 m below ground level, and is typically about 25 m thick beneath the water table.

The relatively shallow nature of the main bedrock aquifer provides a useful moderating feature which protects the aquifer from excessive abstraction. The hydraulic conductivity and the storativity both decrease with depth below the water table because the frequency, size and connectivity of the fractures decrease with depth and pressure of overburden. The effective thickness of the aquifer is rarely greater than 25 m. Consequently, the ability of the aquifer to drain as baseflow or to discharge to pumped boreholes is reduced as the water table falls. Therefore, less water is lost to streams and at the same time borehole yields also begin to decline. This consequential reduction in aquifer properties as the water table falls provides the aquifer with an important self-protection mechanism. However, care needs to be taken at times of water stress to

Table 3. *Groundwater chemistry data*

Parish	Grid ref.	Temp (°C)	pH	SEC ($\mu S\ cm^{-1}$)	Na ($mg\ l^{-1}$)	K ($mg\ l^{-1}$)	Ca ($mg\ l^{-1}$)	Mg ($mg\ l^{-1}$)	Cl ($mg\ l^{-1}$)	SO_4 ($mg\ l^{-1}$)	HCO_3 ($mg\ l^{-1}$; field	NO_3–N ($mg\ l^{-1}$)	$\delta^{18}O$ (‰)	δ^2H (VSMOW)
Forest	300 756	14.3	6.15	607	61.1	2.5	37.0	18.6	71.3	81.3	78.0	11.5		
Vale	316 820	13.4	6.43	718	69.5	8.9	39.2	22.6	78.0	97.3	106	10.6	–5.2	–32
Forest	317 762	12.1	5.68	825	72.0	2.4	51.9	24.8	89.0	81.9	61.0	30.5	–5.9	–31
Forest	312 762	12.4	5.62	805	82.7	2.4	39.6	24.2	134	95.3	67.1	5.2		
St. Saviour	295 763	11.6	6.11	427	50.4	1.6	22.0	10.4	50.3	43.4	77.2	1.4	–5.8	–32
St. Saviour	265 755	12.1	5.94	728	73.3	3.9	40.2	23.1	86.0	117	45.7	15.5		
St. Saviour	286 758	–	–	–	67.3	3.8	31.7	17.9	72.1	106	–	8.7	–5.6	–30
King's Mills	293 786	11.1	–	611	49.6	6.1	48.5	15.1	64.2	44.6	–	5.6		
St. Peters	266 754	11.2	6.33	745	80.9	18.3	27.5	14.8	99.0	81.5	60.4	8.1		
St. Saviour	295 768	13.2	6.51	877	60.8	6.5	57.3	36.6	66.9	106	58.5	48.6		
St. Saviour	293 768	13.3	6.55	800	64.2	3.5	54.3	29.4	73.0	125	100	18.9		
St. Peters	276 762	12.8	5.87	559	72.3	2.4	22.5	7.2	103	40.2	27.4	3.2	–5.6	–30
Castel	279 787	10.0	6.41	650	62.8	7.8	33.8	18.7	93.0	49.4	57.9	15.5	–5.3	–29
St. Sampson	342 813	–	–	–	103	12.0	106	52.6	178	301	–	2.2		
St. Peter Port	327 798	13.8	7.00	1387	76.0	5.2	66.2	31.4	93.0	104	184.1	32.8		
St. Sampson	339 812	14.3	6.65	978	81.6	3.1	64.0	31.5	74.4	161	108.5	27.3		
Vale	355 833	13.6	6.78	858	82.3	7.2	49.4	26.1	96.0	114	119.5	14.3	–3.2	–24
St. Sampson	333 808	11.8	6.68	1578	120	7.0	130	53.3	144	350	246.7	8.0	–5.3	–32
St. Peter Port	333 792	12.0	6.95	548	51.9	2.2	31.5	13.3	50.5	51.0	82.9	12.3	–5.8	–36
St. Sampson	345 810	11.2	6.55	869	83.7	2.8	47.7	21.3	140	66.7	79.9	11.8		
St. Peter Port	337 779	–	–	–	57.2	2.6	36.2	15.2	77.2	53.2	–	4.8	–5.3	–31

VSMOW, Vienna Standard Mean Ocean Water.

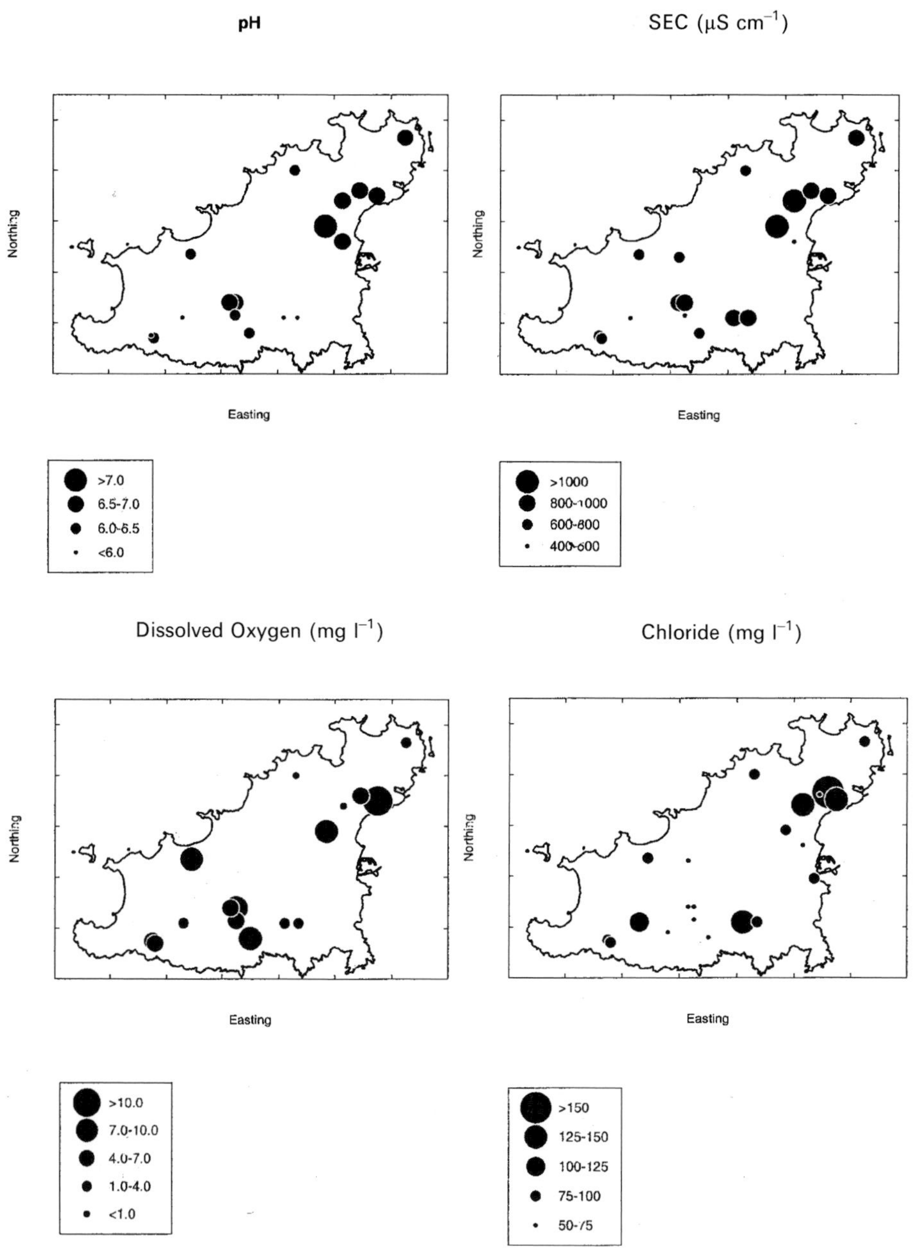

Fig. 5. Island-wide distribution of pH, specific electrical conductivity (SEC), dissolved oxygen and Cl.

ensure that the hydraulic gradient is not reversed near the coast causing saline contamination of at least part of the aquifer.

Deeper groundwater is tapped by a few boreholes which have been drilled to depths up to 100 m. For example, just to the north of the airport are two boreholes which are only 50 m apart; one is 18 m deep with an estimated yield

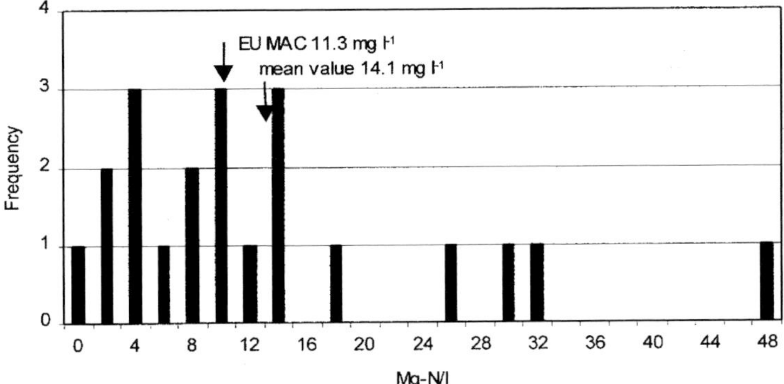

Fig. 6. Occurrence of nitrate as N.

of 60 m^3 d^{-1}, whereas the other is 55 m deep but can only maintain a supply of 17 m^3 d^{-1}. The favourable water bearing fractures intersected at shallow depths by the first borehole were not penetrated by the second borehole which intersected deeper but tighter and lower-yielding fractures at depth.

Anecdotal evidence suggests that artesian conditions have, on occasion, been encountered during drilling, most notably in the vicinity of the airport. Some of these boreholes and wells are relatively shallow and may relate to small perched, and partly confined, groundwater bodies. Others relate to the confining pressure of water contained in a fractured aquifer but recharged from some point of higher ground elevation.

Groundwater chemistry

A sampling campaign was carried out in November 1999 from 21 wells and boreholes distributed across the island. Wherever on-line sampling was possible, wellhead measurement of redox potential (Eh), dissolved oxygen (DO) and pH was made in a flow-through cell, but otherwise pH only was measured in a standing sample. Groundwater temperature, specific electrical conductivity (SEC) and alkalinity were also measured at the wellhead.

Duplicate samples were collected from each location and filtered to 0.45 μm, one sample being acidified with concentrated nitric acid (1% HNO_3) to stabilize cations. A third unfiltered sample was collected in a glass bottle for stable isotope (δ^2H, $\delta^{18}O$) analysis. If the source was not in regular use prior to sampling, water was pumped to waste for approximately three borehole volumes or until such time as stable readings were recorded for SEC. At four borehole sites, 500 ml samples of unfiltered water were collected for analysis of chlorofluorocarbon (CFC) species in immersed glass bottles to prevent atmospheric contamination. The bottles were then sealed within water-filled steel cans to provide further protection against atmospheric ingress prior to measurement (see Oster *et al.* 1996).

Major cations, SO_4 and trace metals were analysed on filtered and acidified samples by inductively-coupled-plasma atomic emission spectrometry (ICP-AES). Analysis of nitrate and chloride was performed by automated colorimetry on filtered but non-acidified samples. Stable isotope analysis was carried out by dual inlet mass spectrometry, and CFC analysis by purge-and-trap gas chromatography using an electron capture detector (ECD).

The analytical data are presented in Table 3. SEC ranges from 427 to 1578 μS cm^{-1} with an average of 809 μS cm^{-1}. SEC is highest in the north-eastern part of the island. The pH varies from 5.6 to 7.0 with an average value of 6.4 and is generally highest in the northern and north-eastern part of the island. There is a wide range of DO from 0.4 to 10.2 mg l^{-1}, with a mean of 5.0 mg l^{-1}.

The significance of the sampled distribution of major ions (in the form of SEC), chloride, dissolved oxygen and pH (Fig. 5) is limited by the number of sample points, but they are compatible with the groundwater flow paths described above with actively recharged and least mineralized waters occurring beneath the higher ground in the south of the island. Very few Guernsey groundwaters are depleted of

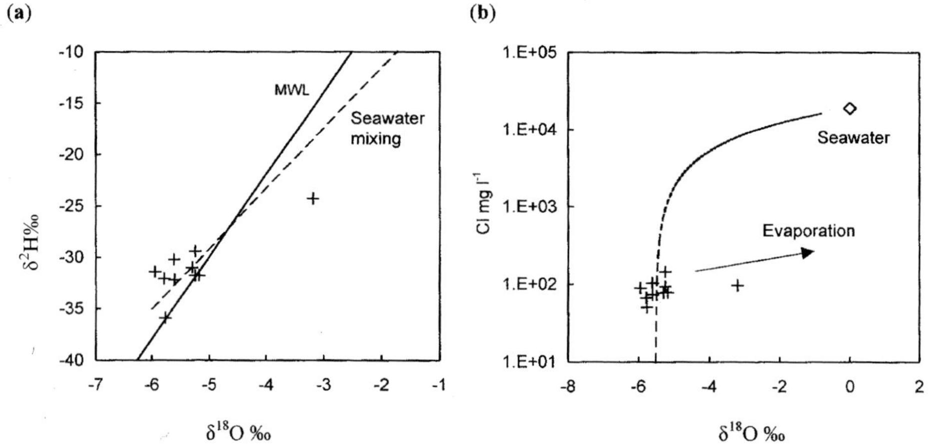

Fig. 7. (**a**) δ-plot with world meteoric water line (MWL) and seawater mixing line, (**b**) δ^{18}O-Cl plot with seawater mixing line.

Table 4. *CFC concentrations*

Sample number and grid reference	Sampling temperature (°C)	Dissolved oxygen (mg l^{-1})	CFC-11 (pmol l^{-1})	CFC-12 (pmol l^{-1})	CFC-113 (pmol l^{-1})
1 Airport [339812]	12.4	3.5	15 ± 2[1]	8.6 ± 0.9[1]	0.50 ± 0.1[1]
2 Airport [317762]	12.8	3.1	8.4 ± 0.9	8.3 ± 0.9	0.40 ± 0.1
3 North-east Guernsey [293786]	13.6	2.1	2.9 ± 0.2	1.5 ± 0.1	0.21 ± 0.1
4 East Guernsey [342813]	11.2	10.2	8.4 ± 0.9	2.9 ± 0.2	0.55 ± 0.1

[1]Data are mean values ± 1 standard error.

oxygen. There is a marine influence reflected in concentrations of Na and Cl in all the groundwaters which derives principally from the local rainwater chemistry and sea spray. Cl concentrations range from 50.3 to 178.0 mg l^{-1} with a mean of 92.0 mg l^{-1}, the higher concentrations tending to be clustered in the low-lying northern part of the island where saline intrusion may occur from time to time.

A histogram of nitrate occurrence (Fig. 6) indicates a polluted groundwater body in which ten out of the 21 samples exceeded the EU MAC. The mean for the sample set is 14.1 mg l^{-1} NO_3-N. The likely nitrogen source is nitrogenous fertilizer leached from the soil horizon, and leakage from cesspits. The higher NO_3-N concentrations do not generally coincide with higher Cl concentrations (which may in any case be of marine influence). This suggests a parallel with findings in Jersey that indicated that sewage is unlikely to be the main source of NO_3-N (Green *et al.* 1998).

Figure 7a shows the stable isotope data for Guernsey groundwaters in relation to the meteoric water line (MWL) of Craig (1961) and a freshwater–seawater mixing line. While most of the samples are quite tightly clustered, there is evidence of a slight evolutionary trend probably due to the presence of small quantities of sea water; the δ^{18}O-Cl plot in Figure 7b supports this theory with most samples falling on or near the freshwater-seawater mixing line. One sample, from the north-eastern tip of the island, appears to contain a proportion of water which has undergone evaporation during surface storage in a lake or reservoir.

CFC results are given in Table 4. Samples 1 and 2 were collected in the south of the island, while 3 and 4 were collected in the north-east and east respectively. CFCs have built-up in the atmosphere at known rates since the 1940s, reaching a peak in the mid-1990s (Oster *et al.* 1996). Given that the maximum modern air-equilibrated concentrations for recharge at an annual average recharge temperature of 11°C are approximately 5.3 (CFC-11), 2.8 (CFC-12) and 0.5 (CFC-113) pmol l^{-1}, it is apparent that three of the sites show at least some evidence of contamination. This may come from a variety of sources and may not affect each species equally. While such contamination raises groundwater CFC concentrations, they can be lowered by degradation under low-oxygen conditions (Oster *et al.* 1996). Generally, this is not serious above concentrations of about 0.5 mg l^{-1} O_2, so none of the sites appears likely to have suffered from this in their overall CFC balance.

The two most contaminated sites (samples 1 and 2) are adjacent to the island airport where solvents are in use, but the species most likely to accompany such pollution (CFC-113) is actually present at natural concentrations which indicates recent recharge. Water from a site in the north-east of the island (sample 3) has relatively consistent CFC concentrations suggesting a time since recharge of 15–20 years. At the most easterly site (sample 4) there is contamination by CFC-11, but the remaining species CFC-12 and CFC-113 are present within measurement error of their mid-90s peaks, indicating very recent recharge.

Conclusions

Historically, the available surface and groundwater resource of Guernsey has sustained the island community without significant problem. The future sustainability of the resource is not guaranteed because: rainfall has been declining locally; demand will increase to overtake the 1970 peak and will place the resource under renewed stress; and intensive use of nitrogenous fertilizer has raised the concentration of nitrate in water supplies to unacceptable levels.

Much of the groundwater in Guernsey is modern and actively recharged. It discharges as baseflow to streams to sustain low flows. A total of 6.5 Mm^3 of surface and groundwater is used on Guernsey per annum, a volume equivalent to a depth of 103 mm over the whole area of the island. The long-term average values for runoff (90 mm) and groundwater recharge (128 mm) suggest that there is normally considerable surplus over demand. However, during periods of successive dry winters when little or no recharge occurs (e.g. 1991–1992, see Fig. 3) there is insufficient groundwater storage to sustain baseflow and surface water storage is depleted rapidly. The incidence of dry periods may be increasing as the long-term average rainfall over the island declines.

There is also concern regarding water quality. The average concentration of nitrate in 21 samples of groundwater collected in November 1999 was 14.1 mg-N l^{-1} compared with the EU MAC of 11.3 mg-N l^{-1}. More rigorous control of land use and resource protection will be essential for the future sustainability of the resource. Control of horticultural leachate from hydroponic culture has already been tackled successfully, but additional controls on the use of nutrients will ultimately be required to reduce nitrate concentrations in the shallow groundwaters.

The authors are grateful to the States of Guernsey Water Board who commissioned this study. The tolerance of the people of Guernsey towards our questions, and their willingness to share information is also gratefully acknowledged. In particular we thank Nigel Jee for permission to reproduce hydrograph data, Dave Jehan at Stan Brouard Limited and the staff of the Guernsey Airport Meteorological Office. The authors are also grateful to the Chief Executive, Public Services Department, St Helier, for permission to use the Jersey recharge model. The analytical chemistry was done at Wallingford; the CFC analyses were by Spurenstofflabor, in Wachenheim, Germany. The paper is published by permission of the Director, British Geological Survey (NERC).

References

Collonette, A. 1916. The Pleistocene period in Guernsey. *Report and Transactions Guernsey Society of Natural Science and Local Research*, **7**, 337–408.

Craig, H. 1961. Isotopic variations in meteoric waters. *Science*, **133**, 1702–1703.

Green, A. R., Feast, N. A., Hiscock, K. M. & Dennis, P. F. 1998. Identification of the source and fate of nitrate contamination of the Jersey bedrock aquifer using stable nitrogen isotopes. *In*: Robins, N. S. (ed.) *Groundwater Pollution, Aquifer Recharge and Vulnerability*. Geological Society, London, Special Publications, **130**, 23-35.

Guernsey Airport Meteorological Office. 1998. Climatological Report for 1997. Report of the Guernsey Airport Meteorological Office, La Villiaze.

Hawksley. 1977. *Report on water resources in the south and west*. Report T&C Hawksley, Aldershot.

Jehan, N. D. 1993. *Groundwater in Guernsey, a geochemical pollution study*. BSc thesis, University of Leicester.

Meteorological Department. 1991. *Jersey Weather 1990*. Report States of Jersey Airports and Harbours Committee, St Helier.

OSTER, H., SONNTAY, C. & MUNNICH, K. O. 1996. Groundwater age dating with chlorofluorocarbons. *Water Resources Research*, **32**, 2989–3001.

ROACH, R. A., TOPLEY, C. G., BROWN, M., BLAND, A. M. & D'LEMOS, R. S. 1991. Outline and guide to the geology of Guernsey. *Guernsey Museum Monograph*, **3**.

ROBINS, N. S. & SMEDLEY, P. L. 1994. Hydrogeology and hydrogeochemistry of a small, hard-rock island – the heavily stressed aquifer of Jersey. *Journal of Hydrology*, **163**, 249–269.

ROBINS, N. S. & SMEDLEY, P. L. 1998. *The Jersey groundwater study*. British Geological Survey Research Report **RR/98/5**.

Sustainable groundwater development in arid, high Andean basins

MARK ANDERSON[1], ROB LOW[1] & STEPHEN FOOT[2]

[1]*Water Management Consultants Limited, 2/3 Wyle Cop, Shrewsbury, Shropshire SY1 1UT, UK (e-mail: dmarkanderson@compuserve.com)*

[2]*Minera Escondida Limitada, Avda de la Mineria 501, Antofagasta, Chile*

Abstract: Mining is an important part of the economy of Chile. A high proportion of mines are located in remote, high altitude, extremely arid environments in northern Chile. The demands of the mining industry for potable and ore-processing water, along with existing longer term demands, mean that water has a high commercial value in this region. Set against this is the desire to conserve the unique flora and fauna, highlighted by the existence of a number of conservation sites of international importance. A typical case study charting the investigation of an aquifer in this region [the Monturaqui–Negrillar–Tilopozo (MNT) aquifer] and the development of a plan for groundwater use observing the requirement for sustainability is presented. The important aspects of the geology and hydrogeology of the aquifer are presented, and a description is given of the techniques used to arrive at a sustainable groundwater development strategy. The unusual characteristics of the aquifer meant that the use of spatially distributed time-variant numerical models to identify a sustainable groundwater abstraction strategy was necessary. The abstraction strategy will enable sustainable abstraction of a significant volume of groundwater from the MNT basin, taking advantage of the high storage to recharge/discharge ratio of the aquifer.

Chile is one of the most rapidly developing countries in Latin America. Revenues from mining (particularly for copper and gold) contribute between 10 and 25% of gross domestic product and 25 to 50% of foreign exchange, therefore representing an important part of the economy. Most mining takes place in northern Chile, with a high proportion of mines bordering the Andes and the Atacama Desert. Precipitation in this region is very low (*c.* 10–250 mm a^{-1}), and is primarily altitude-dependent. Surface water is scarce. Most of the easily accessible water resources are allocated to specific uses, predominantly to the agricultural sector (Karzulovic 1991). The demands of the mining industry for both potable and ore-processing water serve to increase the overall demand for water in this remote, high altitude and extremely arid environment.

Set against the high commercial demand for water is the desire to conserve the flora and fauna of the region. The combination of low moisture conditions and geographical isolation has led to the development of animal and plant communities of conservation importance.

This paper is a case study of the development of a strategy for sustainable groundwater development in a recently discovered aquifer in the region. The various techniques employed to investigate the aquifer are discussed and the key characteristics of the aquifer are described. The relative complexity of the aquifer in relation to assessment of sustainability is then explained, and criteria defining sustainability are presented. Finally, the techniques used to identify a groundwater development strategy to fulfil these criteria are explained.

The Monturaqui–Negrillar–Tilopozo aquifer

Setting

The Monturaqui–Negrillar–Tilopozo (MNT) aquifer is located at the foot of the Andes in

From: Hiscock, K. M., Rivett, M. O. & Davison, R. M. (eds) *Sustainable Groundwater Development*. Geological Society, London, Special Publications, **193**, 133–144. 0305-8719/02/ $15.00 The Geological Society of London 2002.

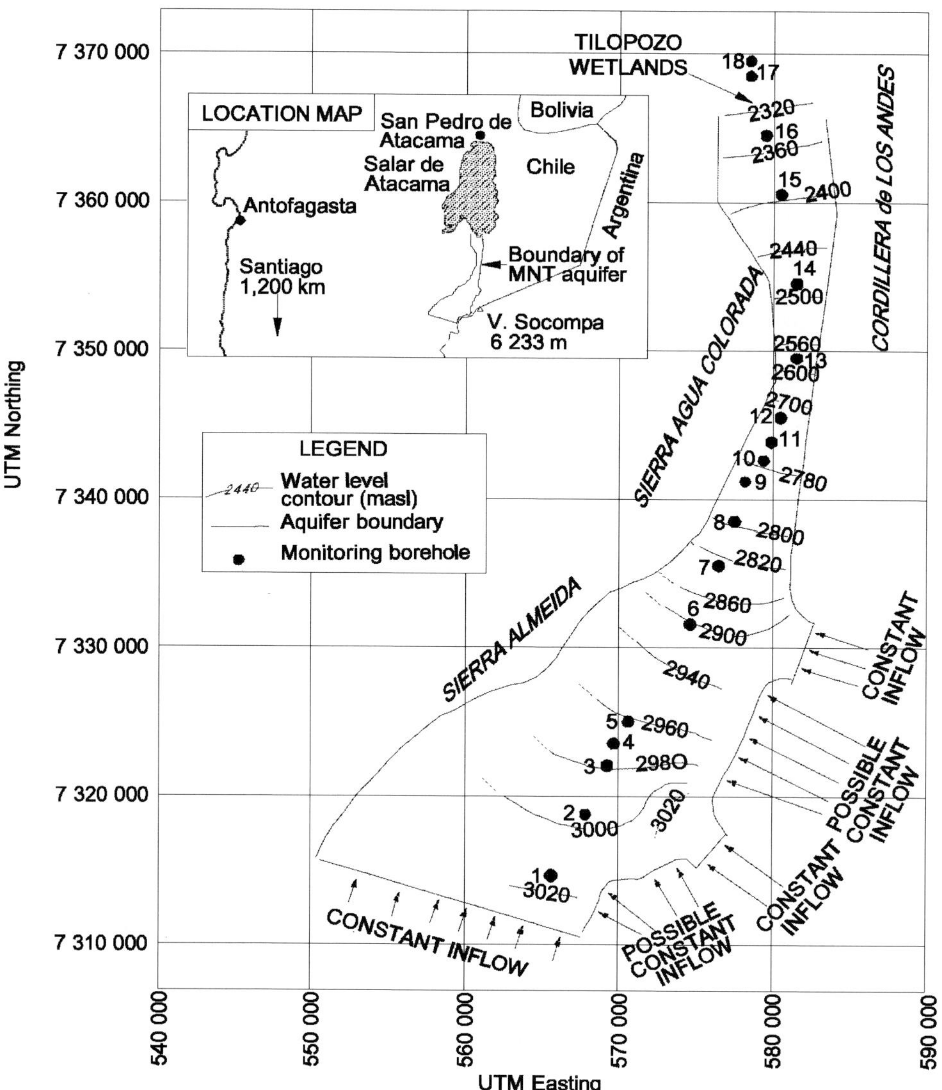

Fig. 1. Location map and plan of the MNT aquifer.

Chile's Region II. It occupies a structural trough approximately 60 km long, with surface elevations falling from around 3200 m above sea level (asl) in the southern (Monturaqui) area to 2300 m asl in the north (Tilopozo wetland), where the aquifer discharges at the south-eastern margin of the Salar de Atacama (Fig. 1). The aquifer is covered in the Monturaqui area by extensive ash and debris deposits resulting from the violent eruption and partial collapse of the Socompa volcano between 10 000 and 500 years ago (Francis *et al.* 1985). The Tilopozo wetland is developed where the aquifer discharges on the Salar margin as it flows over a saline interface, terminating in highly saline, shallow brine lagoons. Vegetation in the Tilopozo wetland varies from dry brush to lush grasslands, depending on the position of the water table in relation to the ground surface.

The Escondida mine, located 90 km SW of the

MNT aquifer, has the largest output of any copper mine in the world, and is operated by Minera Escondida Limitada (MEL). To facilitate both ongoing production and expansion of the mine, the development of new sources of water is required. MEL commenced abstraction from the Monturaqui wellfield in 1998, and the Zaldivar mine (owned by Placer Dome), located close to Escondida, commenced abstraction from a wellfield located in the Negrillar area in 1995. Total combined abstraction is proposed to be approximately 1800 l s^{-1}. The companies cooperate and share most information with each other, and also with the Direccion General de Aguas (DGA), the national authority responsible for regulation of the water environment and abstraction licensing.

Investigation history

Investigation of the Monturaqui aquifer commenced with its discovery in 1983, during a regional assessment of potential water resources for the then-planned Escondida copper mine. Early investigations focused on determining the water resource potential of the aquifer. Subsequent programmes have been designed to provide the information and physical infrastructure required for the development and maintenance of a sustainable groundwater development strategy, and to assess environmental impact. Field investigations are carried out within the constraints imposed by remoteness, the large size of the basin and difficult access. Investigation methods employed to determine the extent and hydraulic properties of the aquifer include:

(a) geological mapping combined with satellite image and structural interpretation;
(b) surface and downhole geophysical studies including ground magnetics and time-domain electromagnetic (TEM) profiling, with associated forward and inverse modelling;
(c) drilling and testing of exploration and production wells, and piezometers;
(d) core drilling and laboratory testing for porosity, permeability and mineralogy;
(e) groundwater sampling and analyses including isotope studies;
(f) evaporation studies in the vicinity of the Tilopozo discharge zone; and
(g) long-term monitoring and sampling of the aquifer and the Tilopozo lagoons.

Associated work has included a baseline environmental study to provide an inventory of flora, fauna, archaeological sites and socio-economic use of the Tilopozo area (CEAL 1994; RESCAN 1996). The scale of the field investigation can be appreciated by the fact that 50 km of TEM and over 180 km of ground magnetic geophysical investigations were completed and that over 80 boreholes have been drilled and completed.

Hydrogeology

The MNT basin is hosted by Palaeozoic basement rocks and is the combined result of numerous tectonic episodes. The most important of these resulted in normal faulting, leading to the development of a N–S oriented graben. This feature was filled in by alluvial and volcanic sediments to depths of over 400 m, mainly during the Tertiary period. The stratigraphy of the basin is summarized in Table 1.

The Tertiary sediment infill to the basin forms the MNT aquifer. Palaeozoic basement on the west, and a N–S basement high on the east, serve as gross controls on the spatial extent of the aquifer. The aquifer has a maximum width of 12 km in the Monturaqui area, narrowing to 3 km to the north of Negrillar, and widening to approximately 8 km in the Tilopozo area. The saturated thickness of the aquifer ranges from around 150 m in Monturaqui to 450 m in Negrillar. The aquifer gradually wedges out towards the Tilopozo discharge area, as it flows over the denser brines of the Salar de Atacama.

The hydrostratigraphy of the aquifer is also summarized in Table 1. The primary aquifer, the Salin Formation, consists of a sequence of interbedded silts, sands and gravels. Pumping tests show a response characteristic of a semi-confined (leaky) aquifer. Across the southern part of the Monturaqui wellfield, investigations have revealed several distinct lower permeability silty layers which can be correlated on a hole by hole basis but, on a basin-wide scale, the leaky behaviour is attributed to the semi-confining properties of discontinuous lower permeability layers.

To facilitate investigation of the vertical distribution of aquifer properties in the Monturaqui wellfield, nested piezometers were installed alongside three of the production wells. Each production well/piezometer installation was monitored during a 3 d constant rate pumping test. Given the established vertical heterogeneity of the aquifer and the complex monitoring provision, it was decided to use a numerical model to analyse results of the pumping test rather than any of the available analytical solutions. The model code used was a

Table 1. *Stratigraphy and hydrostratigraphy*

Unit (age)	Thickness (m)	Stratigraphy	Hydrostratigraphy
Alluvium, volcanic ash and debris (Recent)	0–50	Reworked sediments from other formations. Rhyolitic tuff and lava blocks from eruption of Socompa Volcano	Unsaturated, except near Tilopozo wetland
El Negrillar volcanics (Pleistocene)	0–100	Lava flows overlying significant areas of the Salin, particularly on the basin margins, and formingthe topographic boundary between the Monturaqui and Negrillar areas	Low permeability vents and dykes form only local barriers to groundwater flow
Tucúcaro ignimbrite (Pliocene)	10–50	Strongly welded white-grey ignimbrite forming part of a suite of undifferentiated volcanic rocks. Overlies the Salin in the northern part of the basin	Possible confining layer in the northern part of the aquifer where it is not fractured
Estratos de Quebrada Salin (Salin) (Miocene to Pleistocene)	100–200	Poorly consolidated clastic sediments, ignimbrite and ash deposits which unconformably overlie the Purilactis Formation and crop out where not covered by more recent volcanics. In the Monturaqui area, the Salin sediments generally consist of fine to medium grained sands with a significant ash content and local discontinuous gravel horizons. In the Negrillar area a higher fraction of coarse material and stratification is evident	Primary aquifer with large thicknesses of medium to high permeability material. Upper and lower zones are recognized in Monturaqui
Purilactis Formation (late Cretaceous/ early Tertiary)	100–300	Comprises continental sediments unconformably overlying the Palaeozoic basement, consisting predominantly of silt with ash and only occasional thin sand layers. Crops out in a limited area to the west of Monturaqui	Low permeability
Palaeozoic rocks	Basement	Metamorphosed marine sediments, volcanics and intrusives subject to all four structural phases. Host to the MNT basin.	Extremely low permeability

radial flow pumping test model (the RZ model) developed at the University of Birmingham, UK (A. Spink, pers. comm.). This model has linear and logarithmic space discretization in the vertical and horizontal (radial) directions respectively. The capabilities of the model in relation to the pumping well include representation of pumping from a specific screened interval or intervals, consideration of well losses with an associated seepage face in the pumping well and consideration of well storage. In the wider aquifer, the hydraulic response at any depth and radial distance can be generated, and the model includes an accurate representation of the mechanisms controlling the movement of the water table.

Figure 2 shows the vertical locations of the screened interval of production well MPW-19 and its associated piezometers, located above, within and below the screened interval. To simulate the pumping test response of the monitoring network, the vertical distributions of parameter values in the model were varied in accordance with downhole geophysical and lithological evidence. Figure 3 shows the observed and modelled drawdown responses. It can be seen that despite the marked variation between the observation points in their magnitude of response, the model was successfully configured to reproduce closely each of the

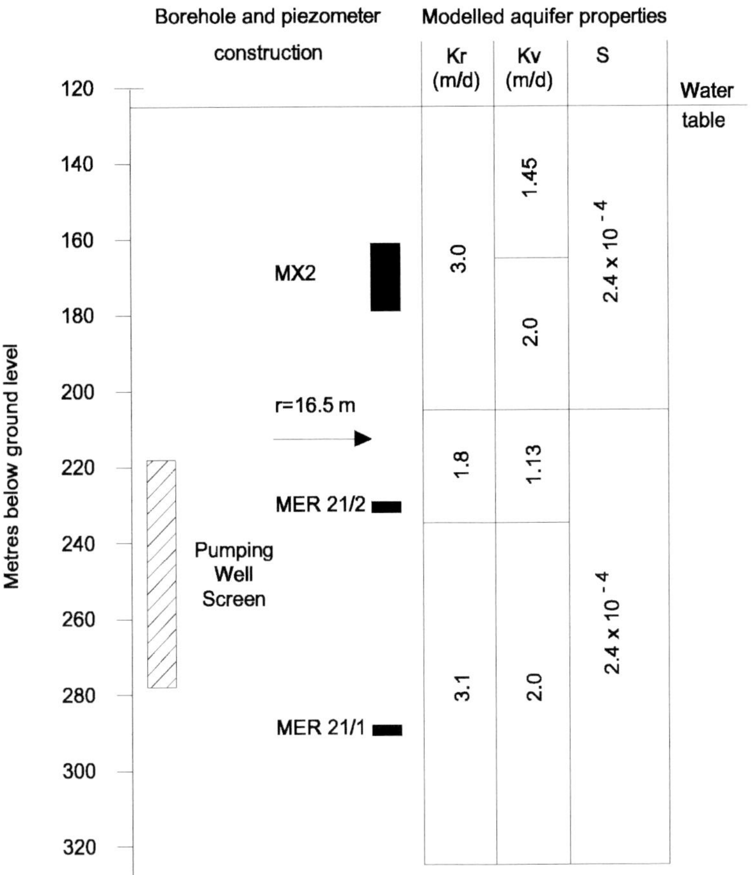

Fig. 2. RZ model construction.

drawdown profiles. Figure 2 also includes the vertical distribution of parameter values in the calibrated model. It should be noted that the model response was insensitive to the value adopted for specific yield within the pumping period, indicating that an insufficient volume was abstracted during the 3 d test to initiate lowering of the water table. The detailed nature of the interpreted aquifer property distribution compared to that available through the use of analytical solutions highlights the added value available from appropriately monitored pumping tests through interpretation using numerical models.

Consideration of all available aquifer test results shows that aquifer transmissivity values range between 500 and 2500 $m^2 d^{-1}$ in Monturaqui, and between 3000 and 4500 $m^2 d^{-1}$ downgradient in Negrillar. Limited data in the Tilopozo area indicate a transmissivity of around 1500 $m^2 d^{-1}$. Hydraulic conductivity values range between 1.6 and 10 m d^{-1} throughout the length of the basin. Confined storage values are around 1×10^{-3}, whilst specific yield has been estimated to be around 10%. An empirical determination of specific yield will only be possible after prolonged pumping.

Numerous techniques have been employed to estimate recharge rate and distribution, aquifer throughflow and aquifer discharge, and thus to develop a conceptual model of aquifer dynamics. Groundwater flow is from south to north, with hydraulic gradients ranging from 0.005 in Monturaqui to 0.05 north of Negrillar. This distribution mainly reflects the geometry of the system with, for example, a steepening of the hydraulic gradient through the constriction in the basin north of Negrillar. Simple calculations

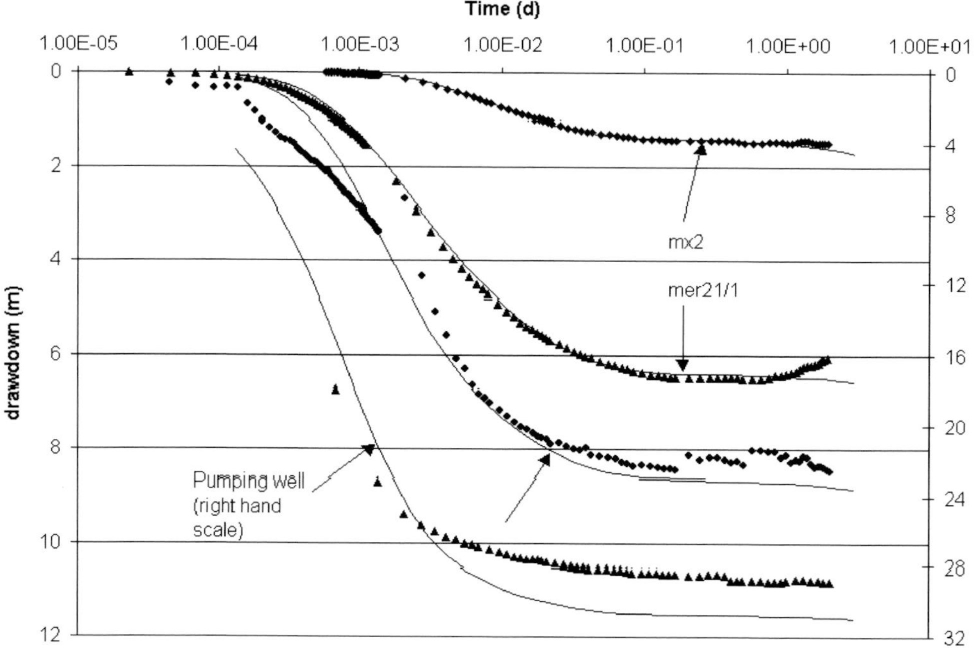

Fig. 3. Observed and modelled (RZ) responses to pumping test.

using Darcy's Law suggest a throughflow of around 400 l s^{-1} in the Monturaqui area, increasing to around 900 l s^{-1} in the Negrillar and Tilopozo areas. This implies that around 40% of inflow occurs in the Monturaqui area, around 60% between Monturaqui and Negrillar, and none to the north of Negrillar.

Some confirmation of this interpretation is offered in the distribution of groundwater chloride concentrations. Alpers and Whitmore (1990) showed that in the Punta Negra Basin, 60 km SW of the MNT basin, high altitude precipitation has very low chloride concentrations (<1 mg l^{-1}), and that after recharge groundwater chloride concentrations increase with residence time through the dissolution of pre-existing evaporite salts and weathering of Neogene volcanic rocks. Using this model for the MNT basin, the markedly lower groundwater chloride concentrations along the central eastern margin of the aquifer, between the Monturaqui and Negrillar wellfields, indicate significant recharge in this area.

Stable isotope (^{18}O and ^{2}H) concentrations were determined for samples from four boreholes in the Monturaqui wellfield. The isotopic composition of the water ranged from -7.5 to -8.0 $\delta^{18}O$ (‰, SMOW) and from -61 to -67 $\delta^{2}H$ (‰, SMOW). These data lie to the right of the Global Meteoric Water Line, and also to the right of the Regional Meteoric Water Line ($\delta^{2}H = 7.3 \times \delta^{18}O + 8$) given by Fritz *et al.* (1978). The isotopic composition of the groundwater samples was also more depleted (more negative) than others reported for this area (Margaritz *et al.* 1989; Alpers & Whitmore 1990). Under normal ranges of relative humidity, evaporation from a freshwater surface results in enrichment of $\delta^{2}H/\delta^{18}O$ with a slope of 3.5 to 6. However, Margaritz *et al.* (1989) suggest that in the high Andes, as a result of high humidities, the slope is close to that of meteoric water (i.e. 8). Extrapolation from Monturaqui groundwaters to determine source water concentrations was carried out using intermediate estimates of slope between 4.5 and 6.5. This suggested that recharge originated from source water with $\delta^{2}H$ between -100 and -140 (‰, SMOW) and $\delta^{18}O$ between -14.7 and -20 (‰, SMOW), these isotopic compositions being characteristic of rainfall and snow at elevations greater than 4000 m asl (Alpers & Whitmore 1990).

An estimate of recharge to the aquifer was made using the method recommended by the DGA (DGA 1987). This method involves empirical determination of altitude/precipitation and altitude/potential evaporation relationships using data from established local meteorological stations. The catchment is then divided into altitudinal zones, and an effective precipitation found for each (using median values). Total effective precipitation is found by multiplying the respective effective rainfall rates by the spatial area of each altitude zone, and then by summing the totals for each zone. An implicit assumption in using this method is that evaporation takes place at the calculated potential rate at all times. At high altitudes this assumption is reasonable as nearly all precipitation falls as snow which remains on the ground for some time and is therefore available for evaporation for prolonged periods. At lower altitudes, precipitation falls more often as rain which is removed more rapidly through run-off processes. This means that the method will tend to overestimate actual evaporation and therefore underestimate recharge.

Application of the DGA-recommended method produced an estimate for average recharge of approximately 250 l s^{-1}. It also suggested that even if lower altitude (below 4000 m asl) recharge is significantly underestimated by the model, the large majority of effective precipitation probably occurs only over the high mountains surrounding the MNT basin, with recharge reaching the MNT aquifer as shallow groundwater flowing through sediment-filled high altitude valleys. This is consistent with the interpretation of the isotope concentrations of groundwater from the Monturaqui wellfield.

The Tilopozo wetland is the sole natural discharge zone for the MNT aquifer. Discharge is effected by evapotranspiration from the vegetation, springs, saline lagoons and bare ground. The spatial distribution of vegetation and springs suggests that subsurface flow also enters the wetland from other adjacent aquifers. Evapotranspiration of groundwater from the MNT aquifer is thought to take place from a central area of about 38 km^2, equivalent to around half the total surface discharge area. Consideration of the area occupied by each type of ground cover, along with characteristic potential evaporation rates for each, suggests a total groundwater discharge of between 400 and 900 l s^{-1} for the MNT aquifer.

A significant discrepancy exists between the estimates of groundwater flux by recharge estimation (*c.* 250 l s^{-1}) and throughflow and discharge estimation (*c.* 400–900 l s^{-1}). A possible explanation for this is groundwater inflows from outside the topographic catchment of the MNT basin. It is known that the Salin Formation extends eastwards at high elevation, and it is possible that this might represent an extension of the groundwater catchment outside the topographic catchment. This phenomenon is common in northern Chile, where older aquifers are intruded by later volcanic rocks, forming local topographic high points. The presence of the low permeability igneous rocks reduces, but does not destroy, the continuity of the aquifer. Such a barrier exists between the Monturaqui and Negrillar areas in the form of the Negrillar volcanics. These form a topographic divide but geophysical, groundwater head and hydrochemical evidence indicates hydraulic continuity although with reduced transmissivity.

The MNT aquifer system exhibits features common to many aquifers in northern Chile, some of which are relatively unusual in a global context. Firstly, an extremely large volume of water (10^{10} m^3) is stored in the aquifer. Secondly, recharge occurs at relatively low rates through subsurface inflows at the edges of the recognized aquifer system, with very little or no areal recharge. Thirdly, the sole discharge from the aquifer is approximately 25 km from the closest point of abstraction, and approximately 50 km from the main centre of abstraction. The discharge zone forms an environmentally sensitive wetland habitat.

Sustainable groundwater development

Assessment of environmental requirements

The first phase in deciding a sustainable aquifer management strategy is the identification of groundwater-dependent (usually natural) features and assessment of the sensitivity of these features to changes in the hydrogeological environment. In all cases, the *de facto* definition of sustainability is decided in a wide socio-economic context, and the resulting definition is employed in deciding policy towards development. In the case of groundwater, the question of whether any derrogation of groundwater features is permissible and, if so, what degree of derrogation will be tolerated, will define the policy. Finally, for each development, the policy is realised through the identification of one or a number of criteria which must be met if the development is to be deemed sustainable.

For the MNT basin, the Tilopozo wetland is the only groundwater-dependent natural feature. The groundwater requirement of the wetland was assessed during a study into the sensitivity of the natural community to changes in water level (CEAL 1994). This suggested that a decline in water level of approximately 0.25 m would have no measurable impact on flora and fauna, and this was adopted by MEL as a maximum tolerable (sustainable) impact at Tilopozo (RESCAN 1996) and accepted by COREMA (Region II, Chile), the government authority responsible for evaluating and approving environmental impact assessments.

The Tilopozo wetland exhibits complex hydrodynamics, with brine from the Salar de Atacama (TDS = 330 000 mg l^{-1}, density = 1217 kg m^{-3}) underlying the inflowing brackish (TDS = 2000–6000 mg l^{-1}) groundwater. A sharp interface exists between these two fluid bodies, the slope of which is related to the rate of inflow of groundwater, the contrast in density, and the rate and distribution of evaporation across the discharge zone. A spreadsheet-based analytical model was constructed to represent the wetland. It was based on the Glover equation for a horizontal discharge zone, and the Ghyben–Herzberg relationship for a saline interface condition (Badon Ghyben 1888; Herzberg 1961). The model was used to translate the maximum tolerable water level reduction into a maximum tolerable groundwater flow reduction at Tilopozo. A reduction in water levels of 0.25 m at Tilopozo was found to correspond to a reduction in aquifer outflow of approximately 6%. This result was very sensitive to the function used for the variation of evaporation rate with depth. Work is ongoing to refine the model both from an analytical viewpoint, and also to improve field measurement of evaporation rate with depth.

Development of a strategy for sustainable groundwater use

Summarizing the effects of artificial pumping on a natural groundwater system, Bredehoft *et al.* (1982) state that 'water withdrawn artificially from an aquifer is derived from a decrease in storage in the aquifer, a reduction in the previous discharge from the aquifer, an increase in the recharge, or a combination of these changes. The decrease in discharge plus the increase in recharge is termed *capture*'. Applying this simple conceptual understanding to the MNT aquifer, water abstracted initially from the aquifer will be derived from a reduction in storage. Since additional recharge cannot be induced, an increasing proportion of abstracted water will be captured over time from the sole discharge from the system at Tilopozo. To meet the agreed criteria defining sustainability, an abstraction strategy needed to be found which limited capture of discharge such that the net effect of abstraction from the two wellfields was a reduction in groundwater flow at Tilopozo of less than 6%.

The MNT aquifer and the established wellfields have complex dynamics in relation to capture of discharge from Tilopozo. Firstly, the aquifer has two main recharge zones and two centres of abstraction. One of the centres of abstraction is downgradient of only one of the recharge zones whilst the other is downgradient of them both. Secondly, the combined cost-effective rate of abstraction from the two wellfields (1800 l s^{-1}) is at least twice the natural discharge from Tilopozo, meaning that indefinite abstraction would capture most, if not all, of the natural discharge from the system. Given these complexities, it was accepted that a three-dimensional time-variant groundwater flow model was required to explore whether a groundwater abstraction strategy could be found which would meet the criteria defining sustainability.

Numerical modelling

The groundwater flow model was developed using the MODFLOW code (McDonald & Harbaugh 1988). The model extended from the southern limit of the Monturaqui basin north to Tilopozo, and used a 1000 m resolution grid with two layers. Figure 4 shows the domain and boundary conditions of the model. Constant inflows on the southern and eastern boundaries were used to represent system recharge, and the full range of possible total inflows was simulated. The model was refined within the bounds established in the conceptual model so that it reproduced as closely as possible the observed 'steady-state' (pre-abstraction) head distribution in the aquifer. Figure 5 shows the observed and modelled pre-abstraction head distribution along the line of monitoring boreholes shown in Figure 1.

A time-variant model of up to 500 years' duration was developed for predictive modelling of impacts. Groundwater abstraction from both the Monturaqui and Negrillar wellfields, at cost-effective rates in both cases, was represented in the model. The model was used to predict outflow changes with time across the northern boundary of the model (representing the Tilopozo wetland) in response to a range of abstraction durations. Significant uncertainty in the value of some aquifer parameters was addressed at all stages of modelling through sensitivity analyses, with the uniform value for specific yield varied between 5 and 20% and the recharge rate varied between 450 and 1800 l s^{-1}.

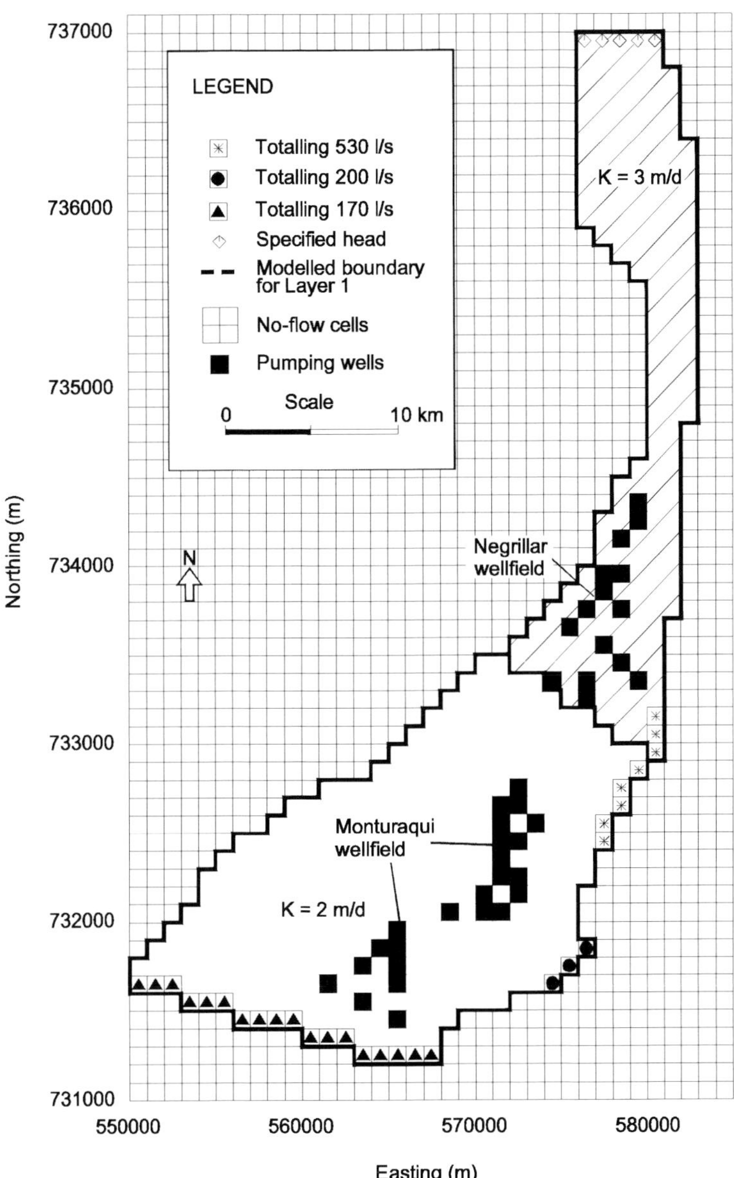

Fig. 4. Domain and boundary conditions of the groundwater flow model.

Analysis of model output revealed a number of interesting system features. Firstly, any abstraction causes a decline in groundwater flow at Tilopozo, i.e. there is no zero impact scenario. Secondly, the impact of abstraction will reach Tilopozo between 20 and 40 years after the start of pumping from either wellfield, with maximum impacts likely to occur 75 to 300 years after abstraction ceases (*c.* 2090–2315). Thirdly, the magnitude of the predicted impact is sensitive to the value used in the model for specific yield, and to the gross volume of groundwater abstracted. Fourthly, the magnitude of the predicted impact is not sensitive to the recharge rate within the

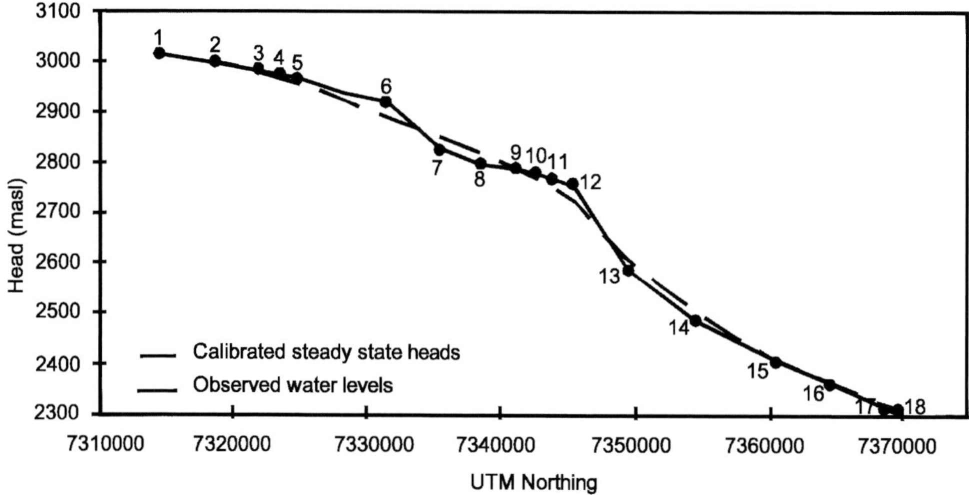

Fig. 5. Observed v. modelled pre-abstraction head distribution.

tested range, although the predicted arrival time of impacts at Tilopozo is sensitive to the recharge rate, with higher rates causing impacts to arrive more quickly. Finally, the location of the Monturaqui wellfield upgradient of approximately 60% of aquifer recharge significantly reduces its relative impact on groundwater flows at Tilopozo.

Interpreting the results of sensitivity analyses, a 'worst case' prediction model was developed which maximized the magnitude of predicted impacts within the established ranges for input parameter values. This model was run numerous times, each time varying the pumping duration (applying to both wellfields) between 10 and 20 years. An iterative approach was used to find a pumping duration, and gross abstracted volume (Fig. 6; 7.38×10^8 m^3), which limited throughflow reduction at Tilopozo to 6% (the maximum tolerable reduction). The resulting model forms the basis for the sustainable groundwater development strategy.

Predicted impacts are moderately sensitive to the inputs to the model which remain uncertain, for example the values adopted for specific yield and the detail of downstream aquifer geometry. The possibility therefore remains that the model could underestimate the critical downstream impacts of groundwater abstraction. With this in mind, a programme of field investigations and model review, concentrated during the early stages of pumping, has been agreed with the DGA. This includes further refinement of a time-variant version of the flow model to the water level changes observed during initial pumping. Numerical modelling of the hydrodynamics of the Tilopozo wetland is also planned. These activities will provide ongoing re-assessment of predicted impacts, and will facilitate revision of abstraction plans if necessary.

The possibility that climate change could influence the sustainability of the chosen strategy was also recognized. Future work will include integration of appropriate outputs from global circulation models under the various emissions scenarios identified by the Intergovernmental Panel on Climatic Change (Younger *et al.* 1997).

Discussion and conclusions

A wide range of hydrogeological investigation techniques has been used to characterize the MNT aquifer, including stable isotope analysis, numerical modelling of a single well pumping tests and geophysics. The conceptual model of the MNT aquifer, which is based on the results of these investigations, shows that the aquifer exhibits some relatively unusual features. Probably the most important of these features in relation to sustainable development of the groundwater is that the aquifer contains an extremely large volume of groundwater (*c.* 10^{10} m^3) compared to the estimated rate of aquifer throughflow (250–900 l s^{-1} *c.* 10^9–10^5 m^3 d^{-1}).

The combined cost-effective abstraction rate

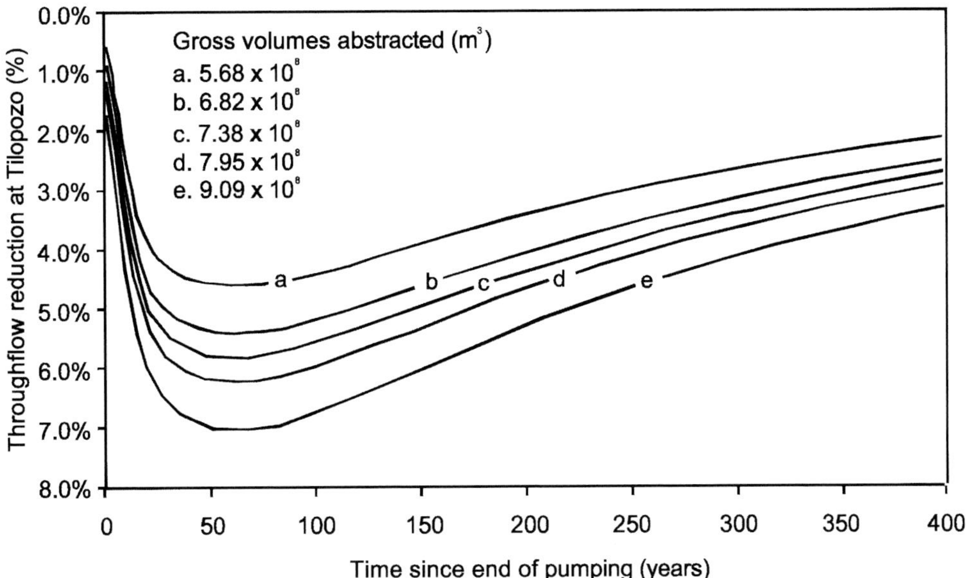

Fig. 6. Modelled groundwater throughflow reductions at the Tilopozo wetland.

from the two wellfields established in the MNT aquifer is at least twice the estimated aquifer throughflow. Judged using the classical notion of 'safe aquifer yield', which equates the quantity of groundwater available for development with the long-term natural (pre-development) recharge rate to the aquifer, the proposed development would clearly be unsustainable. Instead, the sustainability of the proposed development has been explored using a spatially-distributed time-variant numerical model of the aquifer. This has enabled a groundwater abstraction strategy to be developed which is predicted to fulfil the agreed criteria defining sustainability. It has also demonstrated that the magnitude of predicted impacts is not sensitive to the recharge rate to the system within the range of conditions explored.

An interesting feature of the model-derived sustainable groundwater abstraction strategy is that abstraction of a very large volume of water in absolute terms (7.4×10^8 m^3) will be sustainable as defined by the agreed criteria. The fact that this is possible is related simply to the extremely large volume of storage in the aquifer.

Another interesting feature of the proposed development strategy is that only a small percentage (*c.* 5%) of the total storage in the aquifer (and a slightly higher but unknown percentage of the extractable storage) can be abstracted if the sustainability criteria are to be fulfilled. Given the high commercial value of water in the region, the high implied value of conservation, and therefore the high priority afforded to it, is clear.

We thank Minera Escondida Limitada and BHP for assisting in these investigations and allowing publication of this information. We also acknowledge the work on isotope interpretation carried out by A. L. Herczeg and F. W. J. Leaney of CSIRO Division of Water Resources, South Australia, and the collaboration of Andrew Spink of the University of Birmingham, UK, in relation to radial flow pump test modelling.

REFERENCES

Alpers, C. N. & Whitmore, D. O. 1990. Hydrogeochemistry and stable isotopes of ground and surface waters from two adjacent closed basins, Atacama Desert, northern Chile. *Applied Geochemistry*, **5**, 719–734

Badon Ghyben, W. 1888. Nota in verband met de voorgenomen putboring nabij Amsterdam. *Tijdschrift van het koninklyk Instituut van Ingenieurs, 1888*. The Hague, The Netherlands, 8–22.

Bredehoft, J. D., Papadopulos, S. S. & Cooper, H. H. 1982. Groundwater: the water budget myth. *In*: *Studies in Geophysics: Scientific Basis of Water*

Resource Management. National Academy Press, Washington DC, 51-57.

CEAL. 1994. *Linea base ambiental del sector sur del Salar de Atacama.*

Direccion General de Aguas. 1987. *Balance Hidrico de Chile*. Ministerio de Obras Publicas, Chile.

Francis, P. W., Gardeweg, M., Ramirez, C. F. & Rothery, D. A. 1985. Catastrophic debris avalanche deposit of Socompa volcano, northern Chile. *Geology*. **13**, 600–603.

Fritz, P., Silva, C., Suzuki, O. & Salati, E. 1978. Isotope hydrology in northern Chile. *In*: *Proceedings of the Symposium on Isotope Hydrology*. IAEA, Vienna. **2**, 525–543.

Herzberg, A. 1961. Die wasserversorgung einiger Nordseebader, Munich. *Journal Gasbeleuchtung und Wasserversorgung*, **44**, 815–819.

Karzulovic, J. 1991. Analisis y planificacion de los recursos de agua en la segunda region de Antofagasta. ESSAN SA.

McDonald, M. G. & Harbaugh, A. W. 1988. *A modular three-dimensional finite-difference groundwater flow model*. Tech. Water Resources Inventory. Book 6, Chapter A1, US Geological Survey, Washington.

Margaritz, M., Aravena, R., Pena, H., Suzuki, O. & Grilli, A. 1990. Source of groundwater in the deserts of northern Chile: Evidence of deep circulation of groundwater from the Andes. *Groundwater*, **28**, 513–517.

RESCAN. 1996. Environmental Impact Assessment for MEL – '*Lixiviación de Oxidos de Cobre y Aumento de la capaciadad de Tratamiento de Mineral Sulfurado*'.

Younger, P. L., Teutsch, G., Custodio, E., Elliot, T., Sauter, M., Manzano, M., Liedl, R., Clemens, T., Huckinghaus D., Tore C. S., Lamban, J. & Cardosa da Silva G. 1997. Groundwater resources and climate change effects – GRACE. Final Report, January 1997. *EC Framework III, DG XII (Environment and Climate), Contract* CEC EV5V-CT94-0471.

Constraints on sustainable development of arsenic-bearing aquifers in southern Bangladesh. Part 1: A conceptual model of arsenic in the aquifer

W. G. BURGESS[1], M. BURREN[1], J. PERRIN[1,3] & K. M. AHMED[2]

[1]*Department of Geological Sciences, University College London, Gower Street, London WC1E 6BT, UK (e-mail: william.burgess@ucl.ac.uk)*

[2]*Department of Geology, Dhaka University, Dhaka 1000, Bangladesh*

[3]*Present address: Centre for Hydrogeology, Neuchâtel University, 11 rue E-Argand, Neuchâtel 2000, Switzerland*

Abstract: Arsenic is widespread in groundwater of the Holocene alluvial aquifers in southern Bangladesh, yet its concentration is highly variable spatially and with depth. A conceptual model of arsenic in the aquifer is proposed, as a basis for addressing questions concerning sustainability of groundwater development. Patterns and profiles of arsenic distribution in the aquifer have been determined at Meherpur in western Bangladesh, over an area of 15 km^2 and a depth range of 15–225 m. The hydrochemical and hydraulic environments of arsenic occurrence have been established. The conceptual model incorporates the conditions of arsenic release to groundwater, the depth distribution of the arsenic source, likely sedimentological controls on the lateral discontinuity of the arsenic source, and the hydraulic regime imposed by pumping from the hydrogeologically leaky, multi-layered aquifer. Reducing conditions, conducive to arsenic release from sedimentary iron oxyhydroxides, are widespread. The arsenic source occurs at a distinct horizon at a depth of about 20 m, but is laterally discontinuous. The catchments of shallow, hand-pumped tubewells (HTWs) are limited in extent by vertical leakage. Arsenic concentration in water pumped from tubewells depends on the depth separation between the HTW screen and the arsenic source, the overlap between the HTW catchment and the arsenic source layer, and the duration of pumping. Implications are drawn for treatment, tubewell location and design, monitoring, and predictive modelling.

Naturally-occurring high concentrations of arsenic in groundwater are widespread in the Holocene sediments of the Bengal Basin. The objective of this paper is to develop a conceptual model of the hydrochemical context and distribution of arsenic in the aquifer, and the movement of arsenic to tubewells, and to apply the model to questions concerning the sustainability of groundwater development. A companion paper describes numerical models of arsenic concentration in tubewell discharge (Cuthbert *et al.* 2002).

Initial discovery of arsenic in groundwater of the region was made in the alluvial aquifers of West Bengal, India, in the 1980s (Das *et al.* 1994). Arsenic was first detected in groundwater in Bangladesh in 1993, when analysis was prompted by the high incidence of arsenic-related illnesses. Subsequent studies have demonstrated the extensive occurrence of arsenic in groundwater in Bangladesh (Dhar *et al.* 1997; British Geological Survey & Mott MacDonald 1999, and other references therein), at concentrations many times greater than the national regulatory limit for arsenic in drinking water of 50 $\mu g\ l^{-1}$ (Government of Bangladesh 1991) and the internationally recommended guideline for arsenic in drinking water of 10 $\mu g\ l^{-1}$ (WHO

From: Hiscock, K. M., Rivett, M. O. & Davison, R. M. (eds) *Sustainable Groundwater Development*. Geological Society, London, Special Publications, **193**, 145–163. 0305-8719/02/ $15.00 The Geological Society of London 2002.

1994). These regional surveys have demonstrated that aquifers of the Ganges, Brahmaputra and Meghna floodplains are all affected in parts. Groundwater with arsenic in excess of 50 μg l^{-1} is now known to be present over most of the south and south-east of the country. It is estimated that up to 21 million people may be at risk of arsenic poisoning (British Geological Survey & Mott MacDonald 1999).

There have been calls for groundwater to be abandoned as a source of drinking water in the affected areas. Yet the alluvial aquifers provide water for 95% of the population and are also widely used for irrigation. The groundwater is bacteriologically safe, and its increased use since the late 1970s, promoted by international aid agencies, has saved many hundreds of thousands of lives that would otherwise have been lost to water-borne diseases resulting from the use of contaminated surface water sources. Moreover, within the extensive region affected by arsenic in groundwater, considerable variability in arsenic concentration is observed. There are many examples of grossly affected tubewells, pumping groundwater with arsenic at concentrations greater than 1000 μg l^{-1}, being separated by only a few tens of metres from tubewells with arsenic at concentrations less than 10 μg l^{-1} (e.g. Safiullah 1997). This extreme spatial variability is a significant feature of the occurrence of arsenic in groundwater in Bangladesh. It is estimated that across southern Bangladesh, more than 50% of tubewells in the shallow aquifer have arsenic levels that comply with the 50 μg l^{-1} limit (British Geological Survey & Mott MacDonald 1999). Furthermore, it is the uppermost 100 m of sediments that are the most affected; the available data indicate that only 1% of tubewells deeper than 200 m have arsenic above 50 μg l^{-1}.

Sustainable development of the alluvial aquifers may therefore still be possible, and if their safe use is ultimately limited, it is important to consider for how long this readily available and cheap source of water could provide an adequate supply for many millions of people. The key questions concern the mechanisms of arsenic mobilization in groundwater, the distribution of the arsenic source, the patterns of groundwater flow to tubewells, and the processes of water-rock interaction which govern arsenic behaviour on its path towards the tubewells. Only when these questions are resolved can implications for tubewell design and aquifer management be drawn, and the prospects for sustainability with regard to arsenic content be judged. Ultimately, how is arsenic concentration in groundwater pumped from tubewells likely to change with time? No significant time-series monitoring data are available, yet this is a vital aspect in designing a strategy for sustainable groundwater development in Bangladesh.

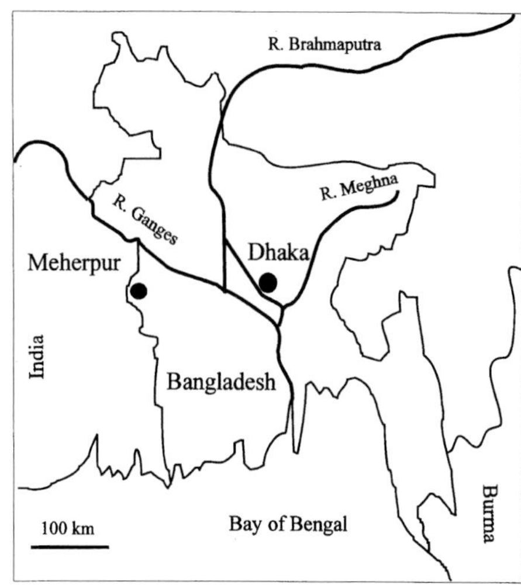

Fig. 1. Location of Meherpur in western Bangladesh.

Field hydrochemical relations observed at a regional scale, supported by limited chemical analysis of sediments, suggest that reductive dissolution of arsenic-bearing iron oxyhydroxides within the sediments releases arsenic to the groundwater (Nickson *et al.* 1998; Nickson *et al.* 2000). However, reducing conditions are widespread in the Holocene alluvial aquifer throughout Bangladesh (Davies 1995), in contrast to the variability of arsenic concentration in groundwater. Patterns of arsenic distribution, in groundwater and the aquifer sediments, are unknown at a level of detail sufficient to address questions of the security of continued groundwater use, and to inform the debate over sustainability.

To support the development of a conceptual model of arsenic in the aquifer, this paper describes the hydrochemical and hydrogeological contexts, and the spatial and depth distributions of arsenic at Meherpur in western Bangladesh (Fig. 1), over an area of 15 km^2 lying within the grossly affected region. Groundwater has been collected from the discharge of tubewells at a spatial frequency of approximately five samples per square kilometre, a scale nearly two orders of magnitude more detailed than previously described for the Bengal Basin. Samples from tubewells ranging in depth from

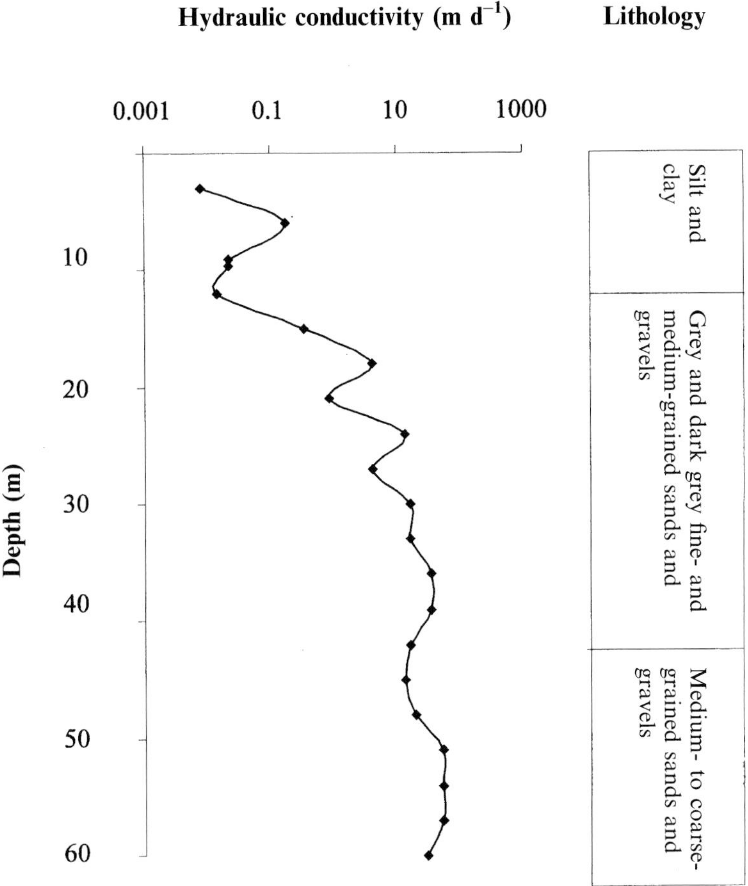

Fig. 2. A simplified lithological log and depth profile of hydraulic conductivity from Ujjalpur, Meherpur district.

15 to 225 m at Meherpur allow a crude but broadly representative description of the distribution of arsenic to these depths. Analysis is also presented of porewater and sediment from cores collected from a borehole drilled to a depth of 60 m close to Meherpur, giving a more precise yet site-specific illustration of the depth distribution of arsenic in the aquifer.

The conceptual model provides a basis for addressing important questions of tubewell location, depth, pumping regime, monitoring and treatment. It enables development of numerical models of arsenic transport in the aquifer. The numerical models, described in a companion paper (Cuthbert *et al.* 2002), simulate arsenic concentration in tubewell discharge, allowing a quantitative assessment of questions concerning the sustainability of groundwater development in southern Bangladesh.

The geology and hydrogeology of the alluvial aquifers at Meherpur

Geology

The Late Pleistocene–Holocene sedimentary sequence of Bangladesh represents a dynamic interaction between sedimentation from glacial meltwaters during regional deglaciation of the Himalayan region subsequent to the last glacial maximum at about 18 000 years BP, when sea level was about 130 m below its present level, and low energy environments of deposition represented by fine-grained estuarine sediments deposited as the sea level rose in response to the contraction of the polar ice-caps (Umitsu 1993). Within the Dhamrai Formation of Late Quaternary age (Davies 1995), fine sands and silts of the alluvial and deltaic floodplains are cut through by an array of palaeo-channels filled with coarse sand and gravel. The principal mineralogical components, determined in a detailed study at Sampta village in Jessore province (N. Shibasaki pers. comm.) are quartz, plagioclase feldspar, potassium feldspars, micas (muscovite, biotite, chlorite), and clays (smectite, kaolinite, illite). Organic matter is present at up to 6% by weight and iron oxyhydroxides occur as grain coatings and fine particulate matter. Pyrite is rare; where observed it is framboidal and apparently authigenic. On a broad scale, dark grey highly micaceous coarse sands and gravels, concentrated at the base of the sequence, fine upwards into fine sands and very fine sands. In the upper part of the sequence, silts and clays predominate, representing conditions of the mid- to late-Holocene, since which time the area has been marshy and poorly drained. The uppermost clays are red-brown in colour, reflecting weathering in an aerobic environment.

Meherpur lies at an elevation of 10–15 m above the Bangladesh Public Works Datum (PWD), 80 km south of the present course of the River Ganges in western Bangladesh, on a currently inactive delta of the River Bhairab, a distributory of the Ganges, in the Bengal Basin (Figs 1 and 4). Clays, silts, sands and occasional coarse sands of the Late Pleistocene and Holocene alluvial and deltaic floodplain sediments deposited since the last glacial maximum achieve a thickness of greater than 330 m in the region. No age determinations have been made, but on sedimentological grounds the sediments below about 100 m depth are ascribed to the Pleistocene, and those above 100 m to the Holocene (C. Bristow pers. comm.). The grey colouration throughout the sequence, apart from the uppermost clays, suggests a pervasive reducing environment.

A partially-cored exploration borehole drilled to 60 m depth at Ujjalpur village 5 km N of Meherpur in 1998 by the Bangladesh Water Development Board (BWDB) penetrated sediments reflecting a predominantly high-energy river channel environment of deposition (Fig. 2). Dark grey and grey, medium to fine-grained sands between depths of 12 and 42 m are channel-fill deposits; coarse to medium-coarse sands and gravels from 42–60 m possibly indicate a braided river-channel environment. The sequence fines upwards. Silts and clays in the uppermost 12 m represent low-energy abandoned channels. Yellow/orange silty, bioturbated clays in the uppermost 3 m demonstrate the effects of weathering on recent floodplain sediments in this inactive delta region.

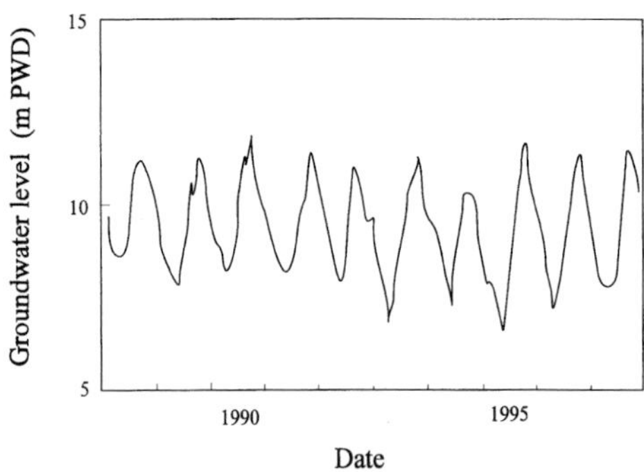

Fig. 3. A borehole water level hydrograph, Meherpur district.

Hydrogeology and groundwater conditions

The Holocene alluvial sediments in Bangladesh form a highly transmissive multi-layered aquifer (Herbert *et al.* 1989). Transmissivity is typically in the range 1000–5000 $m^2\ d^{-1}$, and the aquifers exhibit leaky behaviour. As a result of leakage, equilibrium hydraulic conditions are reached within a few hours of pumping and the radius of influence of pumping tubewells is restricted in relation to yield. For deep production wells yielding up to 3600 $m^3\ d^{-1}$, the radius of influence is typically a few hundred metres; for irrigation wells yielding 240 $m^3\ d^{-1}$ the radius of influence is typically some tens of metres; for hand-pumped tubewells (HTWs) the radius of influence is a few metres only.

Grain size analysis of sediments from the Ujjalpur core can be interpreted as a profile of hydraulic conductivity (Fig. 2). Hydraulic conductivity throughout the sequence spans four orders of magnitude, and in parts it may vary by more than two orders of magnitude between adjacent layers.

The water table is maintained within a few metres of the ground surface by direct rainfall recharge and river flooding resulting from the abundant monsoon rains in June–October each year. Meherpur is in a relatively dry part of Bangladesh; average annual rainfall, concentrated during the monsoon period, is less than 1500 mm. Annual fluctuation of the water table reflects in part the severity of the monsoon, and has ranged from 0.6 m (1975) to almost 6 m (1992) since monitoring began in the Meherpur district. The annual fluctuation may now be significantly influenced by seasonal groundwater abstraction. There is some indication that minimum water levels are currently 2–3 m lower

Table 1. *Modes of groundwater abstraction, Meherpur district*

Well type	Depth (m)	Screen length (m)	Pump	Yield (m^3 d^{-1})
HTW	15–75	3–7	Hand pump	<10
SIW DIW	50–150	15–30 (may be multiple)	Turbine	<250 (seasonal)
DTW (PWS)	100–300	<30	Turbine, electric submersible	<3600

HTW, Hand-pumped tubewell; IW, irrigation tubewell, deep (DIW) or shallow (SIW); DTW, deep tubewell for greater abstractions, e.g. for public water supply (PWS).

than under pre-irrigation conditions (Fig. 3); however there is no overall depletion of the water table, which recovers fully during each monsoon season. The seasonally unsaturated zone is 6 m thick at maximum. Groundwater flow is from north to south across the Meherpur region; the regional hydraulic gradient is approximately 10^{-4}, reflecting the subdued topography and limiting lateral groundwater flow. Differences in elevation between adjacent flooding levels determine a zone of active groundwater circulation, which may be only 5–10 m thick. Below this, groundwater in the Holocene sediments is almost stagnant, a factor that has assisted in the development and maintenance of chemically reducing conditions.

Groundwater hydraulic gradients are likely to have been very low since deposition of the Holocene sediments, and the aquifer would not have undergone the active flushing experienced by the older Pleistocene and Plio-Pleistocene sediments. Also, reactive minerals are abundant. These factors combine to determine the common chemical character of groundwater in the Holocene aquifer. Reducing conditions are pervasive, resulting in a high dissolved Fe content, a moderately high electrical conductivity (EC) of 250–675 $\mu S\ cm^{-1}$, and dominance of a calcium/(magnesium)-bicarbonate type groundwater (Davies 1995; Nickson *et al.* 2000).

Groundwater development in Bangladesh expanded dramatically following independence in 1971 and particularly during the UN Decade for Water in the 1980s. Around Meherpur, HTWs are typically 15–75 m deep, having high density polyethylene (HDPE) casing with a diameter less than 0.1 m and a 3–7 m screened section made of HDPE or galvanized iron. A busy village HTW may be pumped almost continuously for much of the day, giving a typical maximum yield of 10 m^3 d^{-1}; however many HTWs are pumped erratically, and yields are commonly much less. In many parts of southern Bangladesh where HTWs are used for domestic water supply, deeper boreholes, with motorized pumps, are used for irrigation. Irrigation boreholes generally have casing up to 0.4 m diameter, depths of 50–150 m and multiple screened lengths up to 30 m in length. They may be pumped at rates of approximately 250 m^3 d^{-1} for up to 8 h d^{-1} over 2–3 months during the dry season. Public water supply boreholes have been installed in some towns. These are commonly completed to depths of 100–300 m, and may pump several hundred m^3 h^{-1} during much of the day throughout the year, depending on the available supply of electricity. The modes of groundwater abstraction around Meherpur are summarized in Table 1. Two deep municipal public water supply wells have operated in Meherpur since 1978, but this remote region had not benefited substantially from groundwater development prior to 1990. The majority of the population now has access to groundwater for domestic use, and the use of groundwater for irrigation is widespread. Within the Meherpur municipality there are at least 225 municipal HTWs and 1850 private HTWs. On average there is one HTW per family. Irrigation boreholes are spread out every few hundred metres in the fields around the town and surrounding villages.

Arsenic in the alluvial aquifers at Meherpur

Methods and analytical results

In situ hydrochemical measurements were made at 150 HTWs and deeper tubewells in and around Meherpur in June 1998. Arsenic concentration in the tubewell discharge was estimated using a Merck Arsenic Indicator Field Kit at 125 sites. This provided a colorimetric

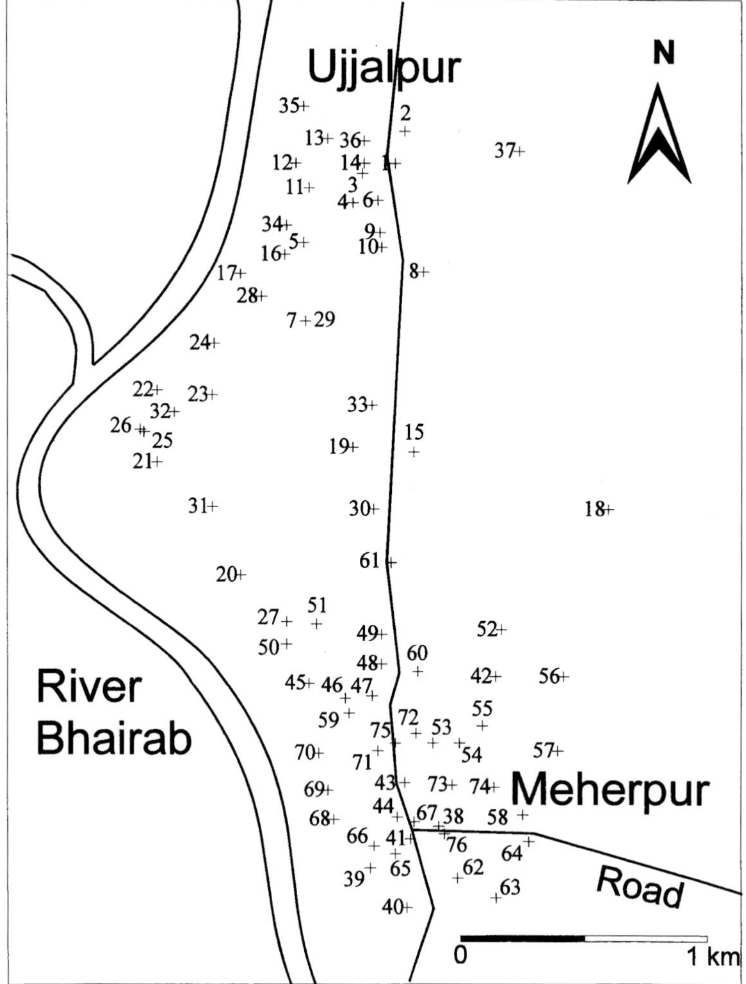

Fig. 4. Location of sampled tubewells, Meherpur district. The cored borehole was situated between site 1 and the road, at Ujjalpur.

indication of the presence of arsenic above 50 μg l^{-1}. Subsequent comparison with laboratory analyses showed the field test to judge exceedence of the 50 μg l^{-1} limit correctly in 80% of cases. The field kit was a valuable guide for fieldwork, but the results have not been used in further interpretation. Seventy-six tubewells (Fig. 4) were selected for sampling and full hydrochemical analysis, representing the full range of arsenic concentration, borehole depth, and pumping regime. Groundwater samples were filtered at well-head using 0.45 μm membrane filters (for anion analysis), or filtered and acidified to pH 2 (for cation analysis). On-site measurements were made of EC, pH, alkalinity, dissolved oxygen (DO) and temperature. Dissolved oxygen measurements were compromised by the erratic discharge of hand-pumps, and are considered overestimates of the true value. Purging of HTWs prior to sampling was limited to approximately 40% of the HTW volume, but irrigation and municipal production wells were purged fully by pumping for at least 10 minutes. Anion analysis was by ion chromatography and cation analysis (excluding arsenic) was by ICP-AES (inductively coupled plasma atomic emission spectrometry). Arsenic was determined using hydride-generation atomic absorption spectrometry (AAS), with a detection limit of 1 μg l^{-1}. On-site measurements and analytical

Table 2. *Location and physico-chemical character of groundwater samples, Meherpur district*

Site no.	Source type*	Depth (m)	T(°C)	DO (% satn.)	EC (μS cm^{-1})	pH	Arsenic (μg l^{-1})
1		30	26.8	3	660	6.97	110
2		18	26.8	10	830	6.90	370
3		43.5	27.2	8	700	7.12	78
4		18	26.8	12	630	7.12	135
5		33	27.2	18	1000	7.02	14
6		18	26.8	0	660	6.96	76
7	ITW	27	26.7	0	470	7.32	24
8	ITW	36	26.7	0	550	7.05	891
9		33	26.8	0	550	6.99	250
10		21.6	26.8	0	600	7.00	1
11		15	27.2	0	1055	6.86	3
12		28.5	27.1	0	750	6.82	11
13		27	27.0	0	690	6.90	470
14		29.4	26.8	0	590	6.99	28
15	ITW	45	26.9	0	550	7.04	243
16		19.5	27.0	0	750	6.95	11
17		24	27.1	0	720	6.99	76
18	ITW	225	27.3	0	540	7.00	63
19	ITW	36	26.9	0	460	7.08	49
20		22.5	27.0	0	560	6.93	55
21		22.5	26.8	1	850	7.03	240
22		21	27.3	0	690	6.98	310
23		22.5	26.4	0	570	6.87	280
24	ITW	33	26.6	0	630	6.91	200
25		21.6	27.2	0	810	6.90	380
26		11.4	27.2	1	820	7.05	1
27		30	26.8	0	570	6.95	106
28		16.5	26.8	0	710	6.99	47
29		66	27.0	0	530	7.06	108
30	ITW	36	26.9	0	530	7.08	49
31	ITW	36	26.9	0	700	6.96	180
32		21	27.2	0	800	7.06	46
33		19.5	26.6	0	420	7.10	2
34		19.5	26.9	0	1120	6.85	19
35		18	26.3	0	740	6.90	17
36		22.5	26.6	0	730	6.90	775
37	ITW	75	27.0	0	550	7.03	126
38		31.5	26.5	4	890	1.60	
39	ITW	27	26.4	1	920		53
40		35.4	26.8	3	680		43
41		21	26.6	4	1050	6.66	100
42		21	26.4	4	590		105
43		15	25.9	4	1100		18
44		19.5	26.4	5	722	6.73	2
45		21	26.9	1	550	6.71	17
46		21	26.9	2	580		2
47		15	26.8	4	1100	6.94	55
48		19.5	26.8	5	1110	6.99	43
49		15	26.7	2	780	7.28	155
50		36	27.0	2	680	8.39	60
51		18	27.0	4	1100	8.31	2
52		16.5	26.8	4	890		29
53		16.5	26.4	4	1060		1
54		15	26.4	3	780		120
55		15	26.8	3	680		190
56		21	27.2	3	680		35
57		24	26.6	2	610		130
58		22.5	27.0	2	650		330
59		13.5	26.9	4	830	6.88	2
60		36	26.7	4	770		2
61		13.5	27.0	4	430	6.70	150
62		14.1	26.3	4	1080	6.70	205
63		35.4	26.4	5	690	6.76	190
64		35.4	26.7	6	720	6.57	131
65		22.5	26.7	5	1360	6.66	2
66		35.4	26.8	1	1060		1
67		22.5	26.5	2	980		29
68	ITW	55.5	26.4	1	820	6.28	100
69		18	26.8	3	1350	6.52	230
70		22.5	26.5	2	640	6.02	155
71		24	26.8	3	1020		4
72		19.5	26.8	3	1310		2
73		15	26.8	3	1030		1
74		28.5	26.9	2	790		29
75	DTW	101	27.0	0	800		102
76	DTW	92	26.8	4	830		80

*All samples are from HTWs, unless otherwise stated (see Table 1 for nomenclature).
DO, Dissolved oxygen; EC, electrical conductivity. Arsenic is the laboratory measurment. A blank indicates no measurement.

Table 3. *Chemical composition of groundwater samples, Meherpur district*

Site	TDS	As	Ca	Mg	Na	K	Fe	HCO_3	NO_3	SO_4	Cl
1	670	110	110	26.5	13.7	4.1	8.5	507	0	2.1	1.0
2	930	370	133	34.1	23.9	5.1	0.9	727	0	0	2.3x
3	860	78	126	26.7	11.7	4.3	3.9	678	0	0	4.6x
4	660	135	99	25.4	24.6	3.8	1.7	503	0	0.9	3.0
5	1190	14	161	52.5	44.5	13.0	0.0	788	7.1	55.2	67.3
6	730	76	114	27.6	15.7	3.8	8.5	561	0	0.1	2.3
7	510	24	96	17.7	7.7	3.9	1.3	381	0.1	1.4	1.6
8	620	891	96	23.3	12.6	4.1	3.0	483	0	0.3	1.7
9	610	250	94	22.4	12.8	4.0	7.9	464	0	0.5	1.9
10	610	1	91	24.5	15.1	4.1	9.7	468	0	0	1.6
11	1060	3	170	56.7	31.7	6.9	0.0	603	48.2	48.2	94.6
12	900	11	141	35.0	14.2	4.0	6.2	693	0	1.5	1.4
13	720	470	125	29.0	15.3	4.4	7.8	498	0	2.4	35.8
14	710	28	122	22.3	9.2	4.3	2.9	542	0	3.4	5.8
15	620	243	91	24.6	11.9	4.6	5.8	481	0	1.0	2.0
16	820	11	135	40.8	13.7	5.6	0.4	608	0.2	5.6	14.9
17	800	76	128	39.5	12.3	5.2	1.3	581	0	2.7	27.2
18	560	63	101	19.9	9.9	3.9	2.8	422	0.1	0.6	4.2
19	480	49	87	19.0	6.9	3.6	1.0	361	0.5	1.8	1.1
20	550	55	102	21.1	12.1	4.0	1.1	407	0	0.8	3.8
21	780	240	143	33.4	22.3	4.5	6.1	493	0	21.1	57.7
22	360	550	42	17.8	17.8	6.9	0.1	261	3.1	10.2	0.2
23	670	310	120	24.3	12.6	5.4	7.7	493	0	0.2	2.2
24	580	280	94	25.6	13.7	3.1	9.6	427	0	0.2	3.3
25	720	200	110	29.3	14.7	4.2	7.9	547	0	0.2	1.9
26	790	380	135	34.2	16.9	5.1	7.9	556	0	9.6	21.3
27	720	1	137	32.9	17.3	5.4	0.0	449	39.7	10.4	31.6x
28	560	106	108	20.0	11.3	4.0	0.4	407	0	0.2	8.2
29	830	47	131	31.2	14.3	4.4	0.8	639	0	3.1	8.0
30	620	108	100	19.2	9.6	3.5	3.5	481	0	0.1	0.9
31	600	49	96	22.2	8.8	3.7	2.3	456	1.3	3.4	4.2
32	830	180	123	29.4	16.9	4.6	8.7	639	0	1.2	3.7
33	820	46	127	38.9	24.1	8.9	3.6	549	0	22.4	44.7
34	450	2	86	15.3	6.6	3.6	0.1	334	4.1	1.6	1.0
35	1150	19	192	49.9	34.8	5.2	0.9	737	0.1	45.5	89.3
36	760	17	134	28.7	24.5	4.0	0.6	522	0	3.5	38.5
37	780	775	122	30.3	24.3	5.3	10.0	578	0	0.1	14.1
38	380	11	30	14.4	47.7	17.8	0.4	212	1.6	6.9	52.5
39	580	126	102	20.4	10.5	4.0	2.6	444	0	0.2	1.1
40	470	570									
41	830	160	135	35.1	26.1	6.0	4.7	573	0	3.1	43.8
42	790	53	142	35.2	22.9	4.4	3.9	481	0	28.3	75.4
43	648	43	113	23.9	19.9	4.0	0.5	469	0	1.4	16.9
44	990	100	161	33.4	50.5	5.1	3.2	622	0.1	30.4	85.2
45	630	105	96	25.1	15.8	4.0	7.0	471	0	0.2	8.4
46	1070	18	131	63.0	57.4	8.3	1.4	695	3.6	36.7	69.5
47	1110	2	140	52.0	69.1	8.8	1.1	650	47.6	48.1	97.4
48	490	17	88	20.9	10.0	3.8	1.6	366	0.2	0.5	3.4
49	550	2	86	24.4	19.3	5.3	0.1	390	0.1	10.0	15.0
50	950	55	125	32.3	95.7	6.3	1.6	505	0.1	46.8	140.7
51	1020	43	127	57.0	60.5	11.5	0.4	605	0	43.9	111.5
52	710	155	109	31.7	23.1	5.0	8.2	505	0.9	0.4	29.8
53	620	60	118	24.5	11.7	4.7	1.4	447	0	1.1	10.4x
54	1120	2	144	49.4	58.9	5.3	0.0	727	58.5	19.3	53.4
55	860	29	155	30.2	22.0	5.5	3.0	593	0.1	6.4	41.5
56	1010	1	124	37.7	33.3	73.4	0.7	581	52.5	27.5	79.1
57	780	120	118	35.8	14.1	6.1	5.2	586	0.2	3.6	12.0
58	650	190	101	26.2	13.8	3.8	9.7	488	0.1	1.2	2.1
59	700	35	117	25.2	13.7	3.1	7.2	525	0.1	0.9	6.7
60	610	130	72	19.0	46.0	3.5	5.4	456	0	0.1	4.6
61	640	330	92	25.0	30.6	4.0	7.4	476	2.0	0.2	3.8
62	1230	2	79	28.5	30.4	15.0	0.0	481	0	14.2	584.0
63	740	2	116	35.5	15.7	11.6	0.0	525	0.3	7.5	26.8
64	430	150	70	17.3	6.2	4.0	3.5	329	0.1	0.5	3.5
65	970	205	158	48.1	25.9	4.7	7.8	605	0.1	5.4	110.9
66	610	190	111	27.2	11.1	4.2	1.6	422	0.1	0.4	29.2
67	690	131	110	29.6	23.3	4.3	7.9	505	0.3	0.2	8.7
68	1270	2	145	40.9	71.4	110.1	0.0	652	59.1	43.8	146.3
69	1000	1	156	40.8	43.4	7.6	0.0	586	26.1	44.2	97.6
70	930	29	152	40.1	32.7	7.0	1.7	578	0.1	42.5	75.0
71	820	100	138	32.3	22.2	6.1	6.3	586	0	2.7	25.8
72	1220	230	167	45.9	104.0	12.9	4.1	671	0.3	44.8	170.3
73	600	155	99	27.4	13.7	4.4	3.3	447	0	0.2	8.6
74	900	4	151	29.1	22.8	49.3	0.1	517	68.8	15.0	49.3
75	1290	2	178	75.9	32.7	8.1	0.0	808	88.2	27.1	67.4
76	960	1	130	52.5	45.7	6.7	0.0	569	82.2	22.3	57.1
77	730	29	140	27.1	8.9	4.0	7.6	495	0.1	17.0	32.9
78	770	102	134	27.0	19.8	5.4	2.5	544	1.4	6.6	29.5
79	780	80	135	29.6	24.6	4.2	3.4	525	3.2	10.0	40.4

Arsenic in ppb. All other analytes in ppm.
Four samples marked x have ionic balance, calculated as {(cations – anions)/(cations + anions)}* 200, < 20%. Remainder are < 10%

Table 4. *Ujjalpur core: As and Fe content of porewater and sediments*

Depth (m)	Arsenic content		Fe content	
	Porewater ($\mu g\ l^{-1}$)	Sediment (whole rock) ($mg\ kg^{-1}$)	Porewater ($mg\ l^{-1}$)	Sediment (whole rock) (%)
3		9.6		3.8
6	10	2	0.2	1.5
9		6.4		2.6
10	8		0.4	
15	9	5.2	2.1	1.4
18	68			
18	537			
21	290	35	6.4	2.7
21	106		0.4	
24	13		17	
27	6	3	2.1	1.2
27	19		17.3	
30	13		17.5	
33	39	1.8		0.8
33	47			
36	14		27.5	
39	13	1.4	21.6	1.2
42	13	2.8	3.3	0.9
45	26		6.3	
46	16	27		3.1
48	35		30.5	
48	23		10.7	
51	27	2.4	41.3	1.1
51	26		26.5	
54	14		11	
57	23	2	18.1	1.6
60	30		17.1	

results are given in Tables 2 and 3 respectively.

Cored sediment samples from the Ujjalpur borehole were preserved on-site by waxing the ends of individual PVC core sleeves. On extraction in the laboratory, approximately 100 g of sediment was mixed with distilled water, stirred for 5 minutes and allowed to settle. The porewater/distilled water mixture was then decanted, and filtered or centrifuged prior to analysis. Results are presesnted in Table 4. Filtration would remove the colloidal fraction. Also, it is probable that some oxidation occurred during porewater extraction prior to analysis. However, dissolved iron is generally high in the porewaters (up to 30 $mg\ l^{-1}$) and apparently has not been oxidized substantially and removed from solution. Porewater calcium concentration is similar to that of local groundwaters, suggesting that calcite precipitation has not occurred to any great extent. In general, porewater composition is similar to that of groundwater in the region. These factors all suggest that the integrity of the porewater has been adequately preserved during treatment, and that the porewater hydrochemical profiles have not been obscured. Nevertheless, arsenic concentrations in the porewater are presented as minimum values.

A fraction of the sediment samples extracted from the wax-sealed PVC core-sleeves was oven-dried, disaggregated and mixed prior to treatment for determination of total arsenic content. The sub-samples were fused with lithium metaborate, and digested in 10 % HNO_3. Total arsenic was determined by hydride-generation AAS. Results are presented in Table 4.

Hydrochemistry of arsenic occurrence

Groundwater at Meherpur is predominantly anoxic, and of calcium/(magnesium) - bicarbonate type. Relatively high background values of total dissolved solids, between 140 and 1290 $mg\ l^{-1}$, are typical of groundwater in young, reactive, alluvial sedimentary sequences (Hem 1989). Spatial distribution of the main hydrochemical components is illustrated in Figure 5. In all but four of the groundwaters sampled, dissolved oxygen was present at less than 6% saturation. In practice, this suggests that anoxic conditions are common, if not pervasive, throughout the

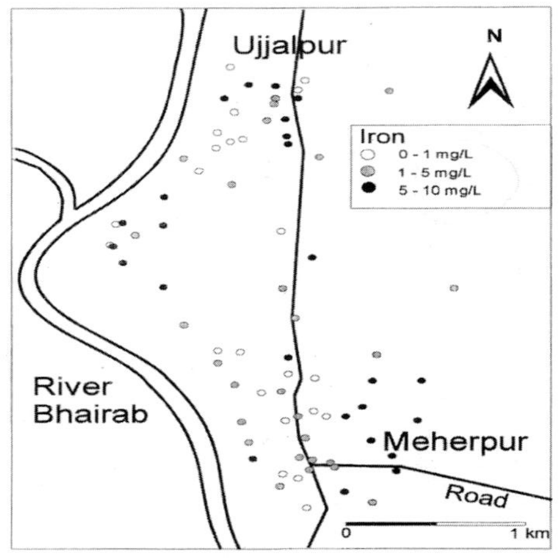

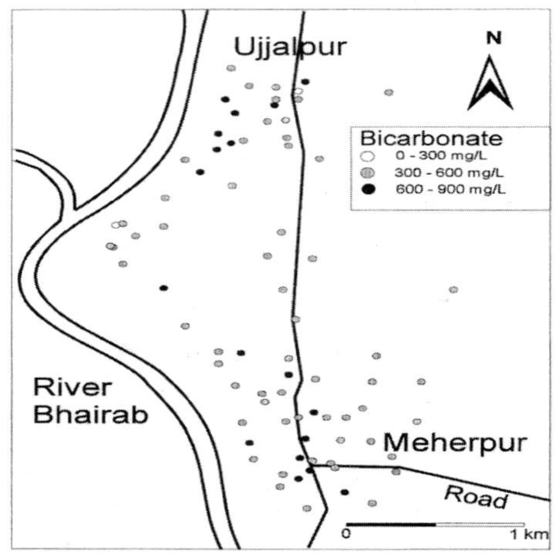

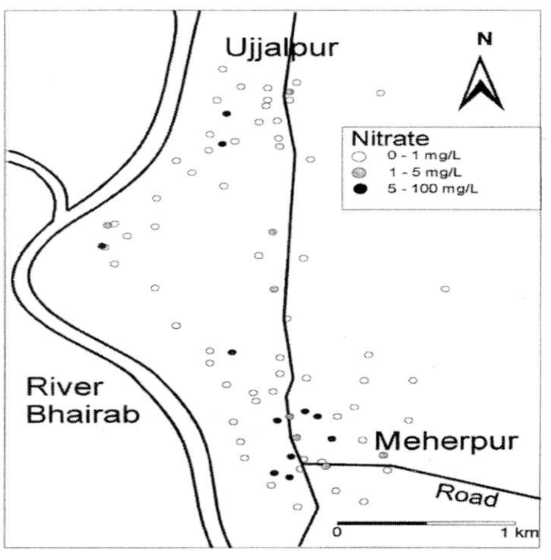

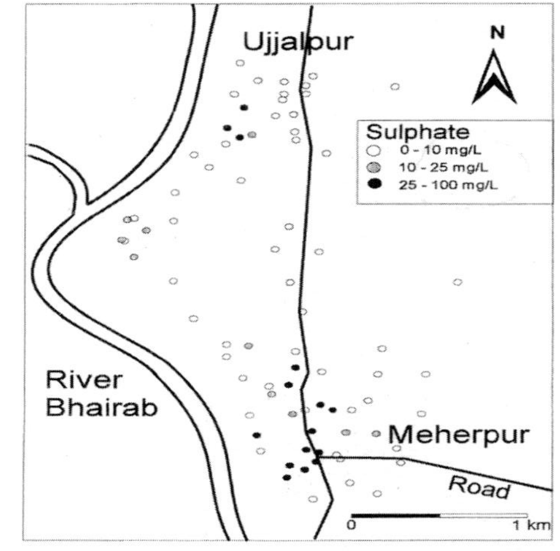

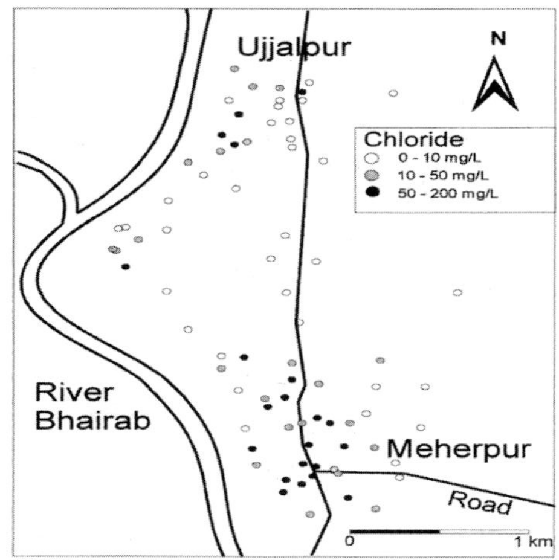

Fig. 5. Spatial distribution of Fe, HCO_3^-, NO_3^-, SO_4^{2-} and Cl^- in groundwater, Meherpur district.

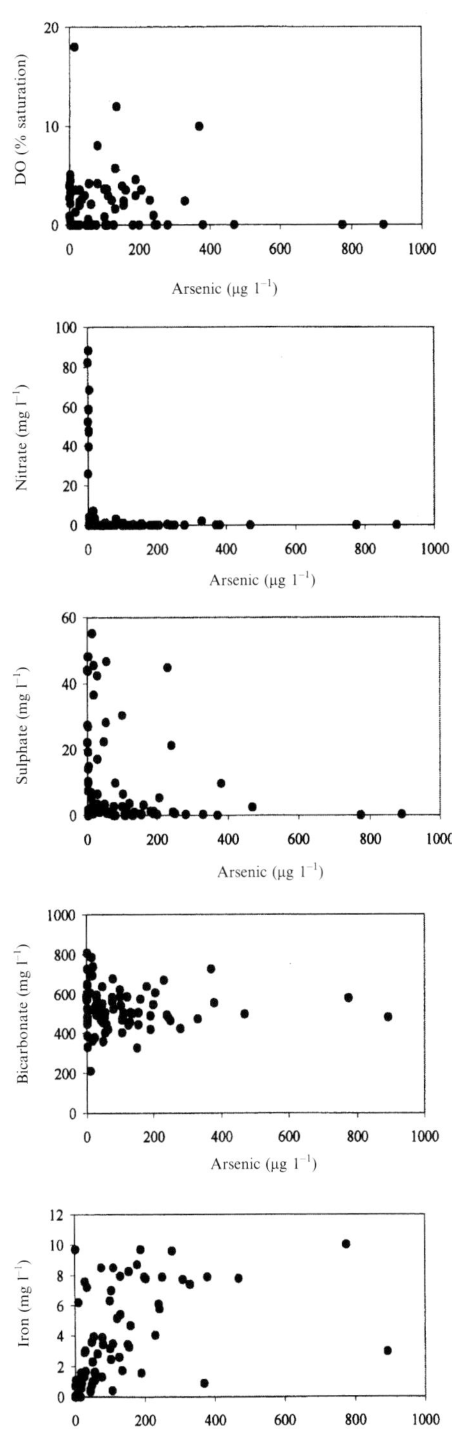

Fig. 6. Hydrochemical relationships between arsenic and dissolved oxygen (DO), NO_3^-, SO_4^{2-}, HCO_3^- and Fe in groundwater, Meherpur district.

aquifer, even at 9 m depth, the shallowest level sampled. Dissolved iron ranges up to 10 mg l^{-1}, reflecting the reducing conditions and the availability of iron in the sediments. Nitrate is generally absent, except at Meherpur town, where nitrate concentration is between 20 and 50 mg l^{-1}, and Ujjalpur village, where nitrate reaches 20 mg l^{-1}. Chloride and sulphate have a distribution similar to that of nitrate in the shallow aquifer. Chloride reaches 150 mg l^{-1} at Meherpur and 55 mg l^{-1} at Ujjalpur, being less than 50 mg l^{-1} beyond the main areas of settlement. Sulphate reaches 40 mg l^{-1} at both Meherpur and Ujjalpur, but is around 5 mg l^{-1} beyond the settlement areas. The distribution of nitrate, sulphate and chloride reflects the density of human settlement. Throughout the aquifer, bicarbonate dominates the anionic component. Median bicarbonate values are around 500 mg l^{-1}; maximum values greater than 700 mg l^{-1} are recorded beneath Meherpur and Ujjalpur. These elevated bicarbonate concentrations, together with the high iron concentrations and other indicators of reducing conditions, demonstrate that the oxidation of organic matter in the sediments (Lovley 1987) is a dominant process in the evolution of the groundwater chemistry.

Relationships between arsenic and dissolved oxygen, nitrate, sulphate, bicarbonate and total iron at Meherpur are illustrated in Figure 6. These confirm that high arsenic concentrations are associated with reducing conditions under which oxygen is limited, nitrate is absent and iron is at high concentrations. Nitrate ranges from zero to 88 mg l^{-1} and has a strongly inverse relationship with arsenic. Where nitrate is above 5 mg l^{-1}, arsenic is less than 10 μg l^{-1} in all but one case (Site 5: NO_3^- 7.1 mg l^{-1}, As 14 μg l^{-1}). In general, arsenic concentrations greater than 50 μg l^{-1} are associated with sulphate concentrations less than 10 mg l^{-1}, but the inverse relationship between arsenic and sulphate is less pronounced than that between arsenic and nitrate. Nevertheless, the positive correlation between iron and arsenic and the pervasively elevated bicarbonate concentrations are similar to those previously recorded by Nickson *et al.* (1998) over a broader geographical area and at lower arsenic concentrations. The results from Meherpur support the hypothesis that desorption of arsenic has accompanied reductive dissolution of iron oxyhydroxides in the aquifer sediments, and suggest this is the principal mechanism by which arsenic is released to groundwater.

It is significant that arsenic is not ubiquitously present in groundwater where oxygen is absent

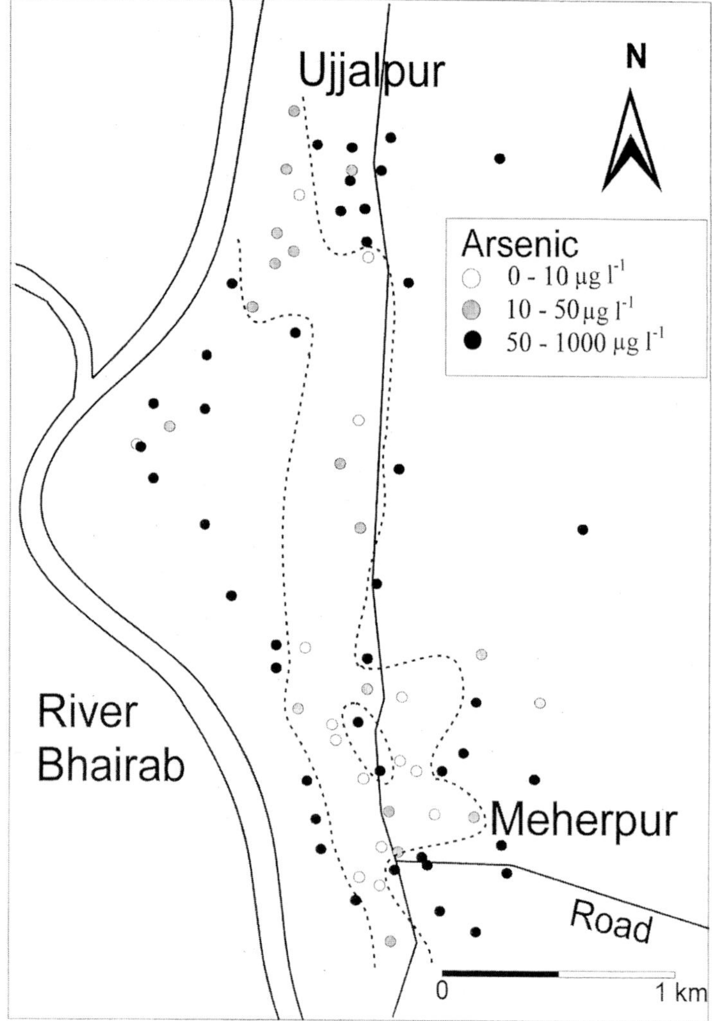

Fig. 7. Spatial distribution of arsenic in groundwater, Meherpur district. Dashed lines indicate the low-arsenic belt within which As is less than 50 µg l^{-1}.

and iron and bicarbonate concentrations are high. A favourable hydrochemical environment is a necessary but not sufficient condition for elevated arsenic concentrations in the groundwater. The results suggest that conditions favourable for arsenic release to groundwater are widespread, but the source of arsenic is more restricted. It is therefore important to establish the distribution of arsenic in the aquifer.

The spatial distribution of arsenic in groundwater

The spatial frequency of sampling in the national delineation of the arsenic-affected area by the British Geological Survey & Mott MacDonald (1999) was approximately one sample per 37 km^2, and one sample per 7 km^2 in each of three special study areas in the western, central and eastern parts of the basin. In the present study around Meherpur, tubewells were sampled at a spatial frequency of five per square kilometres, a scale nearly two orders of magnitude more detailed than previously described in the Bengal Basin.

Arsenic concentration in groundwater around Meherpur ranges from less than 1 µg l^{-1} to nearly 900 µg l^{-1}. Fifty-five per cent of the tubewells sampled have arsenic content greater than 50 µg l^{-1} and only 18% have arsenic at less

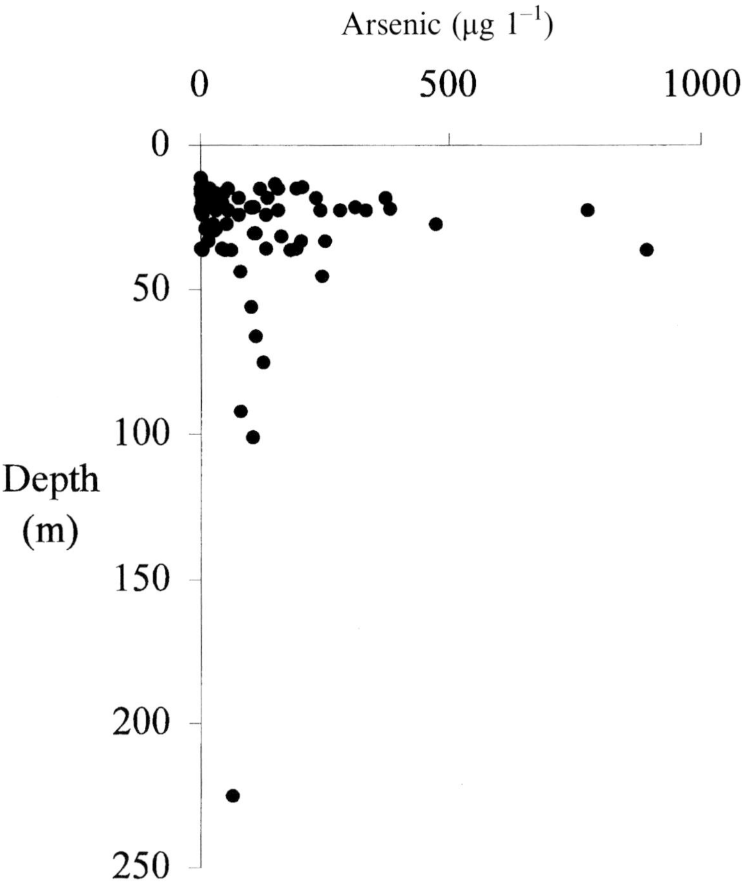

Fig. 8. Depth distribution of arsenic in groundwater, Meherpur district.

than 10 µg l^{-1}. The spatial distribution of arsenic (Fig. 7) reveals a belt of low concentration (less than 50 µg l^{-1}), 500 m wide, crossing the present-day floodplain of the Bhairab River area in an overall N–S orientation between Ujjalpur and Meherpur. Within this belt the average arsenic concentration is 15 µg l^{-1}; beyond it the average is 198 µg l^{-1} (maximum 890 µg l^{-1}). It is defined by the extreme spatial variability along its margins. For example, to the north-east of Ujjalpur one tubewell beyond the margin of the low-arsenic belt pumps water with 250 µg l^{-1} arsenic, just 105 m from a tubewell within the belt with arsenic at less than 1 µg l^{-1}. The pattern of this low-arsenic belt is suggestive of a palaeo-channel. Within the high-arsenic regions on either side of the 'channel', occasional individual tubewells or small groups of tubewells have a more moderate arsenic concentration, less than 50 µg l^{-1}. Available lithological records are insufficient to demonstrate the existence of a palaeo-channel. However, these observations suggest the possibility of linking spatial patterns of arsenic concentration in groundwater to sedimentological features that may have geomorphological manifestation.

The depth distribution of arsenic in groundwater

The short lengths of screen and low discharge of the HTWs at Meherpur make it realistic to approximate the depth distribution of arsenic in the aquifer from a survey of arsenic concentration in the water pumped from different depths in the aquifer. Results show that groundwater with the highest arsenic concentrations, between

200 and 1000 μg l^{-1}, is pumped from depths shallower than 45 m (Fig. 8). Variability is also highest in groundwater pumped from these shallow levels, where the full range of arsenic concentrations, zero to 890 μg l^{-1}, is observed. Arsenic appears to occur at a lower and less variable concentration, commonly around 100 μg l^{-1}, in groundwater pumped from depths greater than 45 m. The deepest tubewell sampled, an irrigation tubewell 220 m deep used since 1975, pumps groundwater with arsenic at 63 μg l^{-1}. The data are too sparse at these greater depths to draw firm conclusions. However, the depth distribution of arsenic in groundwater established at Meherpur is consistent with the overall trends seen across the whole of southern Bangladesh (British Geological Survey & Mott MacDonald 1999). Results from a limited survey of depth distributions in 1997, interpreted as demonstrating increasing arsenic concentrations with depth to 120 m (Nickson *et al.* 1998, 2000), are incomplete and give misleading representations of the depth profiles at the three localities concerned, all close to the convergence of the Ganges and Brahmaputra floodplains west of Dhaka. Nevertheless, lateral variability in the depth distribution of arsenic in groundwater is an issue that needs more attention. Depth distribution of arsenic in groundwater must, in part, be related to the vertical distribution of arsenic within the aquifer sediments themselves, and/or to the vertical distribution of organic carbon as a control of arsenic release by microbial reduction of iron oxyhydroxides.

A depth profile of arsenic in porewater

A more precise yet site-specific view of the depth profile of arsenic in the aquifer has been determined by analysis of porewater eluates from cored sediments recovered from the Ujjalpur borehole, near Meherpur (Fig. 9). The cored borehole was situated between the HTW at site 1 and the road at Ujjalpur, 250 m E of the low-arsenic belt indicated by the HTW samples. There is a single, distinct peak of arsenic in the porewater at 18–21 m depth, where arsenic concentration exceeds 300 μg l^{-1} (range 50–500 μg l^{-1}) against a background concentration less than 50 μg l^{-1}. In a contrasting pattern, chloride has a background concentration of less than 20 mg l^{-1}, with higher values between 80 and 90 mg l^{-1} to a depth of 10 m. The chloride profile reflects the limiting depth of active groundwater circulation, controlled in part by the subdued topography and in part by the occurrence of a silty clay layer just below 10 m depth.

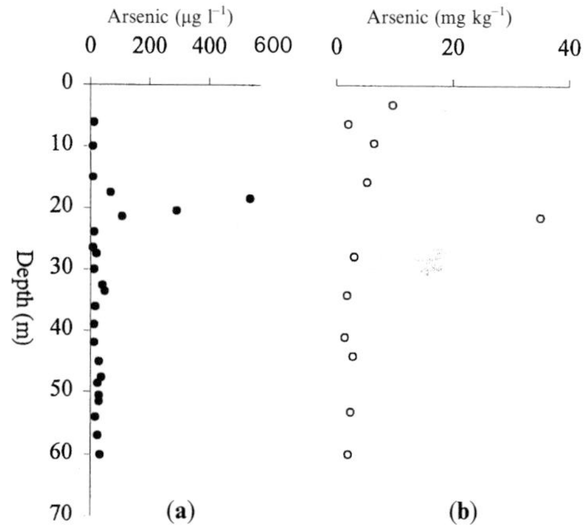

Fig. 9. Profiles of arsenic in (**a**) porewater and (**b**) sediments from Ujjalpur, Meherpur district.

A depth profile of arsenic in sediments

The total arsenic content of sediments in the Ujjalpur core ranges from 1.4 to 35 mg kg^{-1}, comparable with the range 2–11 mg kg^{-1} previously reported for sediments from Chapai Nawabganj in north-west Bangladesh (British Geological Survey & Mott MacDonald 1999). The peak value of total arsenic in the sediment is at a depth of 21 m at Ujjalpur, coincident with the porewater arsenic peak (Fig. 9). The total arsenic content of the sediments is only moderately elevated in comparison to the average crustal arsenic content of 2 mg kg^{-1}. It is likely that other characteristics of the sediment mineralogy, e.g. organic carbon content and type, are important in enhancing the release of arsenic from sediments to groundwater. The correlation between arsenic and iron in the sediments (Fig. 10) supports the hypothesis, based principally on hydrochemical relationships, that iron oxyhydroxide is the source of arsenic in solution. There is also a weak inverse correlation between arsenic and sediment grain size, suggesting that arsenic may be concentrated in the fine-grained, clay-rich sediments beyond the channel sands. Some arsenic may be associated with clay minerals.

How does arsenic concentration in tubewell discharge change with time?

There are no long-term records indicating whether and how arsenic concentration in

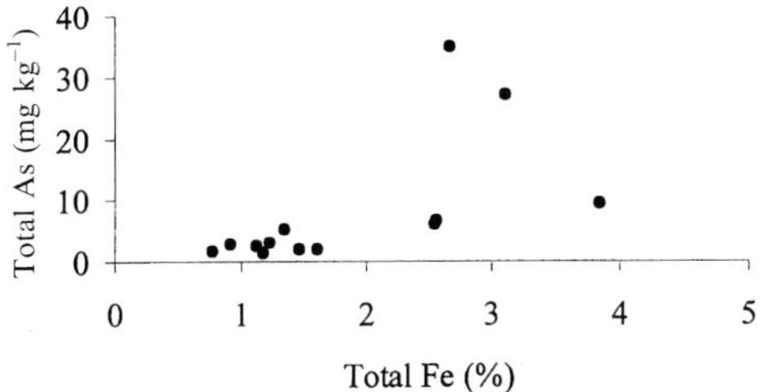

Fig. 10. Relationship between arsenic and iron in sediments from Ujjalpur, Meherpur district.

tubewell discharge changes with time. An indication may be sought by considering arsenic concentration in relation to tubewell age. The data from Meherpur suggest a trend of increasing arsenic concentration with tubewell age (Fig. 11). For the tubewells sampled, less than 50% of those installed in the past 5 years have arsenic levels greater than 50 μg l^{-1}. In contrast, all the tubewells installed more than 15 years ago now pump groundwater that exceeds the 50 μg l^{-1} limit. Similar indications are apparent from the extensive regional database of British Geological Survey & Mott MacDonald (1999). Modelling the movement of arsenic within an individual HTW catchment, from regions of arsenic release within the sedimentary sequence to the HTW screen (Cuthbert *et al.* 2002), enables a clearer evaluation of possible trends with time. Observations of arsenic distribution in the aquifer at Meherpur, together with the hydrochemical and sedimentological associations, suggest a conceptual basis for developing numerical models for this purpose.

A conceptual model of arsenic in the aquifer

The source of arsenic

The source of arsenic in the groundwater has been shown indirectly to be iron oxyhydroxide within the alluvial sediments, concentrated in the finer floodplain deposits beyond the coarse channel sands and occurring presumably as particulate matter and as grain coatings. Arsenic is released to groundwater by desorption from iron oxyhydroxide, associated with its dissolution under the chemically reducing conditions that pervade the aquifer. The data presented here extend the hypothesis of Nickson *et al.* (1998) across a larger range of arsenic concentrations. Within the arsenic source zone, there is a large excess of arsenic in the solid phase over arsenic in solution, and the rate of release of arsenic, although unknown in detail, is considered fast in relation to groundwater flow.

The spatial distribution of arsenic

The spatial distribution of arsenic in groundwater exhibits a distinctive pattern over an area of 15 km^2 in extent, suggestive of sedimentolo-

Fig. 11. Relationship between tubewell age and arsenic concentration in excess of 50 μg l^{-1}, Meherpur district.

gical control and demonstrating extreme lateral variability. At Meherpur, a belt of low arsenic concentration in groundwater (average 15 μg l^{-1}) may delineate a palaeo-channel approximately 500 m wide, which traverses the region. In a separate study at Chaumohani, in south-east Bangladesh (Burgess *et al.* 2000), an arcuate pattern of elevated arsenic in groundwater (above 250 μg l^{-1}) has been identified which may reflect the existence of a backfilled, abandoned meander. Lateral discontinuity of the arsenic source is an essential element of the conceptual model developed to explain the arsenic concentrations observed in groundwater, and must be on a scale comparable with the lateral extent of HTW catchments.

The depth distribution of arsenic

The depth distribution of arsenic in the aquifer around Meherpur has been demonstrated from a profile of arsenic in groundwater pumped from tubewells over a depth range 15–225 m, a profile of arsenic in porewater over a depth range 10–60 m, and a profile of total arsenic in sediments over the same range. The profile derived from pumped groundwater is consistent with the regional database developed by the British Geological Survey & Mott MacDonald (1999), and implies that the zone of arsenic release to groundwater is relatively shallow, within the top 45 m of sediments. The core from Ujjalpur allows a more precise description of the distribution, and shows a single level of arsenic release at 18–21 m depth, associated with laminated, silty sand containing organic material. The arsenic concentration in the porewater from this depth at Ujjalpur averages 300 μg l^{-1} (range 50–500 μg l^{-1}). This detail of the depth distribution of arsenic is another essential element of the conceptual model explaining the arsenic concentrations observed in groundwater.

The highest concentrations of arsenic in groundwater at Meherpur (approaching 900 μg l^{-1}) and the greatest variability (<1–250 μg l^{-1} over approximately 100 m) are coincident over the depth interval 15–45 m. This interval also encompasses the maximum porewater arsenic concentrations observed at Ujjalpur, and the peak value of arsenic content of the sediments. In groundwater below a depth of 50 m, arsenic is present at a lower and more consistent concentration.

The conceptual model summarized

A conceptual model to explain the field observations, combining sedimentological characteristics of the arsenic source zones and hydrogeological characteristics of the aquifer, is illustrated in Figure 12. Within the source zone, arsenic is released to groundwater as a consequence of microbially driven reduction and dissolution of iron oxyhydroxide particles and grain coatings containing sorbed arsenic. The mass of arsenic in the solid phase is much greater than in solution, and the rate of the release to groundwater is sufficient to maintain an arsenic concentration of up to 500 μg l^{-1} in porewater in the source zone. The relatively shallow, depth-specific arsenic source zones principally associated with fine-grained floodplain sediments, are laterally discontinuous and cut through in places by coarse channel sands. The aquifer is multi-layered and hydrogeologically leaky. Under these conditions, HTW catchments are limited to a radius of only a few metres, and higher-yielding, deeper tubewells have catchments of radius a few tens to a few hundreds of metres.

The potential for a range of overlap configurations between the discontinuous arsenic source zone and the limited HTW catchments is an important feature of the conceptual model. Shallow HTWs with screens at the level of the arsenic source could draw directly on groundwater with the highest arsenic concentrations, with minimal dilution, immediately following installation and the start of pumping (HTW A in Fig. 12). Conversely, where the arsenic source and/or the geochemical conditions for microbial reduction are laterally discontinuous, shallow HTWs may draw directly on arsenic-free groundwater, and remain free of arsenic for their entire lifetime (HTW B in Fig. 12). Partial overlap between the arsenic source and the HTW catchment would result in dilution of groundwater from the arsenic source zone, and an intermediate arsenic concentration in the tubewell discharge developing after a period of pumping arsenic-free water (HTW C in Fig. 12). Hence the greatest variability in arsenic concentration is observed in the discharge of HTWs with screens at shallow levels, up to a depth of 45 m at Meherpur. The much larger radii of influence of deep production and irrigation boreholes are more likely to incorporate arsenic source zones, but would also result in significant dilution by arsenic-free groundwater, and possibly a long period of arsenic-free discharge before the arrival of arsenic at the depth of the tubewell screen (ITW and DTW in Fig. 12). This explains the occurrence of arsenic in groundwater from deep production boreholes at more consistent, lower, yet in places still appreciable, concentrations.

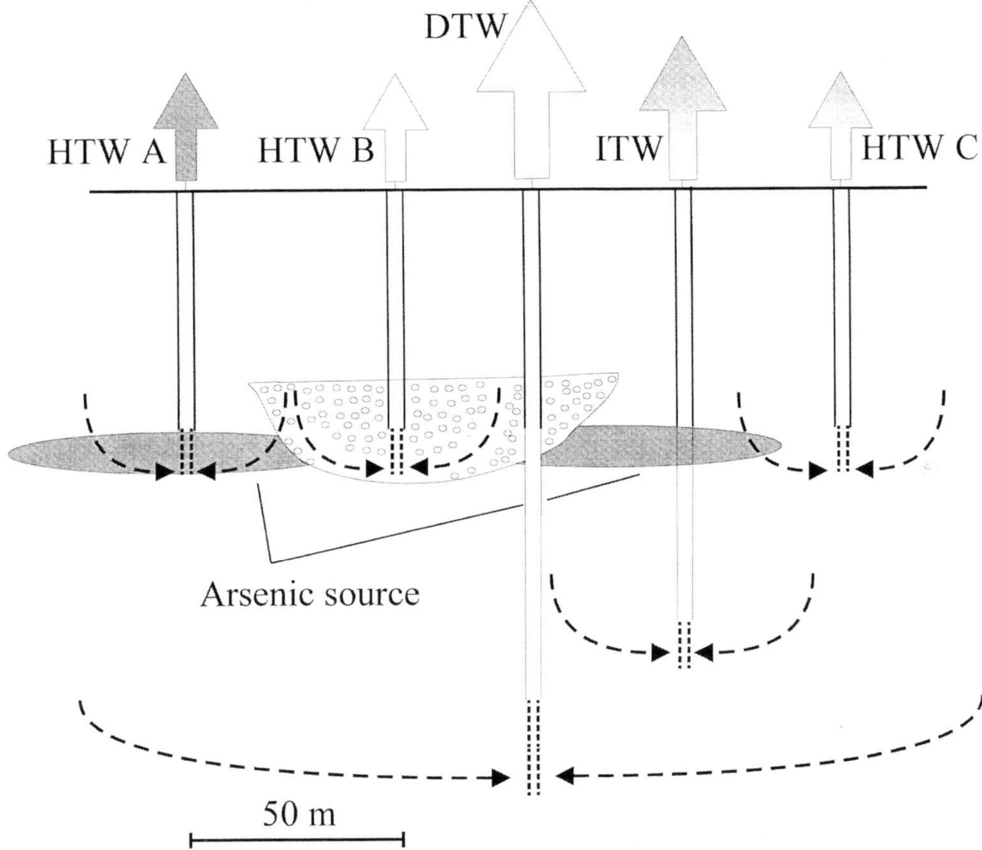

Fig. 12. A conceptual model of arsenic occurrence in the aquifer and variability in HTW discharge. Shading indicates arsenic source zones; ornamentation indicates a channel of coarse sand; screen is indicated at the base of each tubewell; curved dashed lines indicate the extent of tubewell catchments; vertical arrowheads indicate the relative tubewell discharge rates, with shading to indicate trends in arsenic concentration (shading is proportional to concentration, time progresses upwards). Vertical scale is approximately × 2 horizontal scale. DTW, deep tubewall; ITW, irrigation tubewell.

Implications for sustainable development

The conceptual model of arsenic in the aquifer, together with the preliminary numerical models of arsenic transport in the catchments of shallow HTWs (Cuthbert *et al.* 2002), have important implications for sustainable development of arsenic-bearing aquifers in southern Bangladesh.

Treatment

On account of the association of arsenic and iron in the water, aeration and oxidation to remove iron should substantially remove arsenic by co-precipitation. This has occurred fortuitously with iron removal at the Faridpur treatment works (R. Nickson pers. comm.), and the method has been further developed specifically for arsenic removal by Joshi and Chaudhuri (1996), amongst others.

Tubewell location and design

Shallow HTWs located in palaeo-channels containing coarse sands, or screened sufficiently beyond the level of the arsenic source (which is at a depth of around 20 m at Meherpur), should pump groundwater that is essentially free of arsenic for some years, possibly for decades, and in some cases indefinitely. Numerical models can be applied to evaluate these timescales for particular well-screen depths (Cuthbert *et al.* 2002). The challenge is to determine or to be able to predict the depth of the arsenic source zone and the sedimentary sequence at particular locations, sufficiently well to guide HTW placement and screen depth. Remote sensing may have an important rôle to play. High-yielding irrigation and public water supply tubewells with screens significantly deeper than the arsenic source zone, should pump groundwater with only slightly elevated arsenic concentrations, minimizing the need for treatment. Centralized pumping stations and distributed public water supply, where possible, would be advantageous. High-yielding deeper tubewells might significantly enhance the vertical flow of groundwater within the smaller catchments of shallower HTWs, however, and thereby influence arsenic movement to shallow HTWs. It is advisable to locate shallow HTWs beyond the influence of deeper high-yielding tubewells.

Future changes in arsenic at tubewells

The conceptual model has important implications in relation to anticipated changes of arsenic concentration with duration of pumping, illustrated schematically in Figure 12. The potential for evolving arsenic concentrations is an important element in judging the security of existing tubewells, as a guide to monitoring strategies, and as a consideration for design of treatment systems. New boreholes with screens below the depth of the arsenic source in the sediment profile would initially pump groundwater free of arsenic. Arsenic concentration will increase with time in the long term, in line with an increase in that proportion of the tubewell discharge that has flowed through the arsenic source zone and has acquired an elevated level of dissolved arsenic. This should be expected as the long-term trend, regardless of short-term changes that may result from intermittent or variable pumping patterns and seasonal effects. Some time later, the arsenic concentration in the tubewell discharge might be expected to reach an equilibrium. Later still, following depletion of the arsenic source zone, concentration in the tubewell discharge would fall. Monitoring must therefore be a continuing and long-term activity. It should be concentrated on HTWs that are in regular use, and should be performed at consistent times in relation to the pattern of pumping. Numerical models provide a means of evaluating the timescale of the changes in arsenic concentration and the required duration of monitoring, likely to be measured in decades or longer (Cuthbert *et al.* 2002). There remain questions concerning the scale and controls on short-term variations in arsenic concentration at HTWs, which also require evaluation and incorporation into a monitoring strategy.

Guided by the conceptual model of vertical movement of arsenic from relatively shallow source zones, monitoring should also address intermediate levels within the catchments of deeper, high-yielding and strategically important tubewells used for distributed public supply in some towns. This is particularly important given the uncertainty in the scope for retardation of arsenic by re-sorption to aquifer sediments (Cuthbert *et al.* 2002).

A possible 'arsenic-safe' deeper aquifer

The arsenic problem has been shown to be concentrated at a relatively shallow level in the thick alluvial sequence of sediments in southern Bangladesh which is widely used for water supply and irrigation. In the short and medium-term the sustainability of these shallow arsenic-bearing aquifers is a vital issue. However, a longer-term view should consider whether suitable deeper aquifers may exist that could become alternative sources of arsenic-safe water. Deeper aquifers are known to occur in the coastal regions, where they are exploited for their freshwater resources e.g. at Kulna in the south-west and Chaumohani in the south-east (P. Ravenscroft pers. comm.). In these regions the shallow aquifers are saline, and an aquitard protects the deeper aquifer from downward movement of a saline front. North of the coastal area, e.g. at Meherpur, a clay layer 30–65 m thick at 160 m depth may provide equivalent protection for deeper aquifers against the downward movement of arsenic from the shallow aquifer. The lateral extent of this aquitard is unknown. It is thought to be absent in places e.g. at Faridpur (Nasiruddin Ahmad pers. comm.). The deeper aquifers do appear to be almost free of arsenic; the national survey suggests that only 1 % of tubewells deeper than 200 m have arsenic above 50 $\mu g\ l^{-1}$ (British Geological Survey & Mott MacDonald 1999). However, the database for deeper tubewells is much sparser and some 'anomalous' occurrences of arsenic from deeper levels have been reported (Nasiruddin Ahmad pers. comm.). The approach adopted here for the shallow aquifer system, linking an appraisal of the distribution of arsenic and an appreciation of the hydraulic structure of the aquifer, would also be appropriate for the deeper aquifer. However, for the deeper aquifer it is particularly important to develop permeability profiles, especially to evaluate the occurrence of a substantial aquitard, rather than to establish in detail the spatial distribution of arsenic in the shallow sediments. Where the aquitard does not exist, sorption would become a critical issue for the security of the deeper aquifer, as it is for the shallow system (Cuthbert *et al.* 2002). Other questions concern the possible occurrence of arsenic sources and appropriate conditions for arsenic release to groundwater within the deeper sediments.

This paper is an output of the *London-Dhaka Arsenic in Groundwater Programme*. Melanie Burren acknowledges receipt of a Natural Environment Research Council Advanced Course Studentships and Overseas Fieldwork Allowance. We thank the Department of Public Health Engineering of the Government of Bangladesh for provision of facilities during fieldwork, particularly the loan of Merck arsenic field test kits, and access to its tubewell database in Dhaka. We thank the Bangladesh Water Development Board, particularly Mizanur Rahman, for provision of core-samples from the Ujjalpur borehole. Chemical analyses were carried out by the Robens Institute for Public & Environmental Health at Surrey University, UK, at the Environmental Mineralogy laboratory of the Natural History Museum in London, UK, and at the NERC ICP-AES facility at Royal Holloway College, London, UK; grateful thanks to Andrew Taylor, Chris Stanley, Vic Din, Nikki Paige and Tony Osborn. We thank Peter Ravenscroft of Mott MacDonald International for hospitality in Dhaka, and for much invaluable advice on the hydrogeology of Bangladesh in general, and the arsenic problem in particular.

References

British Geological Survey & Mott MacDonald 1999. *Arsenic in groundwater in Bangladesh, Phase 1. Rapid investigation.* Report to UKDFID.

Burgess, W. G., Burren, M., Cuthbert, M. O., Mather, S. E., Perrin, J., Hasan, M. K., Ahmed, K. M., Ravenscroft P. R. & Rahman M. 2000. Field relationships and models of arsenic in aquifers of southern Bangladesh. *In* Sililo, O. (ed.) *Groundwater: Past Achievements and Future Challenges.* Proceedings of the XXX IAH Congress, Cape Town, South Africa, November 2000. Balkema, Rotterdam, 707–712.

Cuthbert, M. O., Burgess, W. G. & Connell, L. 2002. Constraints on sustainable development of arsenic-bearing aquifers in southern Bangladesh. Part 2: Preliminary models of arsenic variability in groundwater. *In*: Hiscock, K. M., Davison, R. M. & Rivett, M. O. (eds) *Sustainable Groundwater Development.* Geological Society, London, Special Publications, **193**, 165–179.

Das, D., Chatterjee, A., Samanta, G., Mandal, B. K., Chowdhury, T. R., Chowdhury, P. P., Chanda, C. R., Basu, G., Lodh, D., Nandi, S., Chakraborti, T., Bhattacharya, S. M. & Chakraborty, D. 1994. Arsenic in groundwater in six districts of West Bengal, India: the biggest arsenic calamity in the world. *Analyst*, **119**, 168–170.

Davies, J. 1995. The hydrochemistry of alluvial aquifers in central Bangladesh. *In:* Nash, H. & McCall,Groundwater Quality (Eds.). Chapman & Hall, London, 9–18.

Dhar, R. K., Biswas, B. K., Samanta, G. S., Mandal, B. K., Chakraborti, D., Roy, S., Jafar, A., Islam, A., Ara, G., Kabir, A. W., Ahmed, S. A. & Hadi, S. A. 1997. Groundwater arsenic calamity in Bangladesh. *Current Science*, **73**, 48–59.

Government of Bangladesh. 1991. *Environmental Quality Standards for Bangladesh.* Department of the Environment, Bangladesh.

Hem, J. D. 1989. *Study and interpretation of the chemical characteristics of natural water.* USGS Water Supply Paper **2254** (3rd edition).

Herbert, R., Barker, J. A. & Davies, J. 1989. *The pilot study into optimum well design: IDA 4000 Deep Tubewell II Project. Volume 6: Summary of the programme and results.* Report by the British Geological Survey for the Overseas Development Administration, UK Government. BGS Technical Report **WD/89/14**.

Joshi, A. & Chaudhuri, M. 1996. Removal of arsenic from groundwater by iron oxide-coated sand. *Journal of Environmental Engineering*, 122, 769-771.

Lovley, D. R. 1987. Organic matter mineralization with the reduction of ferric iron: a review. *Geomicrobiology Journal*, **5**, 375–399.

Nickson, R. T., McArthur, J. M., Burgess, W. G., Ahmed, K. M., Ravenscroft, P. & Rahman, M. 1998. Arsenic poisoning of groundwater in Bangladesh. *Nature*, **395**, 338.

Nickson, R. T., McArthur, J. M., Ravenscroft, P., Burgess, W. G. & Ahmed, K. M. 2000. Mechanism of arsenic release to groundwater, Bangladesh and West Bengal. *Applied Geochemistry*, **15**, 403–413.

Safiullah, S. 1997. *Report on monitoring and mitigation of arsenic in the groundwater of Faridpur municipality.* CIDA Arsenic Project Report, Jahangirnagar Univ. to DPHE, Government of Bangladesh.

Umitsu, M. 1993. Late Quaternary sedimentary environment and landforms in the Ganges delta. *Sedimentary Geology*, **83**, 177–186.

World Health Organisation. 1994. *Guidelines for drinking water quality. Vol. 1: Recommendations. 2nd edn.* WHO, Geneva.

Constraints on sustainable development of arsenic-bearing aquifers in southern Bangladesh. Part 2: Preliminary models of arsenic variability in pumped groundwater

M. O. CUTHBERT[1], W. G. BURGESS & L. CONNELL

Department of Geological Sciences, University College London, Gower St., London WCIE 6BT, UK (e-mail: william.burgess@ucl.ac.uk)

[1]*Present address: Entec UK Ltd, 160–162 Abbey Foregate, Shrewsbury, SY2 6BZ, UK*

Abstract: Numerical models of groundwater flow and arsenic transport to tubewells in southern Bangladesh have been developed, based on a conceptual model derived from field observations. The catchment of a single hand-pumped tubewell (HTW) is incorporated within a model domain 8110 m^2 in area and 60 m thick. Three tubewell specifications represent typical Bangladesh HTW designs. Constant-concentration cells act as a single-layered arsenic source, arranged to represent the observed depth distribution of arsenic in the aquifer and the range of possible patterns of overlap between HTW catchments and discontinuous zones of arsenic release from sediment to groundwater. A variety of sorption regimes is simulated, and sensitivity to sorption is illustrated. Boundary conditions are modified to simulate the effects of deep production wells. The models reproduce the observed scale and range of arsenic concentration in groundwater pumped from HTWs, and demonstrate likely long-term trends. Breakthrough of arsenic to HTWs may occur a few years after the start of pumping, but at many tubewells the concentration of arsenic could continue to rise significantly over tens to hundreds of years. Spatial distributions and depth profiles of arsenic in groundwater from tubewells should be viewed as transient in the long term. These preliminary models allow implications for the sustainability of the shallow alluvial aquifer to be quantified provisionally. The mechanisms and scale of sorption of arsenic by the aquifer sediments remain as significant uncertainties.

Naturally-occurring high concentrations of arsenic in groundwater are widespread in the Holocene sediments of the Bengal Basin (British Geological Survey & Mott MacDonald 1999). Extreme spatial variability in arsenic concentration is a significant feature of this occurrence. It is estimated that across southern Bangladesh, more than 50% of shallow hand-pumped tubewells (HTWs) have arsenic levels below 50 $\mu g\ l^{-1}$, the national limit for arsenic in drinking water. Sustainable development of the alluvial aquifers may therefore still be possible, with the principal operational questions concerning the optimal location and design of tubewells, and future trends in arsenic concentration, which have implications for monitoring requirements and treatment designs as well as for sustainability *per se*.

The nature and patterns of occurrence of arsenic in the Holocene alluvial aquifer at Meherpur in western Bangladesh have been described by Burgess *et al.* (2002), who have proposed a conceptual model of arsenic in the aquifer (Fig. 1). The conceptual model explains the spatial variability and depth distribution of arsenic observed in the discharge of HTWs, and

From: HISCOCK, K. M., RIVETT, M. O. & DAVISON, R. M. (eds) *Sustainable Groundwater Development*. Geological Society, London, Special Publications, **193**, 165–179. 0305-8719/02/ $15.00 The Geological Society of London 2002.

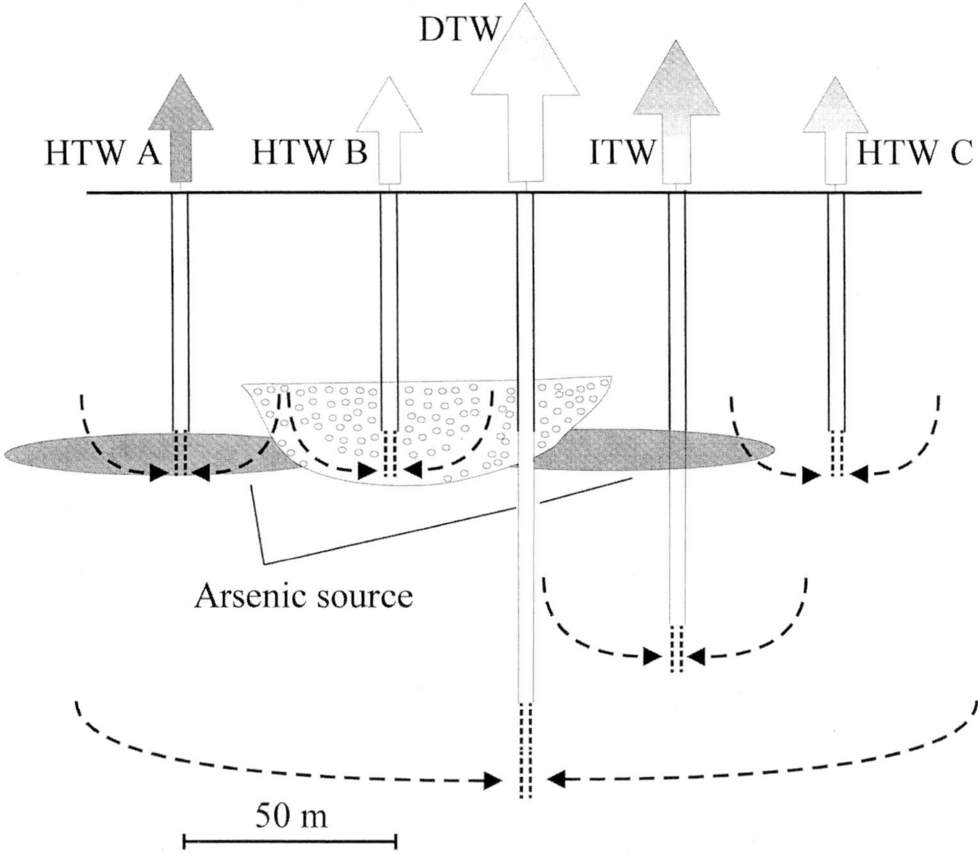

Fig. 1. A conceptual model of arsenic in the alluvial aquifer of southern Bangladesh and its variability in HTW discharge (from Burgess *et al.* 2002). Shading indicates arsenic source zones; ornamentation indicates a channel of coarse sand; screen is indicated at the base of each tubewell; curved dashed lines indicate the extent of tubewell catchments; vertical arrowheads indicate the relative tubewell discharge rates, with shading to indicate trends in arsenic concentration (shading is proportional to concentration, time progresses upwards). Vertical scale is approximately × 2 horizontal scale. DTW, Deep tubewell; ITW, irrigation tubewell.

indicates possible future trends in arsenic concentration in a qualitative manner. It provides a basis for numerical models designed to simulate conditions for typical HTWs in Bangladesh and to provide a more quantitative approach to judging the sustainability of groundwater development from the arsenic-bearing aquifer. The objectives of the modelling are as follows:

(1) to test the conceptual model as an explanation of the observed scale and range of arsenic concentrations observed at HTWs;
(2) to quantify the key issues of tubewell location and screen depth with respect to the arsenic source zone and to deeper, high-yielding irrigation and production wells;
(3) to explore the time-scale of changes in arsenic concentration in HTW discharge in relation to the Bangladesh national limit for arsenic in drinking water (50 $\mu g\ l^{-1}$), as a guide to monitoring policy and design of treatment systems; and
(4) to guide further investigations into the behaviour of arsenic in the aquifer.

The transport models are based on equilibrium groundwater flow conditions, and consider advection, hydrodynamic dispersion and sorption. The models therefore explore the long-term changes in arsenic concentration in HTW discharge. Short-term diurnal and seasonal changes in HTW pumping rate have been ignored. Reversible sorption is the only mechanism considered that is specific to arsenic. The geochemical mechanism of arsenic release to groundwater is not incorporated in these preliminary models, nor are geochemical reactions that may take place along the flow-path, other than sorption.

The conceptual model

The conceptual model of arsenic in the Holocene alluvial aquifer of southern Bangladesh (Burgess *et al.* 2002) is summarized in Figure 1.

The arsenic source

Geochemical relationships have demonstrated that arsenic is released to groundwater from sedimentary iron oxyhydroxides under chemically reducing conditions in the presence of organic matter (Nickson *et al.* 1998, 2000; Burgess *et al.* 2002). Depth profiles of arsenic in groundwater, established to a depth of 225 m from the discharge of HTWs sampled over areas of 10–15 km^2 (Burgess *et al.* 2000), suggest that the arsenic source, or release zone, lies within the top 50 m of the sediment profile. Results from Meherpur in western Bangladesh (Burgess *et al.* 2002) identify it as occurring over a single restricted interval, at a depth of 18–21 m, with the arsenic content of the sediments at a maximum of 35 mg kg^{-1} over this interval and the peak porewater arsenic concentration between 300 and 500 $\mu g\ l^{-1}$. Few speciation studies have been carried out, but arsenic in solution is reported to be approximately evenly distributed between As(III) and As(V) (British Geological Survey & Mott MacDonald 1999). In addition, the detailed spatial distribution of arsenic concentrations in water from HTWs suggests that the arsenic release zone is patchy, forming patterns indicative of sedimentological control (Burgess *et al.* 2000). Furthermore, lateral discontinuity of the arsenic source must be on a scale comparable with the lateral extent of HTW catchments.

The aquifer, and groundwater flow to tubewells

Across the floodplains of southern Bangladesh an upper aquitard of silts and clays overlies a succession of silts, sands and gravels which forms a multi-layered, hydrogeologically leaky aquifer. A deeper aquitard also exists in some parts, limiting the total thickness of the shallower aquifer to about 120 m and separating it from a deeper aquifer, which is largely undeveloped. The regional hydraulic gradient is extremely low; at Meherpur it is approximately 10^{-4}, consistent with the almost flat topography. The shallow aquifer is heavily exploited, with many tens of tubewells per square kilometre.

The significant feature of the multi-layered, leaky, alluvial aquifers is that the groundwater flowing to tubewells is ultimately provided by vertical leakage. Leakage restricts the expansion of the cone of depression, limiting the size of the tubewell catchment (Rushton 1986) to an extent dependent on the permeability of the sediments and the magnitude of pumping. Hand-pumped tubewells in the Holocene alluvial sediments of Bangladesh have very restricted catchments on account of their modest yields and the leaky nature of the aquifer. Limited catchments for the HTWs, and an arsenic source that is depth-specific and laterally discontinuous, combine to explain the high lateral variability observed in detailed field studies of arsenic in groundwater.

At a typical spacing of approximately 100 m between HTWs, an individual tubewell catchment less than 1000 m^2 in area is completely isolated from adjacent HTW catchments. Annual recharge due to precipitation in Bangladesh is on average around 400 –700 mm. Significant recharge may also arise from irrigation losses, typically in the order of 2–3 mm d^{-1} (Rushton 1986) or almost 300 mm a^{-1} for an irrigation period of 3 months. Therefore, the recharge available to supply HTWs by vertical leakage is more than adequate for a HTW pumping only a few $m^3\ d^{-1}$. A deeper well, with a relatively high abstraction rate, in close proximity to a shallow HTW would increase the flux of water through the catchment area of the HTW and modify the HTW catchment.

The conceptual model summarized

In the conceptual model developed on the basis of studies at Meherpur (Burgess *et al.* 2002) and elsewhere (Burgess *et al.* 2000), a discontinuous, single-layered arsenic source zone is present at a discrete depth in the alluvial succession of silts, sands and gravels which forms a multi-layered, hydrogeologically leaky aquifer. Hand-pumped tubewells have catchments which are restricted on account of their modest yields and the contribution of vertical leakage. Flow to the HTWs is provided by vertical leakage from the water table through the upper aquitard, followed by lateral flow through the more permeable alluvial sediments to the tubewell screen. The concentration of arsenic in water pumped from an individual HTW is therefore governed

Table 1. *Hydraulic properties in the groundwater flow model*

Layer	Thickness (m)	Hydraulic conductivity (horizontal) (m d^{-1})	Hydraulic conductivity (vertical) (m d^{-1})	Specific yield S_y	Porosity n
1	6	0.09	0.009	0.05	0.3
2	3	0.02	0.002	0.05	0.3
3	3	0.015	0.0015	0.05	0.3
4	3	0.035	0.0035	0.05	0.3
5	3	4.15	0.415	0.15	0.3
6	3	0.85	0.085	0.15	0.3
7	3	14	1.4	0.2	0.25
8	3	4.15	0.415	0.15	0.3
9	3	16.5	1.65	0.2	0.25
10	3	16.5	1.65	0.2	0.25
11	3	34.5	3.45	0.2	0.25
12	3	34.5	3.45	0.2	0.25
13	3	16.5	1.65	0.2	0.25
14	3	14	1.4	0.2	0.25
15	3	20	2	0.2	0.25
16	3	54.5	5.45	0.2	0.25
17	3	54.5	5.45	0.2	0.25
18	3	54.5	5.45	0.2	0.25
19	3	31	3.1	0.2	0.25

Storativity is taken as 10^{-4} throughout.

by the degree to which the catchment of the HTW overlaps with the discontinuous zone of arsenic release, the depth separation between the source zone and the tubewell screen, and the duration of pumping. Therefore, shallow HTWs exhibit the greatest range and spatial variability of arsenic concentration. The higher-yielding, deeper tubewells have larger catchments that are more likely to encompass a zone of arsenic release but would also lead to greater dilution. Deeper tubewells therefore have arsenic concentrations that are lower and more uniform than shallow HTWs, but which may still be appreciable in relation to national drinking water limits and international guidelines. Where a HTW is located within the catchment of a deeper irrigation or production well, the flux of water through the HTW catchment will be enhanced. A significant increase in the downward vertical flux would increase the rate of arsenic transport to the HTW screen where this is placed below the arsenic source layer.

Modelling groundwater flow

Models have been developed to simulate steady-state groundwater flow to a single HTW pumping at 10 m^3 d^{-1}. This rate reflects the abstraction of a HTW used by a small community, with the pump being in use for much of the day, and is the maximum pumping rate expected for such a well. The use of a constant pumping rate, when in reality pumping is erratic, is justifiable since the steady-state drawdowns for such low abstraction from these permeable alluvial sediments are small, and steady-state conditions are reached very quickly. In addition, the time-scale for arsenic transport is much greater than that of the short-term variations in pumping rate. Therefore, the effects that short-term diurnal and seasonal changes in the groundwater flow regime have on the transport of arsenic over time-scales of decades have been ignored. Modelling was carried out using MODFLOW (McDonald & Harbaugh 1984), and an analytical model for flow to a deep well was constructed within MS Excel (Appendix).

Hydraulic properties

Hydraulic parameter values have been derived as far as possible from field studies specific to the alluvial aquifer of Bangladesh in order to reproduce representative conditions of flow and transport. Where no data specific to Bangladesh are available, realistic values have been derived by comparison with similar hydrogeological environments. Model parameter values for hydraulic properties are summarized in

Table 1. Hydraulic conductivity varies between 10^{-3} and 5 m d^{-1} over the uppermost 60 m of the alluvial sediments, as estimated at Meherpur in western Bangladesh by Burgess *et al.* (2002). Vertical hydraulic conductivity is taken as 10% of the horizontal hydraulic conductivity. Values for storativity and specific yield are based on pumping test analysis of equivalent aquifers in central Bangladesh by Herbert *et al.* (1989). The resulting profiles of hydraulic conductivity and storage properties attributed to 19 horizontal layers in the model represent an upper aquitard, 15 m thick, overlying a multi-layered aquifer.

Boundary conditions and the model domain

Boundary conditions were chosen to represent the conceptual model as closely as possible. The model domain covers an area of 8110 m^2, approximating a circle of radius 50 m. This is sufficient to represent the catchment of a single HTW, within a wider region that can appropriately be delineated by no-flow boundaries without constraining the flow unrealistically. No-flow conditions are therefore imposed on the vertical boundaries defining the limits of the flow domain, and the base of the model. The orthogonal grid spacing ranges from 2.5 m close to the well to 5 m at greater distances. The 19 model layers enable a realistic representation of vertical hydraulic variability within the aquifer. All layers are 3 m thick, with the exception of the uppermost layer which is 6 m thick. This upper layer was modelled as unconfined; the lower layers remain confined throughout all the model runs. The recharge rate was determined so as to balance exactly the abstraction rate of the well, equivalent to approximately 450 mm a^{-1} over the model domain, a simplification justified by the abundance of recharge in Bangladesh. Seasonal effects have initially been ignored. The effects of seasonal pumping from a deep irrigation tubewell have been incorporated by imposing an additional vertical flux estimated from an analytical model.

A hand-pumped tubewell with a constant pumping rate of 10 m^3 d^{-1} is simulated as a negative constant flux boundary condition applied to one cell in a single layer at the centre of the model domain. Since the layers are 3 m thick, this effectively represents a 3 m well screen section. Three HTW specifications have been simulated, based on typical designs in Bangladesh i.e. screened at 18 m (WELL1), 30 m (WELL2) and 45 m (WELL3), all other parameters being held constant.

(**a**) Plan view of layer 9

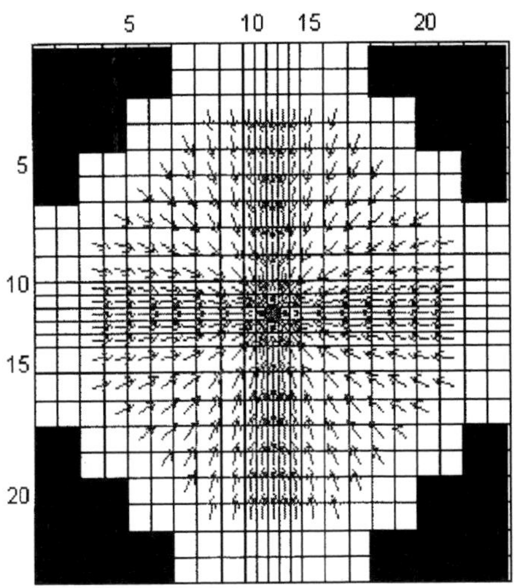

(**b**) Cross-section through the centre of the model domain

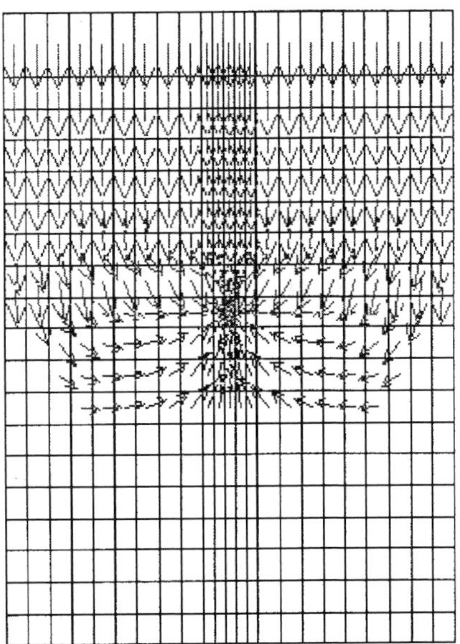

Fig. 2. A plot of equilibrium flow velocity vectors for the model WELL2. Flow velocities are shown on a logarithmic scale; minimum velocity plotted is 0.001 m d^{-1}. (**a**) Plan view of layer 9. (**b**) Cross-section through the centre of the model domain.

Table 2. *Summary of results of the analytical model of vertical flow to a deep tubewell*

WELL	Discharge (m^3 d^{-1})	Capture zone extent (m)	Depth (m)	Average vertical flux at increasing distance from HTW (m d^{-1})			
				50 m	100 m	250 m	500 m
Production well	3600	1600	100	0.045	0.02	0.003	0.0005
Irrigation well	240	500	60	0.003	0.001	0.0004	–

Groundwater flow simulations

The groundwater flow models were run at steady-state with the criterion of change in head for acceptable convergence of the solution set at 0.001 m; the models took 37 iterations to reach convergence. The flow velocity field for the model WELL2 (screened at 30 m depth) is illustrated in Figure 2. The modelled steady-state flow patterns are consistent with the expected hydraulics of the flow system under the specified boundary conditions. They have been taken as the basis for investigating the effects of arsenic transport to individual HTWs. Vertical flow from above the well screen provides water to the lower, more permeable regions of the aquifer in which the screen is set. The head loss across the vertical extent of the domain is approximately 5 m. At the level of the screen the flow is largely horizontal. There is a small upward vertical flux of groundwater close to and immediately below the screen. This is significant only for the transport simulations based on WELL1, when the arsenic source lies below the screen.

Approximating the effects of deep wells

For the purpose of approximating the effects of a deep irrigation or production well on flow to a shallow HTW, an analytical model has been developed (Appendix). The vertical flux of groundwater in the aquifer was modelled for typical production and irrigation wells. The well parameters used and a summary of illustrative results from the analytical model are given in Table 2, as average vertical fluxes through the top 60 m of the aquifer at a range of distances from the deep pumping well. The vertical flux values at specific distances from the deep pumping well (100 m and 250 m respectively) were superimposed on the WELL2 model domain to investigate the effects of deeper production wells on the HTW catchment.

Irrigation wells pump for a restricted duration each day over the 3-month dry-season irrigation period in Bangladesh. To investigate the seasonal effects of irrigation pumping, two steady-state pumping regimes were applied alternately to the WELL2 domain, representing a regular pattern of 3-month irrigation pumping each year. During the irrigation pumping periods an augmented flux in the HTW catchment was applied by modifying boundary conditions to force a background vertical flux across the flow domain. Successive transport calculations took as their initial concentrations the final concentrations of the previous run.

Modelling arsenic transport

Arsenic transport was modelled using MT3D (Zheng 1992). Grid sizes and time-steps were selected to minimize numerical errors as far as possible. Averaged and accumulated mass balance errors in the models were generally acceptably small, less than a few percent.

Representing the arsenic source

The arsenic source is described as a single layer, depth 18–21 m, within which arsenic release maintains a porewater arsenic concentration of 500 μg l^{-1}, typical of highly contaminated groundwater and the highest porewater concentrations observed at Meherpur (Burgess *et al.* 2002). This source is represented as a series of constant concentration cells set to 500 μg l^{-1} within layer 6 of the model. For each of the three modelled well specifications, three different degrees of overlap between the arsenic source zone and the HTW catchment have been simulated, the source being distributed over 100%, 52% and 13% of the HTW domain respectively. A plan view of layer 6 for each contaminant source distribution is shown in Figure 3.

It is emphasized that the mechanism of the release of arsenic to groundwater is not represented in the model. It is implicit that arsenic release from sediments to groundwater is fast in relation to the rate of arsenic transport away from the source, and that in the source zone the concentration of arsenic in porewater is effectively constant throughout the time of the simulations. This is a reasonable initial assumption, considering the low groundwater flow rates in the HTW catchments, in the order of a few mm per day, and the large excess by mass of arsenic in the solid phase.

Hydrodynamic properties and sorption

Effective porosity and dispersivity have been assigned values according to generalized relationships and observations of sediment lithology. Effective porosity is assumed equivalent to total porosity (Table 1) in these unconsolidated sediments. Dispersivity is related to flow distance according to Xu & Eckstein (1995):

$$\alpha = 0.83(\log L)^{2.414}$$

where α is the longitudinal dispersivity and L is taken as the average distance between the contaminant source and the HTW screen. On this basis, longitudinal dispersivity was assigned values between 2.0 and 3.5 m. Transverse dispersivity is taken as one-tenth of the longitudinal dispersivity. Although dispersivity is not well constrained by local data, results were relatively insensitive to changes in dispersivity over the range 0.1–10 m.

The effect of retardation due to sorption has been investigated using published sorption isotherms for As(V) and As(III) sorption to hydrous ferric oxide. Sorption parameters are summarized in Table 3, following the British Geological Survey & Mott MacDonald (1999). These authors indicate that where total arsenic concentration in the groundwater is greater than 100 μg l^{-1}, the proportions of As(III) and As(V) are approximately equal. The sorption parameters represent conditions of neutral pH and available iron concentration in the sediment of 50 mg kg^{-1}. Phosphate restricts As(V) sorption through competition for sorption sites (Dzombak & Morrel 1990). Conditions relevant to low phosphate (0.008 mg P l^{-1}) and high phosphate (0.8 mg P l^{-1}) concentrations have been considered. The flow domain of WELL2 (30 m depth) was used for the model runs which included sorption, with the As(III)/As(V) ratio taken as unity in the source cells. In calculating retardation factors from the sorption parameters, sediment bulk density was taken as 1800 kg m^{-3}, a value typical of alluvial sediments. The effective molecular diffusion coefficient, D^*, was taken as 9.05×10^{-10} m^2 s^{-1} after Lide (1998).

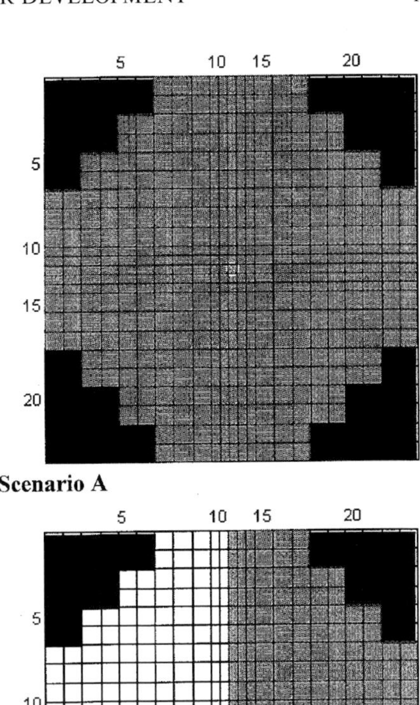

Scenario A

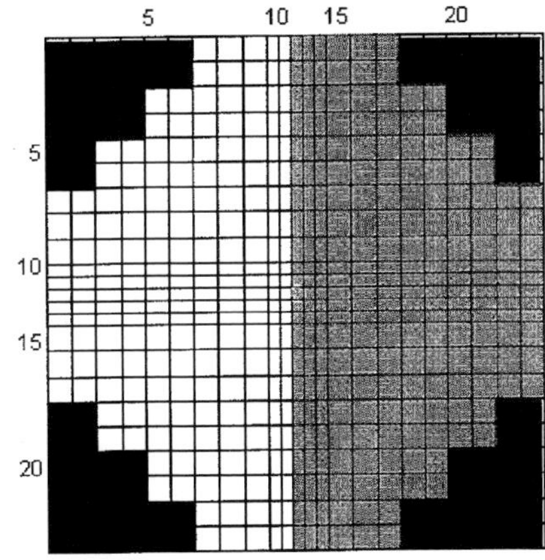

Scenario B

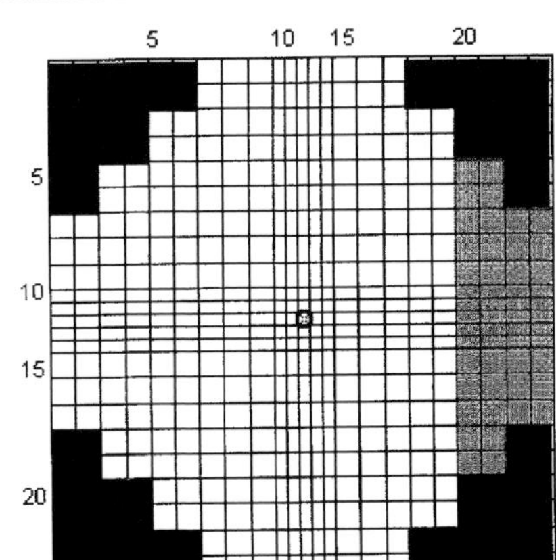

Scenario C

Fig. 3. Contaminant source distributions in layer 6 of the model. Shaded cells are constant concentration cells set at 500 μg l^{-1}. Black areas are outside the model domain.

Table 3. *Sorption isotherm parameters (after British Geological Survey & Mott MacDonald 1999)*

Phosphate concentration ($mg\ P\ l^{-1}$)	Isotherm parameters Freundlich (for arsenate) K_f ($l\ kg^{-1}$)	n	Langmuir (for arsenite) K_l ($l\ kg^{-1}$)	s_m ($mg\ kg^{-1}$)
0.008	3.1	0.18	1.72	13.3
0.8	1.3	0.87	0.417	13.3

These parameters are for conditions of neutral pH and oxalate-extractable iron concentration in the sediment of 50 mg kg^{-1}. K_f and n are the Freundlich isotherm constant and exponent respectively. K_l and s_m are the Langmuir isotherm factor, and the concentration of sorption sites available.

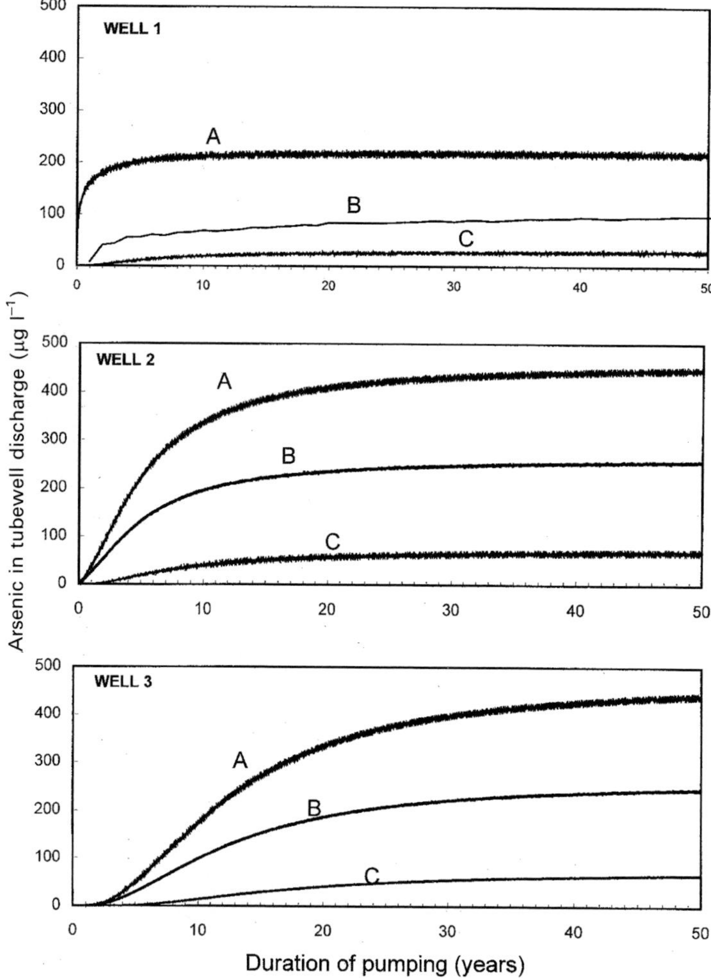

Fig. 4. Modelled evolution of arsenic concentration with pumping duration in WELL1, WELL2 and WELL3, without sorption. A, B and C indicate contaminant source distributions as in Figure 3.

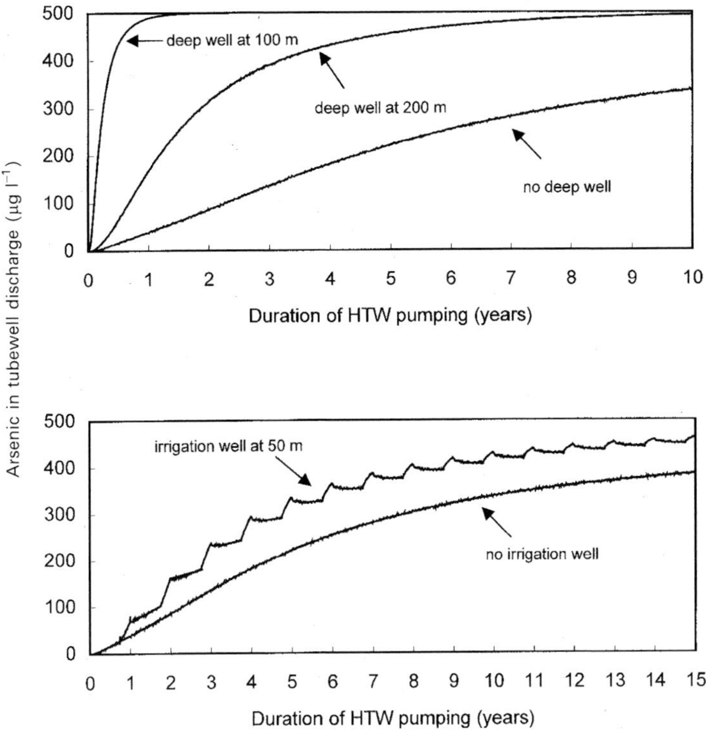

Fig. 5. Modelled effects of nearby deep pumping wells on the arsenic concentration at WELL2A. (**a**) Effect of a continuously pumped water-supply well. (**b**) Effect of a seasonally pumped irrigation well.

Results

Results from the arsenic transport simulations without sorption and in the absence of any deeper high-yielding production or irrigation tubewells are presented in terms of arsenic concentration in HTW discharge v. time (Fig. 4) for the three tubewell specifications (WELL1, WELL2 and WELL3) and for the three arsenic source distributions (A, B and C: 100%, 52% and 13% overlap with the HTW catchment, respectively). The effect of a nearby high-yielding, deeper tubewell on arsenic concentration in the HTW discharge is illustrated in Figure 5, with reference to WELL2 and source distribution scenario A. The possible effects of sorption are illustrated with reference to WELL2 in Figure 6.

Arsenic in the discharge of isolated HTWs, without sorption

At WELL1, with a screen at 15–18 m depth i.e. immediately above the arsenic source layer, arsenic concentration is appreciable from the start of pumping, at about 75 µg l^{-1} for source distributions A and B, and thereafter the concentration builds up to reach an approximately equilibrium condition after 10 years of pumping. At this time, arsenic concentrations have reached 215 µg l^{-1} (scenario A) and 145 µg l^{-1} (scenario B). In contrast, at WELL1 with source scenario C (the arsenic source distant from the HTW), arsenic is absent from the tubewell discharge at the start of pumping; arsenic breakthrough comes after about 1 year, a quasi-equilibrium concentration of 20 µg l^{-1} is reached after 10 years, and the final arsenic concentration is 28 µg l^{-1}.

At WELL2, with a screen at 27–30 m depth i.e. 6 m below the arsenic source layer, breakthrough of arsenic comes almost immediately after the start of pumping for source distributions A and B, and the concentration exceeds 50 µg l^{-1} after about 1 (scenario A) or 2 years (scenario B). Thereafter, the arsenic concentration builds up to reach 335 µg l^{-1} (scenario A) and 195 µg l^{-1} (scenario B) after 10 years of pumping, and quasi-equilibrium is achieved

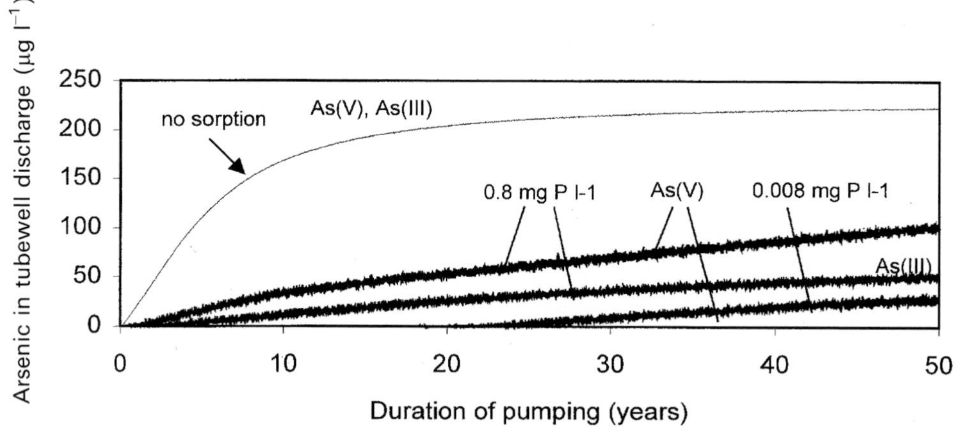

Fig. 6. Effects of sorption on arsenic transport to WELL2A. Results relating to arsenate, As(V), and arsenite, As(III), are indicated separately. As(III) is very strongly sorbed where phosphate concentration is low, and these results have been omitted for clarity.

after a further 20 years of pumping, with concentrations approaching 450 µg l^{-1} (scenario A) and 250 µg l^{-1} (scenario B). For WELL2 with source scenario C, the discharge is free of arsenic for 2 years after the start of pumping. Thereafter it builds up slowly, and exceeds 50 µg l^{-1} only after 15 years. A quasi-equilibrium concentration approaching 60 µg l^{-1} is reached after 20 years of pumping.

At WELL3, with a screen at 42–45 m depth i.e. 21 m below the arsenic source layer, the arsenic content of the tubewell discharge is negligible for at least 3 years after the start of pumping for all source distribution scenarios, and the concentration only exceeds 50 µg l^{-1} after 5 (scenario A) or 7 years (scenario B). The development of arsenic concentration in WELL3 follows the same pattern as in WELL2, but with a delay of approximately 10 years. It takes 20 years of pumping for the arsenic concentration to approach 350 µg l^{-1} (scenario A), 200 µg l^{-1} (scenario B) and 50 µg l^{-1} (scenario C).

The final equilibrium arsenic concentrations for WELL2 and WELL3 are simple reflections of the areal proportion of the model domain which is covered by the arsenic source cells and the arsenic concentration specified at the source cells, thus are 500 µg l^{-1} for source distribution A (100% overlap with the HTW catchment), 260 µg l^{-1} for source distribution B (52% overlap with the HTW catchment) and 65 µg l^{-1} for source distribution C (13% overlap with the HTW catchment). Modelled arsenic concentrations are within approximately 90% of their final values after 50 years of pumping, ignoring sorption. This is due to the HTW screen being located below the arsenic source layer. All the groundwater in this leaky aquifer flows from the water table through the layer in which the arsenic source cells are located, and instantaneous chemical equilibration is assumed in the model with no effective depletion of the source. At WELL1, the final equilibrium arsenic concentrations reflect the placement of the well screen above the arsenic source layer. Under this condition, uncontaminated water drawn from above the screen mixes with arsenic in water drawn through the source layer from below. The resulting arsenic concentrations are always less than the source zone concentration, and equilibrium concentrations are achieved more quickly than if the well screen is placed below the arsenic source layer.

Effects of deeper wells

A deeper tubewell, continuously pumping at a high yield, considerably speeds up the breakthrough of arsenic to WELL2 (source distribution scenario A), even at a distance of 200 m (Fig. 5a). At a distance of 100 m from the deeper pumping tubewell, arsenic breakthrough to the HTW is effectively immediate, and the concentration builds up to the maximum of 500 µg l^{-1} within just 2 years, compared to 75 µg l^{-1} after 2 years where the HTW is beyond the influence of a deeper high-yielding well.

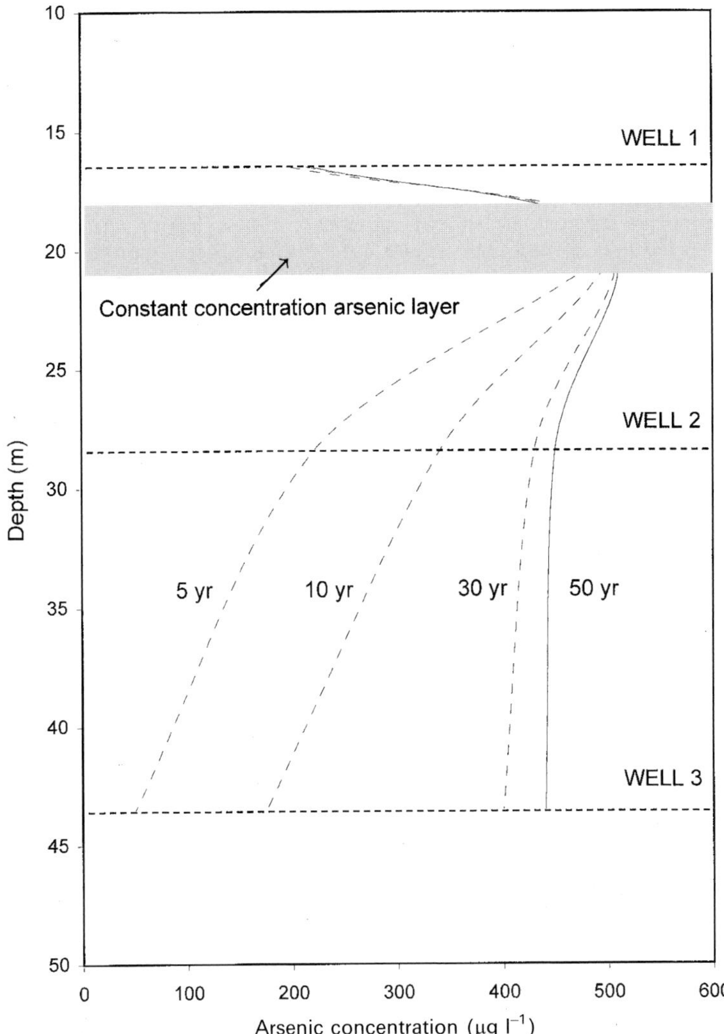

Fig. 7. Development of the depth distribution of arsenic in tubewell discharge: model results for up to 50 years of pumping, without sorption.

Seasonal irrigation pumping from a deeper tubewell has a lesser effect, as shown in Figure 5b. The flux of arsenic to WELL2 is greatly increased during the short period of irrigation pumping. However, after the irrigation pumps are turned off, arsenic concentration at the HTW may actually decrease for a time as the HTW draws more from the relatively less contaminated water below the well screen. The resulting long-term development of arsenic concentration in the HTW discharge is only moderately influenced by the irrigation pumping. For a HTW at a distance of 50 m from the irrigation well, the arsenic concentration is elevated by 20–30% over what would have been in the absence of irrigation pumping.

Effects of sorption

The effects of including sorption in the models of arsenic transport are illustrated in Figure 6, where WELL2 and source distribution scenario A are again used for comparison. Even moderate sorption has a dramatic effect on the development of arsenic concentration in the HTW discharge.

In the event of low phosphate concentration, total arsenic concentration in the WELL2 discharge exceeds 50 μg l^{-1} only after 50 years of pumping, compared to just 1 year when sorption is neglected (or ineffective). As(III) builds up more quickly than As(V) for the first 12 years of pumping under these conditions. After this the situation reverses and by 32 years of pumping As(V) exceeds As(III) in the HTW discharge, illustrating the effect of the different dependencies of sorption on concentration for the two arsenic species.

A higher phosphate concentration reduces the sorption of both arsenic species as phosphate competes for sorption sites. Hence, in the event of high phosphate concentration the total arsenic concentration of the WELL2 discharge reaches about 150 μg l^{-1} after 50 years of pumping. This compares with a concentration of 50 μg l^{-1} under conditions more conducive to arsenic sorption, and to 450 μg l^{-1} in the case of no sorption. At the concentrations concerned, and under the sorption regime applied, As(III) is more strongly retarded that As(V).

Summary and discussion

The model results are consistent with the range of arsenic concentrations in groundwater of the alluvial aquifers of southern Bangladesh and demonstrate how the observed spatial and depth distributions might develop. The models make clear that the arsenic concentrations at HTWs may continue to rise for many years after the onset of pumping, and that the distribution of arsenic in groundwater is transient under the hydraulic conditions imposed on the aquifer by pumping. In detail, the pattern of evolution of arsenic concentration depends on the spatial relationships between the arsenic source and the well screen, the extent of sorption, and the proximity of deeper tubewells.

Implications for HTW placement and recommended screen depths

The spatial distribution of the arsenic sources in relation to the HTW catchment is the main control on the dilution of arsenic-rich groundwater, and hence on the ultimate, maximum concentration of arsenic at an individual HTW. Possible patterns relating to sedimentological control have been reported by Burgess *et al.* (2000), and further research on the sedimentological controls of the geometry, extent and depth of the arsenic source zones is necessary. Where the distribution of the arsenic source is known, the model results can guide the placement of new HTWs and the required screen depths. If screens can be placed above the shallowest arsenic source zone, dilution may be sufficient to maintain an acceptably low arsenic concentration in the HTW discharge. Where this is not possible, a tubewell screened at a depth of 20 m or more below the main arsenic source layer is likely to provide a water supply with acceptable levels of arsenic for many years, and possibly for decades. Extending predictions beyond this, i.e. predicting the arsenic concentration in a new well, is a virtually intractable task given the number of likely uncertainties at a given location. Without substantial site investigation and mapping of arsenic sources, these uncertainties cannot adequately be resolved.

The proximity of HTWs to deeper tubewells

The model results show that deeper irrigation and water-supply boreholes may have a derogatory effect on arsenic concentrations in nearby HTWs because they augment the flux of water through the HTW catchment, increasing the rate of transport of arsenic from the source to the HTW screen. Continuously pumped boreholes, such as those used for public water supply in towns, will have a much more significant effect than seasonally pumped irrigation tubewells. Model results suggest that maximum arsenic concentrations may be reached after only 2 years if the HTW is situated 100 m away from a typical water-supply production well, unless sorption is effective. If sorption is effective, the time-scale for arsenic breakthrough and build up becomes several times greater, in a manner as yet unspecified but dependent on the chemistry of the groundwater and mineralogy of the sediments. Breakthrough times may be more than 10 years and concentrations may continue to build up for 100 years or more, even in the case of moderate sorption. These factors should be taken into account when considering both the spatial and depth placement of new HTWs. It is advisable to locate HTWs intended for drinking water supply beyond the influence of deeper, high-yielding water-supply boreholes. The separation should be in the order of 500 m.

Implications for monitoring and treatment design

The modelling results can be combined to demonstrate the evolution with time of an arsenic depth distribution in a number of wells screened at different depths (Fig. 7). The instance shown is for an arsenic source completely covering the model domain, for advection and dispersion only. The results emphasize that a presently uncontaminated well (or one with an acceptable arsenic concentration) cannot necessarily be relied upon in the future if an arsenic source exists in its proximity. There are important implications for monitoring. The absolute time-scale of arsenic increase is very sensitive to the extent of sorption. It is vital to quantify the effects of sorption in order for models such as these to be useful guides to the required frequency of monitoring. Yet the mechanism and the extent of sorption in the context of the alluvial aquifers of Bangladesh are far from certain.

Due to the time-scales involved in arsenic transport to HTWs, shifting the reliance for safe drinking water onto relatively uncontaminated wells may not provide a truly sustainable long-term solution to the arsenic contamination problem. In some areas, developing a deeper aquifer that is free from arsenic accumulation may be the long-term answer, if this is possible within the usual constraints of groundwater resources development. Where this is not possible, and where the safest and most sustainable strategy may be to use appropriate technology to remove arsenic from drinking water at a household level, it must be appreciated that the main component of the treatment design – the arsenic concentration of the HTW discharge water – is likely to change (increase) with time.

Limitations of the modelling, and research priorities

The models have emphasized the significance of vertical connectivity in the hydraulically 'leaky' sediments that constitute the shallow alluvial aquifer in Bangladesh. The greater the degree of vertical permeability, the more restricted are the catchments of individual HTWs, and the greater the potential spatial variability of arsenic concentrations in HTW discharge (which is also a function of the lateral variability of the arsenic source). The vertical permeability is also a primary control on the rate of migration of arsenic to the level of the HTW screen, governing the time to arsenic breakthrough and the duration of transient arsenic concentrations, up to the time that equilibrium concentrations are established in the tubewell discharge. It is important to establish the vertical permeability in the aquifer over the depth interval between arsenic source zones and HTW screens as one element in an assessment of the quality of the tubewell discharge.

The simulated arsenic concentration in the HTW discharge is more sensitive to the degree of sorption than to any other parameter. Application of sorption parameters derived from the literature suggest a very significant retardation of arsenic, delaying breakthrough and build up for decades and centuries according to the relative position of the arsenic source and well screen. Yet a 'proxy' assessment of the rate of arsenic increase, according to HTW age, has suggested a significant rise in arsenic concentrations over the past 15 years in the Meherpur district (Burgess *et al.* 2002). Sorption mechanisms specific to the sediments and hydrochemical conditions of the alluvial aquifers of Bangladesh, and hence the likely scale of their effects, are virtually unknown. This is a vital area for further research.

Another limitation of the modelling and therefore an additional source of uncertainty is the representation of the arsenic source. In particular, the rate of arsenic release to groundwater in the source zone has been assumed to be instantaneous. The arsenic source, modelled as a constant concentration in porewater within the source zone may, in reality, behave in more complex ways in response to a variety of geochemical factors. Release of arsenic from the sediments to porewater may not keep pace with transport away from the source. The concentration of arsenic in porewater at the source may change according to the mass of arsenic retained in the sediments. Over long periods of time it is likely that depletion of the arsenic source will occur and declining arsenic concentration in the tubewell discharge will eventually result. No attempt has been made to model these effects. In addition, although arsenic in porewater has a simple distribution at Meherpur (Burgess *et al.* 2002), this is not necessarily the case across other parts of Bangladesh. Depositional history and sedimentological controls may combine to produce more complex source patterns elsewhere in the basin. The methodology of our approach is valid to a variety of arsenic source distributions; the validity of the specific conclusions presented is restricted to the depth distribution that we have chosen to represent.

The models have highlighted the long timescales involved in arsenic transport to tubewells in the alluvial aquifers of Bangladesh, and the importance of interpreting field measurements of arsenic concentration in pumped groundwater from HTWs as a transient, evolving condition. Major uncertainties which remain and need to be resolved concern the geometry and depth distribution of the arsenic source zones, the mechanisms and extent of arsenic sorption, and the rates of the geochemical processes governing arsenic release from sediments to porewaters. If these issues can be addressed, numerical modelling has the potential to further the understanding and aid the mitigation of the Bangladesh groundwater arsenic crisis.

This paper is an output of the *London-Dhaka Arsenic in Groundwater Programme*. Mark Cuthbert acknowledges receipt of a Natural Environment Research Council Advanced Course Studentship and Overseas Fieldwork Allowance. John Barker suggested the potential significance of deep wells, and the analytical approach taken to address the magnitude of their influence.

References

British Geological Survey & Mott MacDonald. 1999. *Arsenic in groundwater in Bangladesh, Phase 1. Rapid Investigation.* Report to UKDFID.

Burgess, W. G., Burren, M., Cuthbert, M. O., Mather, S. E., Perrin, J., Hasan, M. K., Ahmed, K. M., Ravenscroft, P. R. & Rahman, M. 2000. Field relationships and models of arsenic in aquifers of southern Bangladesh. *In* Sililo, O. (ed.) *Groundwater: Past Achievements and Future Challenges*, Proceedings of the XXX IAH Congress, Cape Town, South Africa, November 2000. Balkema, Rotterdam, 707–712.

Burgess, W. G., Burren, M., Perrin, J. & Ahmed, K. M. 2002. Constraints on sustainable development of arsenic-bearing aquifers in southern Bangladesh. Part 1: A conceptual model of arsenic in the aquifer. *In*: Hiscock, K. M., Davison, R. M. & Rivett, M. O. (eds) *Sustainable Groundwater Development.* Geological Society, London, Special Publications, **193**, 145–163.

Dzombak, D. A. & Morel, M. M. 1990. *Surface Complexation Modelling.* Wiley & Sons, New York.

Herbert, R., Barker J. A. & Davies, J. 1989. *The pilot study into optimum well design: IDA 4000 Deep Tubewell II Project. Volume 6: Summary of the programme and results.* Report by the British Geological Survey for the Overseas Development Administration, UK Government. BGS Technical Report **WD/89/14**.

Lide, D. R. 1998. *Handbook of Chemistry and Physics. 78th Edition.* CRC Press, Cleveland, USA.

McDonald, M. G. & Harbaugh, A. W. 1984. *A modular three-dimensional finite difference groundwater flow model.* Techniques of Water Resources Investigations of the United States Geological Survey, **06-A1**.

Nickson, R. T, McArthur J. M., Burgess, W. G., Ahmed, K. M., Ravenscroft, P. & Rahman, M. 1998. Arsenic poisoning of groundwater in Bangladesh. *Nature*, **395**, 338.

Nickson, R. T., McArthur J. M., Ravenscroft, P., Burgess, W. G. & Ahmed, K. M. 2000. Mechanism of arsenic release to groundwater, Bangladesh and West Bengal. *Applied Geochemistry*, **15**, 403–413.

Rushton, K. R. 1986. Vertical flow in heavily exploited hard rock and alluvial aquifers. *Ground Water*, **24**, 601–608.

Xu, M. & Eckstein, Y. 1995. Use of a weighted least squares method in evaluation of the relationship between dispersivity and field scale. *Groundwater*, **33**, 905–908.

Zheng, C. 1992. *MT3D: A Modular Three-Dimensional Transport Model, Version 1.5. Documentation and User's Guide.* S.S.Papadopulos and Associates Inc. Bethesda, Maryland, USA.

Appendix

Derivation of the vertical steady-state flow field due to a deep well

Applying Darcy's Law to flow across a spherical surface of radius r around a point source:

$$Q = K4\pi r^2 \frac{\partial h}{\partial r} \tag{1}$$

where Q is discharge (L^3T^{-1}); K is hydraulic conductivity (LT^{-1}); and h is hydraulic head at radial distance r from well (L).
Vertical flux q_z (LT^{-1}), at radial distance r from the well is given by:

$$q_z = K\frac{\partial h}{\partial z} = K\frac{\partial h \partial r}{\partial r \partial z} \tag{2}$$

since

$$r^2 = x^2 + y^2 + z^2$$

$$\frac{\partial r}{\partial z} = \frac{z}{r}. \tag{3}$$

Combining equations gives:

$$q_z = \frac{Qz}{4\pi r^3}.$$

Thus, for the point x, y, z, the total vertical flux (q_{zT}) taking into account the well and all of its images, with coordinates x_i, y_i, z_i is:

$$q_{zT} = \sum \frac{Q(z - z_i)}{4\pi[(x - x_i)^2 + (y - y_i)^2 + (z - z_i)^2]^{3/2}}. \tag{4}$$

The analytical model quantifies the steady-state flow field produced by a deep pumping well. It considers a radial domain centred on the deep well, which is modelled as a negative point source, bounded by a no-flow boundary and with a constant head condition as the upper boundary. The aquifer is assumed to be homogeneous, isotropic and infinitely deep. The boundary conditions are imposed by the use of layers of image wells in a square grid, added in 'shells' around the centrally located pumping well until the solution converges.

Groundwater quality in the Valigamam region of the Jaffna Peninsula, Sri Lanka

L. RAJASOORIYAR[1,2], V. MATHAVAN[3], H. A DHARMAGUNAWARDHANE[2] & V. NANDAKUMAR[3]

[1]*Department of Geography, University of Jaffna, Jaffna, Sri Lanka*

[2]*Present address: School of Environmental Sciences, University of East Anglia, Norwich NR4 7TJ, UK (e-mail: l.rajasooriyar@uea.ac.uk)*

[3]*Department of Geology, University of Peradeniya, Peradeniya, Sri Lanka*

Abstract: The Valigamam region is underlain by a Miocene limestone formation and a highly porous soil cover. The region is totally dependent on groundwater to meet its agricultural, industrial and domestic needs, since other sources of water are seasonal. Recharge from rainfall is limited by high run-off and evapotranspirational losses. The region experiences water supply problems due to high concentrations of chloride, total hardness and nitrate in groundwater. The spatial distribution of chloride varies from year to year, with maximum concentrations experienced during or after the wet season. The major factor explaining high chloride concentrations is the excessive extraction of groundwater that results in saline intrusion from the sea or lagoonal areas. In a large proportion of wells sampled for nitrate, levels exceed the WHO standard due to intensive agricultural practices involving very high inputs of artificial and natural fertilizers and the improper construction of latrine soakaway pits. To improve groundwater quality in the Jaffna Peninsula will require controls on the location of new wells, a revision of existing and future pumping rates and a change in agricultural practices. It is imperative that future work in the region should focus on combining groundwater management and sustainable agricultural practice.

A great deal of emphasis has been given to the study of hydrogeological systems in the dry zone areas of Sri Lanka, as the consumption of groundwater for domestic and agricultural purposes has increased dramatically over the last few decades (Christensen & Dharmagunawardhane 1986). The quantity and quality of groundwater in the dry zones are affected by natural processes that vary according to the geological, hydrogeological and climatic settings in each region, but human activities contribute substantially to the problems of groundwater resources and water quality in these areas.

The Jaffna Peninsula, which is part of the dry zone area in Sri Lanka, is underlain mainly by a Miocene limestone that is considered to be a good aquifer for groundwater storage and discharge. However, the region experiences groundwater problems as the resource is limited and its quality has deteriorated over the years (Arumugam 1969; Nandakumar 1983). Groundwater is the only source of water for the whole peninsula and there are currently no major water supply schemes. The seasonal rainfall is of short duration, and is the only source of recharge. High evapotranspirational loss during the dry season and high run-off loss during the wet season play a major role in determining the limited storage of groundwater in the peninsula.

The Valigamam region, which covers about 50% of the peninsula and which is relatively accessible, was chosen for the present study. This area was selected because severe groundwater quality problems had been identified in previous studies (Arumugam 1969; Nandakumar 1983). In comparison with the other areas of the peninsula, the Valigamam region is moderately highly populated. Intensive agricultural practices occur on thin soils directly overlying the shallow aquifer, from which water is used for both domestic and agricultural purposes. Agricultural practices are increasingly

From: HISCOCK, K. M., RIVETT, M. O. & DAVISON, R. M. (eds) *Sustainable Groundwater Development*. Geological Society, London, Special Publications, **193**, 181–197. 0305-8719/02/ $15.00 The Geological Society of London 2002.

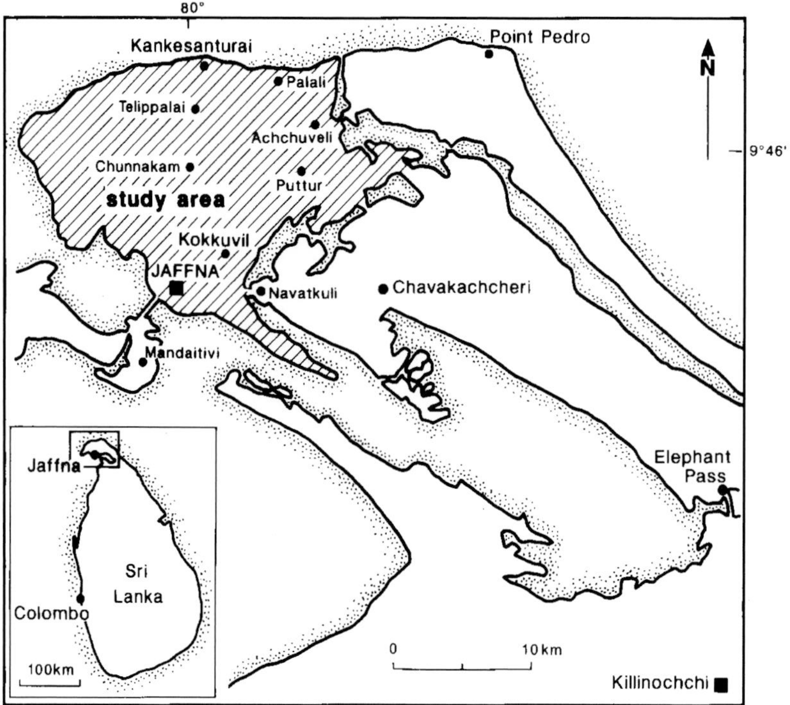

Fig. 1. Location of the Jaffna Peninsula and study area.

dependent on chemical fertilizers and pesticides for greater food production to meet the demands of local markets, and this trend has continued during the past two decades during a period of political instability.

The deteriorating quality of groundwater in the Jaffna Peninsula has justified continued water quality monitoring and investigation. A major water quality problem, identified in the 1950s and highlighted in the 1960s, is seawater intrusion into the groundwater system (Balendran *et al.* 1968). Later concern centred on the high nitrate problems related to high inputs of artificial and natural fertilizers and congested or improperly planned soakaway pit systems. It has been suggested that 80% of the wells in the peninsula are affected by high nitrate concentrations (Gunasekaram 1983).

Since 1979, a number of government and private organizations have initiated a systematic water quality monitoring programme; however, this came to an abrupt end in 1984 due to the unsettled political situation, with the result that the consequences of saline intrusion and the intensification of agriculture on the underlying groundwater are not well understood. It is unfortunate that the vast amount of data collected by government and private authorities from 1972 to 1984 was incomplete and was recorded in a disorderly manner. One of the better systematic surveys was organized by the Water Resources Board (WRB) during the period 1972–1984 in order to determine the levels of chloride and total hardness, including monthly readings, from 543 selected sample wells. In addition, the WRB collected a few borehole and shallow tubewell drilling records. However, much of the unpublished water quality data and water level maps are unavailable at present.

The primary objective of the present study was to carry out a systematic survey of groundwater quality in the Valigamam region by measurement of various parameters (chloride, total hardness, nitrate, sulphate, fluoride, electrical conductivity and pH) and compare variations in their distribution to different geological and geographical environments. The survey reported here emphasizes the distribution and interpretation of chloride in groundwater. The study was carried out between August 1997 and February 1998, to determine seasonal variations during the driest period that prevails before the rainy season (August) and the following wet

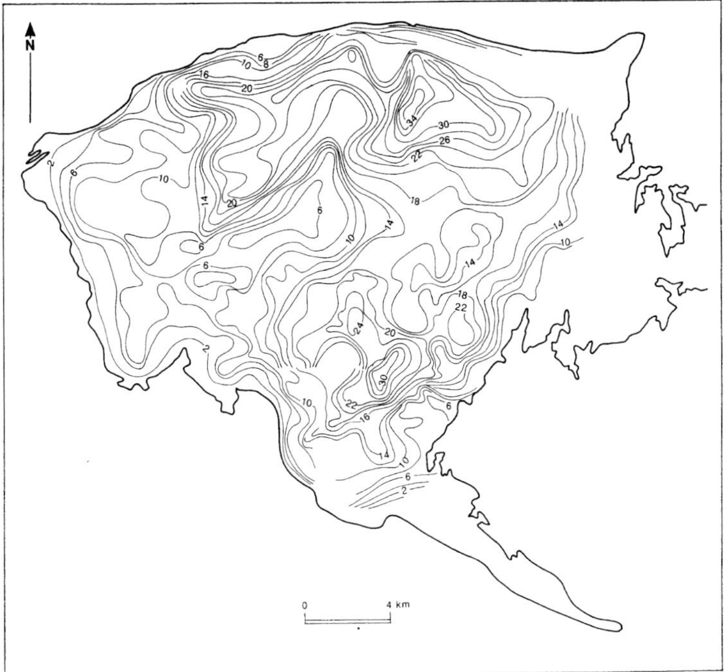

Fig. 2. Topographic map of the study area (contour lines in feet; 1 foot = 0.305 m).

period caused by heavy depressional rains (December). A secondary objective was to compile and reinterpret the groundwater quality data obtained in the previous detailed investigations of the Jaffna Peninsula so as to examine changes during the intervening 15-year period and to suggest remedial actions that should be taken to manage the groundwater resource.

Physiographic setting

Geography

The Jaffna Peninsula is located in the northern part of Sri Lanka. The Valigamam region covers approximately 325 km^2 and is bordered by straits and lagoons. No point within this area is more than 8 km from the sea (Fig. 1). The topography of the area is low and flat. The elevation varies from >35 to <1 m with an average elevation of less than 12 m above mean sea level (MSL) (Fig. 2). The area is devoid of any perennial rivers with the exception of a small intermittent rain-fed stream called the 'Valukai aru', which also drains a few very minor canals in the Jaffna area.

The area experiences a tropical arid climate. The average maximum daily temperature is 30°C and the minimum is 25°C. Winds blowing during the north-east monsoon (December to February) and the south-west monsoon (May to September), at average velocities of 40 km h^{-1}, intensify evaporative losses that amount to about 2000 mm per year (Arumugam 1970; Yogarajah 1991). The mean annual rainfall is approximately 1255 mm and falls mainly during the inter-monsoonal seasons due to depressions moving in from the Bay of Bengal and the Arabian Sea. Surface drainage is minimal under ordinary or light rainfall conditions, but since most precipitation falls during a short period – sometimes 50% within 24 h (Nandakumar 1983), surface run-off is usually very marked. This surface run-off is concentrated in shallow ephemeral channels, the small canals and the Valukai aru mentioned above, and may lead to flooding and overland sheet flow (Balendran *et al.* 1968).

The Valigamam region may be divided into nine major land-use categories that include built-up land and associated non-agricultural land (1.9%), homestead (45.6%), paddy (19.1%), coconut (2.1%), scrub (2.9%), marsh-

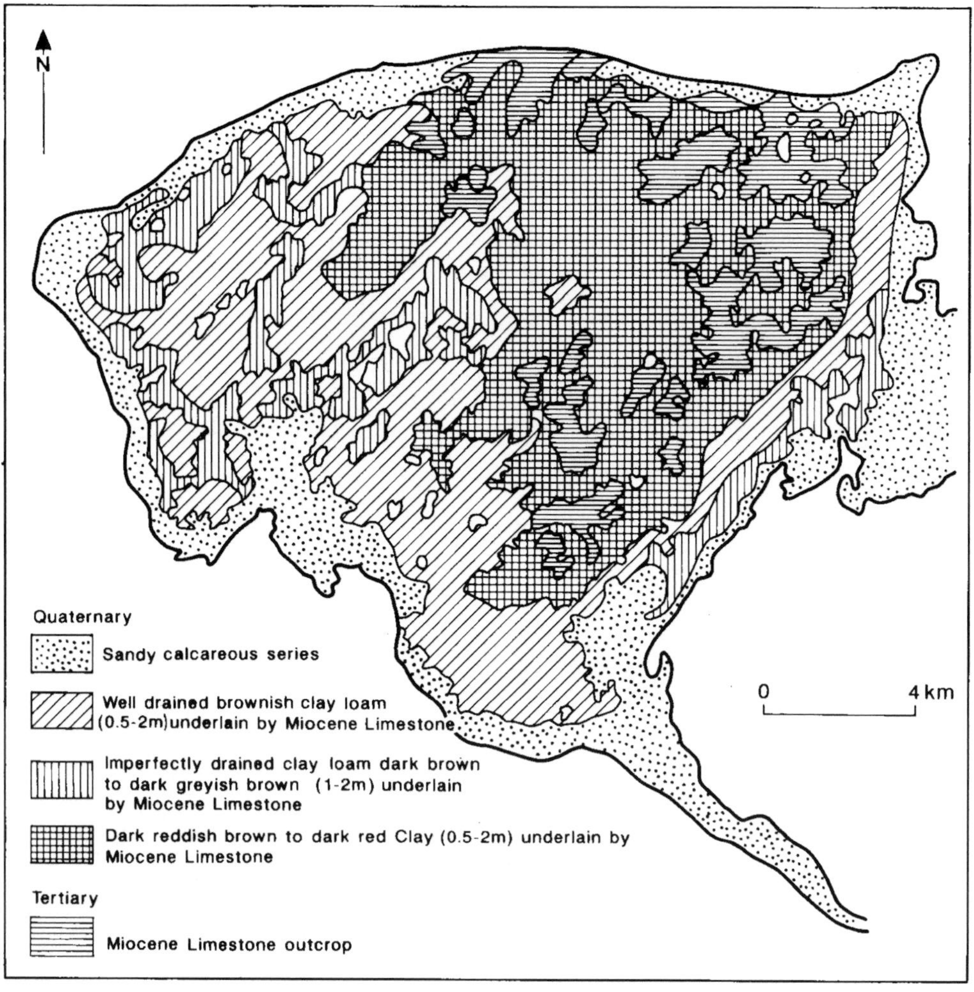

Fig. 3. Generalized soil and geology map of the Valigamam region. The inset map shows identified lineaments based on aerial photographic interpretation.

land (8.1%), sparsely used cropland (8.0%), barren lands (11.1%) and water bodies (1.2%). However, two categories, namely homestead (including gardens), and paddy, cover the major part of the study area. An important feature of the built-up land is the high density of population and housing, particularly in the coastal areas (3000 inhabitants km^{-2}) according to data from the 1981 population census.

Geology

A generalized soil and geology map is shown in Figure 3. The northern and the north-western coastal belt of Sri Lanka (stretching from Puttalam to the Jaffna Peninsula) represents the major sedimentary formation of the island. This formation consists mainly of Miocene limestone (Cooray 1984). In general, this Miocene formation unconformably overlies high-grade pre-Cambrian metamorphic rocks (the Wanni complex, formerly the West Vijayan complex) but in places is underlain by sedimentary layers of Upper Jurassic (Gondwana) age. In offshore drilling programmes, Cantwell *et al.* (1978) recognized sedimentary deposits from Lower Cretaceous to Pliocene age, separated by a number of unconformities (Cooray 1984).

The Miocene limestone of the Jaffna Peninsula is poorly bedded and generally flat, except in

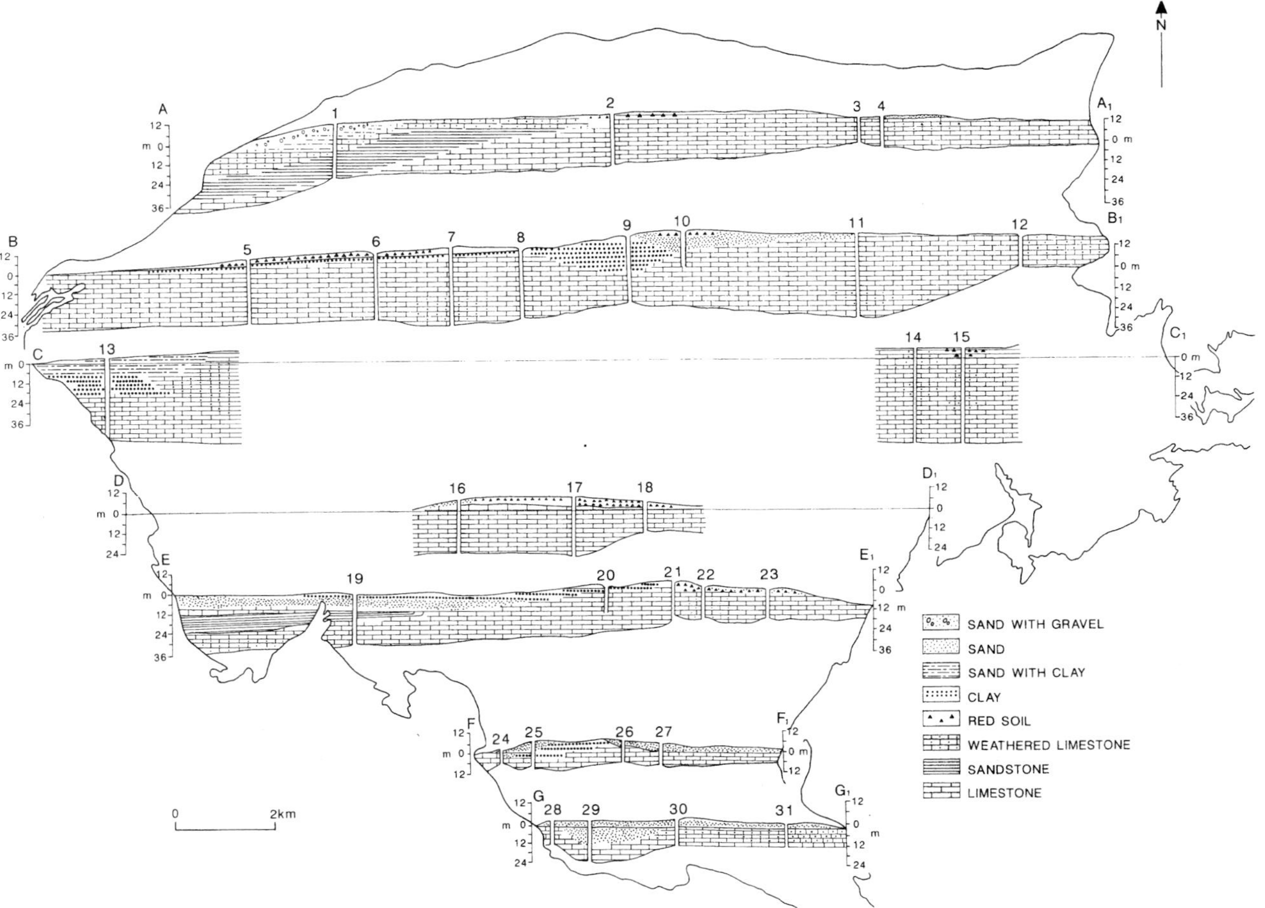

Fig. 4. Generalized cross-sections of the limestone formation in the Valigamam region.

some areas where it shows a slight dip to the west. In places the limestone beds are extremely well jointed and have a marked rectangular pattern of closely spaced joints running in north-west to south-east and north-east to south-west directions (Cooray 1984; Rajeswaran *et al.* 1993) Three lineaments, running in a north-east to south-west direction have been identified in the study area using aerial photographs and borehole information. These are in the Valukai aru basin, Upparu lagoon and Palali (Rajeswaran *et al.* 1993).

Lithologically the limestone is cream coloured varying from white grey to light brown, hard, compact, highly karstic, indistinctly bedded and partly crystalline. It also consists of sandy (siliceous) friable layers with cavities and clastic fossiliferous limestones. The limestone beds are intercalated with coral limestone beds and sandy limestone beds in the northern and southern parts of the study area. A sandstone bed is intercalated with limestone beds at a depth of about 5 m in the north-western part of the study area and this bed extends almost to the surface. The limestone is generally very fine grained but is slightly coarser and more porous where it contains fossils. This fossiliferous limestone contains many interconnected cavities that are mostly filled with calcareous clays. Generally, where the limestone is weathered it produces a yellow calcareous clay.

The vertical thickness of the Miocene limestone exceeds 35 m. In the north-east the limestone scarcely crops out, but there are a number of karstic features including surface depressions (e.g. at Manipay Idikundu), tidal wells (Puthur Nilavarai), cliffs and springs (Keeramalai). The limestone is generally overlain by highly porous thin (maximum 2 m) soil cover (Figs 3 and 4). Coral reefs are deposited around the northern coast and Quaternary red earth (Laterite), gravel and alluvium occur in the mainland (Cooray 1984). The red earth occurs as a thin layer (0–3 m) on the surface of the Jaffna limestone and in cavities, is poorly sorted and contains rounded river pebbles of different rock types. The associated reddish brown soils are deposited along the gentle undulating surface and calcic red yellow latasols occur at lower elevations (Gunesekaram 1983). Low humic clayey soils and alluvium are deposited mainly in the coasts and some parts of the mainland (Yogarajah 1991). The western, southern and eastern coastal regions of the study area are covered by a sandy calcareous soil series while red and grey earths occur inland (see Fig. 3). Beach sand is found as a narrow layer along the northern coast. Red and brown soils overlie a clay formation in the western coastal parts and the central part of the region. This clay formation extends to a depth of 16 m at Telippalai and gradually thins out towards the western coast where it is not mixed with the calcareous sand formation present.

Figure 4 shows generalized cross-sections constructed using the available borehole and tubewell logs. The borehole data reveal that the limestone is highly weathered in the central and the coastal areas at the surface and to shallow depths. In some locations, such as Kalundai, Telippalai and Vasavilan (boreholes 19, 9 and 11, respectively), the weathered limestone layers are intercalated or interbedded with hard limestone beds at different depths and thickness. At Maruthanamadam (boreholes 17 and 18), the weathered limestone beds are interbedded with calcareous sand and clay. Such sand and clay interbedding with the limestone occurs at shallow depths in most areas. The central, inland part of the region has a thick overburden consisting of highly weathered and permeable limestone, while the rest of the area lacks such conditions due to greater variability in the lithology of the interbedded limestone beds and the possibly greater uplift in the central area leading to enhanced weathering.

Hydrogeology

According to investigations carried out in 1997 (176 data points), the maximum water level in the study area in August (at the height of the dry season, i.e. groundwater level minimum) was 4.0 m above msl while in December (following the inter-monsoonal rains) it was about 5.0 m (msl). The groundwater level patterns for the period 1979 to 1997 show differences in water level elevations, especially during the dry season (August). The most significant feature (see Fig. 5) is shown for the year 1997, when the areas experiencing negative groundwater levels (below msl) were relatively large (26% of the study area). Furthermore, the areas with negative water level elevation had migrated inland in 1997 and are separated by areas with positive water levels. In 1997, a large part of the Jaffna Peninsula had negative water level elevations in both the wet and dry seasons. Excessive extraction of groundwater from wells, particularly from agricultural wells using highly efficient electrical pumps for prolonged periods, is the main cause of the reduction in groundwater elevations.

The pattern of groundwater movement in the study area between 1979 and 1997 has also changed in response to increased agricultural abstraction. In general, the direction of ground-

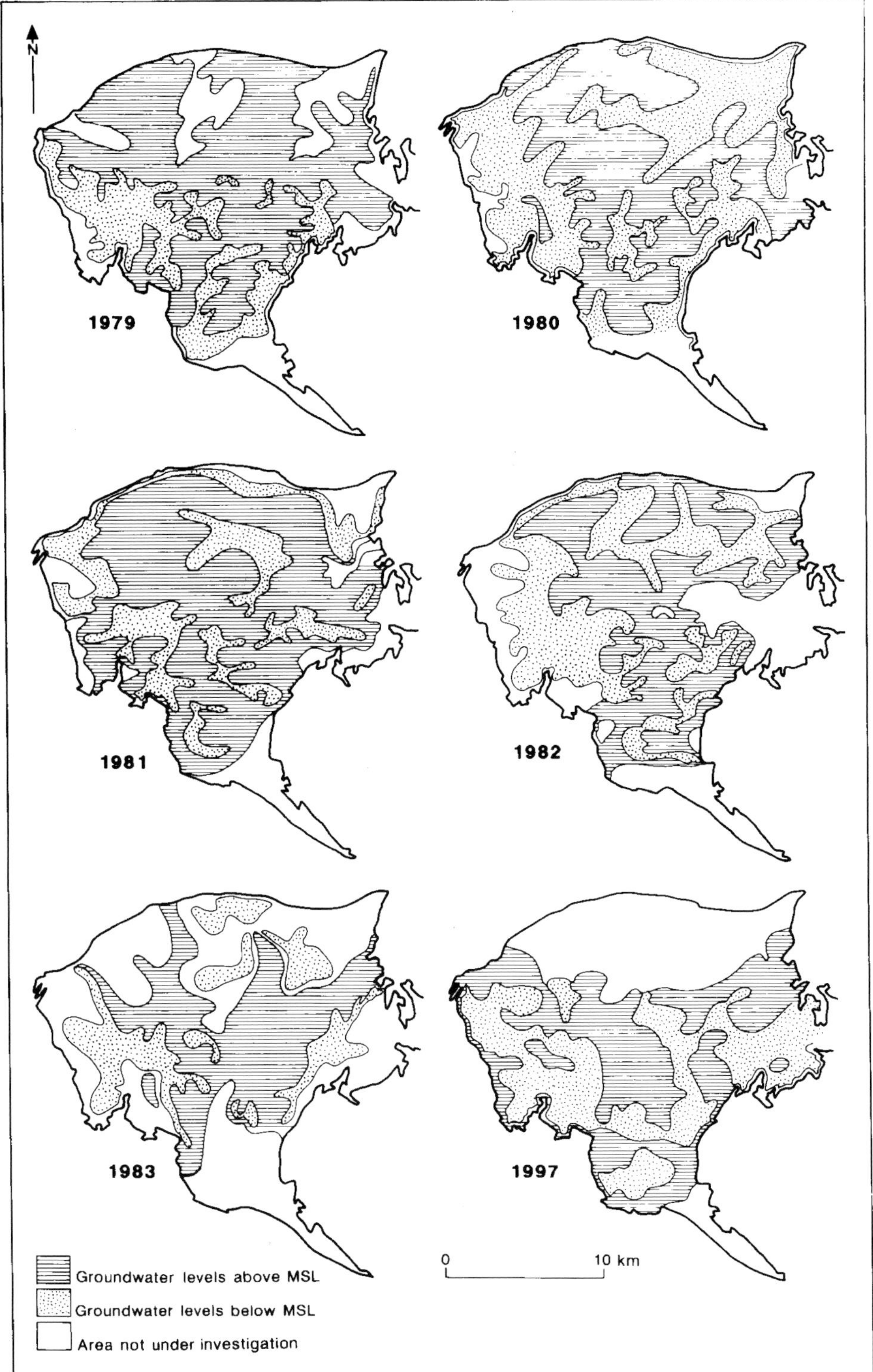

Fig. 5. Spatial distribution pattern of groundwater level elevation (1979–1997 August). Areas of groundwater levels under and above mean sea level (msl) are shown.

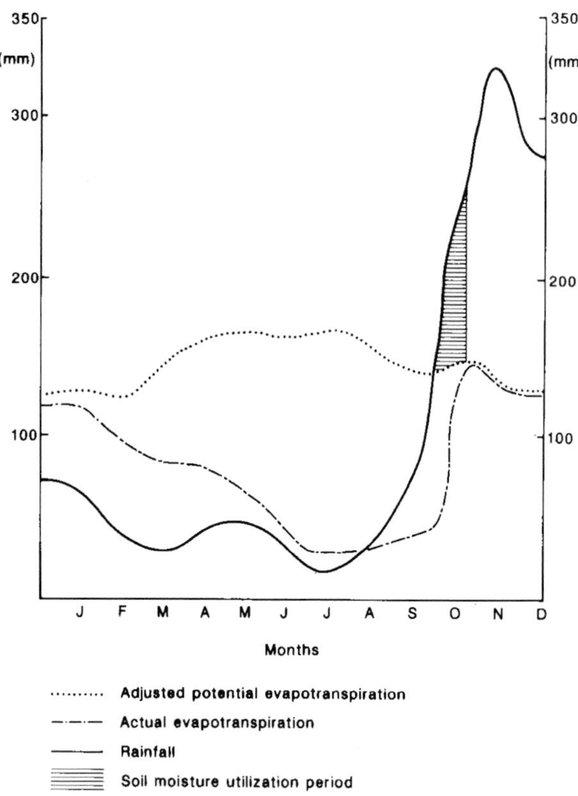

Fig. 6. Trends of annual surplus and deficit of water (rainfall) in the Jaffna Peninsula.

water movement is towards the southern, south-western and eastern coasts, where the negative water level elevations occupy relatively large areas (Fig. 5). It is important to note that the pattern of groundwater movement changes from year to year due to changes in the pumping regime.

Figure 6 shows a simplified water balance based on average monthly temperature and rainfall values for the Jaffna Peninsula over the past 50 years. The graph shows higher rates of average adjusted potential evapotranspiration (1764 mm a^{-1}) than actual evapotranspiration (1034 mm a^{-1}). Thus, an average deficit of 730 mm is typically maintained from January to August each year. The only significant period of soil moisture utilization occurs in August to September, even though the region is under intensive cultivation throughout the year. It is also important to note that the balance of average annual surplus rainfall (about 191 mm) includes both surface and subsurface run-off.

Total catchment recharge estimations have rarely been made for the Jaffna limestone due to the absence of relevant data. For the present study, the estimated volume of recharge was calculated for an area of 185.5 km^2 using estimated values of specific yield and an average increase in groundwater levels of 0.61 m for a few selected wells for the period August to December. This limited approach revealed that assuming a specific yield of 0.18 would furnish an average annual recharge rate for the selected area of 2.0×10^7 m^3 (110 mm a^{-1}). However, this value does not account for the variation in the spatial pattern of recharge in the area, particularly as 25% of the shallow dug wells are located in calcareous sandy and gravely formations whose porosity varies from 0.25 to 0.30.

Field and laboratory methods

For the present study, carried out between 1997 and 1998, seven hydrochemical parameters (chloride, total hardness, electrical conductivity (EC), pH, nitrate, sulphate and fluoride) were

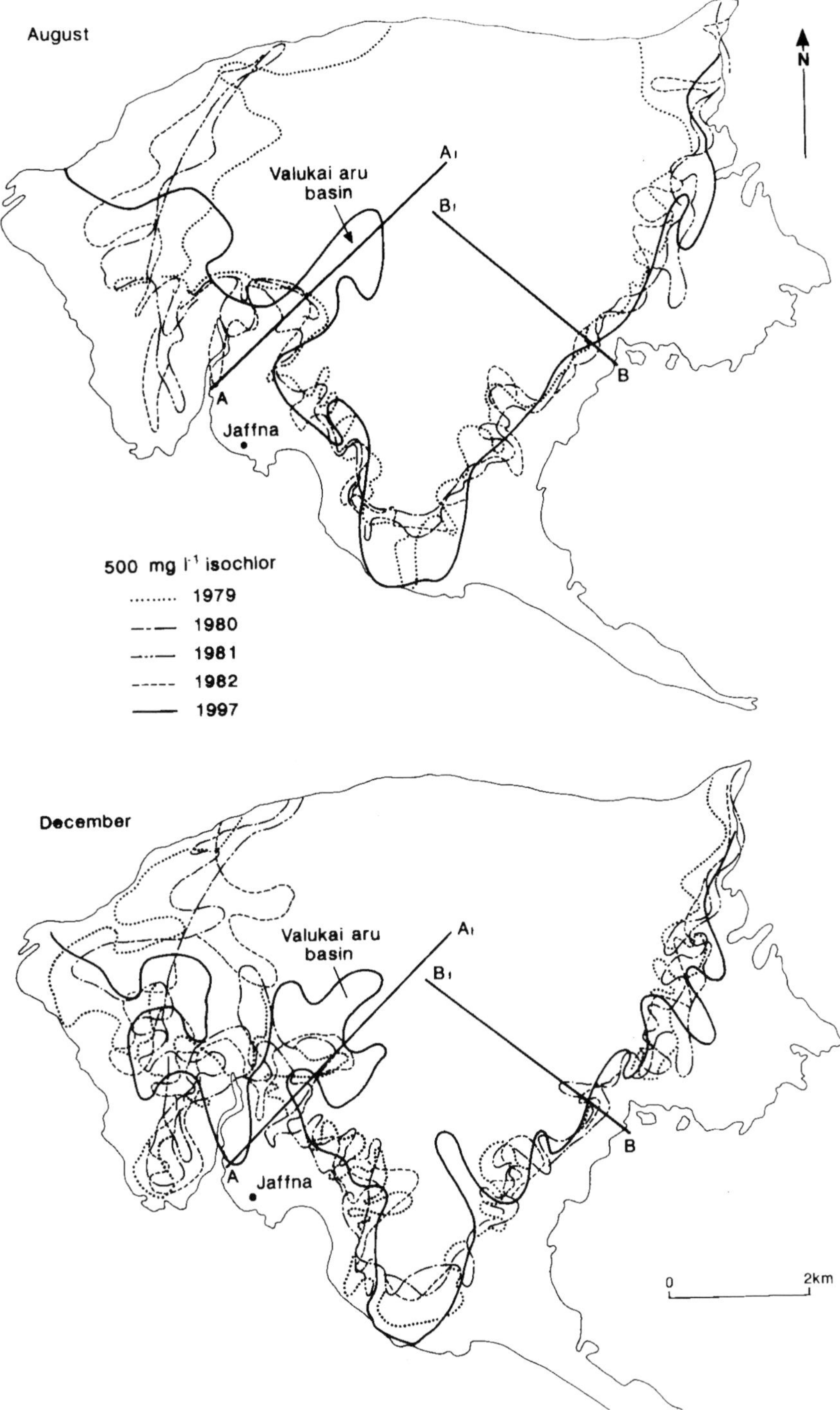

Fig. 7. Spatial distribution of groundwater chloride showing the 500 mg l^{-1} Cl^- contour for the years 1979 to 1982 and for 1997 for both August and December.

determined. Two water quality parameters (chloride and total hardness) are discussed in detail since there is a large number of analyses available, particularly for the period 1979 to 1982 (based on secondary data) and for 1997.

Systematic sampling using a bailer was undertaken at 176 locations during both the dry and wet seasons. Water levels in the sample wells were also recorded for both seasons. Rainfall and temperature data were obtained from the Sri Lanka Meteorology Department for the period 1971–1991 and 1971–1981, respectively. Additionally, rainfall and temperature readings for 1997–1998 were recorded at a single station (University of Jaffna). The tidal variation of the Jaffna lagoon was also measured during the period of this study. A questionnaire was distributed to well owners to obtain information on well use and a qualitative appraisal of water quality.

The analytical work was carried out in three different institutions (University of Jaffna, University of Peradeniya and National Water Supply and Drainage Board). The determination of chloride, total hardness, pH and EC (at 25°C) was completed within 48 hours of sample collection. Chloride was determined by silver nitrate precipitation. Additional chloride measurements were made using an ion specific electrode and a reference electrode. Fluoride was determined using a fluoride ion specific electrode and reference electrode. Water hardness due to Ca^{2+} and Mg^{2+} was estimated by titration with standard EDTA using Eriochrome Black T indicator.

Results and discussion

Electrical conductivity and pH

Electrical conductivity values for groundwaters in the study area range from 300 to >22 000 μS cm^{-1}, with the higher values being found in coastal regions and inland areas under intensive agriculture. All measured groundwaters in the region have pH values above 7.0 reflecting the alkaline nature of the limestone aquifer.

Chloride

The geographical setting and geological environment play a major role in determining the distribution of chloride in groundwater in the study area. Chloride concentrations are very high in the sandy calcareous formations of the coastal regions, increasing from 500 mg l^{-1} (the WHO permissible level for chloride) to >4000 mg l^{-1}. The 500 mg l^{-1} Cl^- isochlors for the years 1979 to 1982 and for 1997, for both August and December, are shown in Figure 7 and reveal a number of important features regarding the spatial variation of groundwater chloride: areas under high salinity (>500 mg l^{-1} Cl^-) are confined to the coastal regions; the 500 mg l^{-1} Cl^- contour shows an irregular, meandering spatial variation, both for the dry and wet seasons; and in the Valukai aru area, the 500 mg l^{-1} Cl^- contour expanded significantly in 1997.

The pattern of chloride concentration is a clear indication of seawater intrusion. Concentrations as high as 20 000–30 000 mg l^{-1} in some selected coastal locations provide categorical evidence of seawater intrusion. A noticeable feature in the areas of high chloride concentration is the influence of fracture zones and lineaments (Fig. 7). For example, the lineament feature of the Valukai aru region always presents a linear pattern of high chloride concentration.

Table 1. *Temporal variation in area of groundwater chloride zones*

Year	Chloride zone		
	Area in km^2		Total area in km^2
	500–1000 mg l^{-1}	>2000 mg l^{-1}	>500 mg l^{-1}
August			
1979	47.8	59.0	155.0
1980	21.8	64.8	129.8
1981	39.1	52.4	199.5
1982	–	–	114.4
1997	53.6	46.9	122.9
December			
1979	29.3	47.5	108.4
1980	–	64.4	132.0
1981	25.8	86.0	134.0
1982	19.1	64.9	123.7
1997	42.1	45.7	112.7

Previous studies have suggested that the areas experiencing high salinity have increased since 1979 (Elankumaran 1994). To test this assumption, areas occupied by different salinity levels (chloride zones) were measured for the period from 1979 to 1982 and for 1997 using Figure 7. Results are given in Table 1, it should be noted that the number of wells studied in 1997 (176 wells) was much smaller than the number of wells studied in the period 1979 to 1982 (543 wells). Therefore, the data given in Table 1 should be treated with caution. In general, the extent of the chloride levels in the range 500–1000 mg l^{-1} and >2000 mg l^{-1} is much less or

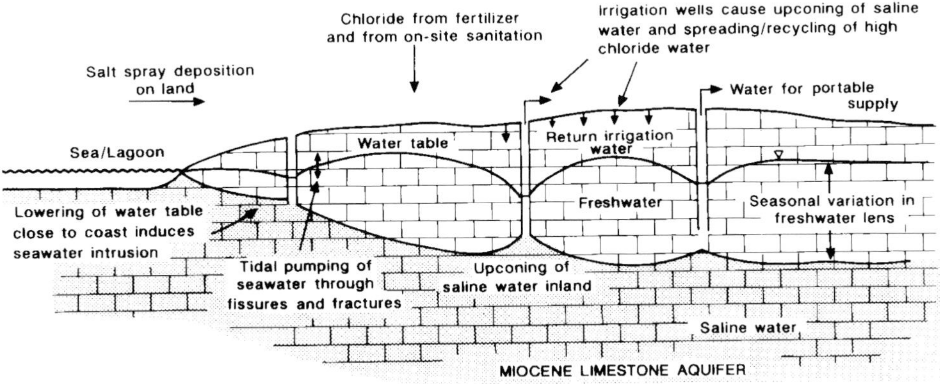

Fig. 8. Conceptual model of the factors controlling the sources and distribution of saline groundwater in the Miocene limestone aquifer of the Jaffna Peninsula.

shows a decline in the wet season (December) compared to the dry season (August). The data also reveal that the total area experiencing groundwater salinity has fluctuated during this period.

The total rainfall data for the corresponding years do not indicate any significant correlation with either the spatial variation of the 500 mg l^{-1} chloride contour or the extent of the areas experiencing saline conditions. A number of factors can be cited that control the chloride pattern, including: changing low and high tides of the lagoons that surround the study area; variation in the thickness of the freshwater lens before and after the inter-monsoonal and the north-east monsoonal rains; changes in the amount and extent of salt spray derived from the surrounding seas/lagoons; pattern of fertilizer inputs on cultivated lands; over-abstraction of groundwater leading to upconing of the saline/freshwater interface and saline intrusion; lineament and fracture position; and improper positioning of latrines.

Figure 8 provides a conceptual illustration of these factors, although whilst each depends to some extent on the location and use of any particular area of land, it is difficult to distinguish between processes due to a lack of data.

In contrast, a gradual increase in chloride concentration with seasonal rainfall has been identified. Figure 9 shows plots of average chloride concentration in the study area against monthly rainfall for the period 1980–1982. Also shown is a well hydrograph for Jaffna (well number 270). Figure 9 reveals a gradual increase in groundwater chloride concentration during

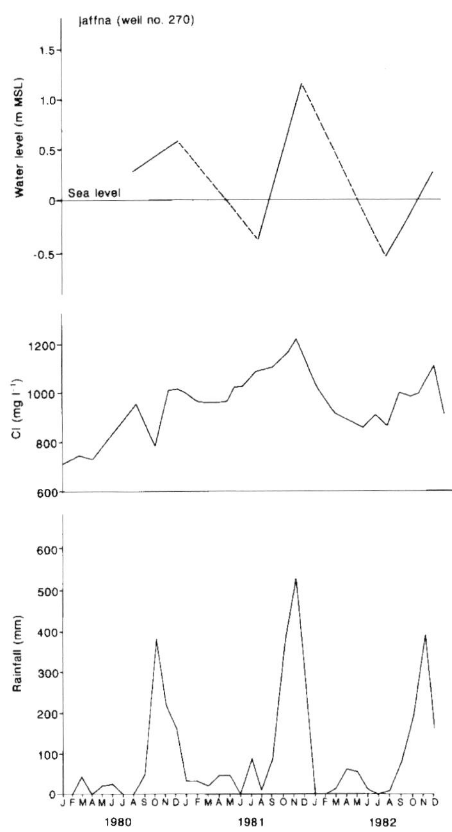

Fig. 9. Relationship between the average monthly rainfall, groundwater level at Jaffna and average chloride concentrations (1980–1982).

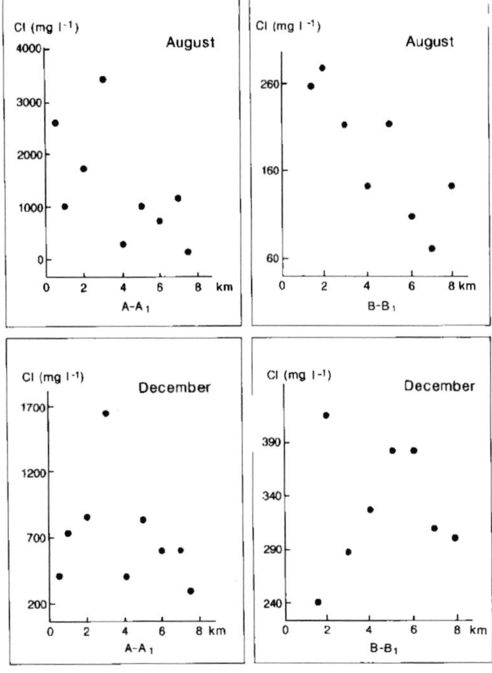

Fig. 10. Graphs showing the relation between distance inland from the coast and chloride concentration for sections A - A_1 and B - B_1 in August and December. The lines of the sections are shown in the Fig. 7.

the dry season, when the groundwater level is near to or below sea level, with the maximum concentration reached soon after the rains commence, followed by a rapid decrease once the rains cease.

In the study area, and as might be expected, the distance inland versus the chloride concentration should maintain a negative correlation as mentioned by Puveneswaran (1986). Figure 10 shows two selected profiles. Profile A-A_1, which includes the Valukai aru basin, shows a weak negative correlation ($r = -0.57$) during the dry season but no correlation during the wet season. In contrast, profile B-B_1 crosses an area that is not influenced by any lineaments. This profile shows a stronger negative correlation between distance inland and chloride concentration ($r = -0.83$) during the dry season, but again shows no trend in December.

Figure 11 shows histograms of chloride levels versus land-use patterns in the study area in August and December for 1979 and 1997. Six land-use types are shown and the histograms suggest that chloride concentrations are relatively low in built-up areas, scrub land, homestead areas, and coconut plantations, while the coastal marsh and barren lands have relatively high chloride levels. As expected, some built-up areas in the coastal zone have higher chloride levels. The irregular chloride distribution in areas under paddy cultivation may be related to the location of these areas. Paddy is cultivated mostly in the coastal region but is also cultivated in some low-lying inland areas. A significant contributory factor to the high levels of chloride in the cultivated land areas is the extensive extraction of groundwater in support of cultivation (Arumugam 1969; Nandakumar 1983) that has caused the lowering of groundwater levels and a rise in the saline water body.

Table 2. *Table of average chloride concentrations identified by cluster analysis for 1983 and 1997*

	Cluster	Number	Dry season ($mg\ l^{-1}$)	Wet season ($mg\ l^{-1}$)
1983	A	16	208	289
(previous	B	11	252	254
study)	C	7	348	404
	D	6	793	513
	E	5	822	685
	F	4	189	263
	G	Not reported	528	854
	H	Not reported	625	547
1997	A	70	190	226
(this	B	7	471	238
study)	C	20	375	506
	D	23	716	571
	E	8	1673	605
	F	15	961	1870
	G	7	2906	662
	H	13	2515	3229
	I	4	5096	4718

The patterns of groundwater chloride concentration inferred on the basis of cluster analysis techniques are shown in Figure 12, with the average chloride concentration for each identified cluster shown in Table 2. Elankumaran (1994) applied the cluster analysis method to study the pattern of chloride and total hardness in the Valigamam region for the years 1979, 1981 and 1983, using data from the Water Resources Board (Jaffna). The clusters labelled A, B and C in Figure 12 for 1983 were considered as zones with chloride concentrations below the WHO permissible limit of 500 mg l^{-1}, although some variation is observed between dry and wet seasons. Elankumaran (1994) suggested that the area affected by chloride has varied every year and therefore the clusters that contain good quality water have been shrinking since

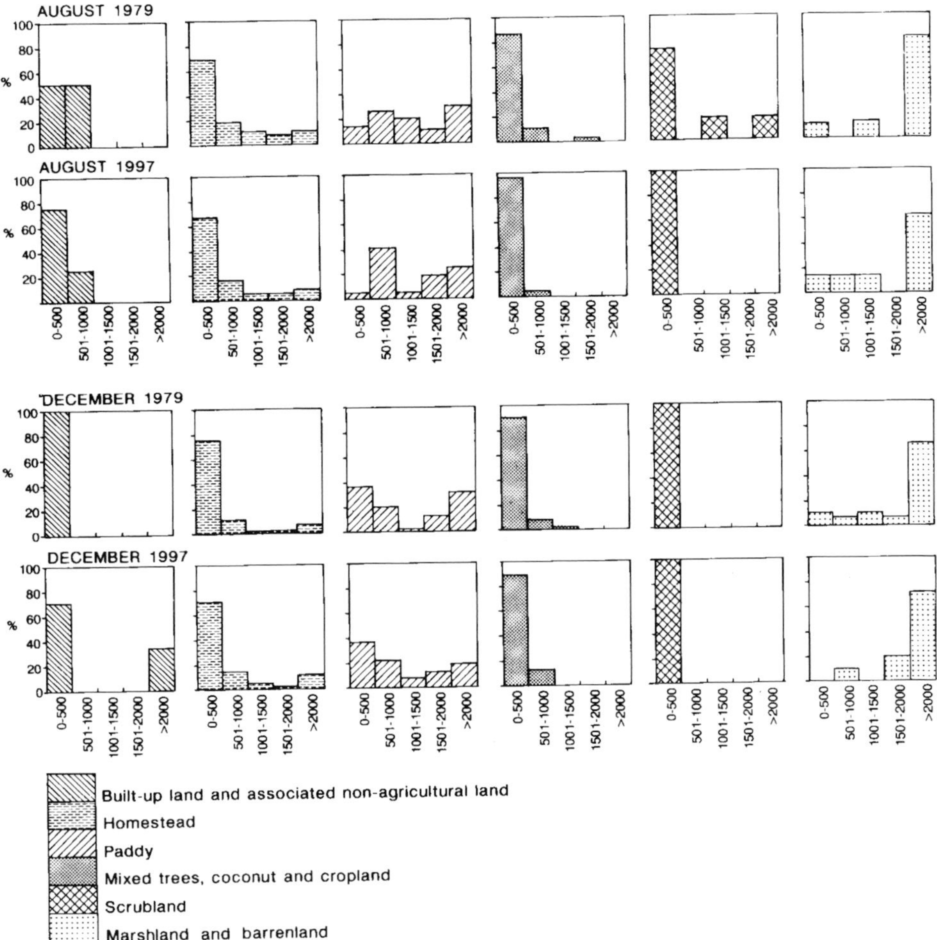

Fig. 11. Histograms showing the relation between land-use pattern and chloride concentrations (1979–1997 August and December).

1979. It is also evident in Figure 12 that the clusters are probably related to the heterogeneity of the limestone aquifer that causes different responses to recharge events, even in wells only 10 m apart.

The cluster analysis performed using data from the present study shows a more complex pattern to that suggested by Elankumaran (1994) for 1983, based on a similar set of site locations. In 1983, the concentration of chloride was higher after the rains when compared to the dry season except in clusters B, D, E, G and I. Higher chloride concentrations were observed in the remaining clusters. Clusters E–I in the 1997 study show a range of chloride concentrations from 961 to 5096 mg l^{-1} (dry season values) that are mainly located in some of the coastal areas in regions of intensified cultivation. Overall, the present study shows no consistent or regular pattern in the extent of groundwater salinity, but there is an apparent decrease in the saline areas by as much two-thirds compared with the earlier studies. In contrast, in the Valukai aru region, there has been an increase of 55% in the area occupied by saline groundwater between 1982 and 1997.

One reason for the observed differences in the spatial distribution of groundwater chloride is the internal migration of people that occurred between 1983 to 1996 as a direct result of the

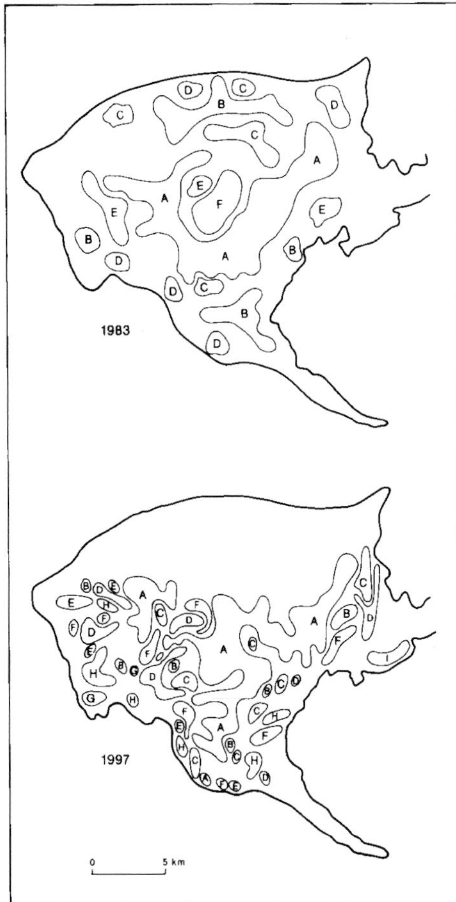

Fig. 12. Pattern of groundwater chloride concentration inferred on the basis of cluster analysis (1983 and 1997 August). See Table 2 for average chloride concentration for each cluster.

political and socio-economic situations. The population density of the Valigamam region underwent a vast change from approximately 4.64 million people in 1981 and 4.86 million in 1993, to 2.42 million in 1997. The population of the northern and north-western parts of the region experienced a sharp decline from about 48 000 to 35 000 in the Valigamam West division and 72 000 to 4600 in the Valigamam North division. These demographic changes may have improved the hydrogeological environment in the northern and north-western coastal areas as reflected in a decline in the total salinity of the area, but the southern area close to the Valukai aru region, which has become densely populated due to internal migration and intensified cultivation, has experienced a deterioration in groundwater quality. The situation was different in 1995 owing to the mass evacuation of people from the entire Valigamam region, a situation that prevailed until the latter part of 1996. Interestingly, even though the population was reduced to almost zero for about a year, there was no improvement in the saline distribution. This is due to the speed with which negative groundwater level elevations were again achieved once pumping restarted and indicates a responsive aquifer, although without more pumping and recharge data it is difficult to determine exactly how responsive.

Total hardness

The spatial pattern of total hardness in groundwaters in the study area, as elsewhere in the Jaffna Peninsula, is related to the hydrochemistry of the limestone aquifer. The distribution of total hardness and its relationship with land use is very similar to the trends observed for chloride.

Table 3. *Temporal variation in area of groundwater total hardness zones*

Year	Total hardness zone		
	Area in km^2		Total area in km^2
	500–1000 mg l^{-1} as $CaCO_3$	> 2000 mg l^{-1} as $CaCO_3$	> 500 mg l^{-1} as $CaCO_3$
August			
1979	55.0	112.4	123.5
1980	39.2	50.8	110.9
1981	26.1	62.5	122.0
1982	–	–	–
1997	–	–	101.5
December			
1979	33.9	29.6	95.3
1980	37.2	50.5	117.1
1981	25.3	45.3	166.7
1982	35.2	56.5	119.3
1997	47.8	–	100.2

Table 3 shows the extent of areas with total hardness between 501–1000 and >2000 mg l^{-1} $CaCO_3$ for the period 1979–1982 and in 1997. The temporal variation of the two selected total hardness zones is again similar to that shown for chloride (Table 1). Bearing in mind that the data are limited for the years 1982 and 1997, they suggest a gradual expansion of the area with total hardness greater than 500 mg l^{-1} as $CaCO_3$

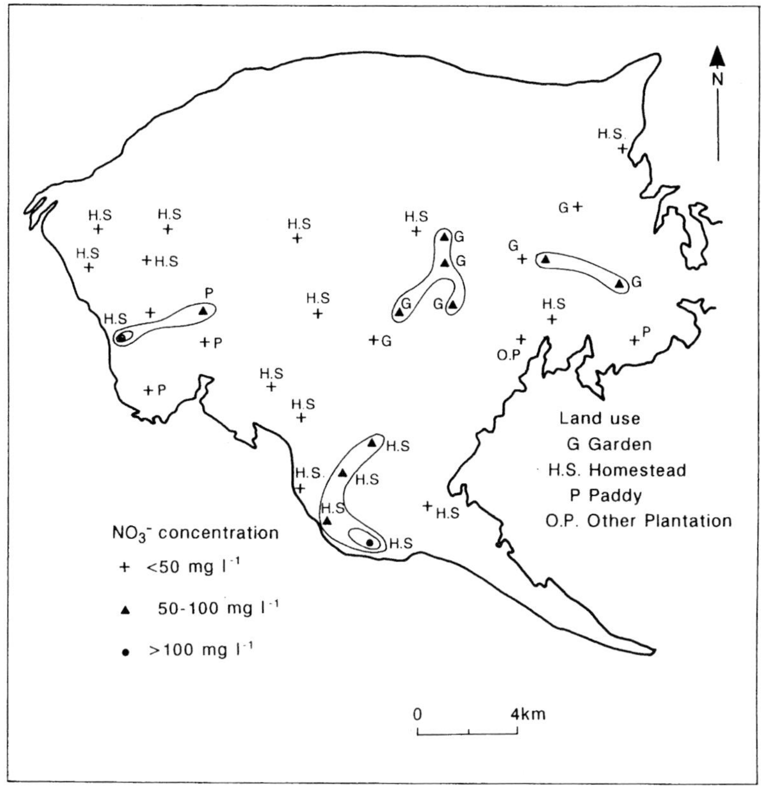

Fig. 13. Concentration of groundwater nitrate in selected locations of the study area for February 1998.

during the wet season (December). A similar trend is seen in the >2000 mg l^{-1} $CaCO_3$ total hardness range. However, in the dry season (August), the 500–1000 and >2000 mg l^{-1} $CaCO_3$ zones show gradual reductions in area. The observed temporary expansion in the total hardness zones shortly after rainfall appears to match expansion in the areas of corresponding chloride concentrations.

Nitrate, sulphate and fluoride

The number of nitrate analyses (35) carried out in the present study is insufficient to properly describe the spatial distribution of groundwater nitrate in the area. A summary of the nitrate concentration data is shown in Figure 13 and it is apparent that the distribution of results is dependent on the source of nitrate, in particular the application of artificial fertilizers and the disposal of human wastes. Nitrate concentrations, particularly in selected agricultural and urban wells, reach values of more than 100 mg l^{-1} NO_3^-. The range of nitrate concentrations is <8 to >165 mg l^{-1} NO_3^- with a mean value of 25 mg l^{-1} NO_3^-.

Previous studies have shown very high levels of nitrate in groundwater in the Jaffna Peninsula (Gunasekaram 1983; Maheswaran & Mahalingam 1983), sometimes up to twice the WHO (1984) standard of 45 mg l^{-1} NO_3^-. In recent years intensive agricultural practices have increased in response to population growth and have resulted in very high inputs of artificial and natural fertilizers, with excessive amounts of manure applied to agricultural land in rotation. Inputs are always above crop requirements, resulting in leaching of the excess to groundwater (Navarathnarajah 1994). In the future, attention should be given to more sustainable agricultural practices that limit this unnecessary loss of nitrate.

Another factor responsible for high nitrate concentrations is the improper planning of soakaway pits and dug wells (Gunasekaram 1983). Distances between latrine pits and dug wells are not maintained as recommended, particularly in highly populated urban areas (Table 4). The distance chosen depends on the site geology and more importantly on the soil type. The Jaffna Municipal Authority recommends a minimum separation of 7.5 m, reflecting a pragmatic approach to the siting of latrine pits in the densely populated coastal areas.

Table 4. *Distances between pit latrines and dug wells for two regions in the Jaffna Peninsula*

1990 Jaffna municipal area		1997 Valigamam area	
Distance (m)	Percentage of dug wells	Distance (m)	Percentage of dug wells
<1.5	5.7	<10	13.6
1.6–3.0	8.0	10.1–20.0	48.2
3.1–4.5	5.7	>20.1	38.2
4.6–6.0	6.8	–	–
>6.1	73.5	–	–

The depths of dug wells in the study area are generally more than 4.0 m, except in areas underlain by sandy formations, and have a shallow lining. The shorter the length of lining the more likely are surface pollutants derived from agricultural and urban inputs to enter the abstracted groundwater (Table 5). In the present study, a single well in Jaffna town demonstrated a high nitrate level (115 mg l^{-1} NO_3^-) but all the other locations in the town showed nitrate levels below the WHO standard. Although the number of samples collected in this study was limited, the current situation may be interpreted in a positive way given that the reduction in the population density of the Jaffna division from about 6000 inhabitants km^{-2} in 1981 to 1800 inhabitants km^{-2} in 1996 may help reduce the threat of groundwater contamination from pit latrines.

Table 5. *Length of lining in dug wells of the Valigamam region, Jaffna Peninsula*

Lining (m)	Percentage of wells
<1.0	7.4
1.1–3.0	41.5
>3.1	38.5
Damaged	12.6

The same limited number of analyses of water samples for sulphate provided values in the range 1.0–1500 mg l^{-1} SO_4^{2-}. The higher values are observed near the coastal regions and are related to seawater intrusion. Groundwater data for fluoride (35 samples) showed that concentrations of this water quality parameter are low, with less than 1 mg l^{-1} F^- measured in 95% of the samples. This is a result of solubility controls causing calcium fluoride precipitation.

Conclusions

The water resources of the Valigamam region of the Jaffna Peninsula depend totally on rainfall recharge to the Miocene limestone aquifer. The study described here confirms previous findings that with respect to chloride and hardness, the quality of groundwater in parts of the Jaffna Peninsula, including the Valigamam region, is poor and deteriorating. Generally speaking, the levels of chloride, total hardness, nitrate and sulphate are high, and the spatial distribution pattern of these parameters has altered as a result of changes in the demography of the region. Human activities, in the form of adverse agricultural practices and careless waste disposal, have compounded the situation over the years. It is imperative that future work in the region should focus on combining groundwater management and sustainable agricultural practice.

A number of measures have or can be implemented to control or reduce the problems of chloride and nitrate contamination of groundwater. To maintain satisfactory levels of chloride, an important scheme was launched in 1952 by the Department of Irrigation to convert the lagoons into freshwater bodies by flushing out salts (Navaratnarajah 1994). This scheme was completed in the early 1970s and the lagoons were recognized to contain freshwater bodies. Farmers who cultivated land adjoining the lagoons derived the benefits, but poor maintenance of the scheme resulted in a return to the previous adverse conditions in the lagoons and adjacent areas. Other suggested measures include augmenting rainfall recharge by natural or artificial means, and the physical control of seawater intrusion by installing artificial subsurface barriers in coastal areas (Nandakumar 1994). Restrictions on pumping rates in existing and new wells should also be considered.

Lastly, and in common with many areas of the world, local educational programmes are required to strengthen public awareness of the reasons for the poor quality and deteriorating nature of groundwater in the Jaffna Peninsula.

We are indebted to: S. Balachandran, Department of Geography, V. Arasaratnam and S. Balasubramaniam, Department of Biochemistry, J. Ganeshamoorthy, Department of Medicine and Pharmacology, V. K. Ganeshalingam, Department of Zoology, University of Jaffna; Atula Senaratne, University of Peradeniya; and J. P. Padmasiri, Regional Chemist, National Water Supply and Drainage Board, Peradeniya for their kind assistance in providing project facilities and supervision. We are grateful to the Chairmen, Water Resources Board, Global Engineering and Technical Services, Geological Survey Department and the Department of Irrigation for their kind assistance in terms of data provision. We also thank members of the Department of Geology, University of Peradeniya and the Department of Biochemistry and Geography, University of Jaffna, for their assistance throughout the project. We are grateful to K. M. Hiscock and A. E. Foley, University of East Anglia, for their help in editing this paper, and to V. Navaratnarajah, University of Jaffna, for his encouragement and assistance.

References

ARUMUGAM, S. 1969. *Water Resources of Ceylon, its Utilisation and Development*. Water Resources Board, Colombo.

ARUMUGAM, S. 1970. Development of ground water and its exploitation in the Jaffna Peninsula. *Transactions of the Institute of Engineers, Ceylon*, **1**, 31–62.

BALENDRAN, V. S., SIRIMANNE, C. H. I. & ARUMUGAM, S. 1968. *Ground Water Resources of the Jaffna Peninsula*. Water Resources Board, Colombo.

CANTWELL, T., BROWN, T. E. & MATHEWS, D. G. 1978. Petroleum geology of the northwest offshore area of Sri Lanka. *Proceedings of the Seapex Conference*, Singapore, 1978.

CHRISTENSEN, H. & DHARMAGUNAWARDHANE, H. A. 1986. *Behaviour of some chemical parameters in tube well water in Matale and Polonnaruwa districts*. Sri Lanka Association for the Advancement of Science, 25–43.

COORAY, P. G. 1984. *An Introduction to the Geology of Sri Lanka (Ceylon). 2nd revised ed.* National Museums Sri Lanka, Colombo, 126–269.

ELANKUMARAN, C. 1994. *Ground Water Patterns of Valigamam Region*. University of Jaffna, Gunesekaran.

GUNESEKARAM, T. 1983. *Ground water contamination and case studies in Jaffna Peninsula, Sri Lanka*. Global Engineering Technology Services, Jaffna.

MAHESWARAN, R. & MAHALINGAM, S. 1983. Nitrate - nitrogen content of well water and soil from selected areas in the Jaffna Peninsula. *Journal of National Scientific Council, Sri Lanka*, **II - 1**, 269–275.

NANDAKUMAR, V. 1983. *Natural Environment and Ground Water in the Jaffna Peninsula, Sri Lanka*. Tsukuba, Japan.

NANDAKUMAR, V. 1994. *Groundwater contamination in the Jaffna Peninsula*. University of Tsukuba, Ibaraki, **305**, 155–164.

NAVARATHNARAJAH, V. 1994. Water problems in the Jaffna Peninsula. *20th WEDC Conference in Affordable Water Supply and Sanitation*, 1994, 191–193.

PUVENESWARAN, K. M. 1986. Spatial temporal variation and the human dimensions of groundwater of Jaffna Peninsula. *Beitrage zur Hydrologie*, **5**, 827-845.

RAJESWARAN, S. T. B., ROBERT, G., RAJASOORIYAR, L. D. 1993. The evolution and the morphological structure of the Jaffna Peninsula. *Sinthanai*, **5**, 58–65.

YOGARAJAH, K. 1991. *De-silting of ponds*. Irrigation Office, Jaffna Division.

WORLD HEALTH ORGANISATION. 1984. *Guidelines for Drinking Water Quality. Vol. I: Recommendations*. WHO, Geneva.

Will reductions in groundwater abstractions improve low river flows?

K. R. RUSHTON

School of Civil Engineering, University of Birmingham, Birmingham B15 2TT, UK

Abstract: It is a commonly held view that a reduction in groundwater abstractions in many heavily exploited aquifers is likely to improve low flows in rivers. In most aquifers, but especially in chalk and limestone aquifers, the improvement in flows may only be small compared with the reduction in abstraction. Using conceptual and numerical models, the relevant flow processes in chalk and limestone aquifers are examined with special reference to the impact of pumped boreholes on stream and river flows under conditions of differing transmissivities resulting from high or low groundwater heads. The key response is that much of the winter recharge is lost to high winter river flows with only a small proportion of the recharge being stored in the aquifer and available in the summer. Consequently, the low summer river flows, which depend primarily on the areas local to the rivers, may not be strongly influenced by pumped boreholes unless these are close to the rivers. Case studies are used to illustrate the discussion.

Abstraction from aquifers almost certainly results in some reduction in river flows. Rarely is it realistic to stop all abstraction from the aquifer to improve river flows, but it may be possible to reduce abstraction from some sites for all or part of the year. Before proceeding with any reduction in abstraction it is essential to estimate what will be gained by such a procedure.

If abstraction boreholes are close to rivers there is likely to be a direct improvement following a reduction in abstraction. However, if abstraction sites are some distance from the river then the impact is more difficult to assess. Analytical solutions are available that predict the impact of an abstraction borehole on river flows. An analytical solution published recently by Hunt (1999) represents the river as being partially penetrating. The value of this type of analytical solution is that it demonstrates that the pumping does not have an immediate effect on the river; in unconfined aquifers it can take many months before the impact reaches the river. Although these analytical solutions do provide insights into the impact of pumping, they focus on the borehole, the aquifer and the river. They fail to include the influence of varying recharge, changing transmissivities with saturated depth, non-perennial rivers, the possibility of a river being perched above (hence disconnected from) the aquifer, and other outlets from or inlets to the aquifer system such as rivers, lakes or the sea.

These effects, which are not included in analytical solutions, occur frequently in limestone and chalk aquifers. To obtain an improved understanding of the impact of any reduction in abstraction on river flows, it is necessary to carry out a detailed hydrogeological study of the aquifer system with particular reference to the variation of aquifer properties between winter and summer. Different responses of pumped boreholes between winter and summer are especially important (Rushton & Chan 1976). It is also necessary to examine and understand non-perennial rivers and streams (Rushton *et al.* 1989). These detailed field studies lead to the development of conceptual models and thence to mathematical models. Once mathematical models have been refined so that they represent field responses, they can be used to explore the effects of changes to the abstraction patterns.

Two case studies are chosen to illustrate the impact of reduced abstractions on river flows.

From: Hiscock, K. M., Rivett, M. O. & Davison, R. M. (eds) *Sustainable Groundwater Development*. Geological Society, London, Special Publications, **193**, 199–210. 0305-8719/02/

The first refers to the Southern Lincolnshire Limestone where various attempts have been made to improve river flows. There were a large number of overflowing uncontrolled 'wild' boreholes in the confined region that were wasting water throughout the year. Controlling or sealing of these wild bores took place in the early 1990s; a numerical groundwater model was used to identify the likely improvements in river flows. The second example refers to the Chalk of East Kent; river flows have been influenced by groundwater abstraction and this particular study is concerned with the potential for improving river flows by reducing summer abstractions.

Southern Lincolnshire Limestone, UK

Flow processes

The Southern Lincolnshire Limestone is an important source of groundwater used for public supply. The confined region has high transmissivities which means that there is little difficulty in pumping out significant quantities of water (Rushton *et al.* 1982). Detailed studies have been carried out for a number of years into the behaviour of the Southern Lincolnshire Limestone catchment. A diagram of the aquifer (Fig. 1) indicates that the limestone is exposed only in limited areas. Elsewhere, Boulder Clay or a sequence of low-permeability strata containing minor aquifers, which will be described by the term Overlying Beds (unshaded in Fig. 1), limit direct recharge; the Overlying Beds can include the Upper Estuarine Series, Great Oolite Limestone, Great Oolite Clay, Cornbrash and Kellaway Beds (Downing & Williams 1969; Bradbury & Rushton 1998). The Boulder Clay and Overlying Beds are shown in the representative cross-sections in Figure 2.

The Boulder Clay and the Overlying Beds have an important effect on the distribution of recharge. Although little water passes vertically through the Boulder Clay (Bradbury & Rushton 1998), run-off occurs which can enter the limestone as run-off recharge when the streams pass onto the limestone outcrop. A good example is the Easton Wood catchment to the west of the West Glen river (Fig. 1). In addition to the immediate run-off, some water is stored in the more permeable material in the Boulder Clay and is released to flow down the streams to the limestone after the rainfall event; significant flows occur for several weeks after major rainfall events. The foresight of those who ensured that gauging stations were provided at locations where run-off recharge is significant has been fundamental to the success of this study.

The run-off recharge from the Overlying Beds is even more important. There is a large catchment of Overlying Beds above Burton Coggles gauging station; run-off from this catchment continues for many weeks after major rainfall events. Most of this water enters the limestone aquifer in the reach of the West Glen immediately below Burton Coggles, although the infiltration into the aquifer is limited by high groundwater heads. For the East Glen, much of the course of the river crosses the Overlying Beds; also the river tends to be very flashy due to surface run-off. However, there are three important Limestone inliers at Irnham, Grimsthorpe Brook and at Toft. At each of these locations run-off recharge does enter the Limestone aquifer. The manner in which both direct recharge and run-off recharge occur in the Southern Lincolnshire Limestone catchment is a key to the assessment of groundwater resources.

Another crucial feature was identified following the low recharge season of 1975–1976 which resulted in serious declines in the confined groundwater heads during spring/summer of 1976 even though water was still stored in the unconfined region of the aquifer. There appeared to be some form of restriction between the unconfined and confined regions; this was eventually identified as a significant decrease in transmissivity due to decreases in the saturated depth and the resultant dewatering of some of the more permeable zones. The reason for the decline in transmissivity with decreasing saturated depth is indicated in Figure 3; Figure 3a is an idealization of the variation of the hydraulic conductivity on a vertical section of the limestone stratum. In the upper part of the section, hydraulic conductivities are higher. In practice, there will not be such a smooth variation in hydraulic conductivity; however, since the hydraulic conductivities and resultant transmissivities are to be included in a groundwater model, the hydraulic conductivity variation of Figure 3 adequately represents the distribution over a grid square or cell. The transmissivity is the sum of the hydraulic conductivities from the base of the aquifer to the water table; as the water table rises there is a significant increase in transmissivity. Transmissivities in winter are appreciably higher than summer transmissivities.

There are many pumping stations in the Southern Lincolnshire Limestone (Fig. 1), most of which have abstraction rates in the range 5 to 20 Ml d^{-1}. All the pumping stations are to the east of the River East Glen where the Limestone is overlain by 20+ m of the Overlying Beds. There are also many wild bores which over-

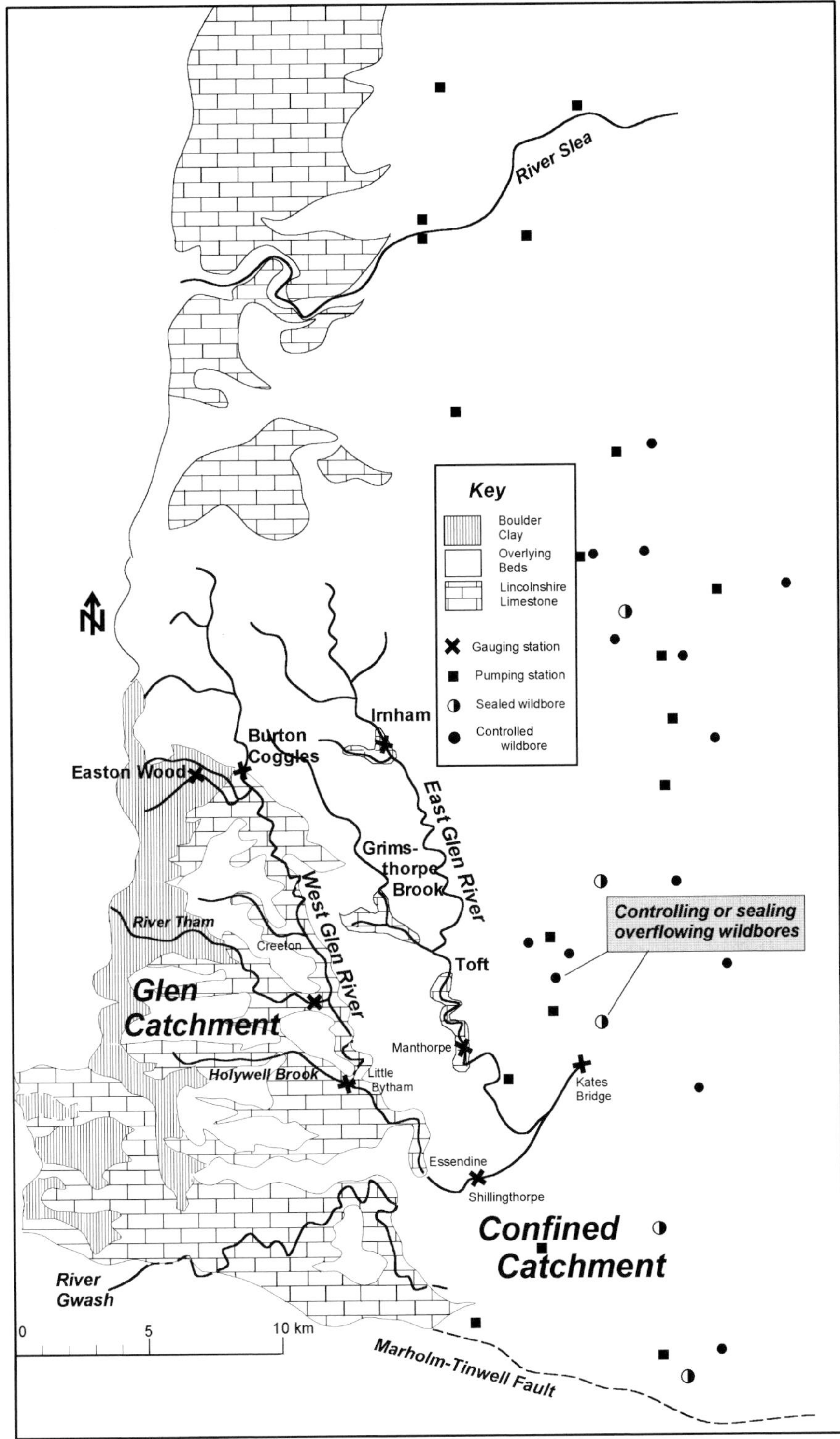

Fig. 1. Plan of Southern Lincolnshire Limestone aquifer.

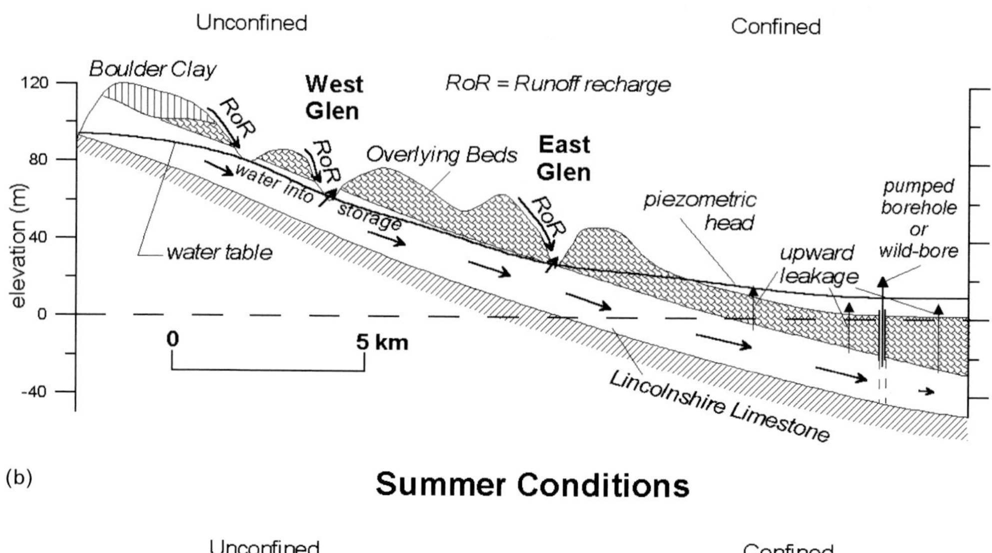

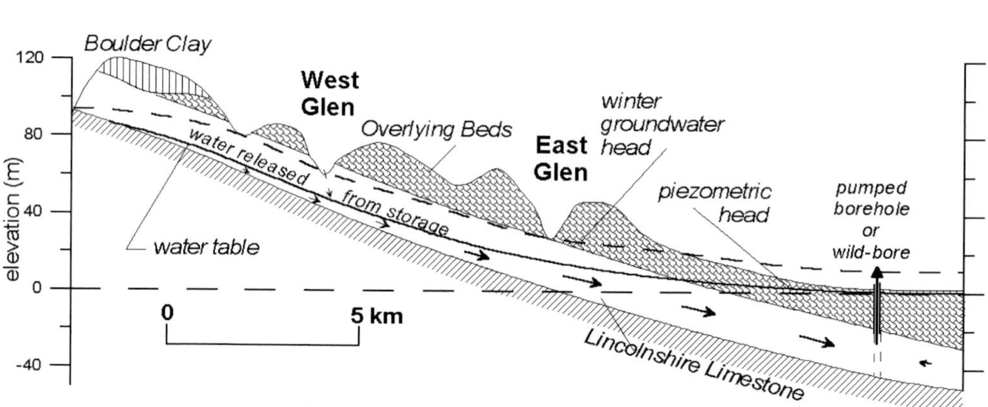

Fig. 2. Schematic diagrams of flow processes in the Southern Lincolnshire Limestone (**a**) during winter and (**b**) during summer.

flowed due to the artesian pressure in the confined region; most of these wild bores were controlled or sealed during the early 1990s (Barton & Perkins 1994; Johnson & Rushton 1999).

Conceptual diagrams

A conceptual cross-section of the important flow processes during a typical winter is presented in Figure 2a. There is substantial recharge, much originating from run-off from the Boulder Clay and Overlying Beds which subsequently enters the limestone aquifer. Some of this water leaves the aquifer from springs and rivers, some is taken into storage in the unconfined region and some becomes groundwater flow from the unconfined to the confined part of the aquifer. Of the flow into the confined part of the aquifer, some is pumped out at abstraction boreholes whilst some is discharged from wild bores. There is also an upwards flow (often called upward leakage) through the Overlying Beds and some

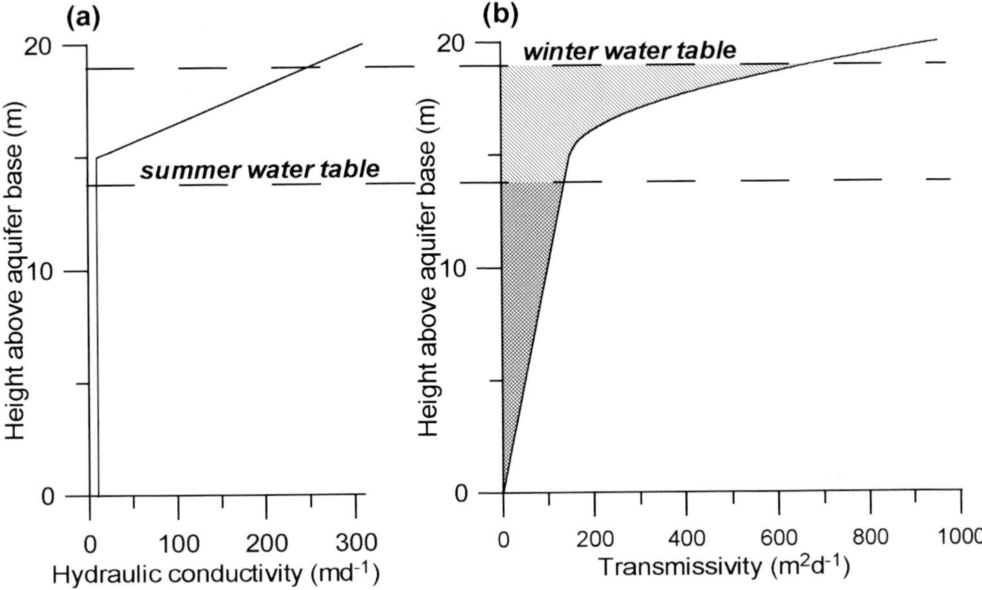

Fig. 3. Variation of hydraulic conductivity and transmissivity with saturated depth between summer (**a**) and winter (**b**) water tables.

water is taken into confined storage.

In the summer (Fig 2b) there is little recharge; some water enters the aquifer from rivers, but the major contribution in the unconfined region is water released from storage as the water table falls. The groundwater gradient between the unconfined and confined regions of the aquifer is similar to that in winter, but the transmissivity is reduced hence there is less flow from unconfined to confined regions. This inflow from the unconfined region supplies water to the confined abstraction sites and the wild bore flows. Due to the reduction in the piezometric heads there is little upward leakage; the balance is supplied by water released from confined storage.

Further insights into the response of the unconfined and confined parts of the aquifer system can be gained from a consideration of the water balance for a typical winter half month and a typical late summer half month; the numerical values for these balances are taken from a groundwater model (Rushton *et al.* 1982). Table 1 summarizes the flow components for the unconfined Glen catchment and also for the 'confined catchment' to the east (see Fig. 1). The sign convention is that inflows to the aquifer are positive; this means that recharge is positive whilst groundwater abstractions are negative. Water released from storage into the aquifer is positive so that when water is taken into unconfined storage during recharge periods this is indicated as a negative component.

Table 1. *Inflows to aquifer system*

	Inflow (Ml d^{-1})	
	2nd half of Jan 1995	1st half of Sept 1995
Glen (unconfined) catchment		
All forms of recharge	801	3
Water from storage	−374	54
Flow from rivers	−328	7
Flow from east	−106	−67
Flow from north	7	3
Confined catchment		
Flow from west	106	67
Other boundary inflows	−3	18
Abstraction	−82	−89
Water from storage	−11	5
Wild bore flows	−6	−1
Water from leakage	−4	0

Considering first the unconfined Glen catchment, the first column of figures in Table 1 relates to the second half of January 1995 when there was a high recharge input; a significant proportion of this recharge either went into storage (resulting in a substantial rise in the unconfined water table) or flowed rapidly from springs or directly into the rivers. Rises in the water table as water entered storage resulted in increased transmissivities, hence a rapid flow to springs and rivers. The other major component of flow is -106 Ml d^{-1} from the east (the minus sign indicates that water actually flows from the unconfined area to the confined area to the east). Considering next the confined catchment, much of the inflow of 106 Ml d^{-1} is used to meet the abstraction demand of 82 Ml d^{-1}. Some water is taken into confined storage whist flow from wild bores (which were controlled or sealed before 1995) and upward leakage from the confined area total 10 Ml d^{-1}.

Values in the right-hand column of Table 1 show that conditions were very different during the first half of September 1995; the quantity of water which flows from the unconfined to confined catchment falls from 106 Ml d^{-1} in January to 67 Ml d^{-1} in September. This reduction is due primarily to the reduction in transmissivity in the east of the unconfined catchment. A detailed examination of the unconfined Glen catchment shows that there is virtually no recharge, 54 Ml d^{-1} is taken from unconfined storage (this results from a further decline in the water table leading to further reductions in transmissivity) and the rivers provide a small flow to the aquifer system. In the confined catchment, the inflow from the unconfined area of 67 Ml d^{-1} does not provide for the abstraction demand of 89 Ml d^{-1}; consequently there is a substantial fall in piezometric head in the confined region, releasing water from storage (the confined storage coefficient is 0.0003). This fall in piezometric head also results in an inflow from the confined region to the north. In the context of considering the impact of abstraction on river flows, this description of the various components of the water balance indicates that there is no direct relationship between the abstraction in the confined region and the flows in rivers in the unconfined Glen catchment although a significant reduction in abstraction would lead to a small increase in river flows.

Impact of controlling and sealing wild bores

Due to the presence of the overflowing wild bores in the confined region, a substantial quantity of water was lost from the confined aquifer. Wild bores which were sealed or controlled are shown in Figure 1. Significant variations in flow from the wild bores between winter and summer were known to occur, but it was anticipated that the average quantity of water which could be saved by sealing or controlling the wild bores would probably be in the range 15–20 Ml d^{-1}.

A total catchment model of the Southern Lincolnshire Limestone was used to assess the impact of sealing and controlling wild bores. Comparisons were made between two predictive simulations, one with no control of the wild bores and the second with sealing and controlling for the whole simulation period. Of the average estimated saving of 15.4 Ml d^{-1} due to controlling or sealing wild bores, the average increases in other flows were calculated as follows (Johnson & Rushton 1999):

East Glen	2.7 Ml d^{-1}
West Glen	2.5 Ml d^{-1}
River Slea (to north)	3.1 Ml d^{-1}
River Gwash (to south-west)	1.7 Ml d^{-1}
Leakage through Overlying Beds	5.4 Ml d^{-1}

Improvement in flows in the West Glen is the desired objective; however, the predicted average increase is only 2.5 Ml d^{-1} with increases in flow also occurring in more distant rivers. The reason for the wide distribution of locations having increased flows is that pumping from the confined aquifer results in water being drawn from a wide area. Figure 4 indicates how the predicted increase in flows in the West Glen changes with time. The graph shows that if the wild bores had been sealed and controlled before the dry periods of 1973–1974 and 1975–1976 the improvements in flows in the West Glen would have been small.

East Kent Chalk Aquifer, UK

Desired improvement in river flows

The second case study refers to the Dour catchment of the East Kent Chalk aquifer; this example focuses on the impact of altered abstractions from pumped boreholes located up groundwater gradient from a river.

The Dour catchment, which is part of the East Kent Chalk aquifer, has a different response in summer to winter (Cross *et al.* 1995). This difference between summer and winter flow conditions is demonstrated by the flows measured at the gauging station at Crabble Mill on the River Dour (Fig. 5). Figure 6 indicates that Crabble Mill gauging station is downstream of a number of lakes; these lakes have a high amenity value. Winter flows are typically five- to ten-

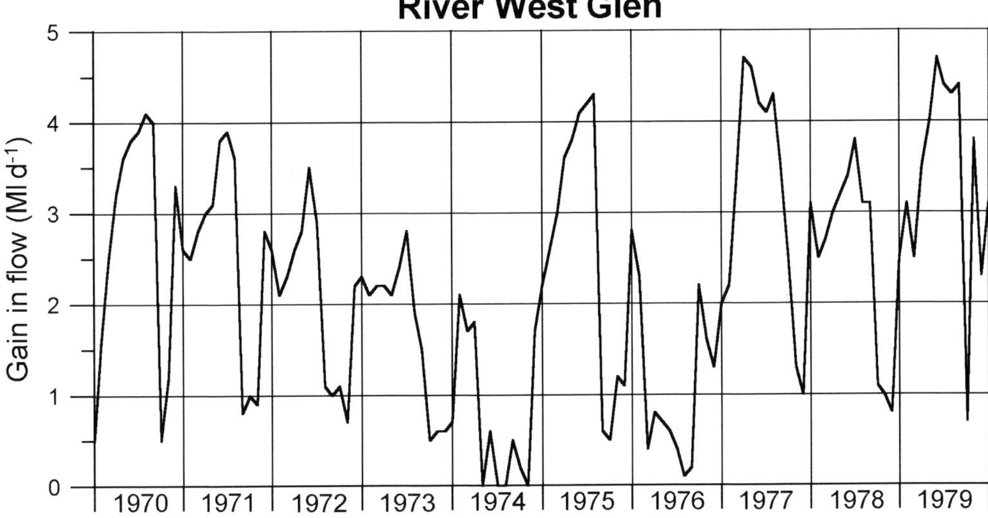

Fig. 4. Predicted increase in flows in the West Glen in the 1970s if wild bores had been sealed or controlled.

times greater than the low summer flows; flows become very low or cease in the tributaries of the Dour and certain lakes may become dry during summer months. Abstraction occurs from a number of pumping stations in the vicinity of Dover; this discussion will concentrate on a multi-borehole pumping station which is more than 3 km up groundwater gradient from the perennial part of the River Dour (Fig. 6). The issue to be explored is whether a reduction in abstraction from this pumping station between April and September will lead to a direct increase in flows in the River Dour. The following discussion is not based on a firm proposal but is intended to explore the likely impact of reductions in abstraction.

Prediction using simple model

The first stage is a predictive analysis using a simplified conceptual model. A first estimate of the effect of the reduction in pumping will be explored using a one-dimensional time-variant idealization of the field conditions; this idealized model is sketched above the graph of Figure 7. The idealized aquifer is assumed to be 6 km long and feeds into a river; the abstraction is positioned 3 km from the river. The object of the simulation is to determine the impact on the river flow of reducing abstractions between April and September. Numerical values were determined from a one-dimensional finite difference numerical model (Rushton & Redshaw 1979)

The shaded part of the graph of Figure 7 shows the reduction in abstraction, and the full line represents the predicted increase in river flow due to this change in abstraction. The main increase in river flow does not occur immediately after the reduction in abstraction due to storage effects of the aquifer system. However, after 6 months the increase in river flow is more than 99% of the decrease in abstraction. When abstraction recommences at the full rate the river flows decrease, but again there is a lag in the response. The area under the graph of the increase in river flow is identical to the shaded area representing the decrease in abstraction; this occurs because the only inflow and outflow locations in the idealized aquifer system are the pumped borehole and the river.

Prediction using a regional groundwater model

A conceptual understanding of the whole of the Chalk aquifer to the east of the River Great Stour has been developed (Cross *et al.* 1995). Valuable insights were gained from detailed studies of the Dour catchment by Reynolds (1947, 1970). He suggested that solution channels had developed along bedding planes and fissures, resulting in swallow holes where substantial quantities of river water can enter the aquifer. In the upper western part of the Dour

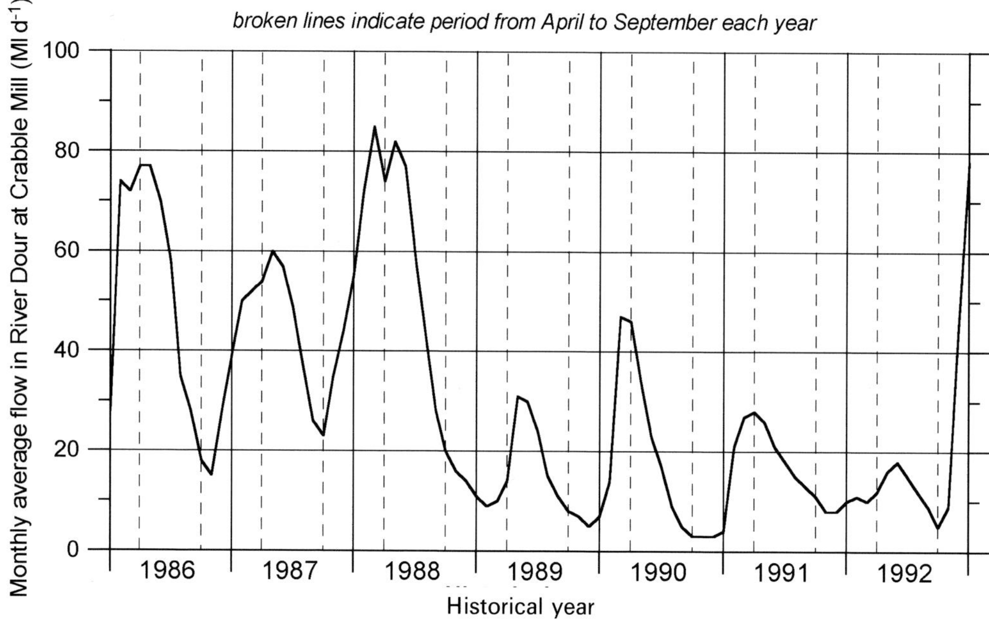

Fig. 5. Flows at Crabble Mill Gauging Station on the River Dour.

catchment there is only a small thickness of the Lower Chalk; hence the ability to store water is limited. Within the Middle Chalk, which is present in all but the upper parts of the catchment, there are persistent marl bands; permeabilities appear to have been developed above these marl bands leading to the occurrence of springs. These features lead to high winter flows (as the aquifer fills) and low summer flows.

This information, together with other hydrogeological insights, allowed the development of a regional groundwater model of the whole of the East Kent Chalk aquifer (Cross *et al.* 1995). Amongst other features, the model represents varying transmissivities of the chalk aquifer between summer and winter together with the effect of all the rivers and all the coastal boundaries associated with the East Kent Chalk. The variation in transmissivity with saturated depth (Fig. 3) is of a similar form to that already described for the Lincolnshire Limestone. However, the difference in elevation between summer and winter water tables is about double that of the Lincolnshire Limestone aquifer. The groundwater model also includes areal variations in transmissivity, with the transmissivities of the interfluves much lower than those in the valleys. The non-perennial nature of the rivers is also simulated; this depends on the variation in recharge which is also included in the model.

The impact of reduced abstraction is explored using the regional groundwater model. Two predictive simulations were carried out using the numerical model, both based on the historical recharge distribution. In the first, the pumping station operates at a constant rate throughout the year, whilst in the second simulation pumping occurs at that constant rate during the winter months but at a reduced rate between April and September. The increase in river flow, which is given by the difference between the two simulations, is plotted in Figure 8. This figure is of the same form as Figure 7 with the shaded area showing the reduction in abstraction, and the full line the predicted increase in river flow. There is an important difference between the predictions of the simple model (Fig. 7) and those of the regional groundwater model (Fig. 8).

A careful consideration of the field conditions provides several explanations for the differences between the response deduced from the idealized model (Fig. 7) and that deduced from the regional groundwater model (Fig. 8). In this discussion reference will be made to the regional groundwater setting (Fig. 9).

(1) The area under the graph of the increase in flow in the River Dour (Fig. 8) is signifi-

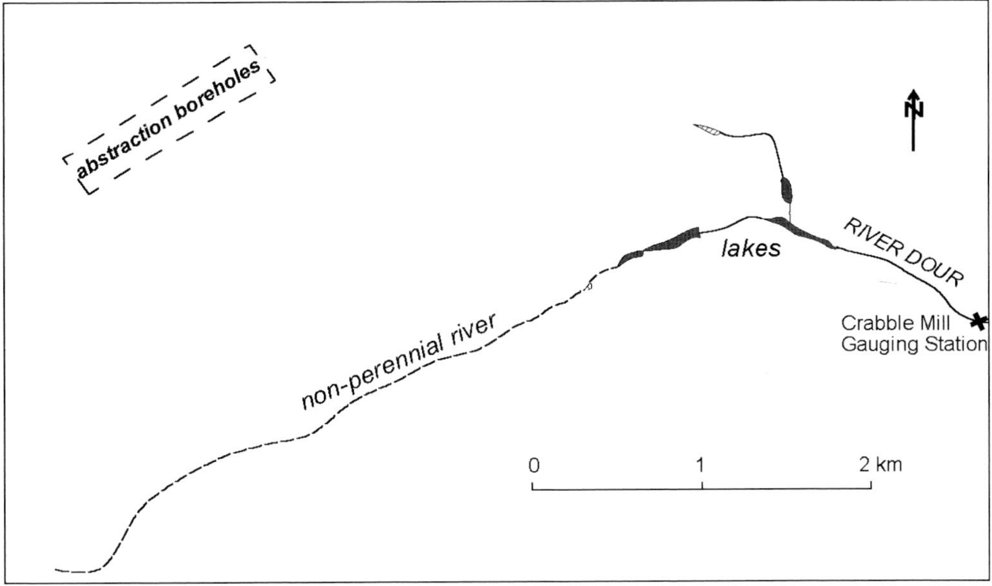

Fig. 6. Features of the upper River Dour and approximate location of the pumping station where reduction in abstraction is considered.

cantly less than the shaded area representing reduced abstraction (it is approximately 25% of the reduction in abstraction). The reason for the small increase in river flows is that there are many other sources of outflow or inflow (Fig. 9), including major rivers and the coastline. The idealized one-dimensional representation fails to take account of these additional flow locations.

(2) The idealized one-dimensional example assumes that the transmissivities at a particular location remain constant with time. However, significant fluctuations occur in the groundwater head in the area surrounding the pumping station; summer to winter differences of up to 10 m have been observed. Between winter and summer, transmissivities typically fall to one-third, with the result that impacts of the changes in abstraction are not transferred efficiently to the River Dour. The regional distribution of transmissivities is also ignored in the simple model.

(3) The one-dimensional example in Figure 7 also ignores differences between annual patterns of recharge. These annual differences lead to varying patterns of groundwater heads and the associated changes in saturated depth and hence in transmissivities. A comparison of the historical flows in the River Dour (Fig. 5), and the predicted increases in flows (Fig. 8) suggests that the smaller increases in flow due to the reduced abstraction occur when the river flows are small. Year 6, which corresponds to 1991, is a good example; the recharge during the winter 1990–1991 was low, leading to smaller than usual flows in the River Dour. The predicted increase in flows is also small, averaging about 20% of the decrease in abstraction.

Discussion and conclusions

Reduction or cessation of pumping is likely to provide some improvement in the flows of the nearest river, but it is unlikely that a reduction in pumping will have a direct and rapid effect. Inevitably there will be a delayed response since flows in aquifers can take a considerable time to change to a new pattern. For instance, due to high specific yields sandstone aquifers may take many years to respond, but even in the faster responding chalk and limestone aquifers, it may take several months before the main effect of any change in abstraction becomes apparent.

The next important issue is that a change in abstraction will not only have a localized impact but it is likely to affect all locations of surface

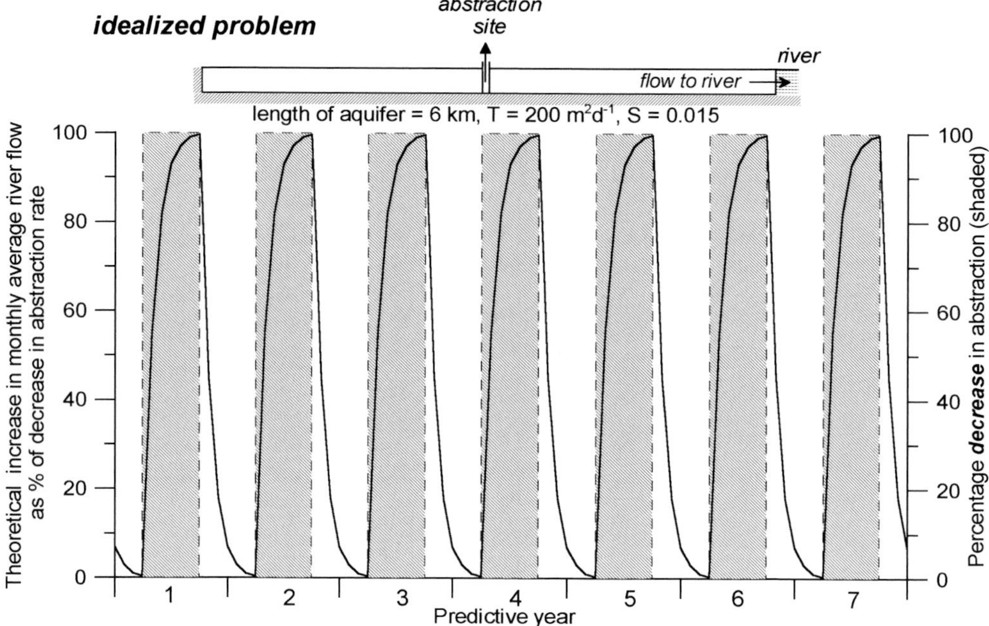

Fig. 7. Prediction using simple one-dimensional model of increase in flows at Crabble Mill due to decreased abstraction during April to September.

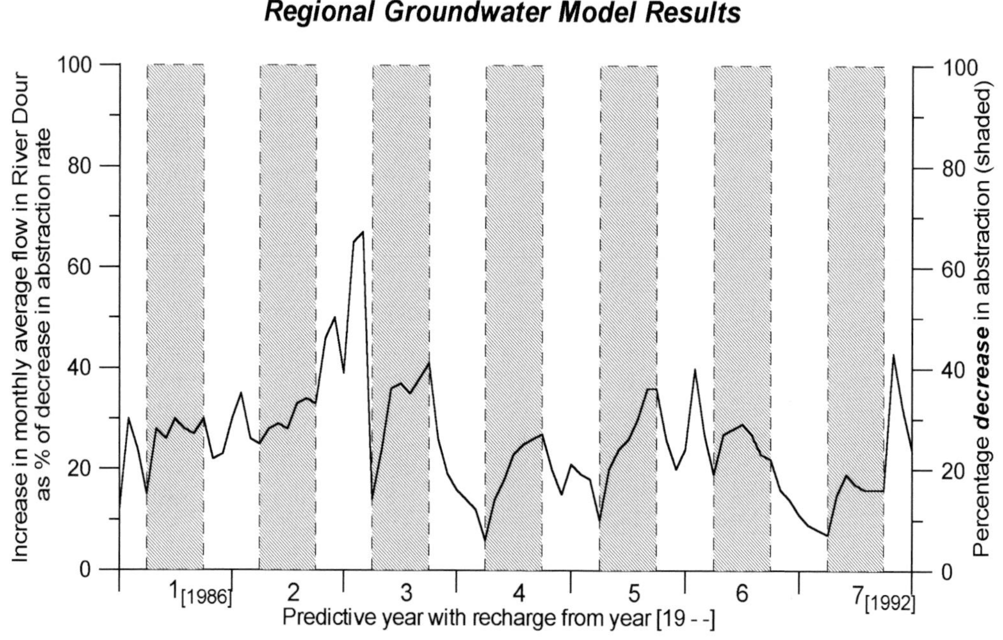

Fig. 8. Prediction using regional groundwater model of increase in flows at Crabble Mill due to decreased abstraction during April to September.

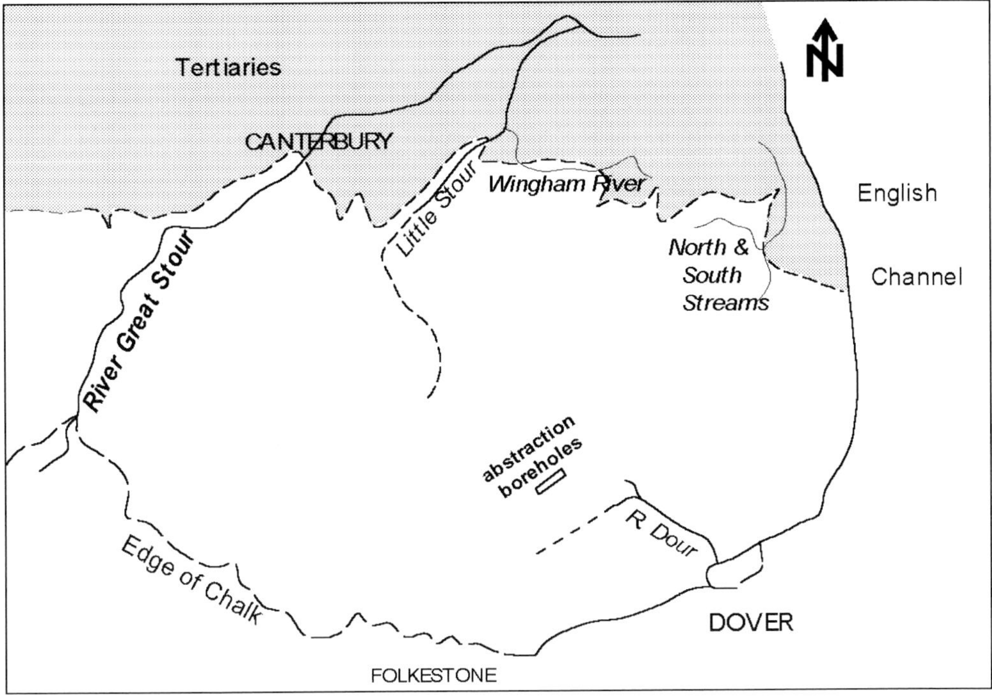

Fig. 9. Details of the entire East Kent Chalk aquifer showing boundaries, rivers etc.

water – groundwater interaction. Both of the case studies in this paper demonstrate convincingly the widespread areal effect of a change in abstraction.

In chalk and limestone aquifers, features such as significant variations in transmissivity with changing water table elevation, intermittent springs, and non-perennial streams all tend to delay or minimize the impact of reduced abstractions. These effects are classified as non-linear behaviour; they inevitably result in more complex responses to changes in abstraction. These non-linear responses become more marked between wet and dry years (or sequences of years). In general, the impacts of changes in abstraction are more marked in wet years. However in dry years, when improved river flows are desirable, the impact of reduced abstractions is usually small.

Unless the abstraction is very close to a river, reducing the amount of water pumped from the aquifer is not a reliable method of improving river flows. Simple analytical solutions are rarely applicable. Instead it is necessary to develop realistic conceptual models of all the flow processes and then devise an appropriate mathematical model. The mathematical model can then be used to explore alternative methods of improving river flows.

References

Barton, B. M. J. & Perkins, M. A. 1994. Controlling the artesian boreholes of the South Lincolnshire Limestone. *Journal of the Institution of Water and Environmental Management*, **8**, 183–196.

Bradbury, C. G. & Rushton, K. R. 1998. Estimating runoff-recharge in the Southern Lincolnshire Limestone catchment, UK. *Journal of Hydrology*, **211**, 86–99.

Cross, G. A., Rushton, K. R. & Tomlinson, L. M. 1995. The East Kent Chalk aquifer during the 1988–92 drought. *Journal of the Institution of Water and Environmental Management*, **9**, 37–48.

Downing, R. A. & Williams, B. P. J. 1969. *The groundwater hydrology of the Lincolnshire Limestone*. Water Resources Board, Publication **No. 9**.

Hunt, B. 1999. Unsteady stream depletion from ground water pumping. *Ground Water*, **37**, 98–102.

Johnson, D. & Rushton, K. R. 1999. The effect of borehole sealing on the Southern Lincolnshire Limestone catchment. *Journal of the Chartered Institution of Water and Environmental Management*, **13**, 37–46.

Reynolds, D. B. H. 1947. The movement of water in the Middle and Lower Chalk of South East Kent. *Journal of the Institution of Civil Engineers*, **29**, 74–108.

Reynolds, D. B. H. 1970. Under-draining of the Lower Chalk of South East Kent. *Journal of the Institution of Water Engineers*, **24**, 471–480.

Rushton, K. R. & Chan, Y. K. 1976. Pumping test analysis when parameters vary with depth. *Ground Water*, **14**, 82–87.

Rushton, K. R. & Redshaw, S. C. 1979. *Seepage and Groundwater Flow*. Wiley, Chichester.

Rushton, K. R., Connorton, B. J. & Tomlinson, L. M. 1989. Estimation of the groundwater resources of the Berkshire Downs supported by mathematical modelling. *Quarterly Journal of Engineering Geology*, **22**, 329–341.

Rushton, K. R., Smith, E. J. & Tomlinson, L. M. 1982. An improved understanding of flow in a Limestone aquifer using field evidence and mathematical models. *Journal of the Institution of Water Engineers and Scientists*, **36**, 369–387.

Assessing the impact of groundwater abstractions on river flows

S. KIRK[1] & A. W. HERBERT[2]

[1]*Environment Agency, National Groundwater and Contaminated Land Centre, Olton Court, 10 Warwick Road, Olton, Solihull, West Midlands B92 7HX, UK (e-mail: stuartkirk@environment-agency.gov.uk)*

[2]*Environmental Simulations International, Priory House, Priory Road, Shrewsbury SY1 1RU, UK*

Abstract: Groundwater abstractions affect the water balance in catchments of rivers in hydraulic continuity with groundwater and may lead to reductions in river baseflow. The Environment Agency issues licences for groundwater abstractions and takes into account the risk of such adverse impacts when deciding whether to grant licences and what conditions to associate with each licence. Assessing the impact of a particular licence is a difficult task, but analytical solutions to idealizations of the complex river–aquifer interaction can help guide such judgements. This paper presents a brief review of available analytical solutions and discusses their applicability to real groundwater systems. The most useful analytical solutions (developed by Theis, Hantush and Stang) have been incorporated into a spreadsheet and a new methodology has been made available to support Environment Agency hydrogeologists working in abstraction licensing. This offers a consistent approach to making an initial evaluation of the impact of groundwater abstraction on river flow that may be applied across the diverse hydrogeological systems found throughout the Environment Agency Regions of England and Wales. The use of analytical solutions in this methodology inevitably represents a significant simplification of what is generally a very complex issue, and the limitations of the new methodology are emphasized.

The Environment Agency (the Agency) is responsible for water resource management in England and Wales. The principal means of management is a formal abstraction licensing system that was first introduced in 1963. Central to the Agency's philosophy and practice of abstraction licensing is the recognition that decisions about groundwater licence applications frequently have implications for surface water flows. However, in contrast to 'run-of-river' surface water abstractions, the impacts of groundwater abstractions on river flows are often difficult to predict. This is due to the complex nature of most river-groundwater interactions and the uncertainties about the nature of the hydraulic connection at a particular site. Therefore, making decisions about groundwater abstractions that are likely to have an impact on surface water flows is technically challenging. Furthermore, the risks associated with the determination of such abstraction licences are often high because unlike 'run-of-river' surface water abstractions, groundwater abstractions cannot generally be controlled by the imposition of 'hands-off-flow' conditions that would otherwise protect a pre-determined river flow (or level). This is primarily due to the inherent time lags associated with the response of the groundwater system and also with the aforementioned uncertainties. Whilst the technical challenges associated with the estimation of the impacts of groundwater abstractions on river flows are great, so too is the need to develop practicable solutions to their estimation. Indeed, our understanding of river-groundwater interactions underpins the Agency's integrated manage-

From: Hiscock, K. M., Rivett, M. O. & Davison, R. M. (eds) *Sustainable Groundwater Development*. Geological Society, London, Special Publications, **193**, 211–233. 0305-8719/02/ $15.00 The Geological Society of London 2002.

ment of surface water and groundwater.

The findings reported in this paper represent the first phase of work being managed by the Agency as part of its programme of work on the 'Impact of Groundwater Abstractions on River Flows' (the IGARF programme). This on-going programme of work is being managed by the 'National Groundwater and Contaminated Land Centre' on behalf of the Agency's 'Groundwater Resources Group'.

Approaches for assessing the impact of groundwater abstraction on river flows

The interaction between rivers and groundwater systems is complex, and a general methodology needs to be able to accommodate a wide range of conditions. This section summarizes the issues that need to be considered, and identifies the restrictions on the system that need to be satisfied for simplified methods to provide useful estimates of the impact of abstractions.

The response of the aquifer to river–aquifer flows will depend on characteristics of the aquifer, namely aquifer transmissivity and specific yield (if unconfined) or storativity (if confined). Confined or leaky aquifers will show the most rapid and most significant head response to pumping and thus can induce the most significant head differences and fluxes at a river.

Rivers may either gain water from or lose water to aquifers, depending on the degree of hydraulic connection and the difference between river stage elevation and the connected groundwater heads. This flux between river and aquifer may vary during the year as groundwater heads and river stage elevation vary, and also with position along the river. It may vary over very short time-scales as flood run-off will lead to rapid variations in river stage elevation on time-scales of hours, and over longer time-scales as groundwater heads change.

The stage of the river is related to the river flow rate. There will generally be a direct relationship between flow and river stage. This may be established directly from river gauging, or may be estimated by one of a number of general hydrological flow equations such as Manning's formula (e.g. Shaw 1994) or more sophisticated hydrograph relations. The aquifer discharge to rivers (baseflow) is determined by groundwater flow arising from hydraulic gradients in the aquifer close to the river and this will generally vary slowly when compared to changes in river flows associated with rainfall run-off. It is likely to rise during periods of recharge in parts of the aquifer providing the baseflow. This is generally followed by periods when the aquifer has reduced or no recharge, during which there is a typical exponential decline in the baseflow as the groundwater system discharges to the river at a rate proportional to the difference between a typical aquifer head and the river stage.

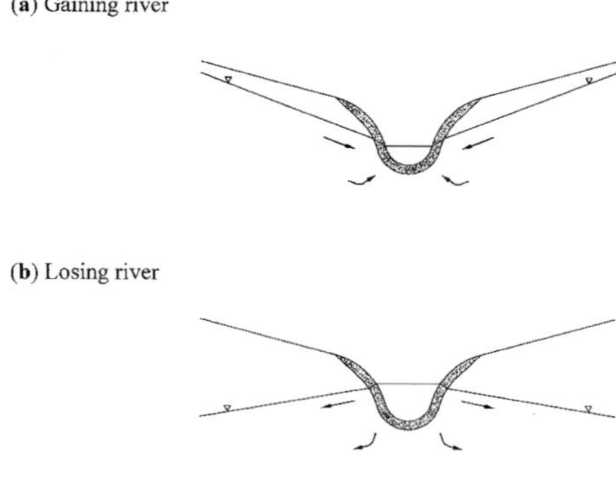

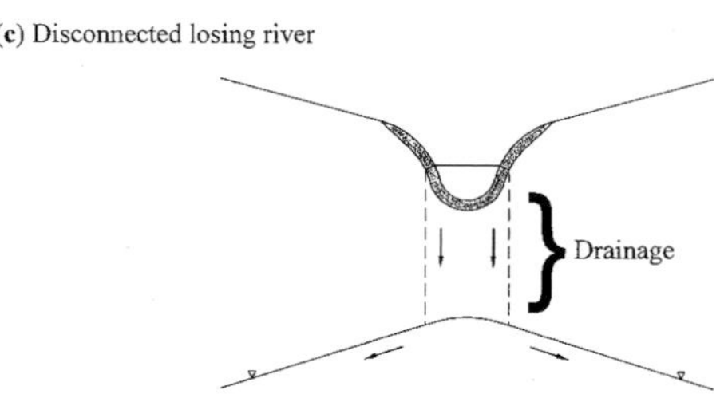

Fig. 1. Influence of river bed/bank sediments on river–aquifer interaction.

The most important determinant of the flux of water between a river and the aquifer is the degree of connection between the river and the aquifer. The degree of connection will depend on the properties of the material comprising the river bed and river bank, the existence of aquitard or unsaturated aquifer material between the river and the aquifer, and the extent to which the channel of the river intersects the saturated part of the aquifer.

The range of circumstances leading to flows between the aquifer and the river may be considered by taking the different configurations of aquifer head and river stage in turn as shown in Figure 1 and described below.

(a) When the aquifer head is above the river stage there is potential for flow from the aquifer to the river as illustrated in Figure 1a. The flux is generally proportional to the difference between the aquifer head and the river stage.

(b) When the aquifer head is below the river stage, then the flow is reversed and the river potentially loses water to the aquifer as illustrated in Figure 1b. Again, this is generally proportional to the difference in elevation of the river stage and the aquifer head.

(c) For a partially penetrating river (one where the aquifer extends beneath the river, as is often the case) the aquifer head may continue to fall below the base of the river. When the groundwater head first falls below the base of the river, the material below the river remains saturated and a mound of groundwater will be formed by the infiltrating river water. As the aquifer head falls further, a column of draining water will form in the aquifer material between the base of the river bed and the water table as illustrated in Figure 1c. This water will drain under gravity with a unit head gradient, and this flux will no longer depend on the elevation of the groundwater head next to the river. This situation corresponds to a limiting infiltration rate and the river losses will not increase as the aquifer head falls further.

The nature of river-aquifer interactions will also depend on the properties of the material lining the river. Many rivers have a zone around their bed and banks where fine sediments have been deposited or have invaded the aquifer material, reducing the permeability below that of the main aquifer. This can result in a very significant resistance to flow between the river and the aquifer, and this aspect is also shown in Figure 1a–c for the gaining river, losing river and disconnected (perched) river, respectively (isolated rivers where there are no losses considered). Where the river is disconnected, the flow may be undersaturated below the river bed since the flux will be limited by gravity drainage at the rate determined by the saturated Darcy flow through the low conductivity sediments. In some situations, the presence of river bed or river bank sediments can have a similar effect on the response to pumping the aquifer as that due to an increased distance between the abstraction borehole and the river.

The rate of change of river flow, and attenuation of short-term river flow changes (flood run-off), can be strongly influenced by the storage provided by the river bank. As river stage increases, the water is able to saturate material along the river bank and access this additional storage. When aquifer heads are reduced then this water-saturated zone will provide an additional source of water and reduce the immediate impact of an abstraction on the river flow. This phenomenon of bank storage is not taken into account in the methodology for an initial assessment, and later discussion only considers the impact of the transmissivity of bank sediments in reducing flow rates.

Typical analysis or approaches, such as the analytical solutions presented in this paper, rely on simplification of the aquifer properties. The extent of river–aquifer interactions will be dependent on the variations in aquifer properties through time and space. Typically, interpretations of the river–aquifer interaction make a simplifying assumption by assigning a representative single value to storage and transmissivity parameters. Indeed, this is generally necessary for analytical solutions. In real systems, the aquifer will be heterogeneous and often highly anisotropic. The aquifer thickness may vary, and there will often be heterogeneous cover restricting recharge and leading to locally varying heads and aquifer characteristics. Similarly, the potential impact of groundwater abstractions will be influenced by the path of the river relative to the abstraction borehole.

In the analytical solutions it is also implicit that the aquifer can be characterized by a single value of head away from the abstraction well. This will not be the case, as the aquifer will be recharged by precipitation and head distributions will result from the interaction of numerous inputs and outputs from the aquifer. In general there will be a reduction of aquifer head towards the river when the river is gaining baseflow from the aquifer. The effect of abstraction will be to create a zone of depression around the well. Using analytical solutions this drawdown can be simplified with assumptions of idealized radial flow to calculate the shape of the drawdown zone that can be superimposed on the distribution of head due to the natural behaviour of the system.

The superposition of a zone of depletion around a pumped well on a typical recharging aquifer head distribution near a gaining river is illustrated schematically in Figure 2. In this figure, the physical distribution of head (Fig. 2c) is obtained by superposition of the drawdown due to abstraction (Fig. 2b) on the natural head distribution without abstraction (Fig. 2a). As a consequence of groundwater gradients which occur towards rivers, the abstraction will usually intercept recharge that would otherwise contribute to baseflow, rather than take water directly from the river. However, the net effect on the water balance of the river is the same,

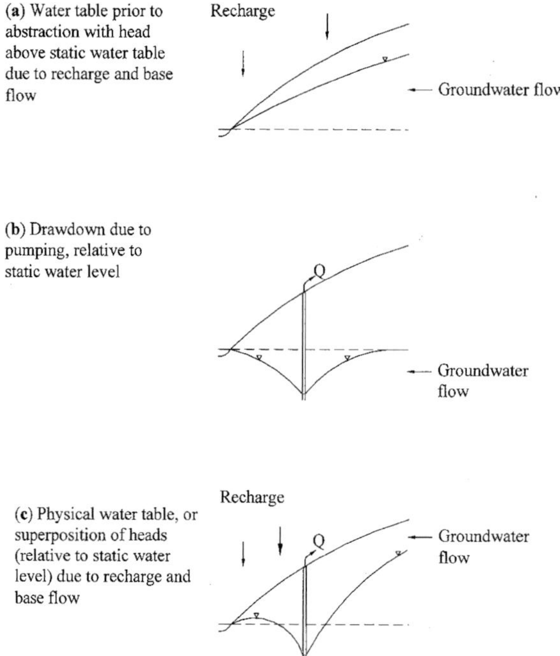

Fig. 2. Illustration of the principle of superposition applied to abstraction close to a gaining river.

with the net flow to the river being depleted as a function of time and position in a very similar way to the depletion due to abstraction from the horizontal rest water level in the aquifer assumed by analytical solutions. Note that the principle of superposition would make this an exact equivalence for an ideal confined aquifer or leaky aquifer. With an unconfined aquifer (or one where the effective properties are influenced by a leaky cover layer and become head dependent), the equations are non-linear and the solution can only be an approximation. Also, the principle of superposition will not apply when the depletion from the stream is at the limiting percolation rate due to the aquifer head falling below the effective stream bed.

The zone of depression will extend away from the borehole and will approach the river. For an ideal, well-connected and fully penetrating river, this will be bounded by the river which will form a zero drawdown boundary; however, for partially penetrating rivers the zone of depression can extend under the river and will be influenced by the storage of the aquifer beyond the river.

The number of possible combinations of aquifer type, aquifer head and river sediment properties, river penetration and river stage distributions, all of which may vary in space and over time, is large. There will generally be insufficient data to identify clearly which situation pertains at any given reach of a river close to a proposed or existing licensed groundwater abstraction. There is often, therefore, no definitive answer to the question of how to evaluate the impact of the abstraction on the river without very significant levels of investigation and modelling, which will not generally be practical. The approach will have to be based on hydrogeological judgement, although the tools presented here will aid and support that judgement.

A key point, however, is that local abstractions will eventually take all their water from the river, or intercept water that would otherwise have discharged into the river, or a combination of these two. Situations where neither of the above apply (and part of the abstraction does not ultimately affect baseflow) are as follows:

(1) the abstracted water would otherwise have discharged to the sea or lake;
(2) the abstracted water would otherwise have discharged to another river;

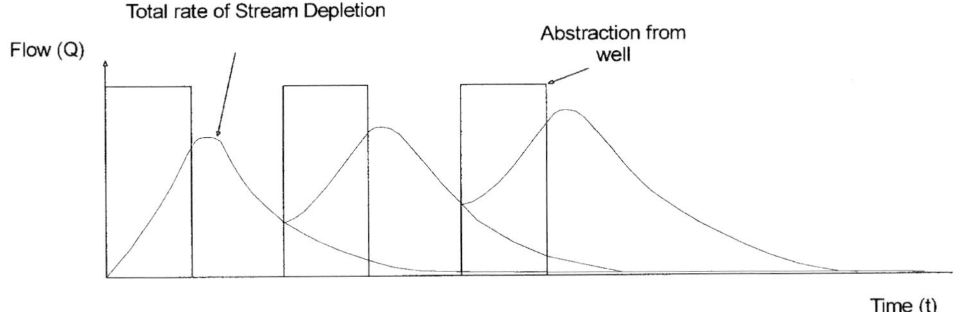

Fig. 3. Illustration of the growth of stream depletion as a result of several years of periodic groundwater abstraction (after Wallace *et al.* 1990).

(3) the abstraction affects other supplies and leads to reduced abstraction of groundwater elsewhere.

These first three possibilities essentially require consideration of a larger system than is considered directly in the scoping methodology described here.

(4) The abstraction affects crops or other plants by removing water that would have been used for evapotranspiration. This is only likely where the water table is particularly close to the ground surface and would be the case, for example, near a wetland, again giving rise to a system that is more complex than can be addressed by this methodology.
(5) If the river is effectively disconnected from the aquifer, limiting percolation applies (although even in this case one is depleting the aquifer and at steady-state the region across which limiting percolation applies will be extended so as to supply the water). For example, rivers perched over impermeable clay drift will be isolated from the underlying aquifer, and will not be affected by nearby abstraction. This condition is not considered by any of the analytical solutions used in the methodology. It is important that a judgement is made as to whether the river might be isolated (no drainage) or disconnected (limiting gravity drainage) over a significant reach of the river under consideration. If so, then the analytical solutions will overestimate the river depletion, and other approaches would be needed to estimate a more realistic impact.

Thus, whilst it is almost inevitable that the abstraction will eventually be taken from the river flows, the timing and distribution along the river of the net depletion will be affected by the aquifer properties, river characteristics and the well location. The position of greatest net river flow depletion will generally be centred on the point closest to the abstraction. This will be the case whenever the equations are nearly linear, even when the groundwater naturally (prior to abstraction) flowing past the borehole location would have contributed to baseflow some distance downstream. The impact will be spread up and downstream affecting a significant reach at steady-state depletion. However, the time to reach steady-state can be long, and may be several years if there is a low permeability barrier between the river and the aquifer. This may not influence the evaluation for long-term steady abstractions for sustainable resource management, but will be important for the assessment of periodic or temporary abstractions. If the water is taken from the aquifer during summer low flow periods, the net depletion of the river may be spread over a longer period (at a correspondingly reduced rate). It may be delayed until winter months when the depletion will reduce higher flows when the river environment is less sensitive to river flow depletion.

If there is a periodic abstraction on a regular annual schedule, then the impact of a given year's pumping period may extend for more than a year. In this case, when the second year's pumping begins its impact will be superposed on the declining residual impact from the previous year. In this way, once periodic abstraction begins, it may take several years for the maximum cycle of river flow depletion to be established, as illustrated in Figure 3 (after Wallace *et al.* 1990).

In the above it has been implicit that the

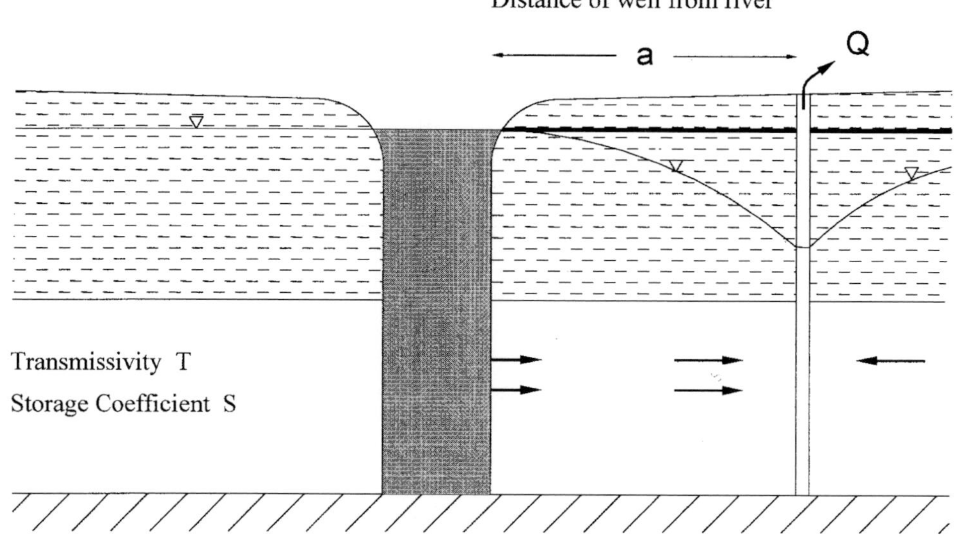

Fig. 4. Conceptual model for river–aquifer interaction used in the Theis solution.

aquifer and river sediments can be represented by effective homogeneous properties. Real aquifers are heterogeneous, and many UK aquifers are fissured. The use of analytical solutions to investigate real river–aquifer interactions therefore requires care. The analytical solutions use single values for key parameters and whilst effective hydraulic conductivity and storage parameters may be estimated, these will only apply on sufficiently large scales. The distribution of the impact on river flows will be strongly influenced by local properties and, for example, the analytical solutions cannot predict what will happen to specific spring discharges in any but the most general sense. However, the general features of river–aquifer interactions will be relevant and, in particular, there will be a water balance over the river catchment and abstracted water will ultimately be debited from the catchment discharges. The methodology below proposes that in many cases, the consequences of heterogeneity may be assessed through expert judgement informed by a broad sensitivity analysis. There will be cases where the methodology cannot be implemented using analytical solutions and other tools will be required.

In assessing abstractions a logical, tiered approach is therefore required. At a preliminary stage it is useful to assess the consequences of simplified representations of the river aquifer system in order to make an informed expert judgement of the consequences for the water table and the river flow. Decisions can then be made as to whether the issues require further tiers of assessment drawing on additional field investigations and numerical modelling techniques.

Review of analytical solutions

As discussed above, the interaction between rivers and groundwater systems is complex. Analytical solutions provide well established and rapid idealizations of reality and are often seen as an important part of a staged investigation process. They are useful scoping tools but inevitably oversimplify the interaction between the aquifer and river. Analytical solutions are usefully deployed in the early stages of any hydrogeological investigation where there is little data and a quick and cheap means of directing further investigation is required. Accurate models will inevitably require numerical methods to be used and will require significant field data to be collected to confirm key parameters and to provide head and flux targets against which the model can be calibrated. These methods require more time and money, resources which are often lacking. Used by hydrogeologists, the analytical models presented here can be useful tools to clarify the need for further, more detailed evaluation. However, the usefulness of analytical equations depends on the user having a clear appreciation of the underlying assumptions of the equations. Only then can the user decide which hydrogeological phenomena are covered by the equations.

The earliest evaluation of the impact of pumping near rivers using analytical solutions was by Theis (1941) as discussed in the classic papers on this subject by Jenkins (1968, 1970). Theis used an analytical solution to calculate the quantity of water supplied from an infinite straight line recharge boundary due to a constant rate abstraction from an aquifer. This result was used by Jenkins to estimate the total depletion of stream flow as a function of time due to nearby abstraction with the following assumptions:

(1) the aquifer is isotropic, homogeneous, semi-infinite in areal extent, bounded by an infinite straight fully penetrating unpuddled stream;
(2) water is released instantaneously from storage;
(3) the well is fully penetrating;
(4) the pumping rate is steady; and
(5) the residual effects of previous pumping are negligible.

The model further assumes that the aquifer is confined or that, for a water table aquifer, the drawdown is negligible compared to the saturated thickness (i.e. the transmissivity does not change). A cross-section, perpendicular to the river, through the idealized model is shown in Figure 4.

The mathematical solution gives the stream flow depletion as a proportion of the abstraction as:

$$\frac{q}{Q} = \mathrm{erfc}\left(\frac{1}{2\tau}\right) \qquad (1)$$

where $\tau = \frac{1}{a}\sqrt{\frac{tT}{S}}$ is a dimensionless length scale for the system; T is the aquifer transmissivity (m^2 d^{-1}); S is the aquifer storage coefficient (specific yield for unconfined aquifer approximations – dimensionless); a is the perpendicular distance of the well to the line of the river (m); Q is the abstraction rate at the well (m^3 d^{-1}); q is the total river depletion rate (m^3 d^{-1}); t is time (days); and erfc is the complementary error function.

This is a very useful analytical result. Although for most applications of interest the assumptions behind the model are oversimplifications, the model can be used to provide rough estimates of the local impact of abstraction on river flow, and of the time-scales over which flow reduction develops. By neglecting stream bed and stream bank sediments and assuming full penetration, the impact of pumping on the stream flow is overestimated and the time delay between abstraction starting and the impact of pumping on stream flow is underestimated. A further assumption behind the result is that prior to pumping the piezometric surface is flat and equal to the constant water level of the stream. In fact this is not a major limitation since the impact of recharge and baseflow can be accounted for by superposing the drawdown predicted by this model on the head distribution without abstraction, and interpreting the stream depletion as a reduction from the baseflow.

The same result is presented by Glover & Balmer (1954). Jenkins (1968, 1970) extends this to the case where the abstraction is taken for a finite period, by using the principle of superposition. Jenkins's report (particularly the expanded version; Jenkins 1970) makes a very useful and applicable summary of the use that can be made of Theis's or Glover & Balmer's result with worked practical examples. Spinazola (1998) has recently implemented Jenkins's (1968) approach in a spreadsheet model and demonstrated it on a case study in Idaho.

Whereas Jenkins considers the impact of a single finite period of pumping, such as might be expected to result from a pumping test, Wallace *et al.* (1990) extend this to investigate the practical issue of regular seasonal abstractions over several years. For cases where the impact develops over long time-scales due to the distance between the well and the river, the aquifer properties and the possible role of river bank/river bed sediments, Wallace *et al.* (1990) show that the impact may develop over several annual cycles. The maximum impact in later years might exceed the first year's maximum river depletion. This is a potentially important feature of sustainable catchment management to avoid low river flow problems.

Boulton (1942) addresses a related problem, evaluating the steady drawdown due to continuous steady pumping with the same assumptions regarding the aquifer and stream properties, but assuming the river flows over the top of the aquifer. He calculates the steady-state drawdown as a function of all three coordinates and also the steady-state location of the water table and the depth-averaged head. This is particularly useful to confirm that the river and aquifer do not become disconnected during the abstraction for cases where Theis's approach is being applied as a bounding estimate of the impact of abstraction near a partially penetrating stream.

A significant development of Theis's result is

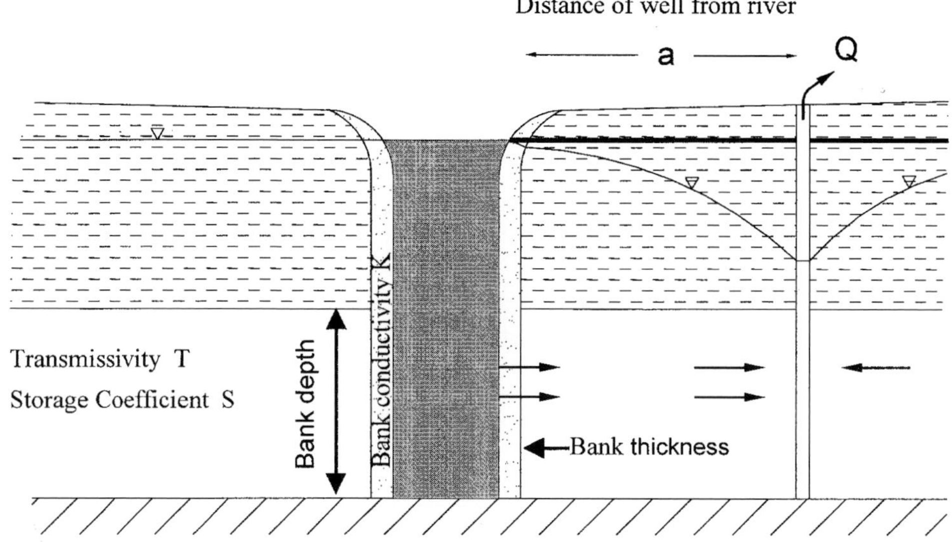

Fig. 5. Conceptual model for river–aquifer interaction used in the Hantush solution.

to consider the influence of low permeability river bank or river bed sediments. The role of these sediments is discussed in detail by Rorabaugh (1963). Hantush (1965) extends the analytical solution to include the impact of semi-permeable sediments adjacent to the river. The corresponding solution to that of equation (1) is given by:

$$\frac{q}{Q} = \text{erfc}(\tfrac{1}{2\tau}) - \exp(\tfrac{1}{2}\lambda + \tfrac{1}{4}\lambda^2\tau^2)\text{erfc}(\tfrac{1}{2\tau} + \tfrac{1}{2}\lambda\tau) \quad (2)$$

where $\tau = \frac{1}{a}\sqrt{\frac{tT}{S}}$ is a dimensionless length scale for the system; $\lambda = \frac{a}{2}\frac{bp'}{Tm'}$ is a dimensionless river bank resistance parameter; b is river bank depth (m); m' is river bank thickness (m); p' is river bank hydraulic conductivity (m d^{-1}); and other parameters are as above. A cross-section, perpendicular to the river, through this idealized model is shown in Figure 5.

Hantush also expresses his result in terms of an effective distance of the pumped well from the river. The two representations are equivalent but the explicit consideration of the river bank properties is conceptually clearer. The difficulty with the conceptual model for calculating the flow in the aquifer is that the river is assumed to be fully penetrating. This corresponds to no vertical components to flow and no account being taken of storage beneath the river or on the opposite side of the river to the well (as also assumed by Theis).

The other useful extension to Theis's solution discussed in the literature is the extension to a partially penetrating river with a semi-permeable base. This is reported in a very detailed paper by Hunt (1999) [although a partial analysis is attributed earlier to Stang (1980) by Bullock *et al.* 1994]. River depletion is given by:

$$\frac{q}{Q} = \text{erfc}(\tfrac{1}{2\tau}) - \exp(\lambda + \lambda^2\tau^2)\text{erfc}(\tfrac{1}{2\tau} + \lambda\tau) \quad (3)$$

where b is now the river bed width (m); m' is river bed thickness (m); p′ is river bed hydraulic conductivity (m d^{-1}); and other parameters are as given for equation (2) above. A cross-section, perpendicular to the river, through the idealized model is shown in Figure 6.

Hunt (1999) also gives an expression for the drawdown in terms of integrals of the well function (i.e. integrals of the exponential integral E_1) which might be integrated numerically. He shows with this solution that the correct approach to dealing with partial penetration is to modify stream bank sediment resistance parameters rather than using an effective distance to the well that is modified from the physical distance (as incorrectly suggested by Hantush). Nawalany *et al.* (1994) suggest a similar correction but do not account for the storage on the far side of the river. Zlotnik & Huang (1998) present a slightly more sophisticated solution that better accounts for the region beneath the river where the confined storage

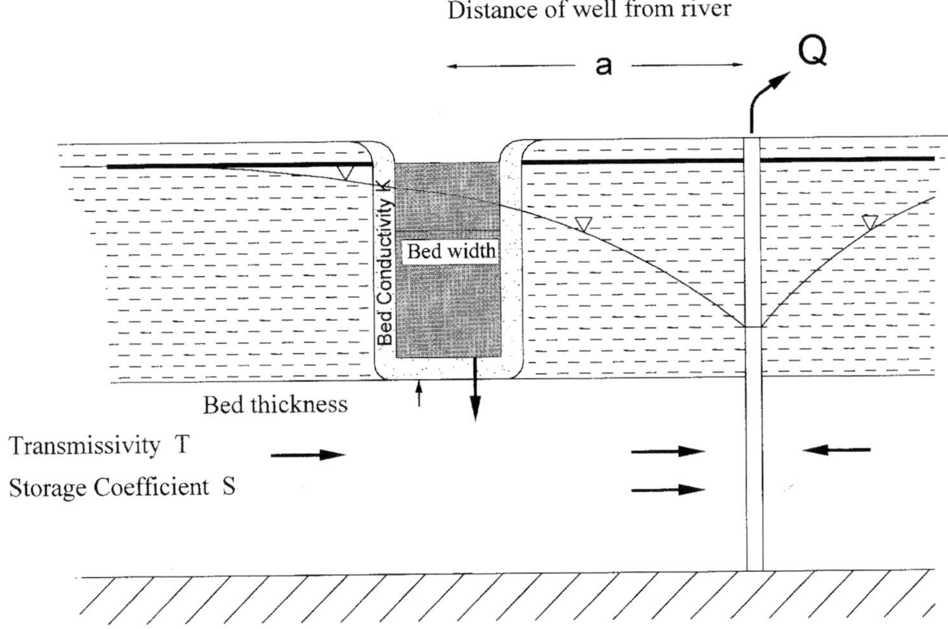

Fig. 6. Conceptual model for river–aquifer interaction used in the Stang solution.

coefficient applies, even in an unconfined aquifer, with continuous contact assumed between the groundwater and the river bed sediments. Their result is for a related problem posed in terms of the response of the aquifer to head variations but it might be modified to address the impact of abstraction on river flow. However, their solution was not used in the methodology developed by this project because it was judged by the authors to be introducing too much detail to the mathematical model used to estimate impacts, given other inherent simplifications to the aquifer properties and river geometry, and given the difficulty in obtaining the required data for the applications where the methodology will be applied.

The above mathematical work develops relatively sophisticated analytical solutions to an idealized river–aquifer interaction. This may often aid an initial assessment of the impact of groundwater abstraction on river flows. However, the authors all simplify the aquifer properties, represent the course of the river as an infinite straight line, ignore transmissivity variation within the zone of depletion, and ignore the possibility of the groundwater becoming disconnected from the river along some reaches. There are a number of useful reviews of the application of these analytical solutions to approximate the river–aquifer interaction. Particularly useful is the report by Bullock *et al.* (1994) which summarizes (in its Appendix B) many of the findings presented here in the context of using and verifying a numerical model. The work reported in this and the companion report by Watts *et al.* (1995) is very similar to that discussed here but is aimed at a more detailed assessment of low flows over river catchments.

Another particularly useful review that compares results from the Theis and Hantush solutions against a numerical model of a partially penetrating river is given by Spalding & Khaleel (1991). Comparison between simplified analytical models and numerical models allows the consequence of conceptual model simplification to be evaluated independently from issues of uncertainty in measurement or transience in real field case studies. Spalding & Khaleel's (1991) paper does not evaluate the Stang/Hunt solution, which was developed later. Considering the other analytical solutions, they show that neglecting the storage beyond the river might lead to significant errors of half an order of magnitude in the time to approach steady-state (or a significant overestimate of

20% or more in peak impact for a 2-month test with the example parameters). In addition, neglecting the role of sediments produced much larger errors (about twice the error for the example parameters). Both of these investigated errors are avoided by the use of the Stang solution.

Sophocleous *et al.* (1995) also contrast results of analytical solutions with those of numerical models (MODFLOW). Again, it is seen that partial penetration significantly increases the time-scale of development of the impact of steady abstraction on river flow (or reduces the maximum impact of finite abstractions). The authors comment that neglecting the role of stream bed sediments can render the use of Theis's equation completely inappropriate (as might be expected!) with very much shorter time-scales for the development of the impact. They also note the role of aquifer heterogeneity, seeing significant differences between numerical model results and analytical estimates. They report errors in the order of 50%, although their numerical model meshes are coarse and their results will be strongly influenced by detailed specification of the structure of the heterogeneity and of the properties adjacent to the well and the river.

Detailed assessment of particular river–aquifer interaction is best carried out with a combination of field work and distributed numerical modelling using models that accurately incorporate the correct conceptual model. This is not always feasible, particularly in an initial assessment; however, it is useful to review case studies of more detailed assessments. Particular case studies within UK aquifers include Morel (1980), Rushton *et al.* (1989), Younger *et al.* (1993), Chen *et al.* (1997), and Chen & Soulsby (1997). A modelling study with a different emphasis is given by Nield *et al.* (1994) who used a numerical model to explore the wide range of possible configurations for river–aquifer interaction.

A new methodology to support licensing decisions

The aims of the new methodology (Environment Agency 1999) were to provide a consistent approach to licensing across the regions of the Agency that also incorporated best current practice. Initial work on the methodology presented here focuses on initial evaluation of the impact of groundwater abstraction on river flows. There are very many applications that must be assessed, and given time and resource constraints, a simplified approach to quantifying the impact is essential. It is not practical to develop detailed numerical models in every case, although for the most critical cases this is likely to be required. Such detailed evaluation is being investigated in further ongoing work, as discussed below. This initial methodology therefore uses analytical solutions to scope the impact. Whilst analytical solutions have often been used previously in the licensing process, the tools have not been readily available to all hydrogeologists involved in licensing, and only the Theis solution has been widely used. This has significant limitations when there are river bed sediments or where wells are close to rivers that do not fully penetrate the abstracted aquifer. The analytical solutions presented above have been integrated into a simple spreadsheet model that is available to all Agency staff and may be obtained by other interested parties. This spreadsheet tool allows users of the methodology to evaluate quantitatively the impact of abstraction on the river flow, but the focus of the methodology is on defining a detailed conceptual model of the physical river–aquifer interaction and developing an understanding of the physical system. This conceptual model is not the same as the simplified mathematical model underlying any analytical or numerical models of the system, which should be chosen to best approximate the conceptual model of the real system. Other tools can be used within the methodology.

The methodology is presented in detail in a User Guide (Environment Agency 1999), but can be summarized by the flowchart shown in Figure 7. It requires iterative cycles of data collation, conceptual model development and quantitative prediction of the consequences of the conceptual model. As the licensing process gives the opportunity of specifying a field pump test of the proposed borehole, this can be used to test and refine the conceptual model and follow through the flow chart a second time to improve understanding of the river-aquifer interaction. In some cases, longer-term monitoring of a time-limited licence may allow for further development of the conceptual model and understanding of the system, corresponding to a further cycle of the procedure. The important feature of this approach is that uncertainty should be estimated as part of the conceptual model. Any further field investigations can then be focused on trying to reduce this uncertainty, as well as confirming other aspects of the conceptual model. The obvious way to reduce uncertainty is to use the pump test to measure local aquifer properties in the usual way. However, by

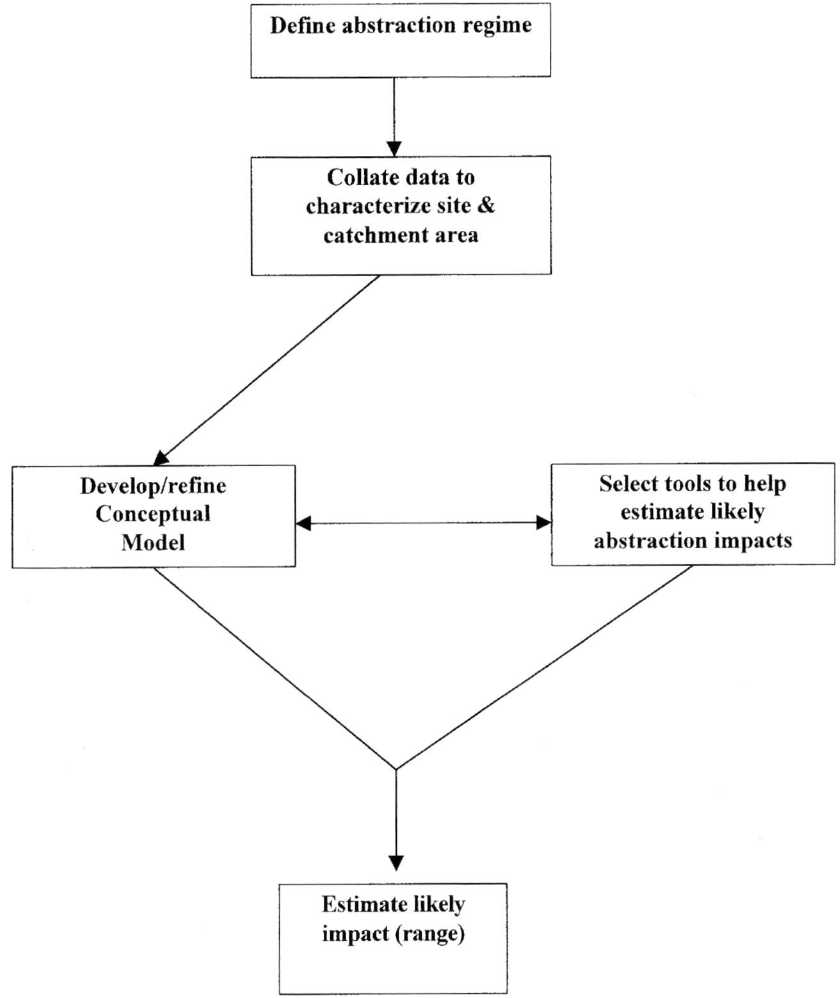

The above procedure should be applied to each of the following sequential stages:

A. To derive an initial estimate of impact for the proposed abstraction and to help design a pumping test for the proposed abstraction.
B. To produce final best estimates of likely impact of the proposed abstraction after analysing the result of the pumping test and revising the conceptual model accordingly.

Fig. 7. Procedure for estimating the range of likely impacts of a proposed groundwater abstraction.

applying the analytical solutions for the river–aquifer interaction one can estimate the expected impact on the river and may be able to design the pump test so that these aspects of the system can be tested. This might involve timing the test to take place during a period where impacts might be more measurable, or extending the period of monitoring to observe longer delayed

Table 1. *Illustrative input parameters used by the spreadsheet tool*

Parameter	Description	Illustrative value
Site	Site identification for plot to produce a quality assured (QA) record. This appears on all output.	Anywell
Run Identification Number	Simulation run number for plot to produce a QA record. This appears on all output.	Base 1
Analytical solution		
Analytical solution selected	A drop-down menu to choose which analytical solution is to be evaluated. The parameters not used by the selected solution are coloured pale yellow and protected to prevent values being entered for them, and a warning is shown if necessary river bank or river bed parameters are given zero values.	Select Hantush
Transmissivity	Transmissivity of saturated thickness of the aquifer	$400\,m^2\,d^{-1}$
Storage coefficient	Storage coefficient for the aquifer (confined) or specific yield for unconfined approximations	0.01 (can only set values between 0 and 1)
Perpendicular distance to stream	Perpendicular distance of well to the assumed straight line of the river course	5 m
Hantush		
Stream bank sediments' conductivity	Hydraulic conductivity of river bank sediments for use in the Hantush solution	$0.2\,m\,d^{-1}$
Stream bank depth	Depth of river bank for use in the Hantush solution	2 m
Stream bank thickness	Thickness of the river bank sediments for use in the Hantush solution. If non-zero, then Hantush solution is invoked.	0.2 m
Stang		Input not required when Hantush selected and cannot be set
Stream bed sediments' conductivity	Hydraulic conductivity of river bed sediments for use in the Stang solution	N/A
Stream bed width	Width of river bed for use in the Stang solution	N/A
Stream bed thickness	Thickness of the river bed sediments for use in the Stang solution. If non-zero, then Stang solution is invoked.	N/A
Periodic annual pumping		
Abstraction rate (by month)	Regular abstraction rate for month for long-term exploitation of the aquifer at the well.	Set to 0 except for May and June for which an average abstraction rate of $3630\,m^3\,d^{-1}$ should be set. This will allow the time lag to be seen in the periodic output
Compensation returns (by month)	Regular rate of return of water to the river for long-term exploitation of the aquifer at the well.	Set compensation to 0 except for $1000\,m^3\,d^{-1}$ in May
Pumping Test Input		
Duration of test	Duration of pumping test	150 d
Pumping rate for test	Constant rate of pumping for the pumping test	$3600\,m^3\,d^{-1}$
Axis limit plotting parameter	The chart for the pumping test will use a time axis drawn so that plot of depletion due to a continuous abstraction would extend to this percentage of the ultimate depletion rate. Set to high value to display long-term impact: a low value to see early impact.	'99.99%' or 0.9999 will display a large proportion of the impact. Lower values will provide more resolution of the early impact

stream depletion or to observe the recharge effect on the time drawdown data.

Thus, the procedure is not a simple linear sequence of tasks. As the various tasks are addressed there will often be a need to reconsider earlier steps and iterate between tasks as an understanding of the behaviour of the hydrogeological system is developed. At all stages, the conceptual model of the river-aquifer interaction should be considered and updated if new information or results suggest it might be improved. Similarly, in assessing the impact of abstraction on river flows, there will inevitably be uncertainty, and parallel assessments considering alternative solutions or alternative parameter scenarios are likely to be needed.

Access to the analytical solutions through the spreadsheet tool

The analytical solutions calculate the river depletion rate which should be interpreted as the net reduction in flow from the aquifer to the river. This may correspond to a reduction in baseflow (reduction in river gains) or to an increase in flow out of the river into the aquifer. The baseflow itself and its evolution over time is not evaluated by these methods. The spreadsheet is designed to estimate the impact in terms of stream flow depletion of regular annual abstraction cycles. Thus the user inputs the proposed abstraction in terms of average abstraction rates for each calendar month and the spreadsheet superimposes the impacts due to this cycle being repeated over many years to obtain the eventual maximum periodic impacts. The spreadsheet also incorporates Jenkins's (1970) model of a single finite period of abstraction to enable the impact of a pumping test to be calculated and to help design such tests to maximize the likelihood of obtaining measurable responses.

The spreadsheet is designed as a 'user-friendly' tool; the inputs are largely self-explanatory, and it is intended to be used quickly and easily. Of course, this is not to say that using the spreadsheet will make the hydrogeological understanding any easier. Indeed, it allows easy access to some new analytical solutions that require additional thought and consideration of the conceptual model of the hydrogeological system. However, the numerical tool itself will aid rather than hinder this process. It is important to investigate ranges of parameter values to see what the sensitivity is and to assess plausible parameter ranges. The spreadsheet can be used to investigate the extreme parameter ranges and to test the consequences of alternative conceptual models.

Table 1 lists the input parameters the user should set in the spreadsheet, as shown in Figure 8, to reproduce the output illustrated in Figure 9. There are three different analytical solutions available, but when one of these is selected from the drop-down menu only relevant parameter input cells will be accessible.

If the Stang solution is used, an evaluation of the drawdown is made on this sheet for the point in the aquifer beneath the closest point of the river to the well. The parameters for this evaluation are taken from the analytical solution parameter list and the time for the drawdown evaluation is taken to be the duration of the pumping test. The layout of this input to the spreadsheet is shown in Figure 8. The following section describes the corresponding output produced by the model as illustrated in Figure 9a and b.

The first pumping scenario in the spreadsheet requires specification of the average pumping rate each month for long-term use of the well over many years. The corresponding output histogram shows the corresponding average river depletion rate in each month. Where there is a significant resistance between the aquifer and the river, there may be a significant delay in the river depletion arising from a given pumping period. Thus, for example, the histogram may show river flow depletion in January when abstraction was only specified for August; the interpretation would be that the January depletion was largely due to the previous year's abstractions. This is illustrated in Figure 9a.

A 'PERCENTAGE EVALUATED' value of 69% (see Fig. 8) is calculated that records the proportion of steady-state river depletion covered by looking back over this 10-year period. Where the river is poorly connected to the aquifer due to low permeability of sediments or underlying aquitards, or for more distant abstractions, this will begin to fall to significantly less than 100%. In these cases, the analytical solution predicts that part of the impact of abstraction is taking more than 10 years to be realized in the river flows. For such cases the remaining fraction of the impact of continually repeated periodic abstractions is approximated as being spread uniformly throughout the year, and this is a reasonable approximation. The percentage evaluated relates to the proportion of the impact evaluated by direct integration of the analytical solutions, as opposed to the proportion that is evaluated more approximately. Note that the analytical solution predicts that once regular period abstractions have begun, the impact may well continue to develop over several years, with the

Input data sheet	Site	Anywell	Run identification number

Analytical solution

Theis Hantush and Stang		
Transmissivity	Trans	400 $m^2 d^{-1}$
Storage coefficient	S or Sy	0.01
Perpendicular distance to stream	a	5 m
Hantush		
Stream bank sediments	K bank	0.2 $m d^{-1}$
Stream bank depth	Depth bank	2 m
Stream bank thickness	Thick bank	0.2 m
Stang		
Stream-bed sediments conductivity	K bed	0.01 $m d^{-1}$
Stream-bed width	width bed	0 m
Stream-bed thickness	thick bed	0 m

Update mode
☐ Auto mode
☐ Manual mode

Analytical solution selected — Hantush — Hantush Modellist

Periodic annual pumping

	Jan	Feb	Mar	Apr	May	Jun	Jul	Aug	Sept	Oct	Nov	Dec
Abstraction rate ($m^3 d^{-1}$)	0	0	0	0	3630	3630	0	0	0	0	0	0
Compensation returns ($m^3 d^{-1}$)	0	0	0	0	1000	0	0	0	0	0	0	0
Cumulative days in year	0	31	59	90	120	151	181	212	243	273	304	334
Volume abstracted (m^3)	0	0	0	0	*112530*	*10890*	0	0	0	0	0	0
Compensation volume (m^3)	0	0	0	0	31000	0	0	0	0	0	0	0
Monthly river depletion (m^3)	5838	5273	5838	5649	26705	77169	31558	9406	5654	5843	5654	5843
Daily river depletion ($m^3 d^{-1}$)	188.3	188.3	188.3	188.3	861.5	2572.3	1018	303.4	188.5	188.5	188.5	188.5

Total abs specified	2.21E+05
Total abs evaluated	1.53E+05
Percentage evaluated	69.22%

Pumping test input

Duration of test	50 days
Pumping rate for test	3600 $m^3 d^{-1}$
Axis limit plotting parameter	0.00% set to high value long-term impact – low value to see early impact

0% sets axis limit to minimum period for display (=2*duration), and 100% sets axis limit to maximum evaluated period

Theis well function to give (over) estimate of max. drawdown below river at end of pumping test (for Stang)

Pumping rate for drawdown calculation	3600 $m^3 d^{-1}$
Transmissivity for drawdown calculation	400 $m^2 d^{-1}$
Calculation	0.01
Drawdown time	150 days
Drawdown distance	5 m

Environment Agency Project NC/06/28

Fig. 8. Illustrative input for spreadsheet tool.

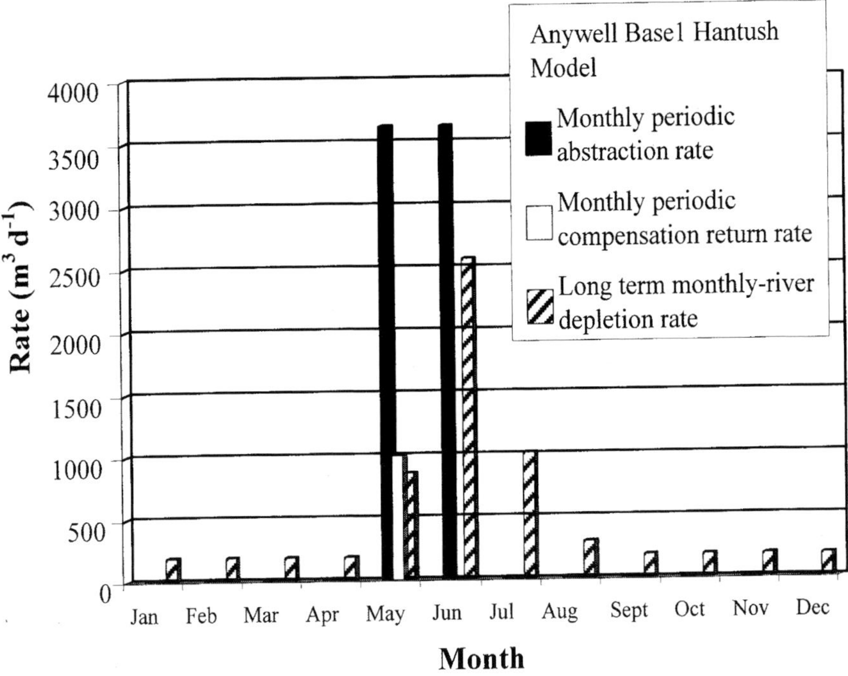

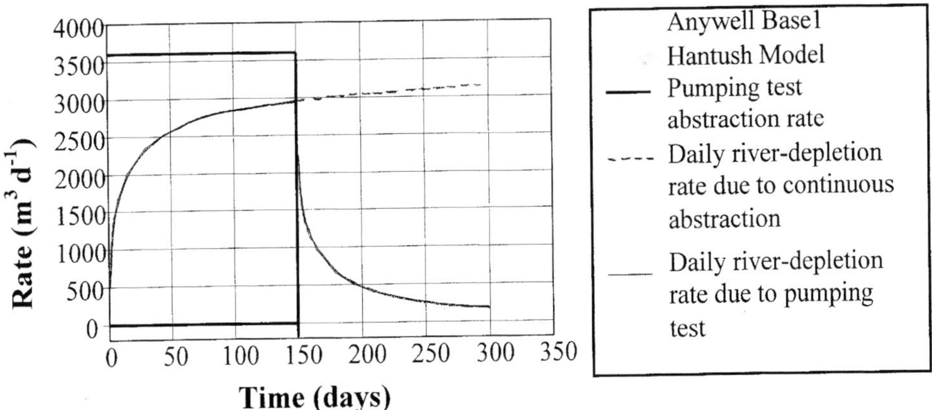

Fig. 9. Illustrative output for spreadsheet tool: (**a**) histogram of monthly impacts due to periodic abstraction; (**b**) the continuous impact due to a finite pumping test.

first year's impact being significantly less than the longer-term steady-state impacts predicted by the spreadsheet tool.

The second pumping scenario allows a single finite period of pumping at a steady rate. The model makes a continuous prediction of the rate of river flow depletion as a function of time measured from the start of abstraction. It is

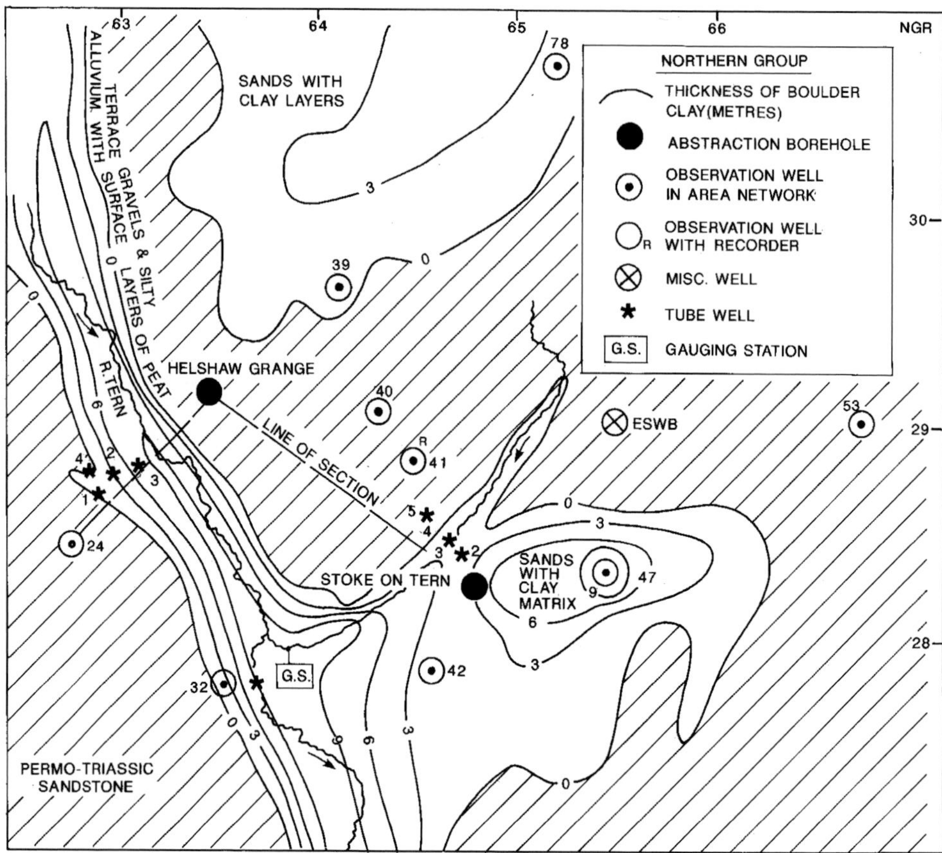

Fig. 10. Generalized base map for Helshaw Grange.

intended to allow estimation of the impact of a pumping test on the river flow. It will also indicate the duration of that impact. The user can optimize the parameters of a pumping test and focus field data collection so as to maximize the likelihood of important impacts being measured. Figure 9b shows the output from the spreadsheet and reference should be made to this output. The output from the pumping test shows three parameters in graphical form. They are:

(1) the pumping rate of the test and the time interval over which the depletion has been evaluated (pumping test abstraction rate);
(2) the impact of the abstraction on the river over the period of the test and any lag that may be occurring prior to that impact. This may be used to identify the length of time for which stream gauging should be performed during the test and to identify the time before impact would be expected (daily river depletion rate due to pumping test); and
(3) the predicted time evolution of the impact of the test, should that test be continued indefinitely. This may also be used in conjunction with hydrometric flow data for the river to predict at what time in the pumping test an impact would be identified from flow gauging measurements (daily river depletion rate due to continuous abstraction).

It is recognized that none of the three analytical solutions can account for the reduced drainage from a river that might occur if the groundwater level falls below the river bed. The solutions will therefore tend to overestimate river flow depletion in such reaches under these circumstances. Note that this implies that

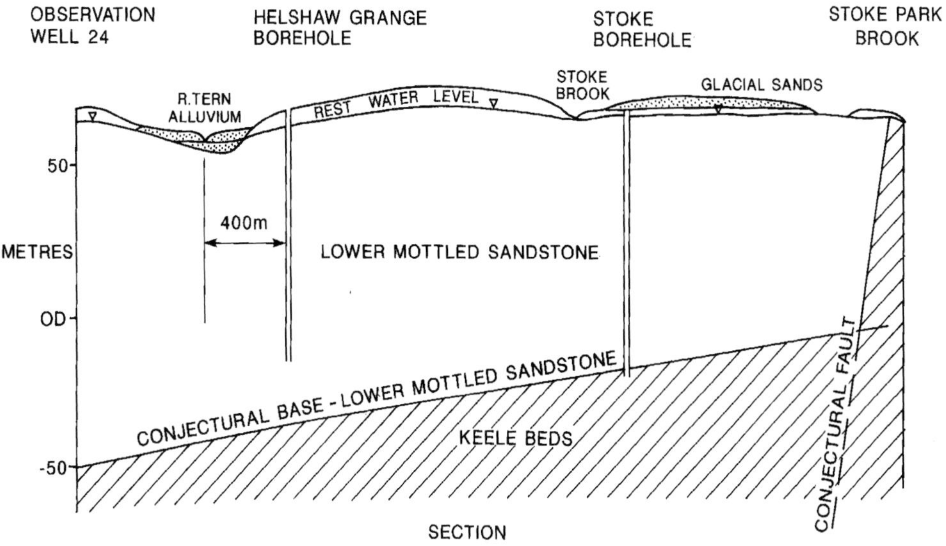

Fig. 11. Generalized geological cross-section for Helshaw Grange.

further from the abstraction, the impact will consequently be underestimated since the model conserves mass. In order to help identify the conditions under which reduced drainage might occur a crude means of estimating the drawdown beneath the river bed at the end of a pumping test (from the point closest to the abstraction) is incorporated in the spreadsheet. The drawdown can then be used to judge whether or not limiting drainage will occur. The drawdown is evaluated for a time corresponding to the end of the pumping test, at the point closest to the well beneath the river bed. The time, distance and pumping rate used are those of the pumping test input.

Illustrative application of the methodology to the Helshaw Grange abstraction

The Helshaw Grange abstraction borehole (NGR SJ63442919) is situated in the River Tern catchment approximately 15 km NE of Shrewsbury in the English Midlands and forms part of the Shropshire Groundwater Scheme. The borehole is constructed in the unconfined Permo-Triassic Sandstone aquifer. Pumping tests confirm that the hydrogeological characteristics of the aquifer are similar to other Permo-Triassic Sandstone aquifers in the region. The geological setting and a cross-section are shown in Figures 10 and 11. Abstraction from the borehole is used for the purposes of river regulation along a short length of the Tern during periods of low flow. The River Tern flows from north to south across the outcrop, passing within 400 m of the borehole and is understood to gain baseflow from the aquifer. The connectivity between the aquifer and the river is understood to be good in places. Pumping test data were obtained during the investigative stages of the Shropshire Groundwater Scheme, and some data on streambed properties are available.

An initial assessment of the impact of groundwater abstractions on river flows was carried out. The assessment was based upon a conceptual model built up from data from literature available prior to the development of the Shropshire Groundwater Scheme. The initial assessment was then used to illustrate how the methodology could help design a pumping test that would test the hydraulic connection between the aquifer and the river. There was no opportunity within this work to undertake new hydrogeological field investigations. Therefore, to illustrate the iterative approach of the methodology, the results of a pumping test carried out during the development of the Shropshire Groundwater Scheme were used to refine the conceptual model of the site. This process is summarized below.

Table 2. *Generic property values used for evaluation of the Helshaw Grange abstraction in the illustrative case study*

Parameter	Value	Notes
Transmissivity (m^2 d^{-1})	495 (range 225–765)	Lovelock (1977)
Storage coefficient	0.01	Low regional value – produces more rapid response
Saturated thickness (m)	90	From Shropshire Groundwater Scheme investigation drawing
Regional hydraulic gradient	0.0008	Groundwater monitoring levels
Distance to borehole (m)	400	Ordnance Survey map
Bed/Bank hydraulic conductivity (m d^{-1})	0.1	Shropshire Groundwater Scheme Investigation, 5th report, 1978
River depth (m)	2	Estimate
River width (m)	2	Estimate

Define the abstraction

The borehole at Helshaw Grange is currently used for the intermittent abstraction of groundwater for river regulation. The abstraction is subject to a maximum licensed rate of 7000 m^3 d^{-1} with an overall maximum rate of 700 000 m^3 a^{-1}.

Collate data to characterize site and catchment area

This illustrative example only considers the impact of abstraction on the River Tern which is assumed to be the only natural discharge for groundwater flow in the vicinity of the abstraction. A full application of the methodology to a licence application would also consider the impact of the abstraction on other tributaries of the Tern and any other susceptible surface water bodies in the catchment. The groundwater contours indicate that locally the groundwater flow is to the river. The distance from the borehole to the river was estimated. Stream bed and bank properties, together with uncertainty introduced through geometrical simplification, were subject to sensitivity calculations, and a range of parameter values was adopted. Hydraulic data relating to the site were collated and are presented in Table 2.

Develop/refine conceptual model and select tools

The conceptual model for this study, which comprises a detailed interpretation of the hydrogeology, is shown in Figures 10 and 11. Some parameters in the model were known (e.g. abstraction rate), others were estimated (e.g. aquifer properties). The Stang model providess the best match to the conceptual model as the majority of losses from the river were thought to be through the river bed. Alternative analytical solutions were also considered, as the bank and bed configuration was not well known. Due to uncertainties in the conceptual model, all three possible solutions are often used to explore the range of impacts that may occur. For example, anisotropy in the aquifer may lead to predominantly horizontal flow; this would make Hantush's model more appropriate than Stang's even if the aquifer extended beneath the river.

The properties for the chosen solutions were then estimated, as shown in Table 2. In this model the bed and bank properties were estimated from the Shropshire Groundwater Scheme investigations and were considered reasonably well constrained. At this stage a decision should be made as to whether the analytical solutions represent an appropriate representation of the river-aquifer interaction. In some cases, more detailed field investigations and modelling techniques would be required, for example for important abstractions in sensitive locations or for particularly complex hydrogeology.

Estimate likely impact

Results from the model may be considered either by use of the pumping test or periodic pumping result sheets. At this site, it is expected that there will be occasional periods of abstraction to regulate the river flows and the pumping test output was considered rather than the results for regular periodic abstractions. In particular, to demonstrate the effectiveness of the river augmentation, the time lag is critical. A pumping rate of 7500 m^3 d^{-1}, and a period of 10 d pumping were specified.

It is necessary to assess the likelihood of the

river and aquifer becoming disconnected as this will significantly affect results. The standard Theis well function may be used to provide an (over) estimate of the impact of pumping on water levels adjacent to the river. In this case after 10 d of abstraction the Theis function estimated a drawdown of 2.4 m in the vicinity of the river. This is only slightly greater than the estimated river depth. It is thus considered unlikely that drainage from the river would be reduced from that estimated using the analytical solutions due to significant reaches of the river becoming 'perched' during the pumping test.

Results

The Theis (river) solution suggested that the depletion of baseflow would reach 5100 m^3 d^{-1} after 10 d. The Stang solution, assuming losses through the low permeability base of the river gave a result of 1100 m^3 d^{-1}, while Hantush's solution gave an impact of only 600 m^3 d^{-1} after 10 d. A degree of lag was noted with both the Hantush and Stang solutions. The results are illustrated in Figure 12 for the three different analytical solutions. This shows that inappropriate use of the Theis solution can give very significant overestimates of the short-term impact.

The amount of depletion can be compared to the 95th percentile of river flow. In the Tern, the Theis solution (extreme) depletion was equivalent to 12% of Q_{95}, whilst the more appropriate solution (Stang) equated to 2.5% of Q_{95}.

To help design a pumping test for the proposed abstraction

When considering an application for a new abstraction licence, the methodology recommends using the above results to help design a pumping test and monitoring scheme to test/confirm the predicted results and constrain the range of uncertainty. In general, the test duration and the nature of the monitoring are the key pumping test variables that may be varied so as to maximize the likelihood of obtaining useful information about the impact of abstraction on river flow. However, there are often practical constraints on these issues.

Following a group pumping test investigation in 1973, the Helshaw Grange borehole was subject to a further pumping test in 1975 to assess the effect of a possible recharge boundary, the River Tern. This 9 d pumping test was undertaken in November 1975 at the Helshaw Grange abstraction borehole (SGS 1978). The pumping rate (7418 m^3 d^{-1}) used in the test is close to the maximum licensed rate. Measurements were taken at nine different sites close to the river and at a nearby river stage recorder site. Abstracted water from the test was disposed of to the river and the river flow gauging undertaken at this time could not be quantified to constrain the interpretation. Water levels in tubewells near the river did suggest a very rapid response of groundwater levels in response to a change in river stage. Type-curve analysis of the pumping test gave a specific yield of 0.025 and a transmissivity of 240 m^2 d^{-1}.

The pumping test values were used as the basis for a refined model and a second iteration of the procedure for estimating the impact. The data available are sufficient for scoping, but not for more detailed modelling. The Stang model is thought to be the most appropriate solution, as in the initial assessment. In this updated assessment abstractions used in the drought year of 1995 were used. This is expected to produce an overestimation of impact compared to average abstractions. The abstractions used were 6233 m^3 d^{-1} during July, 5163 m^3 d^{-1} during August, 3625 m^3 d^{-1} during September and 3360 m^3 d^{-1} during October. All abstracted water was discharged to the Tern for flow augmentation.

The level of drawdown adjacent to the river was again assessed to confirm that there was unlikely to be limiting drainage produced by the river becoming 'perched' over large reaches in the example.

Estimate likely impact (range)

The revised parameters resulted in a reduced impact compared to the initial assessment. The Stang model predicted a depletion of 500 m^3 d^{-1} after 10 d. With continual pumping this will rise slowly to equal the pumping rate. Abstraction rates from a drought year were used in this case study, and in most years abstractions are expected to be lower. This would result in lower depletion levels. A sensitivity analysis was carried out to investigate the response of the models to changes in parameter values. Both the Stang and Hantush outputs are directly proportional to the hydraulic conductivity of the river sediments and inversely proportional to the thickness of these sediments.

The results were entered into an evaluation of the impact of many years of periodic abstraction at this site with the same monthly pattern of abstraction. The results for the central case are shown in Figure 13. It should be noted that the time lag identified means that reductions in flow

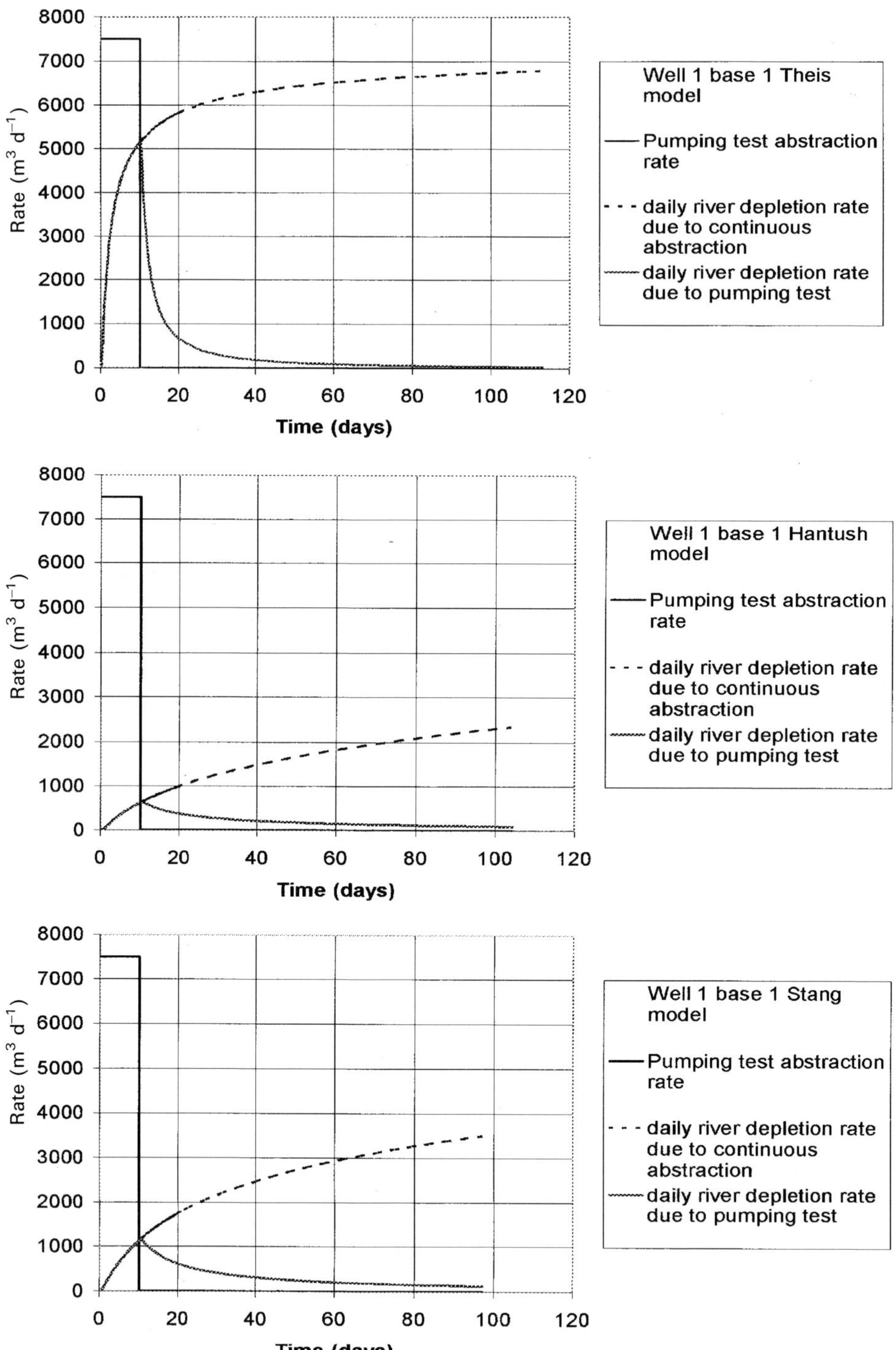

Fig. 12. Results of the pumping test for the general model.

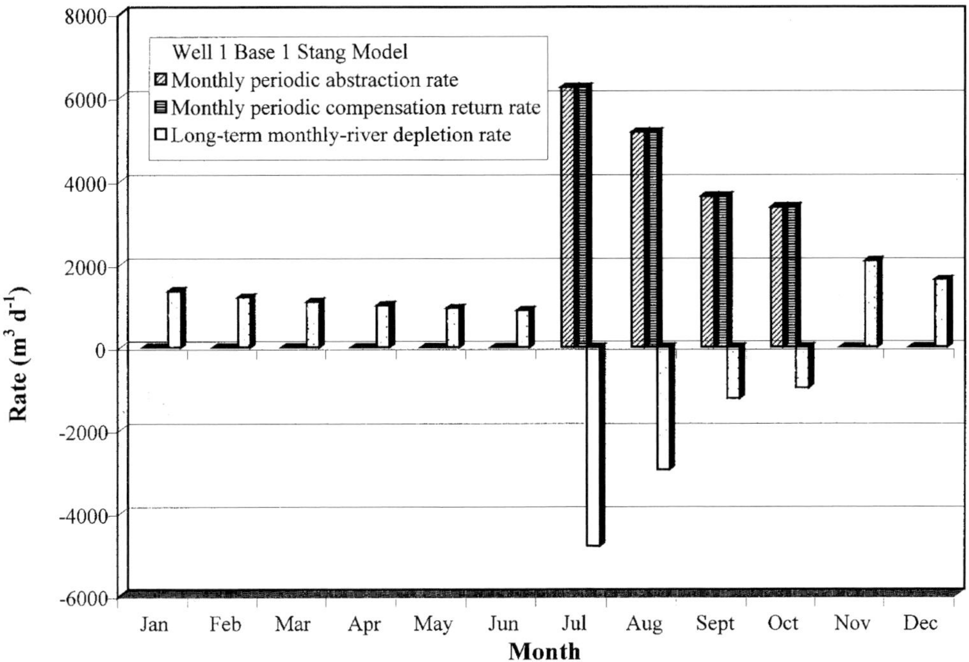

Fig. 13. Monthly periodic abstraction rates and long-term impacts on river flow for Helshaw Grange. The model assumes the 1995 abstraction pattern is repeated annually.

are likely to be seen after pumping has finished in the first year. In later years further abstractions are shown superimposed on the residual depletion from previous year's pumping. Figure 13 shows that river augmentation is likely to be successful, although efficiency is significantly reduced if abstraction continues for prolonged periods, as would be expected for a well-connected river–aquifer system. It also shows that there is an impact extending through the winter recharge period, and subsequent years are predicted to be affected by the use of the scheme through reduced groundwater levels and consequently reduced river baseflow. This would not be suggested by the use of the traditional Theis solution, which would have suggested less efficient augmentation together with negligible impact on baseflow beyond the year of operation.

Future extensions of the methodology

This paper summarizes some of the findings from the first phase of the Agency's programme studying the impact of groundwater abstractions on river flows (Environment Agency 1999). The primary objective of the Agency's programme is to deliver practical guidance to Agency hydrogeologists involved in the licensing of groundwater abstractions that might adversely affect river flows. The first phase of work, described in this paper, has delivered a recommended procedure and a user-friendly spreadsheet tool that incorporates a selection of analytical solutions. This software and procedure have been made available across the Agency, and many hydrogeologists have been trained in their use.

The Agency's IGARF programme is currently in its second phase, with further phases planned over the next 3 years. The second phase of work, which was due to be completed in the summer of 2001, uses a combination of numerical model simulations of generic river–aquifer systems and neural networks to quantify the impacts of groundwater abstractions on river flows. The simulation tool being developed is a user-friendly graphical user interface and embedded neural network software also designed for use in abstraction licensing. This novel approach will provide a modelling tool that is very fast to run but which retains the advantage of numerical methods in being able to represent complex conditions. This will allow the same procedures developed during the first phase of the programme to benefit from a new modelling tool that is capable of simulating a wider range of river-groundwater configurations whilst retaining many of the user-friendly features designed for the spreadsheet tool.

Conclusions

A methodology has been developed for the initial assessment of the impact of groundwater abstraction on river flow. It uses available analytical solutions for river–aquifer interactions. These have been implemented in an easily accessible spreadsheet tool that uses the approach of Jenkins (1968) to consider the distribution in time of the impact of finite or periodic pumping. The methodology has been demonstrated for an abstraction in the Permo-Triassic Sandstone at Helshaw Grange.

It should be emphasized that the approach is based on expert hydrogeological judgement and that the spreadsheet tool should only be used to aid that judgement. The crucial step in the methodology, in common with all quantitative hydrogeology, is the development of an appropriate conceptual model. In the case of this methodology, for the initial assessment of the impact of groundwater abstraction on river flow, conceptual understanding of the real site must then be approximated by selecting from the available analytical solutions each of which is based on a simple underlying mathematical model. The inevitable consequence is increased uncertainty which has been addressed by encouraging consideration of a broad range of uncertainty, and by advocating the application of more than one analytical solution within an assessment, each with a correspondingly different simplified representation of the conceptual model.

To progress to a more detailed and confident assessment, the use of site-specific numerical modelling is recommended where any potential exists for an adverse impact. An important limitation is likely to be the availability of accurate data, in particular on river bed or river bank sediment properties. In such circumstances there will be a great deal of uncertainty and there will be little benefit from the more detailed description of the geometry and heterogeneity afforded by numerical representation; the simple analytical approximations presented are likely to be as useful as numerical methods. The spreadsheet discussed above allows these analytical approximations to be used very easily and enables the consequences of parameter uncertainty to be quantified through sensitivity studies.

This project has benefited from the input of a large number of Environment Agency staff. The authors would like to acknowledge, with thanks, those individuals from the Agency who have given their time to provide input to the project by returning the questionnaire and responding to requests for follow-up discussion. We also thank the staff who provided data and technical inputs for the three trial examples. Finally, we would like to thank the Agency staff on the project board, Steve Fletcher and Dave Burgess who, together with John Aldrick, have provided valuable strategic guidance and detailed review comment during the course of the project.

References

BOULTON, N. S. 1942. The steady flow of ground-water to a pumped well in the vicinity of a river. *Philosophical Magazine Series*, **7**, 34–50.

BULLOCK, A., GUSTARD, A., IRVING, K. & YOUNG, A. 1994. *Low flow estimation in artificially influenced catchments.* Institute of Hydrology, National Rivers Authority, R&D Note **274**.

CHEN, M. & SOULSBY, C. 1997. Risk assessment for a proposed groundwater abstraction scheme in Strathmore, north east Scotland: a modelling approach. *Journal of the Institute of Water and Environmental Management*, **11**, 47–55.

CHEN, M., SOULSBY, C. & WILLETTS, B. 1997. Modelling river-aquifer interactions at the Spey Abstraction Scheme, Scotland: implications for aquifer protection. *Quarterly Journal of Engineering Geology*, **30**, 123–136.

ENVIRONMENT AGENCY. 1999. *Impact of groundwater abstractions on river flow: user manual.* Solihull, National Groundwater and Contaminated Land Centre Report **NC/06/28**.

GLOVER, R. E. & BALMER, G. G. 1954. River depletion resulting from pumping a well near a river. *American Geophysical Union*, **35**, 468–470.

HANTUSH, M. S. 1965. Wells near streams with semipervious beds. *Journal of Geophysical Research*, **70**, 2829–2838.

HUNT, B. 1999. Unsteady stream depletion from groundwater pumping. *Ground Water*, **37**, 98–102.

JENKINS, C. T. 1968. Techniques for computing rate and volume of stream depletion by wells. *Ground Water*, **6**, 37–46.

JENKINS, C. T. 1970. Computation of rate and volume of stream depletion by wells. *In: Techniques of Water-Resources Investigations of the United States Geological Survey.* U. S. Department of the Interior, Ch.D1, 1–17.

LOVELOCK, P. E. R. 1977. *Bulletin 56. Aquifer properties of the Permo-Triassic sandstones in the United Kingdom.* Bulletin of the Geological Survey of Great Britain.

MOREL, E. H. 1980. The use of a numerical model in the management of the Chalk aquifer in the Upper Thames Basin. *Quarterly Journal of Engineering*

Geology, **13**, 153–165.

NAWALANY, M., RECKING, A. & REEVES, C. 1994. Representation of river-aquifer interactions in regional groundwater models. *In*: REEVES, C. & WATTS, J. (eds) *Groundwater-Drought, Pollution & Management*. Balkema, Rotterdam.

NIELD, S. P., TOWNLEY, L. R. & BARR, A. D. 1994. A framework for quantitative analysis of surface water-groundwater interaction: Flow geometry in a vertical section. *Water Resources Research*, **30**, 2461–2475.

RORABAUGH, M. I. 1963. Streambed percolation in development of water supplies. *US Geological Survey Water Supply Paper* **1554-H, H47-H62**.

RUSHTON, K. R., CONNORTON, B. J. & TOMLINSON, L. M. 1989. Estimation of the groundwater resources of the Berkshire Downs supported by mathematical modelling. *Quarterly Journal of Engineering Geology*, **22**, 329–341.

SGS. 1978. *Shropshire Groundwater Scheme Investigation, 5th report*. National Rivers Authority, Shrewsbury.

SHAW, E.M. 1994. *Hydrology in practice* (3rd ed.). Chapman and Hall, London.

SOPHOCLEOUS, M., KOUSSIS, A., MARTIN, J. L. & PERKINS, S. P. 1995. Evaluation of simplified stream-aquifer depletion models for water rights administration. *Ground Water*, **33**, 579–588.

SPALDING, C. P. & KHALEEL, R. 1991. An evaluation of analytical solutions to estimate drawdowns and stream depletions by wells. *Water Resources Research*, **27**, 597–609.

SPINAZOLA, J. 1998. *A spread sheet notebook method to calculate rate and volume of stream depletion by wells in the Lemhi river valley upstream from Lemhi, Idaho*. U. S. Department of the Interior, Idaho, 1–19.

STANG, O. 1980. Stream depletion by wells near a superficial, rectilinear stream. Seminar No.5, Nordiske Hydrologiske Konference, Vemladen. *In*: BULLOCK, A., GUSTARD, A., IRVING, K., SEKULIN, A. & YOUNG, A. (eds) *Low flow estimation in artificially influenced catchments*. Institute of Hydrology, Environment Agency R&D Note 274, WRc, Swindon.

THEIS, C. V. 1941. The effect of a well on the flow of a nearby stream. *American Geophysical Union Transactions*, **22**, 734–738.

WALLACE, R. B., DARAMA, Y. & ANNABLE, M. D. 1990. Stream depletion by circle pumping of wells. *Water Resources Research*, **26**, 1263–1270.

WATTS, G., CREW, R., COOK, S., SCRIVENS, A. & YOUNG, A. 1995. *Predicting artificially influenced flow statistics using micro LOW FLOWS v2.0*. Institute of Hydrology, National Rivers Authority, R&D Note **448**.

YOUNGER, P. L., MACKAY, R. & CONNORTON, B.J. 1993. Streambed sediment as a barrier to groundwater pollution: Insights from fieldwork and modelling in the River Thames basin. *Journal of the Institute of Water and Environmental Management*, **7**, 577-585.

ZLOTNIK, V. A. & HUANG, H. 1998. An analytical model of aquifer response to stream stage fluctuations: effect of partial penetration and streambed sediments. *In*: BRAHANA, J., ECKSTEIN, Y., ONGLEY, L. K., SCHNEIDER, R. & MOORE, J. E. (eds) *Gambling with Groundwater - Physical, Chemical and Biological Aspects of Aquifer-stream Relations*. Proceeding of the Joint Conference of the IAH and AIH, Las Vegas, Nevada, 297–304.

Determination of hydraulic boundary conditions for the interaction between surface water and groundwater

W. MACHELEIDT[1], W. NESTLER[1] & T. GRISCHEK[2]

[1]*Institute for Geotechnics and Water Sciences, University of Applied Sciences Dresden, F.-List-Platz 1, 01069 Dresden, Germany (e-mail: mach@htw-dresden.de)*

[2]*Institute for Water Chemistry, Dresden University of Technology, Mommsenstraße 13, 01069 Dresden, Germany*

Abstract: The main determinants of the interaction between surface water and groundwater are the distribution of areas with different infiltration rates, the thickness of sediment layers and the hydraulic head gradient. These conditions determine the volume and velocity of infiltrating water which, together with the direction of water flow, are required to model the interaction processes. Due to difficulties with measurement, only the direction of water flow is usually determined and boundary conditions are estimated from simplified assumptions. Field techniques have now been developed that help characterize surface water–groundwater interaction. Results from field experiments using a percussion probe and a large-scale laboratory column experiment set up to simulate infiltration processes are presented. Measurements of the ^{222}Rn distribution in the column are used to determine infiltration velocities.

Improved knowledge of the interactions between surface water and groundwater is an important precondition for the solution of various ecological and water management problems. Estimation of infiltration rates and infiltration velocities from surface waters into bed sediments is normally based on water balance calculations, mixing models or geohydraulic simulations. However, exact measurements are necessary to answer questions relating to infiltration regimes during river bank infiltration, floodplain water balances or for the impact assessment of mining areas and hydraulic constructions. In the past the construction of groundwater observation profiles or sophisticated measurement devices has meant that these infiltration measurements could only be made at considerable expense. Thus, the tendency has been towards an insufficient description of exchange processes.

The influence of colmatage layers on infiltration rate

In the German literature the term 'colmatage' is widely used to describe the phenomena of lake and river bed siltation (Busch *et al.* 1993). The exchange between surface waters and groundwater is essentially determined by potential differences consequent upon the hydraulic resistance of colmatage layers. The colmatage process results in the reduction of permeability at the groundwater/surface water interface (Fig. 1). Preliminary studies indicate that colmatage layers, when measuring surface/groundwater interactions, can lead to significant error (Nestler *et al.* 2000).

There are two types of colmatage layer: external and internal. The internal colmatage layer is present in the bed material and consists of pores filled with fine silty and organic sediment. The external colmatage layer is built from the same fine sedimentary material, but lies external to the bed proper. Factors influencing the presence and nature of colmatage layers are illustrated in Figure 1. A colmatage layer may consist of either internal or external colmatage, both types, or may not be present at all. For a more detailed discussion of colmatage layer behaviour refer to Busch *et al.* (1993). This paper will discuss colmatage primarily with

From: HISCOCK, K. M., RIVETT, M. O. & DAVISON, R. M. (eds) *Sustainable Groundwater Development*. Geological Society, London, Special Publications, **193**, 235–243. 0305-8719/02/ $15.00 The Geological Society of London 2002.

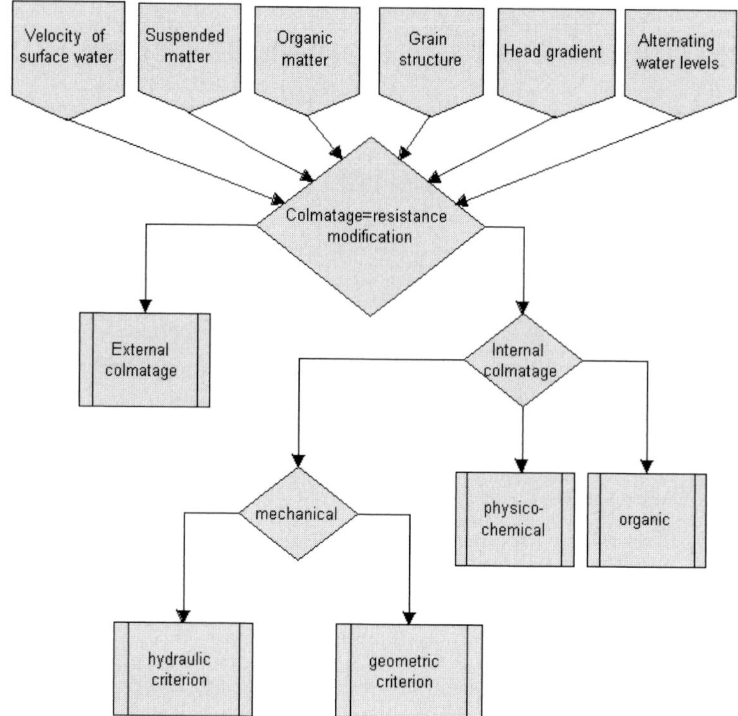

Fig. 1. Influences on the colmatage process.

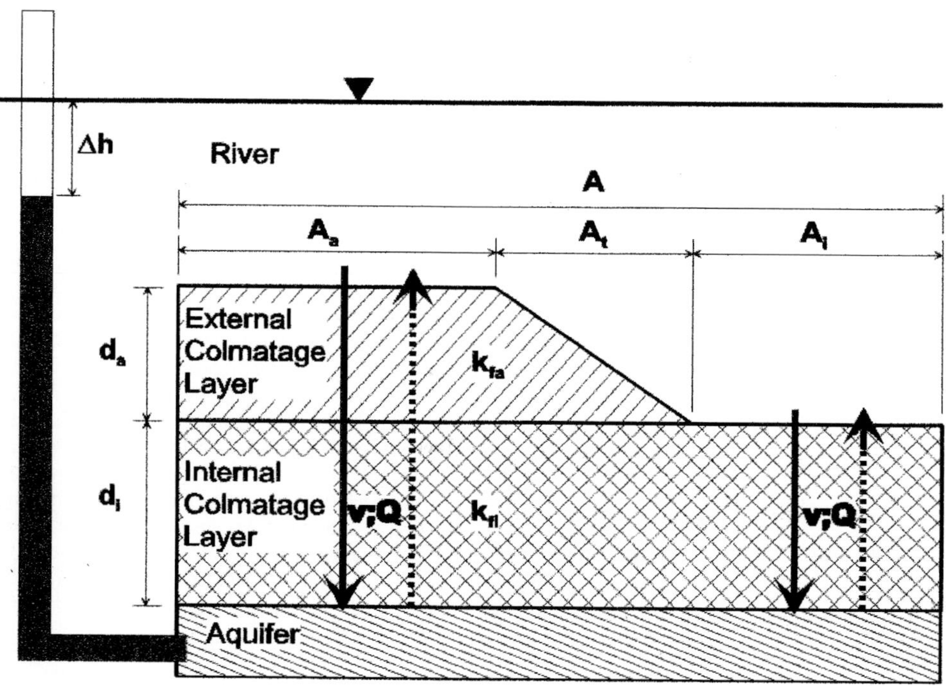

Fig. 2. Conceptual model of surface water–groundwater interaction.

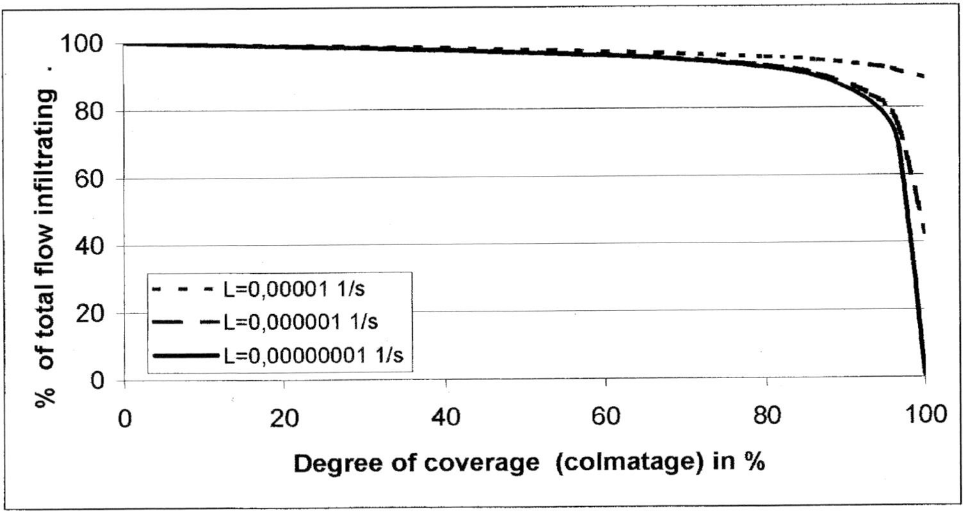

Fig. 3. Effect of the degree of coverage of an external colmatage layer on the infiltration rate. L is the leakage factor.

reference to rivers, as this is the area of greatest importance for water supply in eastern Germany (Grischek *et al.* 1994).

In rivers, the discharge flow and the flow velocity are the primary controls on colmatage layer formation. Other important factors are bed material, pore geometry and the direction of the potential gradient; the higher the potential gradient from the surface water towards the groundwater the more likely it is that a colmatage layer will occur.

The influence of the colmatage layer on the exchange rate between groundwater and surface water is illustrated in Figure 2 and is described by the following equation:

$$Q(t) = \int v \times dA = \int \frac{\Delta h}{d_a/k_{f_a} + d_i/k_{f_i}} dA \qquad (1)$$

where Q is discharge (m s^{-1}); v is velocity of exchange (m s^{-1}); Δh is hydraulic head difference (m); d_a is thickness of the external colmatage layer (m); d_i is thickness of the internal colmatage layer (m); k_{fa} is average permeability of the external colmatage layer (m s^{-1}); k_{fi} is average permeability of the internal colmatage layer (m s^{-1}); A is infiltration area (m^2); A_a is infiltration area with external colmatage (m^2); A_i is infiltration area without external but with internal colmatage (m^2); and A_t is infiltration area between external and internal colmatage (transition area) (m^2).

To determine the infiltration rate either the infiltration velocity related to the river bed or the following parameters need to be known: the area of the external colmatage layer; the thickness and the hydraulic conductivity of the external and internal colmatage layers; the gradient between the hydraulic heads of surface water and groundwater; the temporal dynamism of the parameters; and the direction of the exchange process. Of these, a continuous registration of dynamic changes in the exchange rate is only possible by measuring the head gradient. Otherwise, two methodological approaches may be adopted for the characterization of exchange processes as derived from these conditions (Table 1):

Table 1. *Methods of direct and indirect measurements of exchange processes between groundwater and surface water*

Direct measurement	Indirect measurement
Dating using ^{222}Rn	Compilation of the geometry (d, A)
Seepage meter	Determination of hydraulic conductivity (K)
Tracer tests, e.g. temperature	Water head measurement (h)

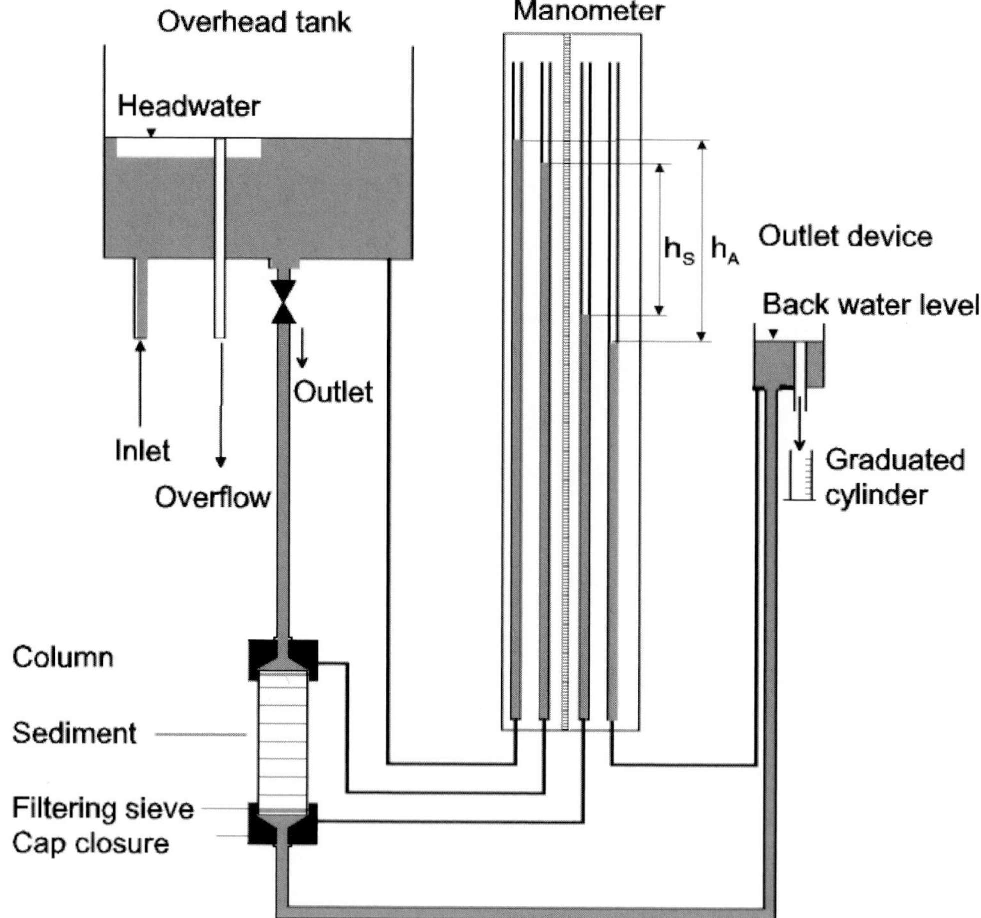

Fig. 4. Operation principle for the determination of hydraulic conductivities in undisturbed sediment samples.

Degree of colmatage coverage and rate of infiltration

The infiltration rate of a losing river depends on the degree of river bed coverage by an external colmatage layer. However, the effects of partial coverage on infiltration are often overestimated. An external colmatage layer with low hydraulic conductivity needs more than 80% coverage to significantly limit infiltration. Modelling simulations show the following correlation between the degree of coverage and infiltration rate (Fig. 3): up to 20% – no effect; between 20 and 60% – only minor effect; between 60 and 80% – low effect (67% coverage causes a reduction of 5%); and more than 80% – considerable effect on the infiltration rate.

These results depend on the leakage factor, L (s^{-1}), for the colmatage layer which is given by:

$$L = \frac{K_f}{M} = \frac{\mathit{Hydraulic\ conductivity}}{\mathit{Thickness}} \qquad (2)$$

Indirect measurement of infiltration rates

Determination of sediment permeability

Indirect measurements of infiltration rates require the determination of the hydraulic conductivity of the colmatage layer. The very fine stratification typical of the colmatage layer results in considerable vertical inhomogeneity. This effect makes it necessary to measure the hydraulic conductivity of undisturbed samples. For this reason a special laboratory-scale experimental system has been developed (Fig. 4). The principle is based on the method of

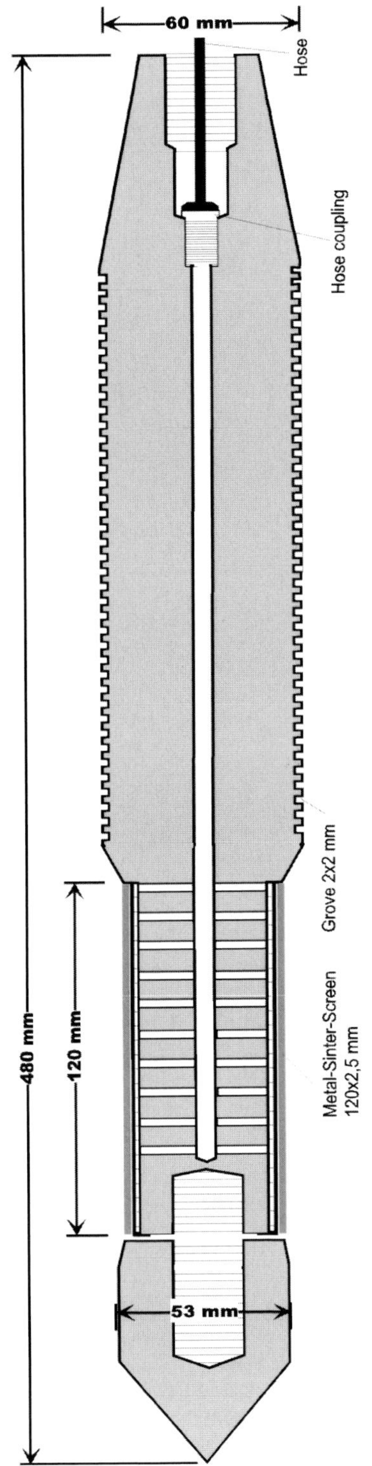

Fig. 5. Stainless steel core probe.

Beyer & Schweiger (1969) for the determination of vertical permeability in undisturbed sediment samples.

For an overview of a river bed, samples have been taken from each different area in the river bed. Sediment samples are collected in stainless steel columns (diameter 0.1 m; length 0.5 m) that are pressed into the upper zone (colmatage layers) of the river bed. After filling with sediment, the liners are withdrawn without loss of material and with only minimal compaction (the degree and effect of which is currently under investigation). Special adapters allow the columns to be installed directly into the experimental system in the laboratory. The apparatus facilitates work with hydraulic head differences up to 1.8 m. The vertical permeability of the sediment sample can be determined as a function of different head gradients. By controlling the flow direction it is possible to examine the permeability of the sediment under either infiltrating or exfiltrating conditions. Tracer tests with transparent columns were undertaken to examine the possibility of bypass flow down the column sides; this was not observed (Pfützner 1998). Furthermore, the technique was compared with results from sieve analyses, the results of which showed a maximum difference of $\pm$ 0.0001 m s^{-1} (Richter 1998).

Measurement of head gradient

For on-site investigations beneath a river bed, a stainless steel core probe has been developed to measure the hydraulic head distribution and to obtain groundwater samples at various defined depths (Fig. 5).

One advantage of the core probe is that it can be driven into sediments using ordinary percussion equipment. The probe enables hydraulic head measurements and infiltration water samples to be taken over vertical intervals of 0.3 m. Important probe features are the exchangeable tip, the exchangeable metal-sinter-screen (which allows the adaptation of the sampling technique to different grain sizes) and the transverse notched shaft. Many tracer and head gradient experiments have been undertaken to ensure that the transverse notched shaft excludes vertical leakage along the sides of the probe (Pfützner 1998; Pätzold 1999).

The hydraulic head at the measurement depth is determined by a pressure sensor connected to the probe. The trend of the head gradient plotted as a function of the measurement depth enables the hydraulically active border of the colmatage layer to be determined.

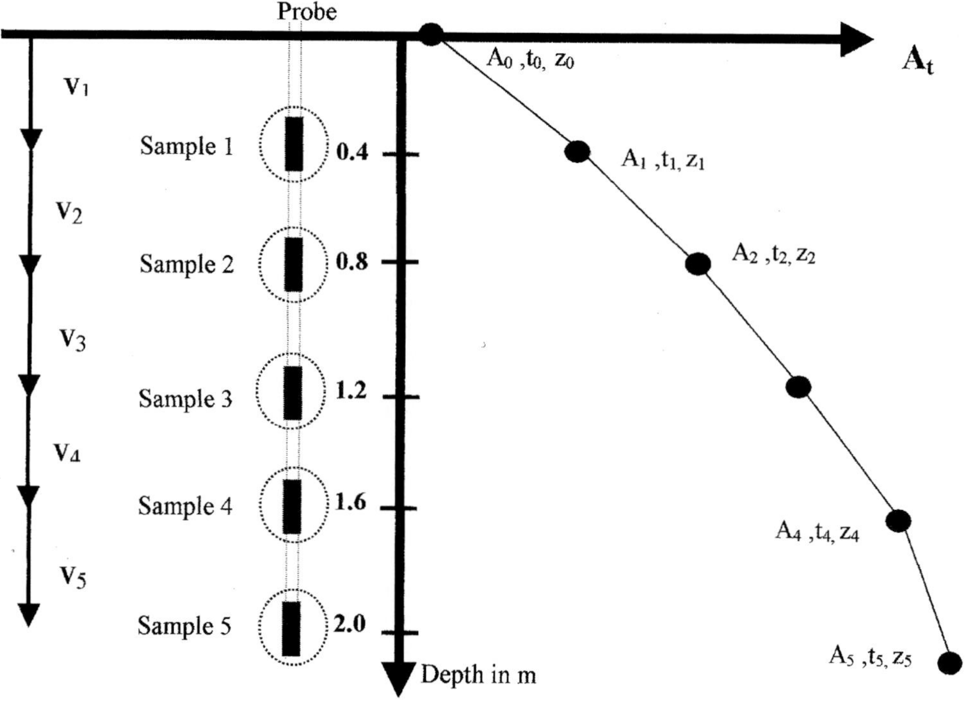

Fig. 6. Theoretical increase of A_t with residence time and sampling depth z_t.

Water samples are obtained using an adjustable low-pressure unit that guarantees minimal disturbance sampling. The sampling is carried out exclusively in a downward movement.

Direct measurement of infiltration rates using ^{222}Rn

Infiltration velocities may be measured by determining ^{222}Rn radioactivity at defined points along a flow path. ^{222}Rn is the natural decomposition product of the radioactive ^{226}Ra, which occurs in soil. ^{222}Rn is an inert gas with a half-life ($T_{1/2}$) of 3.8 d. In stagnant groundwater there is an equilibrium between ^{222}Rn and its parent ^{226}Ra. ^{222}Rn is barely present in surface waters due to its high volatility. However, during infiltration processes, ^{222}Rn is enriched in the aquifer along the flow path until an equilibrium concentration is reached. The increase in radioactivity is described by:

$$A_t = A_e(1 - e^{-\lambda t}) \tag{3}$$

where A_t is concentration of radioactivity at time t (Bq l^{-1}); A_e is equilibrium (background) concentration (Bq l^{-1}); t is retention time (s); and λ is decay constant ($\lambda_{Rn} = 0.18$ d^{-1}) (d^{-1}).

Taking into consideration the radioactivity A_0 of the infiltrating water, equation (3) can be solved for retention time t (Dehnert *et al.* 1998):

$$t = \lambda^{-1} \ln((A_e - A_0)/(A_e - A_t)) \tag{4}$$

Assuming a one-dimensional vertical flow path, the retention time allows the derivation of infiltration velocity averaged as a function of depth z_t (Fig. 6).

The velocity from the surface to the sampling point is calculated by:

$$v_i = \frac{z_i}{t_i}. \tag{5}$$

The first application of this method to determining horizontal groundwater flow velocities was undertaken by Hoehn & van Gunten (1989), with further developments by Freyer *et al.* (1997). The method assumes homogeneity of radon emanation in the soil, and requires, for the calculation, the background equilibrium concentration A_e at a remote observation borehole. However, when applying this method, we

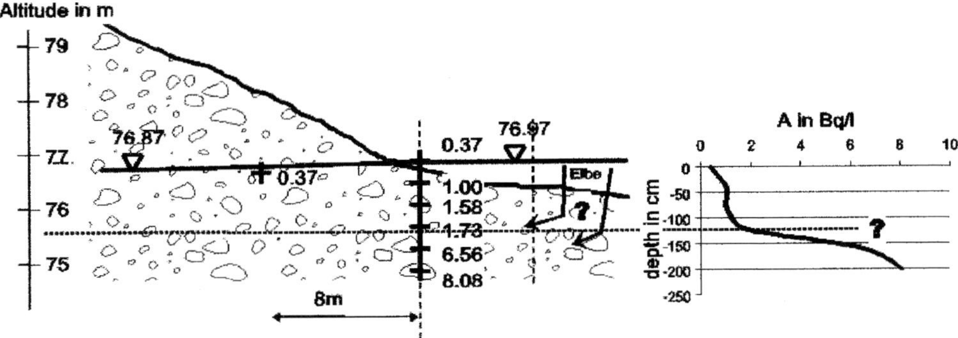

Fig. 7. Example showing ^{222}Rn distribution beneath the bed of the River Elbe at Torgau, eastern Germany.

found the assumptions regarding homogeneity of ^{222}Rn emanation to be inappropriate due to the significantly smaller investigative scale. This effect may be avoided, however, through the combined use of ^{222}Rn and chloride in a mixing model, as shown by Bertin & Bourg (1994). In the following study (Fig. 7), performed at the River Elbe in eastern Germany, the significant observed change in radon concentration at a depth of 1.5 m can be interpreted in different ways. It can be explained by a transverse water flow, by a stratigraphic change, or by alterations in the soil parameters (emanation rate). These different effects influence infiltration measurements in river beds considerably and thus this method could not be used to determine the vertical infiltration velocity properly.

Consequently, reliable measurements can only be made if there is a well-defined, long-term stable infiltrating boundary condition between the surface water and groundwater. Calculation of infiltration velocities requires knowledge of the concentration balance of ^{222}Rn for the investigation area in general (background concentration) and for individual stratigraphic layers. To obtain reasonable results with this method, additional information regarding the stratification of the soil, the background level of radon activity at the sampling point, and the density and porosity of each layer is necessary. Acquisition of this additional information represents the main challenge. At the University of Applied Sciences in Dresden, Germany, a laboratory-scale column test has been set up in an attempt to resolve this problem. The set up of the column is shown in Figure 8.

The initial test of the soil column was to reproduce the theoretical concentration trend of ^{222}Rn, as shown in Figure 9, with the highest possible precision. The column was filled with sediments from the River Elbe. To reproduce field conditions at the River Elbe, the flow velocity through the column was set at 0.1 m d^{-1}. Before the test began, the ^{222}Rn concentration A_0 was determined at every sampling point and, having achieved a time-invariant concentration profile with depth, water samples of 0.5 l each were taken at three different sampling points at every depth. The vertical distance between the sampling points was 0.30 m. As shown in Figure 9, a reasonable correlation was achieved between the theoretical and measured values. Thus, we are able to demonstrate that the use of this method, together with sufficient knowledge of relevant background conditions, permits an accurate evaluation of infiltration velocities in sediments.

Summary

(1) The morphology and hydraulic conductivity of the colmatage layers and the hydraulic gradient are the main determinants for the exchange between groundwater and surface water.

(2) The degree of coverage of the river bed by colmatage layers is important for the evaluation of infiltration processes.

(3) The hydraulic resistance due to colmatage of a river bed first takes significant effect once external coverage reaches 80%.

(4) A probe has been developed to measure hydraulic gradients as a function of depth and to sample water within river bed sediments.

(5) The hydraulic resistance of the external colmatage layer can be measured directly with special column apparatus.

(6) Hydraulic resistance and the hydraulically active thickness of the internal colmatage

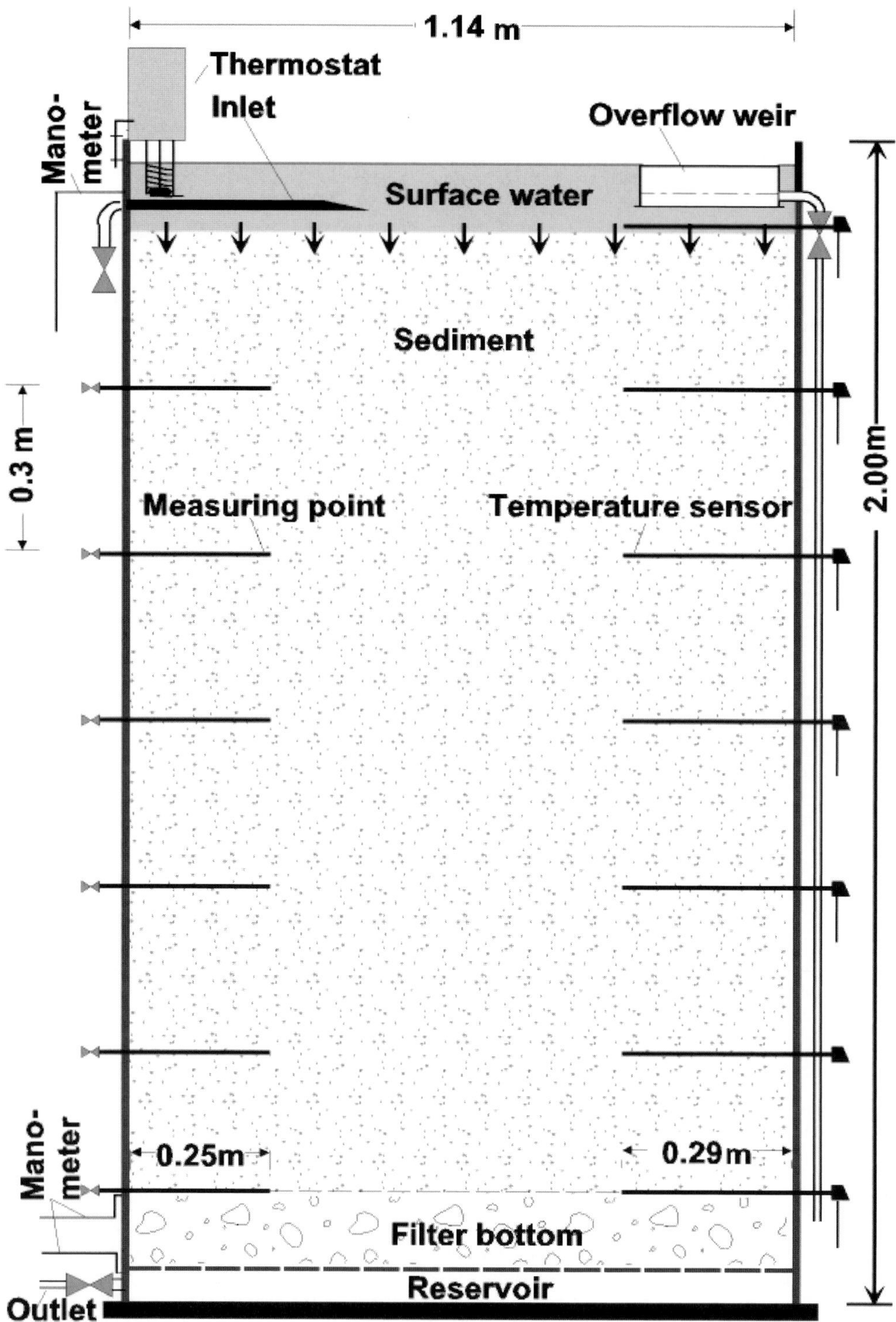

Fig. 8. Schematic diagram of the column used to determine infiltration rates with the distribution of radon concentration.

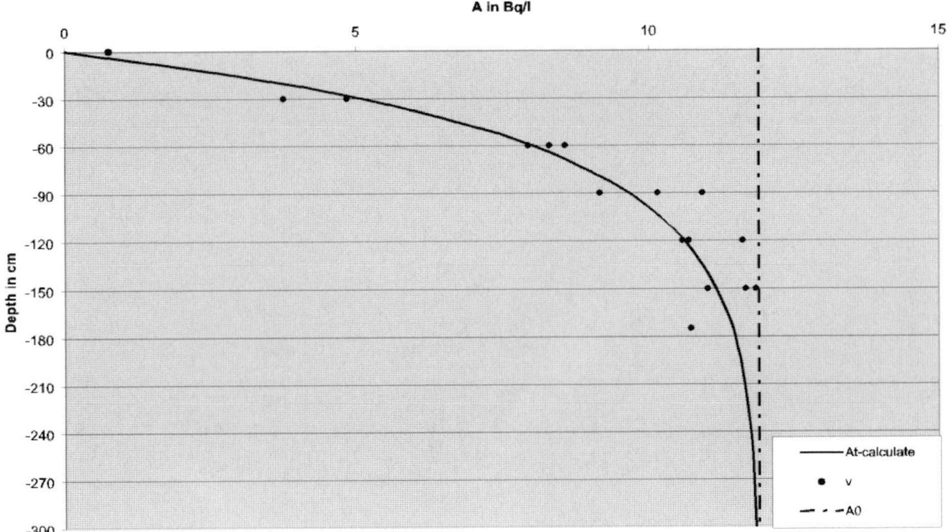

Fig. 9. Depth profile of measured concentrations of ^{222}Rn in comparison with theoretical concentrations. A_t is the calculated ^{222}Rn concentration, v is the measured ^{222}Rn concentration, A_0 is the equilibrium ^{222}Rn concentration.

layer can be measured using the indirect methods of compilation of the geometry (d, A), determination of hydraulic conductivity (K), and water head measurement.

(7) Reliably measured results require additional information, for example, water heads, water quality.

(8) Using ^{222}Rn dating it is possible to measure infiltration velocities in sediments emplaced in laboratory-scale columns.

(9) To use the ^{222}Rn method in fieldwork, methodological improvements are required (equilibrium concentration, stratification, etc.).

References

BERTIN, C. & BOURG, A. C. M. 1994. Radon-222 and chloride as natural tracers of the infiltration of river water into an alluvial aquifer in which there is significant river/groundwater mixing. *Environmental Science & Technology*, **28**, 794–798.

BEYER, W. & SCHWEIGER, K.-H. 1969. Zur Bestimmung des entwässerbaren Porenanteils der Grundwasserleiter. *Wasserwirtschaft Wassertechnik*, **19**(2).

BUSCH, K.-F., LUCKNER, L. & TIEMER, K. 1993. *Geohydraulik, Band 3, Lehrbuch der Hydrogeologie.* Verlag Gebrüder Bornträger, Stuttgart.

DEHNERT, J., FREYER, K., TREUTLER, H. C. & NESTLER, W. 1998. *Wassergewinnung in Talgrundwasserleitern im Einzugsgebiet der Elbe. Teilbericht 5. Radon zur Charakterisierung geohydraulischer Prozesse.* BMBF-Verbundforschungsvorhaben **02WT9454**.

FREYER, K., TREUTLER, H. C., DEHNERT, J. & NESTLER, W. 1997. Sampling and measurement of radon-222 in water. *Journal of Environmental Radioactivity*, **37**, 327–337.

GRISCHEK, T., NESTLER, W., DEHNERT, J. & NEITZEL, P. 1994. Groundwater/river interaction in the Elbe river basin in Saxony. *In*: STANFORD, J. (ed.) *Groundwater Ecology Proceedings of the 2nd International Conference Groundwater Ecology*, Atlanta, 1994, 309–318.

HOEHN, E. & VAN GUNTEN, H. R. 1989. Radon in groundwater: a tool to assess infiltration from surface waters to aquifers. *Water Resources Research*, **25**, 1795–1803.

NESTLER, W., MACHELEIDT, W. & HERLITZIUS, J. 2000. *Grundwasserströmung in der Elbaue bei Falkenberg.* Abschlußbericht UFZ, unpublished.

PÄTZOLD, J. 1999. *Untersuchungen zur meßtechnischen Erfassung der Versickerung aus Flüssen und Seen.* Diplomarbeit, HTW Dresden, unpublished.

PFÜTZNER, R. 1998. *Methodische Untersuchungen zur meßtechnischen Erfassung der Versickerung aus Flüssen und Seen.* Diplomarbeit, HTW Dresden, unpublished.

RICHTER, D. 1998. *Untersuchungen zur laborativen Bestimmung von Durchlässigkeit und Porosität.* Diplomarbeit, HTW Dresden, unpublished.

Direct assessment of groundwater vulnerability from borehole observations

F. WORRALL

Department of Geological Sciences, University of Durham, Science Laboratories, South Road, Durham DH1 3LE, UK (e-mail:Fred.Worrall@durham.ac.uk)

Abstract: A method is developed for measuring the vulnerability of the catchment of a borehole to groundwater pollution based on observation of contaminant in the borehole and the region. The method uses Bayesian statistics to compare the proportion of pesticide detections in the region with that found in the borehole as a measure of groundwater vulnerability of that borehole catchment. Using data from California the method is demonstrated for multi-annual borehole observations of a set of compounds. The method has distinct advantages over present vulnerability assessment methods. The method calculates a continuous measure of borehole vulnerability rather than an index or score. This measure of borehole vulnerability is based solely on observations and not on expert opinion or the application of models. The use of Bayesian methods means that results can be used within a probabilistic risk assessment and can be updated as more information becomes available or as new regional data are considered. The assumptions and disadvantages of the methodology are discussed, but such methods could form the basis of testing the importance of climatic, soil and geological factors in controlling pollutant movement leading to improved management and protection of groundwater resources.

The concept of groundwater vulnerability (Robins *et al.* 1994) has become a central concept in assessing the risk of groundwater pollution. The vulnerability of groundwater constitutes the susceptibility of groundwater to contamination by surface, or near-surface pollutants (Palmer *et al.* 1995). This approach recognizes that the vulnerability depends on the characteristics of the site and that differing soil and hydrogeological conditions will give differing vulnerabilities and afford different degrees of protection. It is also important to note that this concept is independent of the nature of the pollutant. Because borehole vulnerability is a property of a locality and is not dependent upon the type of pollutant it can be mapped across a region as long as the spatial distribution of the properties that control the borehole vulnerability is known.

Several vulnerability assessment systems have been developed. Palmer & Lewis (1995) outline the system developed for the UK that has identified soil type, the presence/absence of drift, and the nature of the aquifer. Groundwater vulnerability relative to each of these factors has been classified and groundwater vulnerability maps of the UK have been produced as part of the UK's national groundwater protection policy (NRA 1992). In the United States the DRASTIC system has been developed for ranking regions with respect to vulnerability of pesticide pollution (Aller *et al.* 1987). As a means of ranking regions by their vulnerability the DRASTIC system relies on weighting seven factors: Depth to water table; net Recharge; Aquifer media; Soil media; Topography; Impact of the vadose zone; and hydraulic Conductivity. The UK system does not include climatic factors, such as net recharge, nor depth to the water table for ease of mapping (Palmer & Lewis 1995). The DRASTIC system has also been applied in New Zealand (Close 1993*a*).

Indices and rankings are the basis of most vulnerability assessment methodologies. Crowe & Booty (1995) focused on crop usage data as a

From: HISCOCK, K. M., RIVETT, M. O. & DAVISON, R. M. (eds) *Sustainable Groundwater Development*. Geological Society, London, Special Publications, **193**, 245–254. 0305-8719/02/ $15.00 The Geological Society of London 2002.

means of highlighting regions at risk of pesticide pollution. A number of authors have used simplified pollutant transport models to develop indices and rankings. Meeks & Dean (1990) used a one-dimensional (vertical movement through a soil to the water table) advection-dispersion transport model to develop a leaching potential index (LPI). The LPI is compound specific and includes adsorption and degradation parameters, similar to other physically-based modelling approaches (Villeneuve *et al.* 1990; Kookana & Aylmore 1994). Britt *et al.* (1992) included chronic toxicity data for compounds with simulations from a groundwater transport model, but the results were classified as an index for interpretation purposes. Compound-specific results will limit the possibility of generating vulnerability maps and prevent risk-based management. Brüggemann *et al.* (1994) took an alternative mathematical approach of using graph theory to assess the pollution status of stretches of the River Elbe, but the system still only developed a relative ranking.

All these systems have a number of significant flaws. Firstly, weightings are chosen arbitrarily and are based solely on expert opinion. Secondly, the indices are not based on observations or measurements and so even when they are based on physically based models they are prone to errors in the assumptions made in the model or in choosing input values for model parameters. Fuest *et al.* (1998) took an empirical approach, examining borehole records across a region and overlying these results with spatial information for the region in a GIS system, but the authors proposed no system based on their observations. Troiano *et al.* (1994, 1997, 1998) took a similar empirical approach based on observations of pesticides in boreholes across California. This work characterized the catchment of each borehole where pesticides were detected and used a series of multivariate techniques to profile regions that would be vulnerable to groundwater pollution. In this way the authors identified five clusters of soil and/or climatic variables that represented vulnerable regions and allowed the assessment of regions not part of the original study. This latter method represents an essentially non-parametric approach to borehole vulnerability as it counts all pesticides as one. Different compounds would be expected to have different mobilities in the environment and this, along with frequency of observation of those differing compounds, is information not used by this approach.

Thirdly, systems based on indices do not fit well with risk assessment which is essentially a probabilistic assessment of an event. The empirical, non-parametric approach given above does not fit with a probabilistic framework for borehole vulnerability assessment. Goss *et al.* (1998) used a logistic regression model to predict the probability of nitrate pollution in domestic supply wells in Ontario based on the well and catchment characteristics. Again this approach represents a simplification of the data as logistic regression predicts the probability of the presence or absence of a contaminant and does not use or predict the frequency of detection or the level of the compound that might be expected. Teso *et al.* (1996) used a similar approach to link soil particle classes to the risk of groundwater pollution from 1,3-dibromochloropropane (DBCP). Kolpin *et al.* (1997) used Spearman's rank correlation coefficient to examine the relationship between nitrate and pesticide concentrations found in boreholes and their surrounding catchment.

Fourthly, indices have to be interpreted against arbitrary classification schemes, i.e. what represents a value for a vulnerable site as opposed to an acceptable situation? The vulnerability of boreholes is a continuous property of an area or borehole catchment and should be measured as such. It is difficult to interpret indices relative to each other and so gain further information regarding what controls a region's vulnerability. If the vulnerability of a borehole could be measured then this information could be compared to borehole properties to assess, for example, the importance of soil properties v. climatic factors. The methods cited above that link contaminant occurrence to properties of the borehole catchment do not give a measure of borehole vulnerability and can therefore only provide a non-parametric analysis and thus only profile vulnerable regions. A measure of borehole vulnerability for a given region would enable the factors important in controlling pollutant transport to be identified and assessed relative to each other.

Finally, the use of indices makes validation of such schemes difficult if not impossible. Close (1993*b*) gave a validation for the DRASTIC method, while Maas *et al.* (1995) showed it to be a poor predictor of vulnerable groundwater regions. The UK vulnerability system remains unvalidated. Risk assessment is a probabilistic problem and so risk assessment methods can only be truly validated by observing the frequency of the event of interest in comparison to predicted risk. Methods that assess borehole vulnerability outside of a probabilistic framework will be of limited use in a larger risk assessment methodology.

This study presents a method of assessing

vulnerability from borehole observations from multi-annual observations of several compounds in boreholes.

Methodology and approach

This study uses the well inventory database maintained by the California Department of Pesticide Regulation (CDPR) (Miller Maes *et al.* 1992). The Pesticide Prevention Contamination Act (Connelly 1986) requires the CDPR to determine whether or not reported detections are due to legal agricultural practice. This means that the database is an extensive resource for understanding the controls on pesticide pollution. The database has been used as part of the development of pre-screening models for pesticide regulation (Gustafson 1989; Worrall *et al.* 1998) and for *ab initio* models of organic pollution (Worrall 2001). The database has been used to profile areas where groundwater has been contaminated with pesticides and to develop a real vulnerability assessment (Troiano *et al.* 1994, 1997, 1998). The geographical extent, i.e. across the range of climate and soils of the state of California, and the number of boreholes included in this database mean that it is ideal for measuring the vulnerability of these boreholes.

To assess the vulnerability of a borehole as a measure of the vulnerability of the catchment area of that borehole it is necessary to examine the occurrence of pollutant in that borehole, taking into account the differing polluting potential of the compounds being analysed. For example, if compound X is found to be above legal limits in a particular borehole, is this because the borehole and its surrounding aquifer are particularly vulnerable or because the compound is highly mobile, and/or highly persistent? Highly mobile or persistent compounds may well be found in the groundwater below the site of application. In this approach the risk of pollution is a function of either the site of application or the borehole being assessed and the compound itself:

$$\text{risk} = f(\text{chemical, site}). \qquad (1)$$

In this case, the site factor, includes all the climatic, soil, aquifer and management practices that could affect the mobility of the compound in that setting. The chemical factor is the innate properties of the compound that mean that it has a high potential to reach groundwater and cause pollution, e.g. it is highly soluble. The approach represented by equation (1) was shown to be statistically valid by Worrall *et al.* (2000) and was used by Worrall *et al.* (1998) to predict the probability of a compound leaching to groundwater. In a symbolic fashion equation (1) can be summarized as: risk = hazard × vulnerability. This approach typifies most vulnerability schemes, i.e. the hazard is the innate risk of the compound and the vulnerability is the risk due to the site. The multiplicative relationship implies that there is no risk without both a hazard and a vulnerability, but also that a high risk compound (high hazard) may still represent a threat to groundwater even in a low vulnerability area.

In borehole studies it is possible to observe pollution directly, i.e. to measure possible pollutant concentrations, and therefore to estimate the actual risk of pollution. If the hazard represented by the chemical could be estimated, then a measure of the vulnerability of the borehole could be calculated. Utilizing the extensive CDPR database it is possible to calculate the overall probability of finding a pesticide in groundwater. This figure represents an average probability of finding a particular compound in the whole of the survey area. Records of individual boreholes can be tested against this figure to discover whether the borehole is more or less vulnerable than the average for the region.

Across the region compound X is analysed n_1 times and of these s_1 are positive, i.e. show a concentration above the specified limit (typically $0.1\,\mu g\ l^{-1}$), and f_1 are negative, i.e. show no detectable pesticide. Equally, for the borehole of interest, the compound is analysed n_2 times of which s_2 are positive and f_2 negative. These two proportions can then be compared. If a real difference exists between the proportion expected (s_1/n_1) and the proportion observed (s_2/n_2) then this difference represents the relative vulnerability of that borehole in comparison to the region as a whole. This method calculates the difference between the proportions and tests its significance.

Comparisons between proportions can be performed by a number of methods, but for reasons discussed below this study takes a Bayesian approach. Bayes theorem (Berry 1996) can be stated as:

$$P(Borehole|Region) = \frac{[P(Borehole)P(Region|Borehole)]}{P(Region)} \qquad (2)$$

where $P(Borehole\,|\,Region)$ is the probability of borehole observations given the regional observations (the posterior probability of the borehole observations); $P(Borehole)$ is the probability of the borehole observations (the prior probability of the borehole observations); *P(Region | Borehole)* is the probability of the regional observations given the borehole observations (the likelihood of the regional observations); and *P(Region)* is the probability of the regional observations. The likelihood of a model is the probability of the observed data given that model. For observations containing s_1 successes and f_1 failures the likelihood of the proportion is:

$$P(Data|p) = p^{s_1}(1-p)^{f_1} \quad (3)$$

where *P(Data | p)* is the the probability of the observed data given proportion p. This can be calculated for both the region as a whole (D_r) and the borehole being considered (D_b). To compare whether the proportions are different the likelihood of D_r given D_b needs to be calculated for p_r and p_b. The probability distribution of such a proportion is a β distribution and is commonly represented as β(a, b) where:

$$P(Data|p) = p^{a-1}(1-p)^{b-1} \quad (4)$$

where the mean of this distribution is $(r) = a/(a+b)$. Initially, values of a and b are taken as 1, which is a uniform distribution. Such a uniform distribution is referred to as a prior distribution. The uniform nature of this distribution indicates a lack of knowledge about the difference between the borehole and the region prior to this calculation.

To understand the difference between the proportion of detections in the borehole (p_b) compared to that in the region (p_r) it is necessary to calculate the probability of the borehole observations (D_b) given the regional observations (D_r). This is a joint likelihood and can be calculated by multiplying the two likelihoods together for all values of both p_r and p_b, i.e. applying Bayes Rule [equation (2)]. The joint likelihood can be converted to a probability distribution by normalizing such that the area under the distribution sums to 1. This distribution is referred to as the posterior probability distribution. The posterior probability distribution is also a β distribution, where the new value of the mean is:

$$r^{+} = \frac{(a+s)}{(a+b+s+f)} \quad (5)$$

Given the values of s successes and f failures of finding the compound in the borehole under consideration.

The joint posterior probability distribution can now be examined to judge the difference between one proportion and the other. The joint posterior distribution approximates to a normal distribution for large a and b (a, b >10). Therefore, the mean of the posterior distribution is the best estimate of the distance between the proportions and the confidence interval can be calculated. The probability of the distance between the two proportions being at least some value x (Probability of distance at least x = Pdalx) can be calculated with reference to z-scores and the standard normal distribution (Miller & Miller, 1993) with respect to the borehole:

$$z = \frac{(d-(r_b - r_r))}{\sqrt{t_b^2 + t_r^2}} \quad (6)$$

$$t_b = \sqrt{r_b(r_b^{+} - r_b)} \quad (7)$$

where r is the mean of the prior distribution; d is the distance between the two proportions; and r^{+} is the mean of the posterior distribution. For the distance between the two proportions to be insignificantly different from 0 (at the 95% level), i.e. no difference at all, the following inequality would have to be satisfied:

$$-1.96 < \frac{(0-(r_b - r_r))}{\sqrt{t_b^2 + t_r^2}} < 1.96. \quad (8)$$

In cases when s or f < 10 the assumption of normality of the posterior distribution does not hold and the above procedure cannot be followed. The probability that the distance between the two proportions is at least a certain value is also the area under the posterior probability distribution where the difference between the two proportions is that value or less. Pdalx can be calculated for both the borehole being vulnerable relative to the region or for being less vulnerable than the average for the region depending upon the portion of the posterior distribution examined.

The above calculation assumes no prior information, i.e. $a = b = 1$, but Bayes theorem allows a prior probability to be set such that information from other sources regarding the difference between the two proportions can be included. Inclusion of prior information will strengthen any inference regarding the difference between the borehole and the region. In the above calculation there is no reasonable source of prior information. However, if further compounds were analysed for the same borehole then the posterior probability distribution for the first compound becomes the prior distribu-

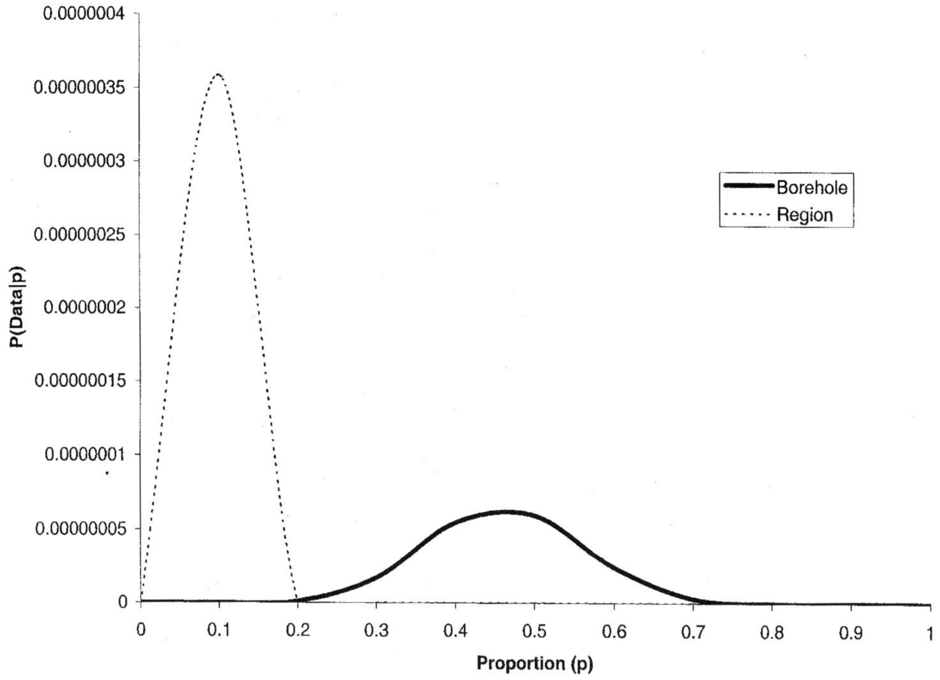

Fig. 1. The prior distribution of positive atrazine samples in the borehole in comparison to the region.

tion for the second compound. In such a case the joint likelihoods calculated from the new likelihoods of the proportion for this compound across the region and for the borehole under study are multiplied by the prior probability distribution. Again the new posterior probability distribution is calculated by normalizing this distribution and is examined as above. This process can be repeated for each of the compounds analysed across the region and in the borehole under consideration. With each new compound the estimate of the difference between the proportions is refined. If it is assumed that the vulnerability of a borehole is independent of the compound under consideration then the use of observations from a range of compounds strengthens the vulnerability estimate.

This methodology can now be demonstrated using data from the CDPR database.

Application to Californian Groundwater

Examining the proportion of positive results to negative results for the whole of the CDPR database shows that a number of compounds are never found in the region; only those compounds showing positive results somewhere in the region were considered. It is also assumed that because a compound is tested for in a borehole it is sufficient reason to believe that it is possible that the compound could occur in this borehole. That is to say, a compound's inclusion in the list of analytes for a borehole is taken as strong evidence that the compound is used in the area to the extent that it might be considered a hazard to groundwater. However, it should be emphasized that when this method is applied to a whole region the assumption that each of the compounds used in the assessment is applied in the catchment of each of the boreholes in the study should be tested. The compounds atrazine, simazine and diuron have each been analysed over 1000 times suggesting that they have been considered a threat over a broad range of conditions across the region. The regional proportions for each of these compounds are: atrazine ($s_1 = 165$, $n_1 = 1791$), simazine ($s_1 = 220$, $n_1 = 1974$), and diuron ($s_1 = 42$, $n_1 = 1241$). Between 1986 and 1988 these three compounds were analysed in one borehole

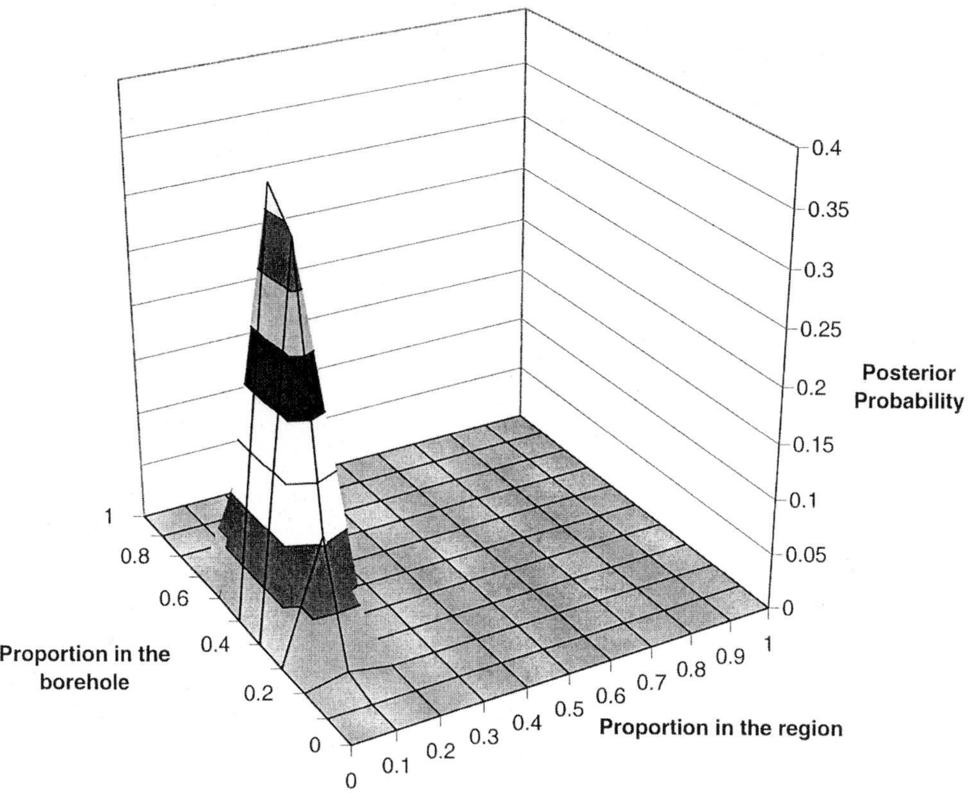

Fig. 2. Posterior distribution for the observation of atrazine in the borehole is true given the regional observations.

(CDPR database ref. – W1119N03W08) and were found in the following proportions: atrazine ($s_2 = 11$, $f_2 = 13$), simazine ($s_2 = 12$, $f_2 = 18$), and diuron ($s_2 = 0$, $f_2 = 5$). So how vulnerable is this borehole?

To ensure independence of the samples the results for the borehole are subtracted from those for the region. The proportions found in the borehole given above can be converted to probability distributions in comparison to the regional proportion (as shown for atrazine in Fig. 1). Examining atrazine first, the posterior probability is given in Figure 2. The mean of the posterior distribution is 0.37 (Table 1), and Pdalx can be calculated from this distribution (Fig. 3). The posterior distribution for the atrazine data can be then be updated using data for simazine and diuron.

The great advantage of the Bayesian approach is that the order in which the data are considered does not matter. Performing the same calculation but using the data in reverse order, i.e. diuron first, followed by simazine and then atrazine, the data converge to the same result (Fig. 4, Table 1).

Table 1. *Vulnerabilities calculated for the example borehole*

Compound	Regional probability	Borehole vulnerability	Confidence interval
After atrazine	0.09	0.37	±0.20
After simazine	0.11	0.37	±0.24
After diuron	0.04	0.34	±0.29

It is far more difficult to estimate the vulnerability of a borehole that is in a catchment that is less vulnerable than the average. Although the calculation is just as straightforward as when compounds occur in a greater proportion than the regional average, the overall

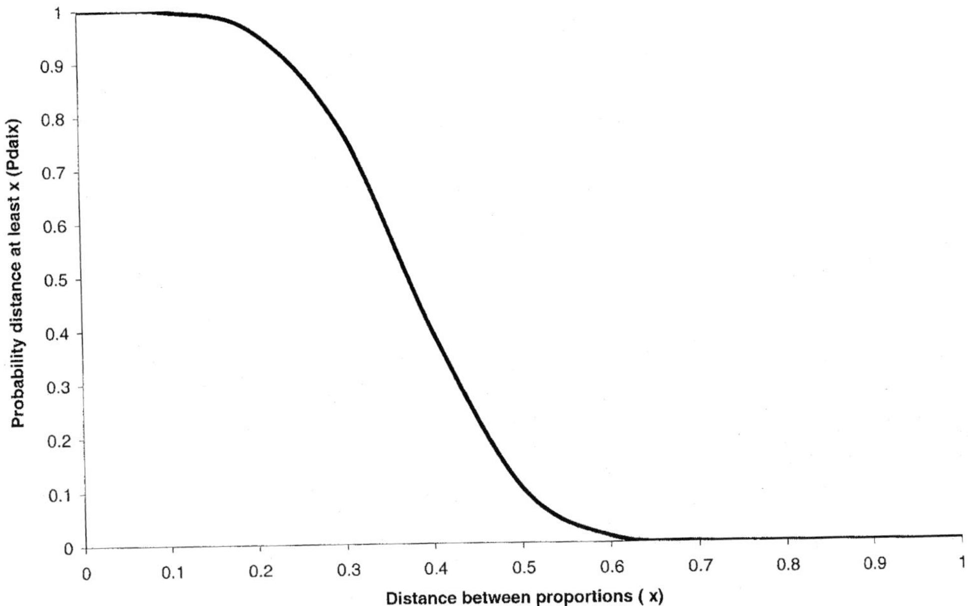

Fig. 3. The probability of distance at least x (Pdalx) for the atrazine observations.

low detection rate in the region as a whole (e.g. approx. 10% for atrazine) makes this technique less sensitive.

Discussion

The method presented above has a number of shortcomings. Firstly, the observations of any contaminant in a borehole are dependent upon the detection limit being applied. However, the techniques rely on relative information, i.e. the number of detections in the borehole in comparison to the number of detections in the region, and so as long as the survey has a consistent limit of detection this is not a problem. Because the probability of finding pesticides in groundwater is relatively low then the lower the detection limit in any survey the more sensitive the measure of borehole vulnerability. The methods given above are insensitive to boreholes that are less vulnerable relative to the region.

In a study of pesticides in 100 boreholes across the midwestern United States, Kolpin *et al.* (1995) found that at the lowest possible detection limits atrazine is detected in 43% of boreholes. Making vulnerability observations relative to a compound with such a high proportion of detections gives measurements approximately equally sensitive to protective and to vulnerable catchments. Thus when studying groundwater vulnerability it is important to focus on the most mobile compounds as these give the most sensitive information for calculating the vulnerability of a borehole. Kolpin *et al.* (1998) included degradation products of atrazine, alachlor, acetochlor, cyanazine and metolachlor; this could be a method of extending the sensitivity of vulnerability assessment as they were found in almost 75% of the wells examined in the study.

The method here works relative to data for a whole regional survey and as such the values as calculated are relative borehole vulnerabilities. The relative vulnerabilities that can be calculated can be converted to an absolute vulnerability by calculating the regional average. The absolute vulnerability rescales the vulnerability on to a probabilistic scale, i.e. 0 to 1. On this scale the vulnerability can be interpreted as the probability of detecting the next compound, no matter what that compound is, at that particular monitoring point. The regional average is easily calculated by calculating the relative vulnerability for a site in which no compounds have ever been observed. This paper does not cover the calculation of vulnerabilities across a whole survey region and so such conversions cannot be executed here.

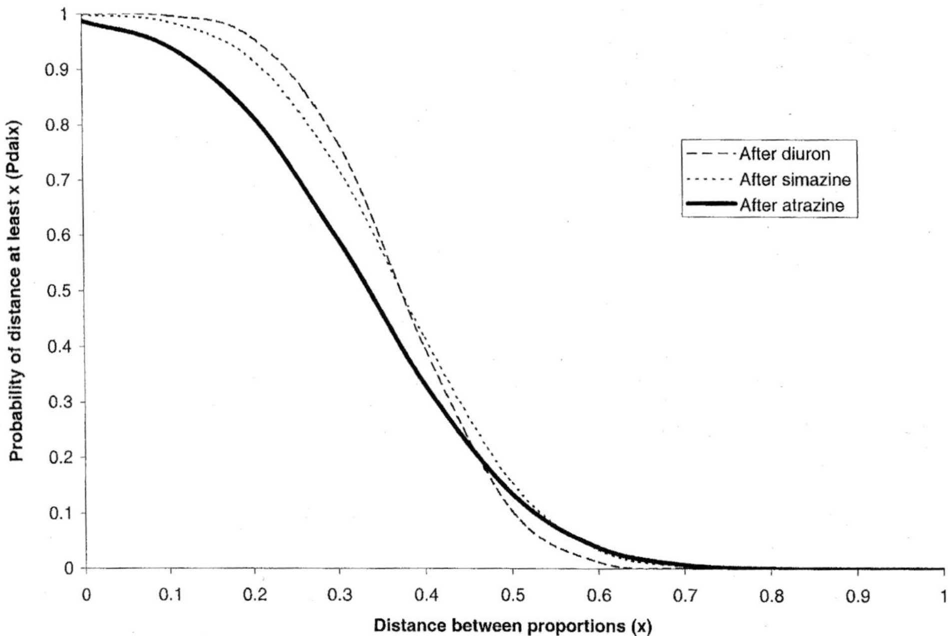

Fig. 4. Pdalx after combining information from atrazine observations with those for simazine and diuron in the same borehole.

The methods used here ignore the information available from the concentration of the pesticides found in the borehole. A more vulnerable borehole would not only be expected to have a greater proportion of contaminant detections, but that these detections would be at higher concentrations. The methods presented here work only on presence/absence data. Future modelling would hope to include all available information. However, it is important to remember, that both the presence and the concentration of a contaminant may vary across the year because pesticides are applied at different times. Thus, the sampling time of any groundwater survey may be critical and some compounds being screened for in any groundwater survey may well be present at other times of the year from that of the particular survey. Several elements of the proposed method militate against this problem. Firstly, the concentration of a compound will be more affected by the timing of any survey than the presence or absence of the compound. In this sense the presence/absence information is a more robust measure of vulnerability than concentration. Secondly, the methodologies work relative to the region and so if the whole survey is conducted at approximately the same time, or at least within the same season with respect to the predominant types of agriculture, then there is still useful vulnerability information in any observations. Such information is useful even if it is not the optimum time of year for detecting any particular compound because the vulnerability is measured from the difference between the proportion of detections in the region and the proportion of detections in the borehole. This difference could be expected to be more robust against temporal change as the use of any particular compound would be expected to be approximately coincident across the region.

In the introduction to this work, vulnerability was defined as a property that was independent of the type of contaminant being considered. Indeed, it is an assumption in the methods given above that observations for one compound can be combined with those from another. However, it is conceivable that one particular combination of site parameters would make a borehole more vulnerable to one compound than to another. For example, if the soils surrounding a borehole were particularly acidic then the behaviour in that soil of an acidic v. a basic compound, or a compound whose degradation is pH dependent, could be dramatically different from that predicted by the methods outlined above. Understanding the interaction effects between compound and site would be the aim of further statistical modelling. Equally, within the catchment of any borehole the areas where different compounds are applied may be widely dispersed. One compound may be applied in the field immediately surrounding the borehole whereas another compound in the survey might only be applied near the edge of the borehole's zone of influence. It is important, therefore, to emphasize that these methods also mean that the uncertainty in the vulnerability of the borehole can be calculated.

The disadvantages outlined above also apply to many of the vulnerability assessment methods reviewed in the introduction, but this method overcomes many of the flaws also outlined therein. In addition, these measures of vulnerability can be used in a predictive sense and be interpreted within a risk assessment context. Measures of borehole vulnerability can be related to properties of the borehole catchment, such that vulnerability could be predicted in areas where no observations or boreholes yet exist. The predicted borehole vulnerability can be easily tested or validated as values calculated by the above method need only to be added or subtracted from the regional average to give a prediction of the proportion of detections in the new borehole. At present the approach needs to be relative to the specific region where a groundwater survey has been conducted and so prediction would also be limited to that region. The data used in this study come from a large regional survey and this could be extended to other large areas from the many studies available in the literature (e.g. Maas *et al.* 1995). The Bayesian framework, within which these methods were developed, allows results to be used in a flexible manner with results from one region or at one time forming prior information for the basis of work in another region.

Conclusions

This work presents a method for calculating a measure of groundwater vulnerability directly from borehole observations. This approach has the following advantages:

(1) it calculates vulnerability as a continuous variable not as an index or score;
(2) the measure of vulnerability is based on observations and not on opinion or the application of models;
(3) the method works for multiple observations of a single compound but can also be adapted to single observations of multiple compounds;
(4) the method works within a probabilistic framework that allows direct interpretation as part of pollution risk assessment;
(5) the method is testable; and
(6) the method is flexible and estimates can be updated as more information becomes available.

Due to the low probability of detecting pesticides in aquifers the method provides a more sensitive measure of high vulnerability regions compared to less vulnerable regions. The method is also susceptible to changes in the temporal pattern of pesticide occurrence in the aquifer relative to pesticide usage, and the assumption of the independence of vulnerability from the compounds screened for within a groundwater survey. However, the possibility of measuring vulnerability of boreholes as part of a larger regional survey means that the importance of hydrogeological, geological, climatic and soil factors etc. can be assessed and measured. Effective protection of groundwater resources for future generations will only be achieved with accurate risk assessment methods.

References

ALLER, L., BENNETT, T., LEHR, J. H., PETTY, R. J. & HACKETT, G. 1987. *DRASTIC: A standardised system for evaluating groundwater pollution potential using hydrogeologic settings.* U.S. Environment Protection Agency, Ada, Oklahoma, **EPA/600/2-87/035**.

BERRY, D. A. 1996. *Statistics: A Bayesian Perspective.* Duxbury Press, London.

BRITT, J. K., DWINELL, S. E. & MCDOWELL, T. C. 1992. Matrix decision procedure to assess new pesticides based on relative groundwater leaching potential and chronic toxicity. *Environmental Toxicology and Chemistry*, **11**, 721–728.

BRüGGEMANN, R., MÜNZER, B. & HALFON, E. 1994. An algebraic/graphical tool to compare ecosystems with respect to their pollution the German River Elbe as an example I: Hasse diagrams. *Chemosphere*, **28**, 863–872.

CLOSE, M. E. 1993*a*. Assessment of pesticide contamination of groundwater in New Zealand 1. Ranking of regions for potential contamination. *New Zealand Journal of Marine & Freshwater Research*, **27**, 257–266.

CLOSE, M. E. 1993*b*. Assessment of pesticide contamination of groundwater in New Zealand 2. Results of groundwater sampling. New Zealand Journal of Marine & Freshwater Research, **27**, 267–273.

CONNELLY, L. 1986. *AB2021-Pesticide contamination prevention act. Article 15, Chapter 2, Revision 7.* Food and Agriculture Code, California.

Crowe, A. S. & Booty, W. G. 1995. Multi-level assessment methodology for determining the potential for groundwater contamination by pesticides. *Environmental Monitoring & Assessment*, **32**, 239–261.

Fuest, S., Berlekamp, J., Klein, M. & Matthies, M. 1998. Risk hazard mapping of groundwater contamination using long-term monitoring data of shallow drinking water wells. *Journal of Hazardous Materials*, **61**, 197–202.

Goss, M. J., Barry, D. A. J. & Rudolph, D. L. 1998. Contamination in Ontario farmstead domestic wells and its association with agriculture: 1. Results from drinking water wells. *Journal of Contaminant Hydrology*, **32**, 267–293.

Gustafson, D. I. 1989. Groundwater ubiquity score: a simple method for assessing pesticide leachability. *Environmental Toxicology and Chemistry*, **8**, 339–357.

Kolpin, D. W. 1997. Agricultural chemicals in groundwater of the midwestern United States: relations to land use. *Journal of Environmental Quality*, **26**, 1025–1037.

Kolpin, D. W., Goolsby, D. A. & Thurman, E. M. 1995. Pesticides in near-surface aquifers: an assessment using highly sensitive analytical methods and tritium. *Journal of Environmental Quality*, **24**, 1125–1132.

Kolpin, D. W., Thurman, E. M. & Linhart, S. M. 1998. The environmental occurrence of herbicides: the importance of degradates in ground water. *Archive of Environmental Contamination & Toxicology*, **35**, 385–390.

Kookana, R. S. & Aylmore, L. A. G. 1994. Estimating the pollution potential of pesticides to ground water. *Australian Journal of Soil Research*, **32**, 1141–1152.

Maas, R. P., Kucken, D. J., Patch, S. C., Peek, B. T. & Van Engelen, D. L. 1995. Pesticides in eastern North Carolina rural supply wells: land-use factors and persistence. *Journal of Environmental Quality*, **24**, 426–431.

Meeks, Y. J. & Dean, J. D. 1990. Evaluating groundwater vulnerability to pesticides. *Journal of Water Resource Planning & Management*, **116**, 693–707.

Miller, J. C. & Miller, J. N. 1993. *Statistics for Analytical Chemistry* (3rd Edition). Ellis Horwood, Chichester, UK.

Miller Maes, C., Pepple, M., Troiano, J., Weaver, D., Kinmaru, W. & SWRCB Staff. 1992. *Sampling for pesticide residues in California well water 1992 well inventory data base, cumulative report 1986-1992*. Environmental Monitoring and Pest Management Branch, Department of Pesticide Regulation, **report EH 93-2**.

National Rivers Authority. 1992. *Policy and practice for the protection of groundwater*. National Rivers Authority, Bristol.

Palmer, R. C. & Lewis, M. A. 1998. Assessment of groundwater vulnerability in England and Wales. *In*: Robins, N. S. (ed.) *Groundwater Pollution, Aquifer Recharge and Vulnerability*. Geological Society, London, Special Publications, **130**, 191–198.

Palmer, R. C., Holman, I. P., Robins, N. S. & Lewis, M. A. 1995. *Guide to groundwater vulnerability mapping in England and Wales*. National Rivers Authority, Bristol.

Robins, N. S., Adams, B., Foster, S. S. D. & Palmer, R. C. 1994. Groundwater vulnerability mapping: the British perspective. *Hydrologeologie*, **3**, 35–42.

Teso, R. R., Poe, M. P., Younglove, T. & McCool, P. M. 1996. Use of logistic regression and GIS modeling to predict groundwater vulnerability to pesticides. *Journal of Environmental Quality*, **25**, 425–432.

Troiano, J., Johnson, B. R., Powell, S. & Schoenig, S. 1994. Use of cluster and principal component analyses to profile areas in California where ground water has been contaminated by pesticides. *Environmental Monitoring & Assessment*, **32**, 269–288.

Troiano, J., Nordmark, C., Barry, T. & Johnson, B. 1997. Profiling areas of ground water contamination by pesticides in California: phase II evaluation and modification of a statistical model. *Environmental Monitoring & Assessment*, **45**, 301–318.

Troiano, J., Nordmark, C., Barry, T., Johnson, B. & Spurlock, F. 1998. Pesticide movement in groundwater: application of areal vulnerability assessments and well monitoring to mitigation measures. *In*: Ballatine, L. G., McFarland, J. E. & Hackett, D. S. (eds) *Triazine Herbicide: Risk Assessment*. ACS Symposium Series 638, American Chemical Society, 239–251.

Villeneuve, J-P., Banton, O. & Lafrance, P. 1990. A probabilistic approach for the groundwater vulnerability to contamination by pesticides: the vulpest model. *Ecological Modelling*, **51**, 47–58.

Worrall, F. 2001. A molecular topology approach to predicting pesticide pollution of groundwater. *Environmental Science & Technology*, **39**, 2282–2287.

Worrall, F., Wooff, D. A., Seheult, A. H. & Coolen, F. P. A. 1998. A Bayesian approach to the analysis of environmental fate and behaviour data for pesticide registration. *Pesticide Science*, **54**, 99–112.

Worrall, F., Wooff, D. A., Seheult, A. H. & Coolen, F. P. A. 2000. New approaches to assessing the risk of groundwater contamination by pesticides. *Journal of the Geological Society of London*, **157**, 877–884.

A simple analytical solution for unsaturated solute migration under dynamic water movement conditions and root zone effects

L. D. CONNELL

Department of Geological Sciences, University College London, Gower Street, London, WC1E 6BT, UK (e-mail: l.connell@ucl.ac.uk)

Abstract: Vulnerability mapping involves an assessment of the potential for solute migration to the water table. Existing procedures rely on simple qualitative indices to calculate this migration potential. As an alternative to these vulnerability indices this paper presents a simple, analytical solution to the advection–dispersion equation. This solution is physically representative, in that it allows for the effects of root zone processes and climate-induced water movement variation on solute transport. The basis of this analytical solution is a spatially transformed advection–dispersion equation. The root zone is treated in a lumped fashion where a relation is developed for the solute concentration of water that has passed through the root zone that includes retardation, degradation and transpiration. This lumped root zone treatment is used in an existing analytical solution that is adapted so that it can be used for a transformed unsaturated advection–dispersion equation. The procedure is tested against hypothetical simulations with dynamic water movement conditions calculated using SWIMv2. The accuracy of the approach is found to be a function of the soil permeability. Where the dynamic variation is averaged out by the water movement process (such as in soils with moderate permeability) the method can accurately represent the transport process. For soils where surface conditions lead to significant variation in water content throughout the profile the technique is more approximate.

For a solute to pose a risk to groundwater it must first migrate through the unsaturated zone. The rate of migration will be driven by the net effect of the surface water processes, but is also a function of the water movement properties of the unsaturated zone. Surface water inputs, through infiltration from rainfall and irrigation, and outputs, from evapotranspiration, mean that water movement is highly dynamic. Vertical unsaturated water movement can be a complicated process, for many situations being a mixture of matrix and macro-pore flow.

A wide range of procedures has been developed to relate the risk from unsaturated solute migration. The simplest are the vulnerability indices used in vulnerability mapping. These indices are intended to be a generalized reflection of the potential for migration (Foster 1998) and have proved popular, in a variety of forms, for mapping because of their ease of application. Another approach is to represent the effect of unsaturated zone solute transport by a transfer function (Jury 1982). This method describes the breakthrough of solute at a depth using a probability distribution for the unsaturated zone travel time.

With aquifer vulnerability mapping the risk of unsaturated zone contaminant transport is estimated over an area. As mentioned above, existing approaches use relatively simple, qualitative, indices to relate the potential for contaminant migration. While these have proved to be useful tools and are in widespread use, their highly qualitative basis means that they are an extremely poor representation of the transport process. Two problems exist in representing migration through the unsaturated zone; the first is developing relationships between physical (site and contaminant specific) properties and the mechanisms that affect the transport process.

From: HISCOCK, K. M., RIVETT, M. O. & DAVISON, R. M. (eds) *Sustainable Groundwater Development*. Geological Society, London, Special Publications, **193**, 255–264. 0305-8719/02/ $15.00 The Geological Society of London 2002.

For example, vulnerability indices use conceptual models. Like all conceptual modelling this is an inherently flawed approach since it is totally reliant on human perception of the transport mechanisms and how they operate. The second obstacle to representing transport is the estimation of site and contaminant transport properties.

Deterministic theory, in principle, provides a mechanistic basis for the relationships between site and contaminant properties and the transport process. However, the production of a vulnerability map could involve a large number of simulations with potentially long durations. An obstacle to the use of deterministic theory for vulnerability mapping has been the computational burden and complexity of solving the coupled flow and transport partial differential equations. This paper presents a simplified deterministic approach that has a low computational burden and avoids having to solve the unsaturated flow equation numerically. The focus of the paper is therefore the first of the obstacles mentioned above, i.e. developing tractable relationships between physical properties and the transport process. Since the approach presented in this paper is very tractable it could prove suitable for vulnerability mapping.

Mathematical background

The deterministic approach for solute transport is based on the advection–dispersion equation:

$$\frac{\partial(c\theta)}{\partial t}+\rho\frac{\partial m_s}{\partial t}=\frac{\partial}{\partial z}\left[\theta D\frac{\partial c}{\partial z}\right]-\frac{\partial(qc)}{\partial z}+\mu\theta c \quad (1)$$

where c is solute concentration (M 1^{-3}), m_s is the adsorbed mass (M solute/M soil); θ is volumetric water content (1^3 water 1^{-3}); q the water flux (1^3 water $1^{-2}T^{-1}$); μ the degradation rate (T^{-1}); ρ is the soil bulk density (M soil 1^{-3}); D is the dispersivity (1^2 T^{-1}); and t and z are the time and space dimensions.

With a linear adsorption isotherm ($m_s = K_d\, c$, where K_d is known as the distribution coefficient) and continuity of water mass, which can be written as:

$$\frac{\partial\theta}{\partial t}=-\frac{\partial q}{\partial z}+S_w \quad (2)$$

equation (1) becomes:

$$K_a\frac{\partial c}{\partial t}=\frac{\partial}{\partial z}\left[\theta D\frac{\partial c}{\partial z}\right]-q(z,t)\frac{\partial c}{\partial z}+(\mu\theta-S_w)c \quad (3)$$

where $K_a = \theta + K_d\rho$, and S_w is the sink/source of water (1^3water 1^{-3} T^{-1}).

To solve equation (3) requires the water content, flux and transpiration (S_w) as functions of position and time. One approach would be to measure these properties, however more typically these are obtained by a solution of Richards' equation (Richards 1931) for unsaturated water movement:

$$\frac{d\theta\partial\psi}{d\psi\partial t}=\frac{\partial}{\partial z}\left[K(\psi)\left(\frac{\partial\psi}{\partial z}-1\right)\right]-S_w \quad (4)$$

where ψ is the matric potential (l); $K(\psi)$ the unsaturated conductivity (l T^{-1}); and $\theta(\psi)$ is the moisture retention relationship.

Richards' equation assumes laminar porous media flow. While it is possible to solve more complex unsaturated flow systems mathematically (i.e. Germann & Beven 1985) the overwhelming problem is characterizing these, particularly for routine field applications. Multiple porosity media can be represented approximately using one-dimensional Richards' equation representations for each porosity class, such as with the SWIMv2 model of Verburg *et al.* (1996). However, since even this is still a significant approximation, the approach needs to be tested against field data to evaluate its validity.

There are a number of packages available that solve the combined advection–dispersion and Richards' equations. The SWIMv2 model uses the efficient numerical procedure of Ross (1990) for Richards' equation and provides a flexible approach for solute transport under field conditions of plant transpiration. However, this approach is significantly more complicated to apply than the approximate procedures presented above. Another approach has been to use analytical solutions to the advection-dispersion equation (ADE) with simplistic water movement conditions. In Elrick *et al.* (1994), salt transport with capillary rise from a saline water table was represented using an analytical solution for a transformed ADE with a steady-state solution for water movement. In Elrick's model, as with many other analytical solutions, the soil profile was considered to be uniform. For vertical migration of a reactive solute, the transport properties of the root zone can be different from the rest of the profile and can play a key role in solute concentration evolution. In addition, transpiration acts to concentrate solutes as they move through the root zone.

The ease of use of analytical solutions has meant that they have found application as screening tools. However, the physical accuracy of the analytical solutions is unknown under the assumptions associated with their derivation. One of the key approximations is related to the nature of unsaturated water movement. At best, water movement is assumed to be at steady state. In reality, in a field soil exposed to climate, water movement will be highly dynamic responding to infiltration and evapotranspiration. This paper investigates the use of analytical solutions for unsaturated solute transport and presents procedures that attempt to minimize the impact of the physical approximations.

Uniform soil profile

Most soil profiles are layered, with a biologically active zone towards the soil surface within which rates of degradation and adsorption are higher than those deeper in the profile. However, a common simplifying assumption has been to assume that the transport parameters are uniform throughout the soil profile. This approximation means that a range of existing analytical solutions, originally developed for the constant coefficient advection–dispersion equation, can be used (i.e. Carslaw & Jaeger 1959). This is the basis of the model of Elrick *et al.* (1994) where an existing analytical solution was used with a transformed advection–dispersion equation. The transformation used came from Bond & Smiles (1983) and consists of the following variable:

$$Q = \int_0^z \theta \mathrm{d}\bar{z} \quad (5)$$

which is the cumulative water volume within the soil profile. This can be extended to:

$$Q = \int_0^z K_a \mathrm{d}\bar{z} \quad (6)$$

appropriate for solute transport with adsorption. Substitution of equation (6) into equation (3) leads to:

$$\frac{\partial f}{\partial t} = \frac{\partial}{\partial Q}\left[K_a \theta D \frac{\partial f}{\partial Q}\right] - \left[q(0,t) + \int_0^z S_w \mathrm{d}\bar{z}\right]\frac{\partial f}{\partial Q} + \frac{(\mu\theta - S_w)}{K_a} f \quad (7)$$

where $f(Q,t) = c(z,t)$.

Equation (7) is an important transformation since the water flux, q, is now only required at the soil surface and not through the profile, as in equation (3). However, to find the concentration at a given location and time, the vertical water content profile is required at that time. For a soil exposed to climate, water movement, particularly towards the soil surface, is highly dynamic and driven by discrete infiltration events and time-varying transpiration. The effect of these processes may become smoothed with depth, but the nature of this smoothing will depend on the hydraulic properties: the lower the hydraulic conductivity the slower the water movement and thus the slower the response to events.

However, the water movement parameters in equation (6) can be simplified significantly. The basis for this is the following transformation,

$Q' = Q - \int_0^t q(0,\bar{t}) + S_w(\bar{t}) z_{av} \mathrm{d}\bar{t}$, that with equation (7) leads to

$$\frac{\partial f'}{\partial t} = \frac{\partial}{\partial Q'}\left[K_a \theta D \frac{\partial f'}{\partial Q'}\right] + \frac{(\mu\theta - S_w)}{K_a} f' \quad (8)$$

where S_w is constant over the profile and z_{av} is a constant, a depth within the profile.

This transformation removes the advective term from equation (7) and redefines the problem in a moving boundary form. This moving boundary problem can be solved, as shown by Barry & Sposito (1989), but is presented here to demonstrate the role of parameters describing water movement in the transformed equation . It is the cumulatives of infiltration and transpiration and not the time evolution of these processes that determine net advective migration in the Q space equation. The infiltration and transpiration fluxes in equation (7) can thus be replaced by constants defined by

$$q_{oc} = \int_0^t q(0,\bar{t}) \mathrm{d}\bar{t}/t \text{ and } S_{wc} = \int_0^t S_w(\bar{t}) \mathrm{d}\bar{t}/t.$$

Assuming the other transport properties in equation (7) can also be described as constants leads to the following analytical solution

$$c(Q,t) = \frac{c_o}{2} \exp\left(\frac{q_{oc} Q}{2[K_a \theta D]}\right) \times \left\{ e^{-\nu Q} erfc\left[\frac{Q - t\sqrt{q^2{}_{oc} - 4\mu\theta^2 D}}{2\sqrt{[K_a \theta D]t}}\right] + e^{\nu Q} erfc\left[\frac{Q + t\sqrt{q^2{}_{oc} - 4\mu\theta^2 D}}{2\sqrt{[K_a \theta D]t}}\right]\right\} \quad (9)$$

where

$$\nu = \sqrt{\left(\frac{q_{oc}}{2[K_a\theta D]}\right)^2 - \frac{\mu}{K^2{}_a D}}$$

where c_o is the fixed surface solute concentration, or with plant water use

$$c(Q,t) = \frac{c_o}{2}\exp\left(\frac{q_a Q}{2[K_a\theta D]}\right)\times$$

$$\left\{e^{-\nu Q} erfc\left[\frac{Q - t\sqrt{q^2{}_a - 4(\mu\theta - S_{wc})\theta D}}{2\sqrt{[K_a\theta D]t}}\right] + \right.$$

$$\left. e^{\nu Q} erfc\left[\frac{Q + t\sqrt{q^2{}_a - 4(\mu\theta - S_{wc})\theta D}}{2\sqrt{[K_a\theta D]t}}\right]\right\} \quad (10)$$

where

$$\nu = \sqrt{\left(\frac{q_a}{2[K_a\theta D]}\right)^2 - \frac{(\mu\theta - S_{wc})}{K_a{}^2\theta D}}$$

and $q_a = q_{oc} + S_{wc}z_{av}$.

Root zone effects

Travel time through the root zone

In equation (10), plant water uptake, S_w, is considered a constant with depth. This approximation allows the term

$$\int_0^z S_{wc} d\bar{z}$$

to be integrated to $S_{wc}z$ and, in a second approximation, replaced by the constant $S_{wc}z_{av}$, where z_{av} is a depth within the profile and S_{wc} is estimated using the procedures presented above. The effect of this second approximation is that advective transport is overestimated above z_{av} and underestimated below it. z_{av} could be chosen to ensure that the travel time through the root zone is correctly preserved in the model. This procedure is based on the following expression for a water profile at steady state:

$$\frac{dz}{dt} = \frac{q_{oc} + S_{wc}z}{K_a}. \quad (11)$$

Equation (11) can be integrated to give a travel time, t_T, for a profile of z_D depth:

$$t_T = \frac{K_a}{S_{wc}}\ln\left[\frac{q_{oc} + S_{wc}z_D}{q_{oc}}\right]. \quad (12)$$

This travel time estimate can be used to find z_{av} through the following, appropriate for a constant velocity

$$\frac{q_{oc} + S_{wc}z_{av}}{K_a} t_T = z_D \text{ or } z_{av} = \frac{1}{S_{wc}}\left[\frac{K_a z_D}{t_T} - q_{oc}\right]. \quad (13)$$

A difficulty with using equations (10) or (11) for modelling vertical migration is that they are based on the assumption of a uniform profile. Most field soils will have vertical variation in transport and flow properties. Physically, a more realistic approximation would be to consider the soil to be composed of two layers: a biologically active root zone and a sub-root zone. An alternative, less complicated procedure, would be to assume that solute transport through the root zone (for a constant source) rapidly approaches a quasi-steady state. While this is clearly a coarse approximation for the root zone, the impact of this should decrease with depth where the concentration will tend to a state determined by the average of the surface conditions. With this assumption mass balance leads to the following relation:

$$q_{co} - q_{ci} = M_{loss} \quad (14)$$

where q_{ci} is the mass flux at the surface; q_{co} is the mass flux at the bottom of the root zone; and M_{loss} is the solute mass lost to degradation. Equation (14) can be reorganized to give:

$$c_o = \frac{q_{ci} + M_{loss}}{q_{oc} + S_{wc}z_r} \quad (15)$$

where the effect of dispersion is neglected and the definition $q_{co} = c_o\ (q_{oc} + S_{wc}z_r)$ has been introduced where z_r is the root zone thickness and c_o the concentration of solute leaving the root zone. The following definition

$$q_{ci} + M_{loss} = q_{ci}\exp(\mu t_T) \quad (16)$$

for the mass lost to degradation can be substituted into equation (15) leading to:

$$c_o = \frac{q_{ci}\exp(\mu t_T)}{q_{oc} + S_{wc}z_r}. \quad (17)$$

Equation 17 accounts for the root zone effects of degradation, retardation and concentration of solutes due to plant water uptake. Substituting equations (17) and (12) into equation (9) gives:

$$c(Q,t) = \frac{q_{ci}}{2(q_{oc} + S_{wc}z_r)}\left[\frac{q_{oc} + S_{wc}z_r}{q_{oc}}\right]^{\frac{\mu K_a}{S_{wc}}}$$

$$\exp\left(\frac{q_{oc}Q}{2[K_a\theta D]}\right)\times$$

$$\left\{ e^{-\upsilon Q} erfc\left[\frac{Q - t\sqrt{q_{oc}^2 - 4\mu\theta^2 D}}{2\sqrt{[K_a\theta D]t}}\right] + \right.$$

$$\left. e^{\upsilon Q} erfc\left[\frac{Q + t\sqrt{q_{oc}^2 - 4\mu\theta^2 D}}{2\sqrt{[K_a\theta D]t}}\right]\right\} \quad (18)$$

where

$$\upsilon = \sqrt{\left(\frac{q_{oc}}{2[K_a\theta D]}\right)^2 - \frac{\mu}{K_a^2 D}}.$$

Tests of the analytical approach

In the subsequent sections of this paper equation (18) is used to estimate the solute migration to a depth will be investigated. The bases for these investigations will be a series of hypothetical simulations calculated using the SWIMv2 water and solute transport model of Verburg *et al.* (1996).

Constant infiltration and transpiration rate

In this section simulations will be conducted for solute transport into a vertical column of 10 m depth with constant infiltration, q_{oc}, and transpiration, S_{wc}, rates. The solute will be applied to the top surface of the column at a fixed flux, equivalent to the solute being dissolved in the infiltrating water at a fixed concentration. These simulations will test the effect of using profile averaged water contents in the solute transport solution where, in the simulations, the water content varies in time and with depth.

SWIMv2 solves coupled water and solute migration. To characterize water movement the moisture retention relationship, $\theta(\psi)$, and unsaturated hydraulic conductivity functions, $K(\theta)$, are required. The moisture retention relationship used in the analysis presented here was the van Genuchten function (van Genuchten 1980),

$$\theta = \theta_r + \frac{\theta_s - \theta_r}{[1 + (\psi/\psi_a)^n]^m} \quad (19)$$

where θ_s is the saturated water content, θ_r the residual water content, ψ_a the matric potential scaling parameter and $n = 1/(1\text{-}m)$, where m is an empirical constant.

The unsaturated hydraulic conductivity function used was also from van Genuchten (1980):

$$K(\theta) = K_s S_e^{1/2}[1-(1-S_e^{1/m})^m]^2 \quad (20)$$

where $S_e = (\theta - \theta_r)/(\theta_s - \theta_r)$ and K_s is the saturated hydraulic conductivity.

Two sets of published measurements were used for the soil hydraulic properties; Guelph loam (van Genuchten 1980) and Grenoble sand (Fuentes *et al.* 1992). The soil hydraulic properties for these soils are presented in Table 1.

The infiltration rate for the simulations was 0.04 cm h^{-1} and the transpiration rate was 0.01 cm h^{-1} spread over the root zone. A 10 m deep profile was used in the SWIMv2 simulations. Equation (18) assumes a semi-infinite profile.

Equation (18) defines the concentration variation in Q space. The moisture content profile is required in order to invert from Q to z space. In the analyses presented here the SWIM water content profile results are used to invert from Q to z space.

Figure 1 compares solute concentration profiles from SWIMv2 and equation (18). For results below the root zone there is excellent agreement between the accurate numerical results and the approximate analytical approach represented by equation (18). Within the root zone the two procedures differ considerably, a result of equation (18) being accurate for sub-root zone transport and the root zone effects being lumped. The use of constant, vertically averaged water contents for the root and sub-root zone has not had an appreciable impact on the accuracy of the analytical results.

Figure 2 presents the solute concentration profiles for Grenoble sand hydraulic properties where advective transport occurs at a much higher rate than in Guelph loam, a result of the higher flow rates leading to lower profile averaged water content. As with Figure 1, the results from SWIMv2 and equation (18) are in close agreement, indicating the general accuracy that can be obtained using constant parameters in equation (18).

Table 1. *Soil hydraulic properties used in SWIMv2 model simulations*

	θ_s (cm^3 cm^{-3})	θ_r (cm^3 cm^{-3})	K_s (cm h^{-1})	φ_a (cm)	m
Grenoble sand	0.312	0.0	15.32	−16.39	0.2838
Guelph loam	0.434	0.218	1.32	−50	0.275

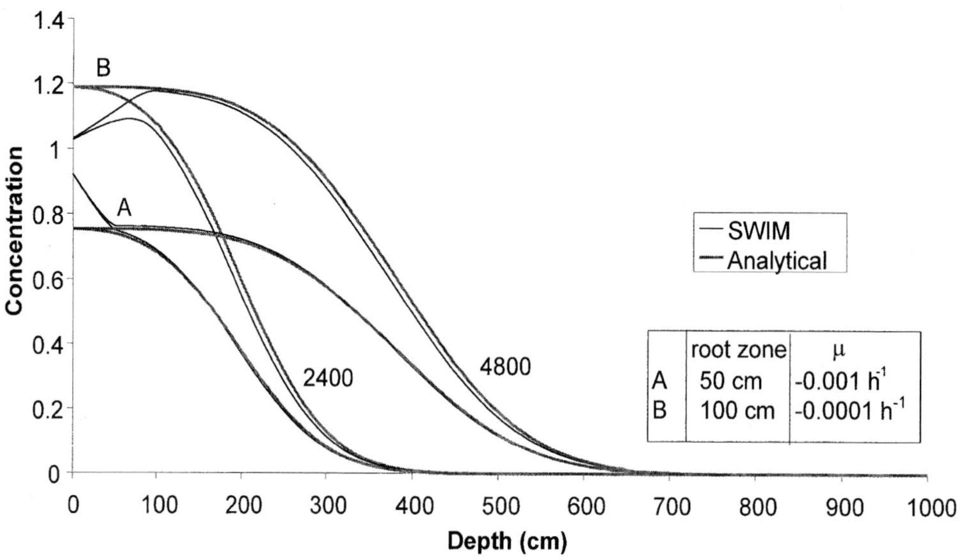

Fig. 1. Vertical solute concentration profiles at two times and for two sets of root zone depth and degradation rate under constant infiltration and transpiration rates calculated with SWIMv2, for Guelph loam hydraulic properties, and equation (18).

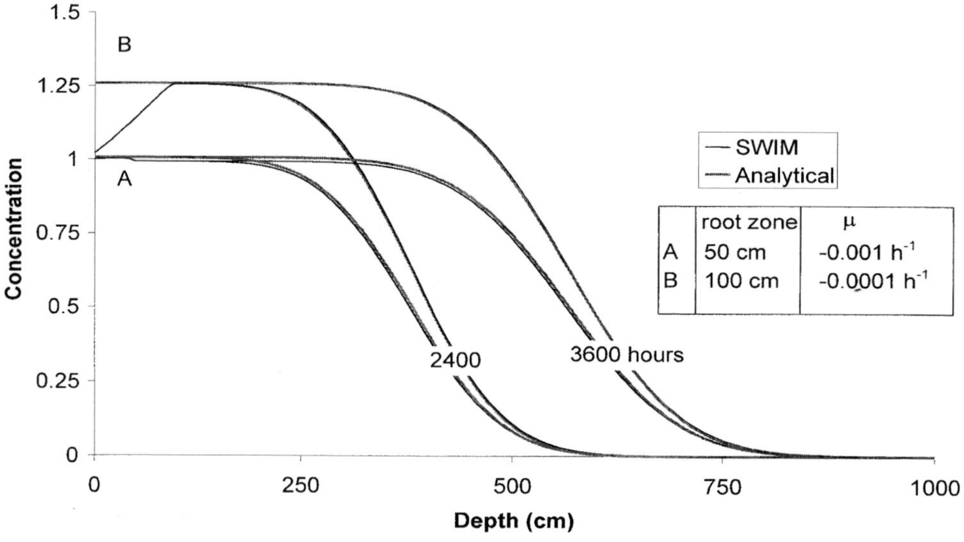

Fig. 2. Vertical solute concentration profiles at two times and for two sets of root zone depth and degradation rate under constant infiltration and transpiration rates calculated with SWIMv2, for Grenoble sand hydraulic properties, and equation (18).

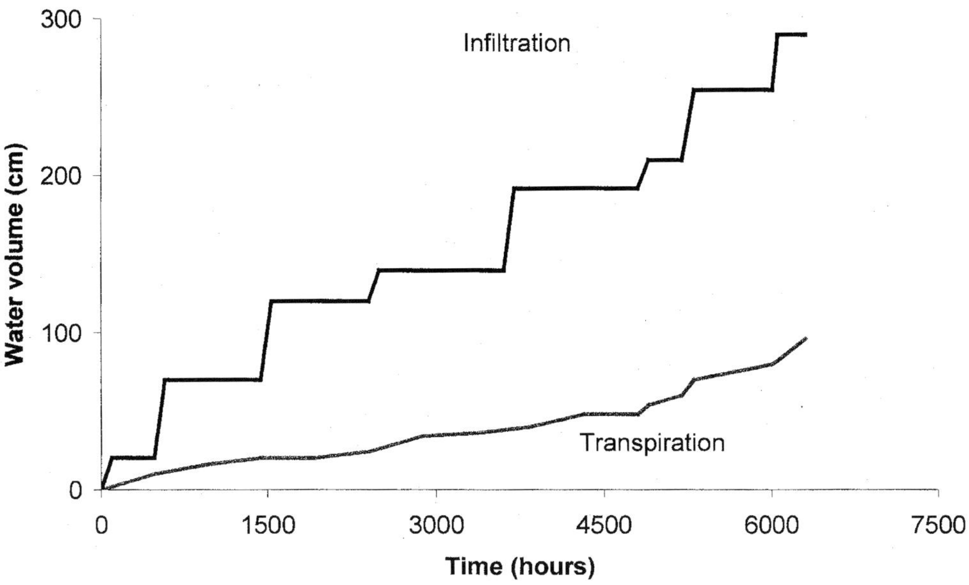

Fig. 3. Cumulative infiltration and transpiration rates imposed in the SWIMv2 simulations of solute transport.

Dynamic variation in infiltration and transpiration rates

A key question identified in the introduction to this paper is the applicability of analytical solutions where there are dynamic water movement conditions. A common assumption made in order to derive analytical solutions to the advection–dispersion equation is that it has constant coefficients. In the transformed equation used as the basis for equation (18) this means that the infiltration and transpiration rates are replaced by constants. It was shown above that these constants could be calculated as the averages of the dynamic rates over a time period. An underlying assumption with the lumped root zone treatment that is part of equation (18) is that the effects of the dynamic variations in infiltration and transpiration rates are smoothed with depth and that the concentration approaches an average of the surface conditions. This smoothing will be a function of the soil hydraulic properties, with lower conductivity material allowing greater smoothing than higher conductivity material.

In this section the accuracy of equation (18) for the simulation of unsaturated solute migration under dynamic variation in infiltration and transpiration rates will be investigated. As with the previous section, this investigation will comprise a series of hypothetical simulations with results from the SWIMv2 coupled flow and transport code being the basis of comparison. The imposed surface boundary conditions are presented in Figure 3. The infiltration time series presented in this figure is composed of a number of 'events' of varying magnitude separated by periods of no infiltration. Transpiration also varies through time. The fluxes used in equation (18) are the averages up to the time of interest and so are different for each point calculated over the simulation.

Figure 4 presents a comparison of SWIMv2, with Guelph loam properties, and equation (18) concentration results through time at several depths within the unsaturated profile. The effect of variation in surface infiltration rate is apparent in these results where solute is seen to move in a series of pulses down the profile. Each solute pulse is shown to have a sharp front as it passes the observation location with successive pulses raising the concentration to a maximum. The agreement between the accurate numerical and approximate analytical results is excellent, bearing in mind the degree of approximation involved in the analytical solution. With this, water movement effects on solute transport are treated by constant, time averages of the dynamic behaviour. Water movement varies

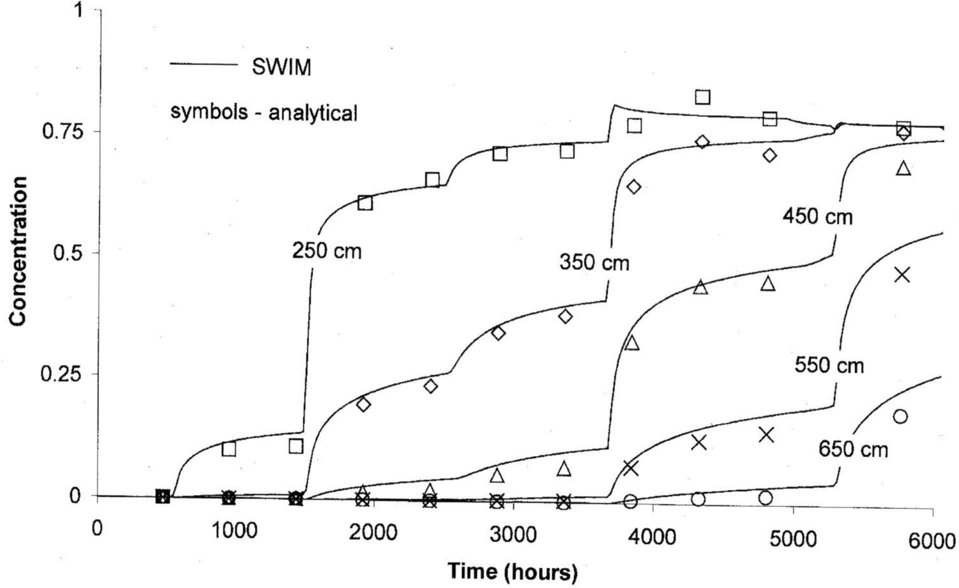

Fig. 4. A comparison between SWIMv2 (with Guelph loam hydraulic properties) and equation (18) solute concentration results at several depths within the soil profile.

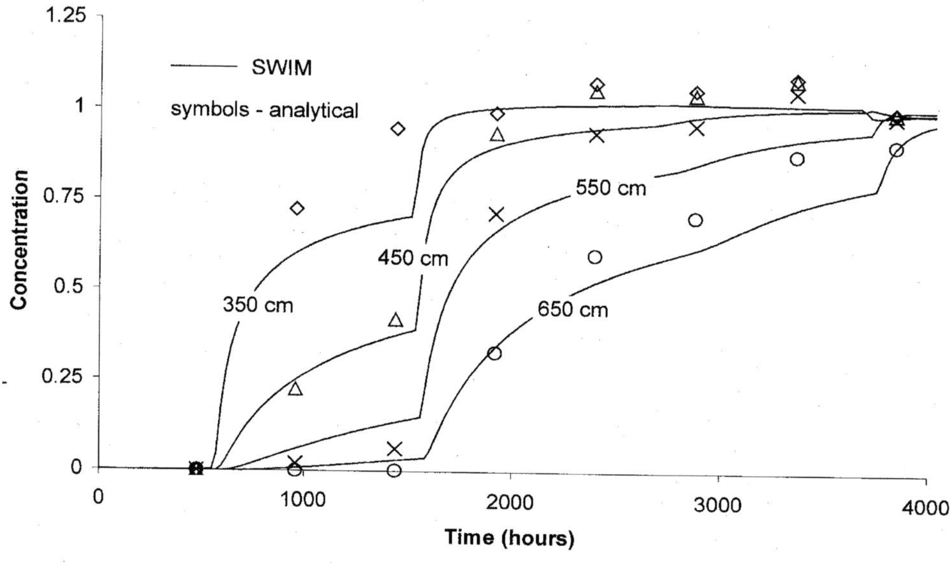

Fig. 5. A comparison between SWIMv2 (with Grenoble sand hydraulic properties) and equation (18) solute concentration results at several depths within the soil profile.

dramatically in response to the infiltration events, with successive wetting fronts progressing through the profile. Thus, water content and flow rate have a temporal and spatial behaviour not represented in the analytical solution; however, the agreement is still close.

Figure 5 presents the concentration results for Grenoble sand hydraulic properties for the two procedures with the infiltration and transpiration time series presented in Figure 3. As with Guelph loam properties, pulses of solute are apparent passing the observation depths; however, in sharp contrast, the rates of migration are significantly higher. While the analytical results do represent the arrival of each solute pulse relatively well, the differences between the two procedures increases behind the solute front with the analytical solution overestimating the solute concentration. This is a consequence of the profile average water contents used in the analytical solution while in the numerical calculation the water content varies substantially as each wetting front passes.

Conclusions

Analytical solutions to the advection–dispersion equation can offer considerable computational efficiencies over numerical approaches and have therefore become popular as screening tools or when large numbers of calculations are required, as in Monte Carlo uncertainty analyses. However, significant physical approximations are involved in the derivation of the analytical solution. This is particularly the case for unsaturated solute migration where there is usually vertical structure in transport parameters and dynamic variation in water movement. This paper has presented methods where approximate results can be provided by a transformed analytical solution under dynamic water movement conditions and where there is vertical structure in transport properties.

A key part of the approach involves a spatial transformation to the transport equation. This transformation redefines the problem from the spatial domain to the spatial cumulative of K_a, the product of water content and retardation. One result of this transformation is that the water flux is now only required at the soil surface. In a second transformation it was shown that advective transport could be represented by the averages of the surface water flux and the transpiration rate, properties that are readily measurable. This is a significant simplification over the standard approach used with numerical solutions where the time evolution of these quantities is necessary.

A complication in the use of the averages of the infiltration and transpiration rates is that the water content profile is required to invert from the spatial transform space. Water content is determined by the water movement history, and thus to resolve accurately the water content profile at a particular time the temporal behaviour prior to the time of interest must be described. Of course the importance of the event history to the current state will be a function of the magnitude of flow events and the length of time that has elapsed since the event occurred. Another factor is that flow associated with rainfall events will be smoothed with depth by the unsaturated flow process. This smoothing will be a function of the soil hydraulic properties: the lower the permeability the greater the smoothing. Thus, in some situations, it may be a safe approximation to use steady-state solutions to describe the water content profile at depth.

The approach described in this paper was shown to work very well for the Guelph loam soil hydraulic properties where results calculated using equation (18) agreed closely with those of SWIMv2. For this soil type, soil water contents at depth were relatively uniform. For the highly permeable Grenoble sand hydraulic properties the two procedures were not as close, a result of the large variations in water content within the soil profile after rainfall and drying events.

This paper has demonstrated that analytical solutions with constant coefficients do have a physical basis under dynamic water movement conditions.

This work was supported by Engineering & Physical Sciences Research Council Grant GR/N33119.

References

BARRY, D. A. & SPOSITO, G. 1989. Analytical solution of a convection-dispersion model with time-dependent transport coefficients. *Water Resources Research*, **25**, 2407–2416.

BOND, W. J. & SMILES, D. E. 1983. Influence of velocity on hydrodynamic dispersion during unsteady flow. *Soil Science Society of America Journal*, **47**, 438–441.

CARSLAW, H. S. & JAEGER, J. C. 1959. *Conduction of Heat in Solids*. Oxford University Press, London.

ELRICK, D. E., MERMOUD, A. & MONNIER, T. 1994. An analysis of solute accumulation during steady-state evaporation in an initially contaminated soil. *Journal of Hydrology*, **155**, 27–38.

FOSTER, S. S. D. 1998. Groundwater recharge and pollution vulnerability of British aquifers: a critical overview. *In:* ROBINS, N. S. (ed.) *Groundwater Pollution, Aquifer Recharge and vulnerability*. Geological Society, London, Special Publications, **130**, 7–22.

FUENTES, C., HAVERKAMP, R. & PARLANGE, J.-Y. 1992. Parameter constraints on closed-form soil water relationships. *Journal of Hydrology*, **134**, 117–142.

GERMANN, P. F. & BEVEN, K.J. 1985. Kinematic wave approximation to infiltration into soils with sorbing macropores. *Water Resources Research*, **21**, 990–996.

JURY, W. A. 1982. Simulation of solute transport with

a transfer function model. *Water Resources Research*, **18**, 363–368.

RICHARDS, L. A. 1931. Capillary conduction of liquids through porous mediums, *Physics*, **1**, 318–333.

ROSS, P. J. 1990. Efficient numerical methods for infiltration using Richards equation. *Water Resources Research*, **26**, 279–290.

VAN GENUCHTEN, M. Th. 1980. A closed form equation for predicting the hydraulic conductivity of unsaturated soils. *Soil Science Society of America Journal*, **44**, 892–898.

VERBURG, K., ROSS, P. J. & BRISTOW, K. L. 1996. *SWIMv2.1 User Manual*. CSIRO Divisional Report No 130, CSIRO, Canberra.

A probabilistic management system to optimize the use of urban groundwater

R. M. DAVISON[1], P. PRABNARONG[1], J. J. WHITTAKER[2] & D. N. LERNER[1]

[1]*Groundwater Protection and Restoration Group, Department of Civil and Structural Engineering, University of Sheffield, Mappin Street, Sheffield S1 3JD, UK (e-mail: R.M.Davison@Sheffield.ac.uk)*

[2]*Environmental Simulations International, Priory House, Priory Road, Shrewsbury SY1 1RU, UK*

Abstract: Urban groundwater is an underused resource mostly due to the perceived risk of contamination; conversely rural groundwater is being over-exploited. To enable urban groundwater to be utilized effectively a probabilistic water management tool has been developed. The management tool combines three models to identify the best use for water pumped from a user-defined location. The probabilistic catchment zone model determines the spatial distribution of the probability that water originating from a given point in the aquifer reaches the pumping well. The land-use model then identifies the potential contaminant sources in this region from a large set of GIS-based coverages and databases. The contaminant source data are passed to a pollution risk model that calculates the probability distribution for the concentration of a contaminant at the pumped borehole. Initial model validation using contaminant concentrations from two sites shows the model fits the available field data. These case studies show there is risk from chlorinated solvents but little risk from the BTEX compounds.

Urban groundwater is a largely underused resource and the potentially productive Triassic Sherwood Sandstone aquifer beneath the City of Nottingham in the English Midlands is no exception. The perceived risk of contamination and potentially high clean-up costs have in the past made the resource unfavourable to suppliers. The underuse of urban groundwater has led to rising water tables, with consequent basement and tunnel flooding and geotechnical problems (Greswell *et al.* 1994; Lerner & Barrett 1996; Lerner & Tellam 1997). Conversely the overuse of rural groundwater has led to environmental problems including depleted aquifers, and rivers commonly experiencing low flow conditions. The fragility of England's water supply has been highlighted with the acute water shortages experienced in recent times. To enable a sustainable city to develop, a balance needs to be attained to reduce the rising water tables, alleviate water shortage and to reduce the pressure on rural groundwater.

Cities, however, have complex interactions with their underlying groundwater. New recharge sources such as leaking water mains, sewers and soakaways ensure that recharge is at least equivalent to rural areas (Lerner 1986). The long industrial history of most cities leads to complex patterns of multipoint pollution sources that may impact groundwater quality (Rivett *et al.* 1990; Burston *et al.* 1993; Ford & Tellem 1994). If urban groundwater is to be used efficiently then borehole positioning is critical. The position of the borehole will invariably influence the water quality, therefore benefit may be gained by abstracting water for different uses from different areas of the city. Established boreholes are also at risk from contamination, but to what degree is difficult to predict. Due to the inherent uncertainties and limited data, a single conceptual model cannot represent the complex interaction between the city and its groundwater. With the aim of helping to convert urban groundwater from a problem to a

From: Hiscock, K. M., Rivett, M. O. & Davison, R. M. (eds) *Sustainable Groundwater Development.* Geological Society, London, Special Publications, **193**, 265–276. 0305-8719/02/ $15.00 The Geological Society of London 2002.

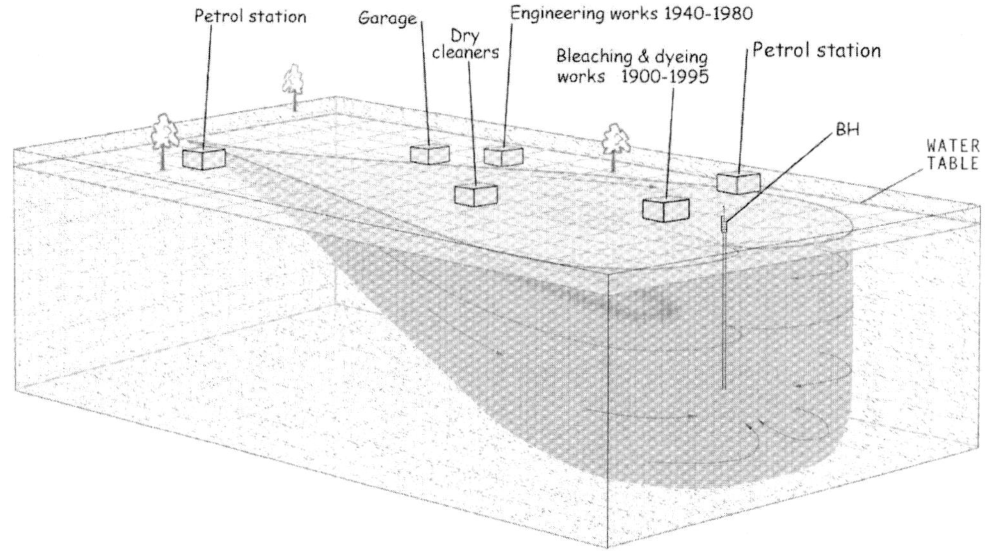

Fig. 1. The target-pathway-sources conceptual model.

resource, a risk-based management tool that builds in the inherent uncertainty is under development.

The risk-based management tool consists of three main components: a probabilistic catchment zone model to identify the land area that may impact the water quality; a GIS (geographic information system)-based land-use model to identify the potential contaminant sources within that area; and a pollution risk model to estimate the threat that these sources may pose. The management tool could be expanded to include a fourth component, the economic model, which could compare the value of water for different end-users, including the cost to treat the water to the required standard. The management tool will ultimately be used to identify the best use for urban groundwater between public supply, industrial supply or river augmentation, in order to provide an economically viable method of moving towards sustainable water management for cities.

The conceptual model is slightly different from the conventional source-pathway-target analogy. Instead, we begin with the target and retrace the flowlines of the capture zone, to identify the multiple potential sources (target-pathway-sources; Fig. 1). The following sections present this novel methodology in the context of two boreholes in Nottingham.

Hydrogeology and groundwater quality

The study is based in the City of Nottingham, which lies above the Sherwood Sandstone aquifer. The Sandstone unit begins in the west along the River Leen and increases in thickness to over 150 m in the north-east. The relatively impermeable Mercia Mudstone Group confines the Sherwood Sandstone to the east. The aquifer is unconfined over much of the study area, with little drift protection away from the valley bottoms. The River Leen runs along the western border of the Sherwood Sandstone at its junction with the Lower Magnesian Limestone. The regional hydraulic gradient of the Sherwood Sandstone is steepest in the north-west, from where it drives flow south-easterly to discharge as baseflow to the River Trent.

The groundwater quality beneath Nottingham has been studied by Barrett *et al.* (1996) and was found to be of poorer quality than in nearby rural areas. The deterioration is not great for inorganic species, except for localized pollution incidents, and no appreciable concentrations of trace metals were found. Nitrate concentrations are similar in urban and rural areas, and frequently exceed the drinking water limits. Chlorinated solvent and BTEX pollution is widespread, originating from fuel and solvent spills.

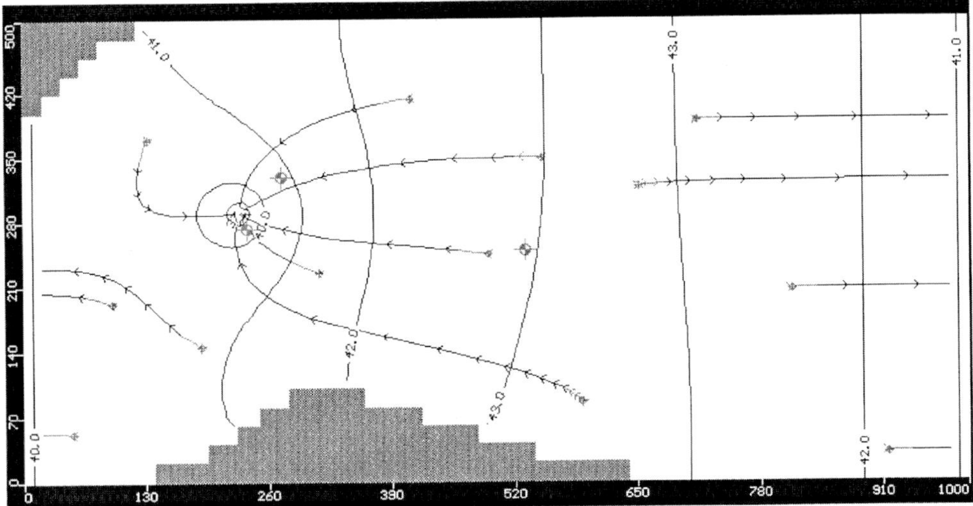

Fig. 2. The technique used in the catchment zone uncertainty model. For each simulation a particle is placed in each model cell, those that reach the borehole are allocated a '1' those not captured by the borehole are allocated a '0'.

The risk model

GIS technology has been used to link the four individual model components to provide an integrated management tool. During the last few years the GIS approach has emerged as an extremely effective tool for analysing and prioritizing natural resource management alternatives (Tim *et al.* 1996). Natural resource management problems are typically spatial in nature, therefore GIS technology provides an ideal tool for defining a problem and facilitates the design and implementation of alternative management strategies (Tim *et al.* 1996). In this data management system the GIS is used as a data storage centre for the inputs to the catchment zone model and the pollution risk model as well as a graphical user interface (GUI). The GUI prompts for the selection of a new or existing borehole location where the user requires information regarding the potential water quality. The interface then guides the user through a series of wizards that instigate the three models described in the following sections.

Catchment zone uncertainty model

The first modular component of the urban groundwater management system is an uncertainty-based catchment zone model. The objective of this component is to calculate the spatial distribution of the probability that water originating from a given point in the aquifer reaches the user-selected pumping well.

The basis of the catchment uncertainty model is a groundwater flow model constructed using MODFLOW. Thus it can include all the geometric boundaries, hydrological features and physical processes that can be represented in MODFLOW. The catchment uncertainty model simulates the effects of uncertainty in the key model parameters: recharge, hydraulic conductivity, porosity, storage coefficient and specific yield. These parameters can be defined as homogeneous model values or in terms of property zones. The uncertainty of recharge with time may also be considered: it may be defined as being constant for a given time (a stress period in MODFLOW) within the total time span.

In addition to the parameter choice, the uncertainty model allows for a choice of parameter distribution types. The basic distribution is a uniform distribution ranging between a minimum and a maximum value. The triangular parameter distribution is defined in terms of a minimum, a maximum and a most likely value. The third option available is a normal distribution, defined by the parameter mean and variance. The choice of the different parameter distributions depends on the level of data available and on the desired weighting of extreme values.

The uncertainty model undertakes a given

Table 1. *Details of the coverages stored within the GIS*

Coverage category	Details
Land-use	The spatial distribution of land-use in Nottingham over the past century, including the locations of each industry that existed on the Triassic Sherwood Sandstone outcrop (Fig. 3).
Contaminated land	The spatial distribution of known contaminated land was acquired from Environment Agency records.
Solid geology	This coverage was used to delineate the area of interest for this study, as all efforts are concentrating on the Sherwood Sandstone outcrop.
Drift cover	The drift cover is not currently required as an input for the pollution risk model but may be added as a feature in the future.
Recharge zones	The recharge zones are required as an input to the pollution risk model in the calculation of the contaminant flux entering the unsaturated zone.
Depth to water table	The depth to water table is a direct input to the pollution risk model, where it controls the time the contaminants spend in the vadose zone and therefore the amount of first order decay.

Table 2. *Details of the databases stored within the GIS*

Database	Details
Specific industry	This forms a central data store for all information relating to a specific industry identified on the land-use coverage.
General industry	This database contains information about generic industry types, for example the type of contaminants associated with bleach works, and the volumes used depending on the facility size.
Contaminant properties	The contaminant properties include data on degradation potential, retardation, solubility, all of which are direct inputs to the pollution risk model

number of simulations, each based on a parameter set selected randomly, but such that the ensemble fulfils the specified distributions. For each parameter set, a flow field is calculated. If the flow field is accepted under the calibration criteria, the advective transport of a hypothetical contaminant originating from the midpoint of each cell of the flow model is simulated using particle tracking in MODPATH. The catchment probability distribution P[x] is defined as the probability that a particle starting at position x arrives at the well. For a transient model the capture zones are also transient. For a single simulation, P[x] at a given position is either 0 (the particle does not reach the well) or 1 (the particle reaches the well). After n simulations the probability distribution is given by P[$x \mid n$], the probability over n simulations of a particle starting at x reaching the well (Fig. 2). The catchment zone model provides a second output, a probability density function (PDF) of travel time for each cell to the borehole. The travel time information is required as a direct input to the pollution risk model. For the Monte-Carlo method in its simplest form, all simulations are accepted with equal weighting. A modification of the method allows equal weighting only for those simulations meeting the calibration criteria; the rest are neglected. The Generalized Likelihood Uncertainty Estimation (GLUE) method weights the individual probability distributions according to how well the flow simulation fulfils the calibration criteria. The weights are termed likelihood functions. So, for example, Bevan & Binley (1991) suggested a likelihood, L, based on the sum of the squared residuals (σ_e^2):

$$L = (\sigma_e^2)^{-N} \quad (1)$$

where the residual is the difference between the simulated and observed heads and N is a number chosen by the user. When N is zero all simulations receive equal weighting, as before. As $N \rightarrow \infty$ the single best simulation receives a weighting of 1 and all other simulations are discarded from consideration. Other likelihood functions can be constructed, for example, to take into account both head and flux calibration targets.

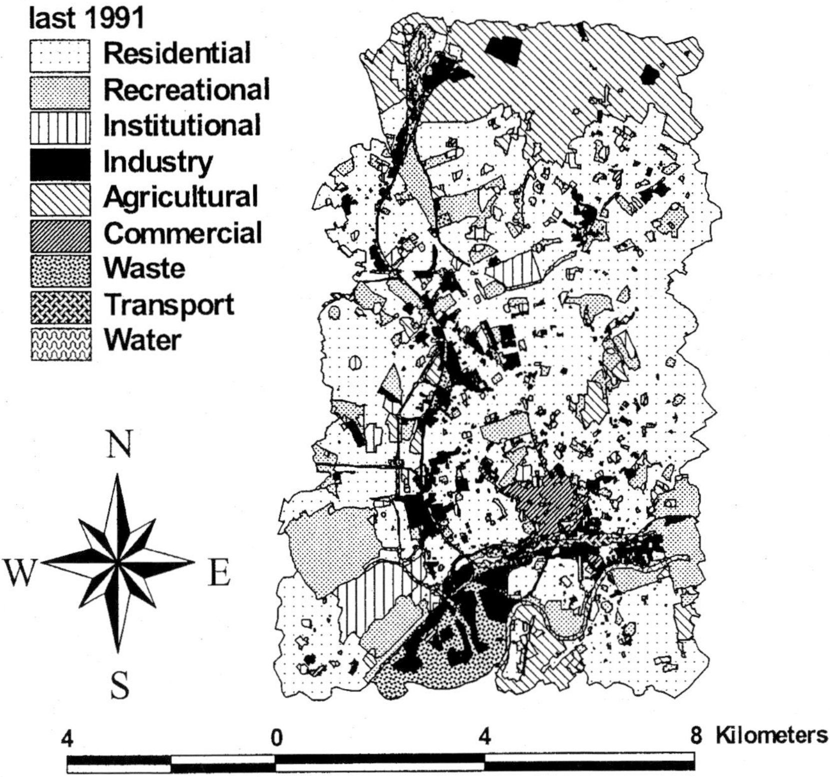

Fig. 3. The land-use coverage for the outcrop area of the Sherwood Sandstone in Nottingham for 1970.

The main output from the catchment zone uncertainty model is a set of contours that enclose areas of varying certainty as to whether they lie within the borehole's catchment zone.

Land-use model

The catchment zone model identifies the approximate area of interest for the chosen borehole. The land-use model can then be used to identify the past and present surface activities that may affect the groundwater quality now and into the future.

The aim of the land-use model is to establish all the potential contaminant sources in a given catchment area. Two types of data were collated for Nottingham: spatial data that were stored within the GIS as coverages, and details of the individual features shown on the coverages, stored in databases. The coverages included in the GIS are described in Table 1 and the databases in Table 2.

The surface activity coverages and their associated databases involved intensive data collection and hence will be described in more detail than the entry in Table 1. The surface activity coverages categorized the area shown in Figure 3 into the following groups: residential, recreational, institutional, industrial, agricultural, commercial, waste, transport and water.

The surface activity data were traced from 1 : 10 000 Ordnance Survey maps, with reference to 1 : 1250 maps to provide detail. Coverages for the years 1900, 1920, 1939, 1954, 1972 and 1991 were digitized to account for historical variations. The specific industry details from the Ordnance Survey maps were supplemented with additional references from historical texts, industrial directories, local records and the Nottingham Industrial Archaeology Society (NIAS). This data collection exercise involved six tracings for each square kilometre, with reference to 24 individual maps, some grid squares containing over 50 industries. All the coverages

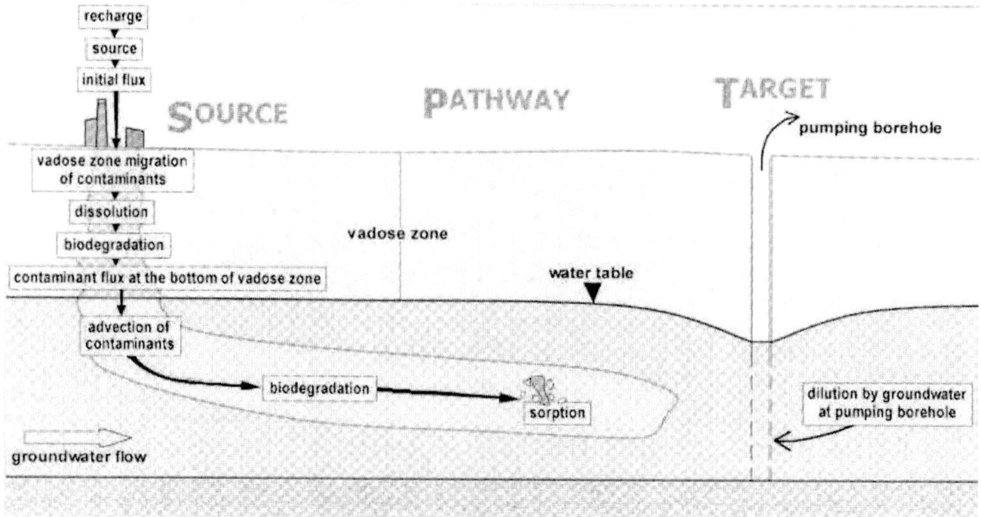

Fig. 4. Processes included within the pollution risk model.

have been digitized using ARC/INFO. The databases are stored in Microsoft Access and link to the GIS to allow easy editing. Each industry is associated with a site-specific database entry using a code. The specific industry database is used as a central store to collate all the data about the industry from two lookup databases: general industry and contaminant properties. The central database includes the following headings: area, perimeter, land-use identification code, industry number, distance from borehole, industry start date, industry closure date, recharge rate, industry type, depth to groundwater, associated contaminants, contaminant properties, minimum volume of each contaminant used, and maximum volume of each contaminant used.

The results from the data collection in Nottingham show that the textile industry is the most common; however, most of the firms are very small and have a low potential for causing significant contamination. Other common industries likely to cause contamination are bleaching/dye works, foundries and petrol stations.

The graphical user interface uses the land-use maps as the base screen. The user may select any industry in any time period in order to display the information shown in Table 2.

The pollution risk model

An analytical solute transport model has been developed to assess risk of contamination at a pumped borehole. The model incorporates a one-dimensional solution to simulate advection, dilution, retardation and biodegradation processes [equation (2)]. The model consists of two interconnected components: an unsaturated (vadose) zone model and a saturated (groundwater) zone model (Fig. 4). The two components are connected by the condition that the contaminant flux through the bottom of the unsaturated zone equals the contaminant flux to the top of the saturated zone. This model is different from many solute transport models as it includes contaminant dilution by groundwater at a pumping well. The simplifying assumptions for transport in the model are: (1) the unsaturated zone has predominantly vertical flow; and (2) the saturated zone has horizontal flow.

$$C_{\mathrm{w}} = C_{\mathrm{s}} X \frac{RA}{Q} \exp\left(-\lambda R_{\mathrm{f}} \left(\frac{\theta w}{R} + \frac{L}{V} \right) \right). \quad (2)$$

Terms are defined in Figure 5.

The simple solute transport model, based on equation (2), has been solved analytically in a spreadsheet to ascertain the concentration of contamination at the pumping borehole (Fig. 5). Due to the uncertainty in many of the input parameters a probabilistic approach was taken using the software package Crystal Ball (Decisioneering 1997). Instead of assigning a single representative value to each input, PDFs were used to represent the range of possible values.

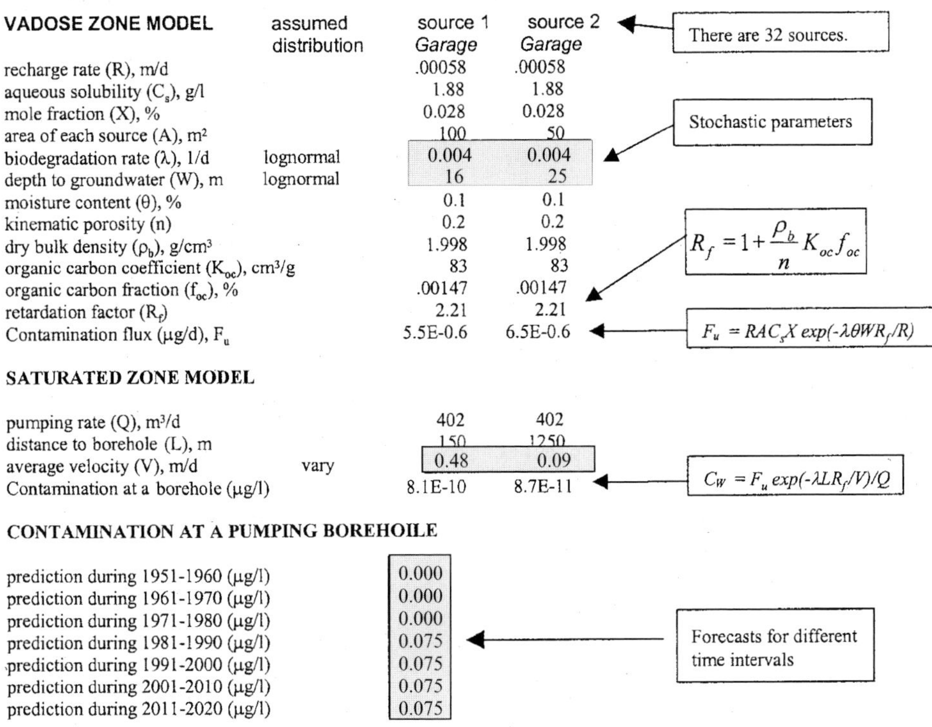

Fig. 5. The probabilistic spreadsheet model.

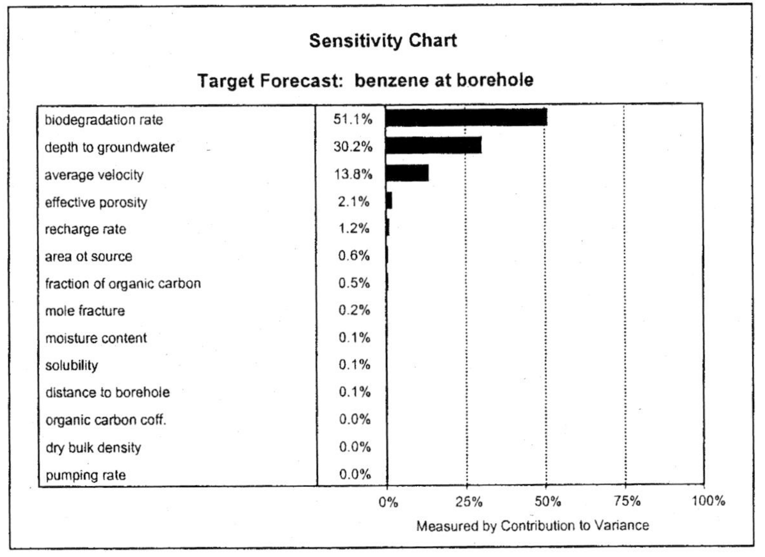

Fig. 6. The contribution of each model parameter to the model variance.

Table 3. *Summary of input data for the public water supply borehole case study*

Parameter	Value	Data source	Data type
Degradation rate of benzene (d^{-1})	0.001–0.004	Mace *et al.* (1997)	Model estimation
	0.002	McNab & Dooher (1998) Mace *et al.* (1997)	Mean values of specific sites
	0.0026	Bjerg *et al.* (1994)	Laboratory data
	0.0027	Reid *et al.* (1999)	Laboratory data
	0.003	Chapelle *et al.* (1996)	Field and laboratory
Recharge ($m\ d^{-1}$)	0.00058	Yang *et al.* (1999)	Total recharge
Organic carbon coefficient ($cm^3\ g^{-1}$)	83	Fetter (1994)	Laboratory data
Organic carbon fraction (%)	0.147	Thornton *et al.* (1996)	Laboratory data
Dry bulk density ($g\ cm^{-3}$)	1.988	Jones (1997)	Field data
Moisture content (%)	10	Barrett *et al.* (1996)	Estimated data
Solubility of benzene at 23.5°C ($g\ l^{-1}$)	1.88	The Merck Index (1990)	Laboratory data
Pumping rate ($m^3 d^{-1}$)	402	Severn Trent Water (1995)	Raw data
Area of sources (m^2)	vary*	Field data and Ordnance Survey map	Raw data
Distance to borehole (m)	vary	Ordnance Survey map, field data, and groundwater model	Model data
Groundwater velocity ($m\,d^{-1}$)	vary	Groundwater model	Model data
The mole fraction of benzene (%)	2.8	McNab & Dooher (1998)	Mean data
Kinematic porosity	0.2	Davison (1998)	Model calibration
Depth to groundwater (m)	vary	The actual sites and water table map from groundwater model	Calculated data

*The value of a parameter is varied depending on sources.

The data collection effort for the model inputs concentrated on the most influential parameters identified from the results of a sensitivity analysis. The sensitivity analysis was performed on the uncertain parameters to quantify the effect of their variation. To quantify sensitivity, input parameters and the results are rank-correlated. Rank correlation involves assigning ranks to both the forecast and the input parameter and performing a linear regression on the corresponding rank sets. The resulting correlation coefficients are then tallied for the forecast and normalized (Decisioneering 1997; Fig. 6). This yields the relative contribution to variance of each input parameter (e.g. biodegradation rate, depth to groundwater) to the forecast (concentration of benzene at the borehole).

Uncertainty in three parameters appears to dominate the impact on the forecast value in this example: the biodegradation rate, the depth to groundwater, and the average velocity (Fig. 6). The uncertainty in biodegradation rate is perhaps the most significant factor in the forecast of benzene concentration. While updating the parameter PDFs for inclusion in the land-use databases particular attention was paid to these three most sensitive parameters.

In the stochastic model, the PDFs representing variability or uncertainty are sampled N times using Latin-Hypercube sampling and the model is run N times. Model output provides cumulative distribution functions of total risk.

Case study

The management tool has been applied to a public water supply borehole in the Basford area of Nottingham (SK 4566 3428). The borehole fully penetrates the Sherwood Sandstone aquifer and has an average abstraction rate of $402\,m^3\,d^{-1}$. The catchment area extends to the north-west of Nottingham city centre and its shape is largely controlled by surrounding abstractions. The Basford area of Nottingham is one of the most industrial areas of the city. The land-use in 1900 was largely agricultural with small clusters of industry. By 1997 the land-use was largely residential with large numbers of industrial sites positioned around the catchment including 13 garages, four printers and one metal working site.

The input parameters to the pollution risk model were all assumed to be constants except

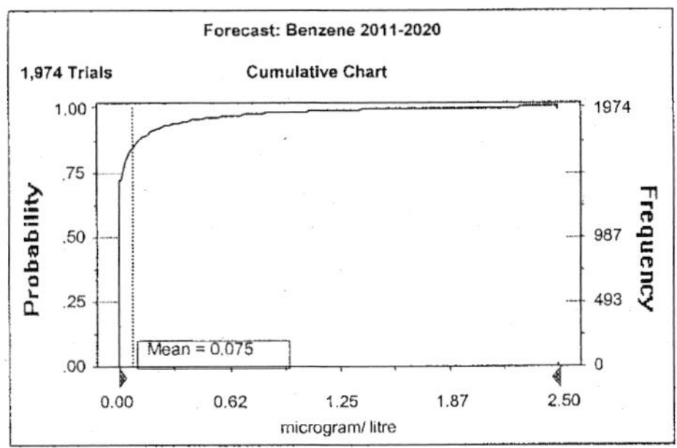

Fig. 7. Cumulative probability distribution function for benzene between 2011 and 2020.

the biodegradation rate, depth to groundwater and average groundwater velocity. The chosen constants are listed in Table 3 with the data sources. The three most sensitive parameters were given probability density functions to represent their uncertainty. The distribution of biodegradation rates is believed to be a lognormal distribution (Mace *et al.* 1997; McNab & Dooher 1998). To characterize the lognormal distribution, mean and standard deviation values are required. The studies of Mace *et al.* (1997) and McNab & Dooher (1998) showed that the mean biodegradation rate for benzene was 0.002 per day. The standard deviation value was estimated by a trial and error method. Using the mean value of 0.002 per day, a standard deviation of 0.004 per day was found to produce a lognormal curve capable of covering all values in the literature review. The range of the biodegradation rate of benzene in this study was between 0.001 and 0.004 per day. At this stage in model development the same decay rate was used for the unsaturated and saturated zones; however, this could easily be changed in the future. The probability distributions for depth to groundwater and average groundwater velocity were model-derived parameters and

Table 4. *A comparison of the drinking standards, the field data, and the model predictions for the time period 2011–2020 at the public water supply borehole*

Parameter	The UK Drinking Water Standard* ($\mu g\ l^{-1}$)	Field data† ($\mu g\ l^{-1}$)	The model prediction in 2000 (mean value, $\mu g\ l^{-1}$)
Trichloroethene (TCE)	30	140	230
1,1,1-trichloroethane (TCA)	30	4	9
Tetrachloroethene (PCE)	10	260	360
Benzene (B)	1	bql‡	0.07
Toluene (T)	< 10	bql	0.03
Ethylbenzene (E)	< 10	bql	0.05
Total xylenes (X)	< 10	bql	0.001

*Source: Nicolson (1993).
† Source: Barrett *et al.* (1996)
‡ bql, below quantification limit.

Table 5. *Comparison of drinking water standards, the field data and the model predictions for the time period 2011–2020 at the industrial borehole*

Parameter	The UK Drinking Water Standard* ($\mu g\ l^{-1}$)	Field data† ($\mu g\ l^{-1}$)	The model prediction in 2000 (mean, $\mu g\ l^{-1}$)
Trichloroethene (TCE)	30	100	220
1,1,1-trichloroethane (TCA)	30	5	13
Tetrachloroethene (PCE)	10	320	400
Benzene (B)	1	bql‡	0.0
Toluene (T)	< 10	bql	0.0
Ethylbenzene (E)	< 10	bql	0.1
Total xylenes (X)	< 10	12	0.1

*Source: Nicolson (1993).
†Source: Barrett *et al.* (1996).
‡bql, below quantification limit.

probability distribution functions were automatically fitted to the data by Crystal Ball.

There are numerous sources of benzene in the catchment, both current and historical including garages and petrol stations. The first stage of the case study investigated the threat that benzene may pose to the borehole. The probabilistic prediction of benzene concentration at the borehole was calculated for different time intervals from 1951 to 2020. The forecasts predicted that there would be no measurable benzene at the borehole during 1951 to 1980. During 1981–1990 the mean benzene concentration was predicted to rise to 0.075 μg l^{-1} (median concentration 0 mg l^{-1}) and to remain steady until 2020, which was the last time interval (Fig. 7). Field data collected by Barrett *et al.* (1996) showed that no benzene was detected in the borehole in 1995 (detection limit = 0.1 μg l^{-1}). Our model showed that the frequency distribution of benzene concentration was highly skewed to the right, where most of the benzene predictions were nearly the minimum value, 0 μg l^{-1} (Fig. 7), and the probability of contaminant concentrations lower than 0.1 μg l^{-1} was 85%. This prediction is consistent with the field data. The risk of the benzene concentration exceeding the UK water drinking standard of 1 μg l^{-1} (Environment Agency, 2000 pers. comm.) was only 2%. Therefore, it is predicted that benzene does not pose an unacceptable risk to water quality in this example.

The risk analysis method was also used to study the impacts of toluene, ethylbenzene, xylene and some chlorinated solvents such as 1,1,1-trichloroethane (TCA), trichloroethylene (TCE), and tetrachloroethylene (PCE). Table 4 shows a comparison of the model prediction with the UK drinking water standards and the field data collected by Barrett *et al.* (1996). The mean predicted values for all BTEX compounds are lower than the drinking water limit, while the mean predicted values for PCE and TCE are higher and it is considered that these contaminants pose the main risk to water quality in this borehole.

A second site at an industrial borehole was used to check that the model is also applicable at other locations. This second borehole is sited in the grounds of a bleaching and dyeing works in the area to the north of Nottingham city centre (SK 4553 3433). The borehole's catchment contains numerous potentially contaminating industries such as engineering works, chemical works, and garages. Measured concentrations of TCE, TCA and PCE at this industrial site are within the interval of the model prediction (Table 5). This agreement with the field data was similar to that found at the public water supply borehole and increased confidence that the model could be applied in different locations. The only contaminant to show a poor calibration at this borehole was total xylenes; this could be due to an unknown contaminant source in the catchment area, which is always a problem in this type of model.

Discussion

Probabilistic risk tools are valuable for interpreting the complex transient interactions between a city and the underlying groundwater. Predicting risk of contamination to a new or established borehole involves uncertainty at all stages, from estimating the catchment zone through to finding the contaminant sources and predicting the pumped concentrations. The risk model presented in this paper allows the user to predict probability density functions for contaminant concentration at a pumped borehole, instead of a single, unreliable estimate. The volume of data required for this type of analysis is large and maintenance is required to keep the data up to date. However, the recent implementation of Part IIa of the Environmental Protection Act (1990) means these data will be readily available for most UK cities over the next few years.

The risk-based management tool could be extended in the future to produce an economic assessment of the viability of the abstracted water for various uses. The viability of the options for water use would be established from the balance between the value of the water and the cost to achieve the necessary quality. Three sets of economic information would be required: the value of the water for each use, the concentration limits for contaminants relevant to different water uses, and the costs to reduce the contaminant concentration to below the limits. The result is a probabilistic indication of the usage involving the lowest financial risk for the abstracted groundwater.

In its current form the risk-based management tool would enable new groundwater abstractions to be appropriately positioned based on their water quality needs. Existing borehole owners could also use this model to assess the risk of future contamination. The GIS environment is ideal for storing the land-use and contaminated land information that is currently being collated and the pollution risk model can be used to identify risk from the contamination. Other models could be added to this framework for additional objectives, such as identifying contaminant risk to groundwater or rivers.

Conclusions

The combination of borehole catchment analysis, detailed land-use and contamination information and a pollution risk model provides a powerful tool for addressing issues regarding the best use of urban groundwater given the inherent uncertainties of natural systems. The GIS-based decision making tool may be applied in order to protect existing boreholes and to determine the lowest risk locations for new wells.

The management tool has been used to predict concentrations of BTEX and chlorinated solvents at two pumped boreholes; the simulated results are consistent with the available field data. In this case study the two example borehole locations were unlikely to be adversely impacted by BTEX contamination but chlorinated solvents pose a real risk.

The research presented in this paper was sponsored by an EPSRC grant. The authors would like to thank Don Morley of NIAS for collecting the land-use data for Nottingham, Charles Vulliamy for digitizing the data, Nigel Tait for setting up the graphic user interface, and Catalist for providing the petrol station information.

References

BARRETT, M. H., LERNER, D. N., FRENCH, M. J. & TELLAM, J. H. 1996. A comparison of the impacts of the UK cities of Birmingham and Nottingham on underlying groundwater quality. *Proceedings of the American Institute of Hydrology Annual Meeting. Hydrology and hydrogeology of urban and urbanising areas*, 1–17.

BEVAN, K. & BINLEY, A. 1991. The future of distributed models: model calibration and uncertainty prediction. *Hydrological Processes*, **6**, 279–298.

BJERG, P. L., BRUN, A., NIELSEN, P. H. & CHRISTENSEN, T. H. 1994. Modelling the fate of organic compounds in in situ microcosm experiments. *In*: DRACOS, T. H. & STAUFFER, F. (eds) *Transport and reactive processes in aquifers*. Balkema, Rotterdam, 131–136.

BURSTON, M. W., NAZARI, M. M., BISHOP, P. K. & LERNER, D. N. 1993. Pollution of groundwater in the Coventry region (UK) by chlorinated hydrocarbon solvents. *Journal of Hydrology*, **149**, 137–161.

CHAPELLE, F. H., BRADLEY, P. M., LOVLEY, D. R. & VROBLESKY, D. A. 1996. Measuring rates of biodegradation in a contaminated aquifer using field and laboratory methods. *Ground Water*, **34**, 180–189.

DAVISON, R. M. 1998. *Natural attenuation and risk assessment of groundwater contaminated with ammonium and phenolics*. PhD thesis, Department of Civil and Environmental Engineering, University of Bradford.

DECISIONEERING. 1997. *Crystal Ball Manual: Forecasting and risk analysis for spreadsheet users*. Decisioneering Inc., Aurora, Colorado.

FETTER, C. W. 1994. *Applied hydrogeology*. Prentice Hall, Englewood Cliffs, New Jersey.

FORD, M. & TELLAM, J. H. 1994. Source, type and extent of inorganic contamination within the Birmingham urban aquifer system. *Journal of Hydrology*, **156**, 101–135.

GRESWELL, R. B., LLOYD, J. W., LERNER, D. N. & KNIPE, C. V. 1994. Rising groundwater in the Birmingham area. *In*: WILKINSON, W. B. (ed.) *Groundwater Problems in Urban Areas. Proceedings of Conference of the Institution of Civil Engineers*, London, 1993, 64–75, discussion 355–368.

JONES, I. 1997. *Field site report: Working paper 2 to 'Processes underlying remediation of creosote contaminated groundwater in fractured sandstone'*. EU contract number **EV5V-CT94-0529**.

LERNER, D. N. 1986. Leaking pipes recharge groundwater. *Ground Water*, **24 (5)**, 654–662.

LERNER, D. N. & BARRETT, M. H. 1996. Urban groundwater issues in the United Kingdom. *Hydrogeology Journal*, **4**, 80–89.

LERNER, D. N. & TELLAM, J. H. 1997. The protection of urban groundwater from pollution. *Journal of the Institution of Water and Environmental Management*, **6** , 28–37.

MCNAB, W. W. & DOOHER, B. P. 1998. Uncertainty analyses of fuel hydrocarbon biodegradation signatures in groundwater by probabilistic modeling. *Ground Water*, **36**, 691–698.

MACE, R. E., FISHER, R. S., WELCH, D. M. & PARRA, S. P. 1997. *Extent, mass, and duration of hydrocarbon plumes from leaking petroleum storage tank sites in Texas*. Geological Circular 97-1, The University of Texas at Austin, Austin, Texas.

NICOLSON, J. 1993. *An introduction to drinking water quality*. Institution of Water and Environmental Management, London, 38-50.

REID, J. B., REISINGER, H. J., BARTHOLOMAE, P. G., GRAY, J. C. & HULLMAN, A. S. 1999. A comparative assessment of the long-term behavior of MTBE and benzene plumes in Florida, USA. *In:* ALLEMAN, B. C. & LEESON, A. (eds) *Natural Attenuation of Chlorinated Solvents, Petroleum Hydrocarbons, and other Organic Compounds. The Fifth International* in situ *and on-site Bioremediation Symposium*, San Diego, California, 97–102.

RIVETT, M. O., LERNER, D. N., LLOYD, J. W. & CLARK, L. 1990. Organic contamination of the Birmingham aquifer. *Journal of Hydrology*, **113**, 307-323.

SEVERN TRENT DATA. 1995. *Licence details: Nottingham district*. Severn Trent Water Plc., UK.

THE MERCK INDEX. 1990. (12th ed). Merck & Co., Inc., USA, 1101.

THORNTON, S. F., LERNER, D. N. & TELLAM, J. H. 1996. *Laboratory studies of landfill leachate Triassic Sandstone interactions*. Department of the Environment, Report number **CWM035A194**.

TIM, U. S., JAIN, D. & LIAO, H. 1996. Interactive modeling of groundwater vulnerability within a geographic information system environment. Ground Water, **34**, 618–627.

YANG, Y., LERNER, D. N., BARRETT, M. H. & TELLAM, J. H. 1999. Quantification of groundwater recharge in the city of Nottingham, UK. *Environmental Geology*, **38**, 183–197.

Groundwater pollution at a pulp and paper mill at Sjasstroj near Lake Ladoga, Russia

D. SCHOENHEINZ[1], T. GRISCHEK[1], E. WORCH[1], V. BEREZNOY[2], I. GUTKIN[2], A. SHEBESTA[3], K. HISCOCK[4], W. MACHELEIDT[5], & W. NESTLER[5]

[1]*Institute for Water Chemistry, Dresden University of Technology, 01062 Dresden, Germany (e-mail: hydroche@rcs.urz.tu-dresden.de)*

[2]*North-West Polytechnical Institute, Millionaja ul. 5, 199178 St. Petersburg, Russia*

[3]*State University St. Petersburg, 10 Linaja ul. 33/35, 190008 St. Petersburg, Russia*

[4]*School of Environmental Sciences, University of East Anglia, Norwich NR4 7TJ, UK*

[5]*University of Applied Science, Friedrich-List-Platz 1, 01069 Dresden, Germany*

Abstract: Lake Ladoga, situated in north-west Russia, is the main source of drinking water for more than 6 million people in the region of St Petersburg. During recent years eutrophication of Lake Ladoga has increased, with pollution of the lake water and sediments by heavy metals, phenols and chloro-organic substances. This paper reports the influence of effluents from a pulp and paper mill at Sjasstroj situated on the southern shore of Lake Ladoga. At Sjasstroj, contamination of a wide area of land, and the production of large volumes of wastewater requiring treatment and disposal, threaten the sustainability of surface and domestic well supplies and lake ecosystems. A tiered approach to risk assessment was adopted in this investigation beginning with a survey of historical data and an initial field reconnaissance followed by more detailed, targeted hydrogeological and hydrochemical investigations. Concentrations of dissolved organic carbon (DOC) as high as $44\,mg\,l^{-1}$ and adsorbable organic halogens as high as $130\,\mu g\,l^{-1}$ were found in groundwater samples. The detailed field reconnaissance survey contradicted toxicity index values for the wastewater, sludge and groundwater samples calculated using a Russian method of risk characterization, which would otherwise predict a high potential environmental pollution risk. In reality, there is good evidence for natural attenuation of organic compounds in groundwater, with as much as 40–50% removal of DOC. However, the residual DOC can be problematic when chlorination of raw surface and groundwaters creates organo-chlorine by-products in treated water.

The catchment areas of Lakes Ladoga and Onega (see Fig. 1) are characterized by high industrialization (economic output based on oil, wood, pulp, paper and chemical industries) and intensive land use. The resulting air, water and soil pollution is associated with increasing biomass production in the lakes, a decrease of biodiversity and the occurrence of pathological changes of water organisms (Drabkova *et al.* 1996). From the 1960s to the 1980s, the ecosystem of Lake Ladoga changed from oligotrophic to eutrophic conditions. The present self-purification capacity of the lake is evaluated to be very poor (Frumin *et al.* 1996). According to Vorobieva *et al.* (1996), the use of lake water for potable water is a frequent reason for diseases of the digestive system and of stomach cancer.

In this study, a pulp and paper mill (PPM) in north-west Russia was characterized with respect to its environmental impact and consequent potential risk to the population, principally caused by surface water and groundwater pollution. The investigation site at the Sjasstroj PPM is on Europe's largest lake, Lake Ladoga. The River Neva is the only outlet from

From: HISCOCK, K. M., RIVETT, M. O. & DAVISON, R. M. (eds) *Sustainable Groundwater Development*. Geological Society, London, Special Publications, **193**, 277–291. 0305-8719/02/ $15.00 The Geological Society of London 2002.

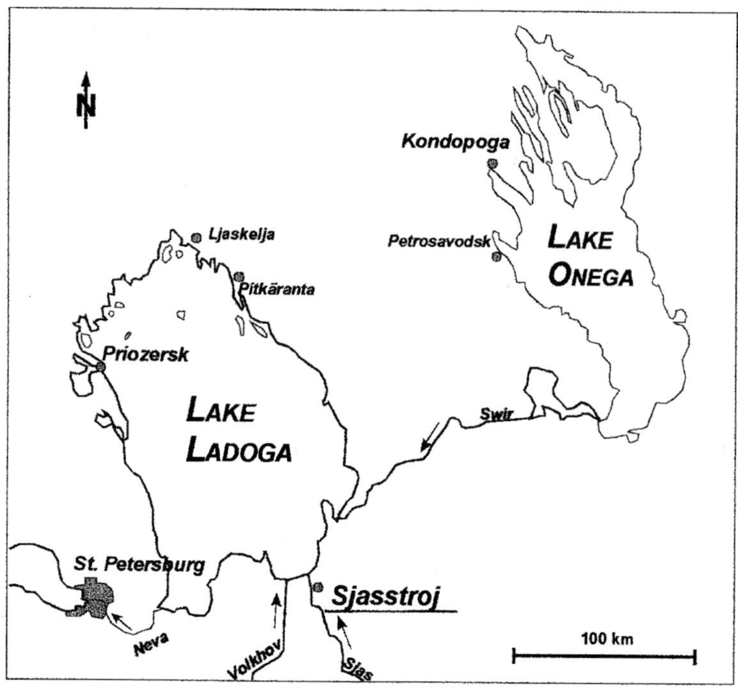

Fig. 1. Location of the pulp and paper mill investigation site at Sjasstroj (underlined).

Lake Ladoga and provides the entire raw drinking water supply for St Petersburg. Hence, the water quality of Lake Ladoga is of importance to the 6 million inhabitants of St Petersburg.

The aim of this study was a risk assessment of groundwater pollution at the Sjasstroj PPM site in north-west Russia as part of an evaluation of the future sustainability of local and regional surface water and groundwater supplies. The objectives were to carry out a tiered risk assessment (Fig. 2). The first level of risk assessment focused on a reconnaissance of historical data for source-pathway-target identification and initial hydrogeological and hydrochemical investigations. The second level of assessment focused on a detailed field reconnaissance, including continuous sampling and analysis of groundwater, surface water, wastewater, soil, sediment and bottom deposits, and risk characterization.

Currently, different approaches to risk characterization are being compared to provide an overall risk assessment methodology. Biksey & Bernhardt (2000) distinguished between three principal methods for characterizing risk: (1) Hazard Quotient (HQ); (2) Line-of-Evidence/Weight-of-Evidence; and (3) Probabilistic. The Russian method of a Hazard Quotient type has been applied to the PPM site at Sjasstroj and is evaluated in this paper.

Scope of investigations

Field methods

Samples of mainly wastewater and groundwater, but also surface water, drinking water and raw water were collected. To allow load calculations of possible impacts on groundwater and surface waters, wastewater samples and sludge water were taken more frequently than groundwater samples.

For the groundwater investigations, eight permanent monitoring wells with screened intervals of 1 m were drilled and developed. Spottests from domestic wells were also taken. Surface water samples from the Rivers Sjas and Valgomka were collected as well as from Lake Ladoga close to the wastewater outflow from the PPM.

According to the planned determinations,

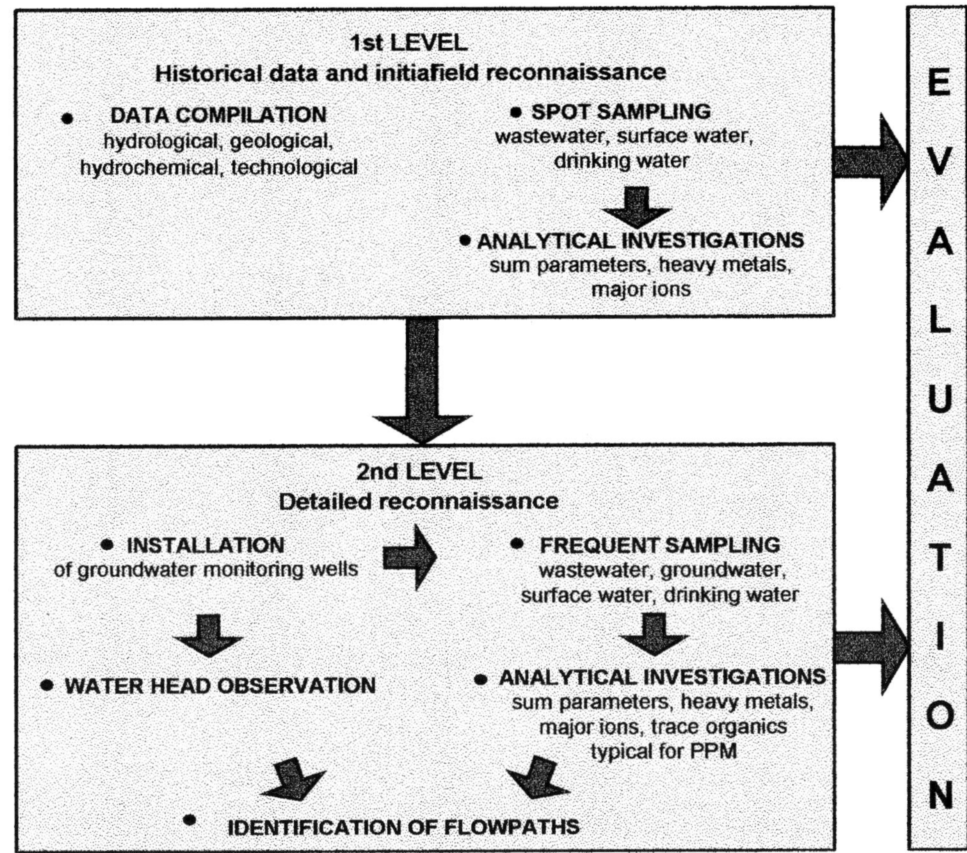

Fig. 2. Scheme of the risk assessment approach.

samples for analysis of dissolved organic carbon (DOC), adsorbable organic halogens (AOX) and phenol were preserved by adding concentrated HNO_3 to pH < 2. Samples for analysis of resin acids were preserved by adding concentrated NaOH to pH > 9. In general, analyses were carried out according to German guidelines and DIN methods. Prior to analysis, water samples were filtered through a 0.45 μm cellulose-nitrate filter.

Analytical methods

Sum parameters. For DOC concentrations, samples were analysed with a Carbon Analyzer (model TOC-5000, Shimadzu). The spectral absorption coefficient at 254 nm (SAC-254) was measured with a UV/VIS-spectrometer (model Spekol, Zeiss Jena). AOX concentrations were determined using an AOX-Analyzer (model Coulomat 702 CL, Stroehlein). For AOX analysis, 100 ml of the water sample was filtered and shaken with 50 mg activated carbon for 1 h. The carbon was washed to eliminate inorganic halogens, burnt in an oven at 950°C, and the gaseous halogens determined by a micro-coulometric method after absorption in water.

Organic pollutants. To identify organic pollutants a gas chromatograph (HP 5890, Hewlett Packard) coupled to a mass spectrometer (HP 5970, Hewlett Packard) was used. The organic compounds were separated with an HP-5-MS column. The water samples to be analysed were filtered through a 0.5 m glass fibre filter. A solid phase extraction with four different reversed phase (RP) materials as stationary phase and ethylacetate as eluent were applied. The combined eluates were reduced in volume to 1 ml under a stream of nitrogen gas, giving an enrichment factor of 2000. A spectra library was used to identify the main peaks.

Ions. The anions NO_3^-, SO_4^{2-} and Cl^- were determined using an ion chromatograph (model DX 100, Dionex). The separation column was AS12A. For the elimination of the organic matrix an OnGuard-RP cartridge was used. A titrimetric method with EDTA as titrant was used for the determination of the cations calcium and magnesium (DIN 38406 T3). Potassium, sodium and heavy metals were analysed with an atomic absorption spectrometer (AAS 5 FL and AAS 5 EA, Zeiss Jena). Depending on the sample concentrations, either the flame technique or the graphite furnace technique was applied. Ammonium was determined using a photometric method (Merck test kit).

Sulphonic acids. The substances naphthalene-2,6-disulphonic acid, 1-hydroxy-3,6-disulphonic acid, naphthalene-2,7-disulphonic acid, 2-naphthalene sulphonic acid, and naphthalene-1,3,6-trisulphonic acid were found. For these, 500 ml of water samples were adjusted to pH 6.5 and 20 µg of anthrachinon-2-sulphonate and 0.25 mmol ion-pair agent (tetrabutylammonium bromide TBA-Br) were added. Prior to enrichment, solid phase cartridges (Merck LiChrolut EN) were conditioned with 10 ml TBA-Br solution (0.1 mmol). The sample was aspirated through the solid phase, which was then dried under a stream of nitrogen gas. Sample material was eluted from the solid phase with 3 ml of acetonitrile/dichloromethane (50 : 50) and concentrated to dryness under vacuum in a rotavapor. The sample was then dissolved in 1 ml HPLC-grade water and analysed using an HPLC with UV-detector.

Resin acids. Pimaric acid, sandarapimaric acid, isopimaric acid, dehydroabietic acid, 8(14)-abietic acid, abietic acid, neoabietic acid and 7-oxo-dehydroabietic acid were analysed. For these, 500 ml of the water sample was adjusted to pH 8–9 by adding 0.1M sodium hydroxide and 50 ml (10%) of trace analytic grade methanol (Merck Suprasolv). Prior to enrichment, solid phase cartridges (Merck LiChrolut EN, 200 mg) were conditioned with 5 ml of methanol (Suprasolv) and 4 ml of distilled water adjusted to pH 8–9. The sample was aspirated through the solid phase, which was then dried under a stream of nitrogen gas. If not eluted immediately, cartridges were stored in a freezer at $<0°C$. Sample material was eluted from the solid phase with 8 ml of analytical grade methanol and concentrated to dryness under vacuum in a rotavapor. Esterization of resin acids was accomplished through addition of 200 µl pentafluorobenzylbromide (PFBB) and 10 µl triethylamine; the solution was then kept at 90°C for 1 h. The sample was concentrated to dryness under a gentle stream of nitrogen gas and, following dissolution in 1 ml toluene, analysed with a Hewlett-Packard Gas Chromatograph, HP GC 5890 series II with mass-selective detector 5971. GC parameters were as follows: column (DB-17) length, 30 m; inner diameter, 0.2 mm; film thickness, 250 m; flow rate, 0.995 ml min^{-1}; injector temperature, 250°C; and oven programme from 70 to 210°C with 30°C min^{-1} ramp, followed by 210 to 260°C with a ramp of 2°C min^{-1}. The mass-selective detector was used with a detector temperature of 300°C and a solvent delay of 20 min.

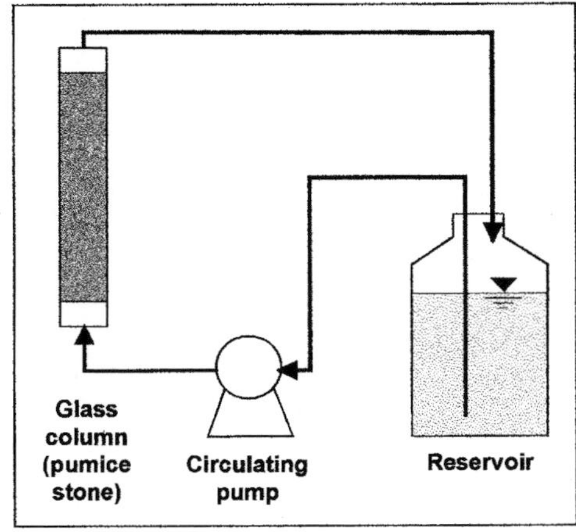

Fig. 3. Test filter experiment set-up.

Laboratory studies

Biological degradation of organic compounds can be studied in the laboratory using a biologically active test filter. The test filter concept was developed by Sontheimer *et al.* (1988), mainly to investigate the degradation of dissolved organic matter (DOC) and single organic compounds. Figure 3 shows a typical test filter set-up. The glass filter column (diameter, 42 mm; height, 90 mm) is filled with an inert solid material. To exclude adsorption processes the solid material should have no or very low adsorption capacity. In our investigations we used the commercial product Hydrofilt (Akdolit GmbH), which is a pumice stone material. The effective porosity of the column

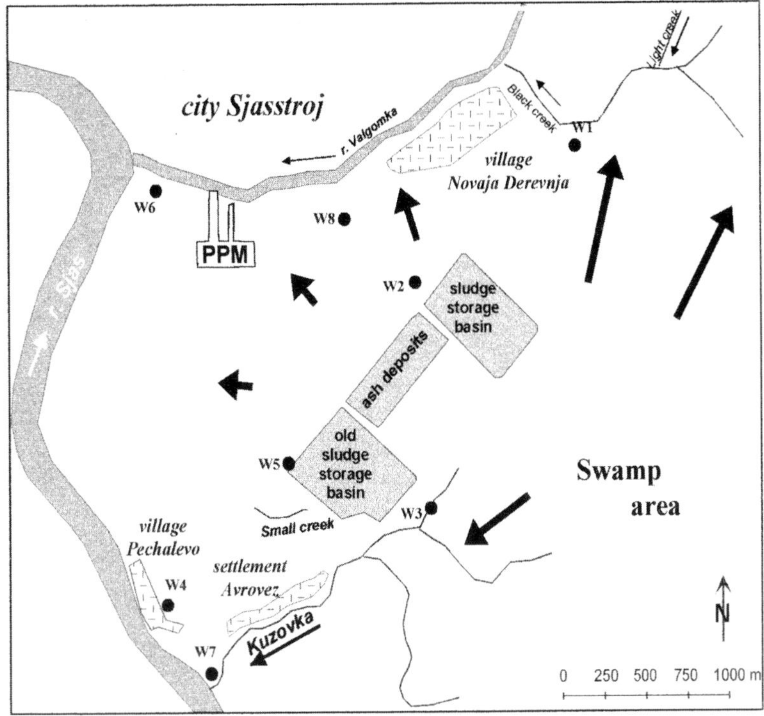

Fig. 4. Direction of groundwater flow (arrows) and installed monitoring wells (W1 to W8)

Fig. 4. Direction of groundwater flow (arrows) and installed monitoring wells (W1 to W8).

pack was determined by tracer experiments to be 0.82.

The experiments were executed in darkness at a temperature of 20°C. Stable aerobic conditions were proven by measurements of dissolved oxygen in the water from the outlet of the column. The pumping rate was 7 ml min^{-1}. After packing the columns, unfiltered water from the River Elbe (Germany) was circulated for a minimum of 1 month, and replaced by fresh river water every week to set up a biofilm. During the following experiments, the water was replaced every 30 to 50 days by a mixture of PPM wastewater from Sjasstroj and filtered Elbe river water (membrane filter of 0.45 μm) in the ratio of 1 : 10.

The first set of test filter runs was to investigate the biodegradability of DOC. The initial DOC concentration of the PPM wastewater was about 320 mg l^{-1}, and the DOC of the Elbe river water about 5 mg l^{-1}. The experiments were carried out with the mixed waters to avoid toxic effects of the PPM wastewater and to simulate natural conditions (discharge of wastewater to Lake Ladoga). Furthermore, the wastewater contained only low concentrations of the nutrients nitrogen and phosphorus. By mixing river water with the wastewater, the provision of sufficient nutrients was assumed.

The decrease in the volume of circulating water due to sampling was taken into account by calculating the relative contact times in the filter. For each test filter, experiments were conducted on a control sample which was unpumped and maintained under the same experimental conditions in an open bottle. In total, four test filter experiments were carried out.

Results of historical data collection and initial field reconnaissance

Site characterization

Sjasstroj is about 120 km NE of St Petersburg. The investigation site is characterized by a high abundance of surface waters: the River Valgomka in the north, flowing into the River Sjas in the

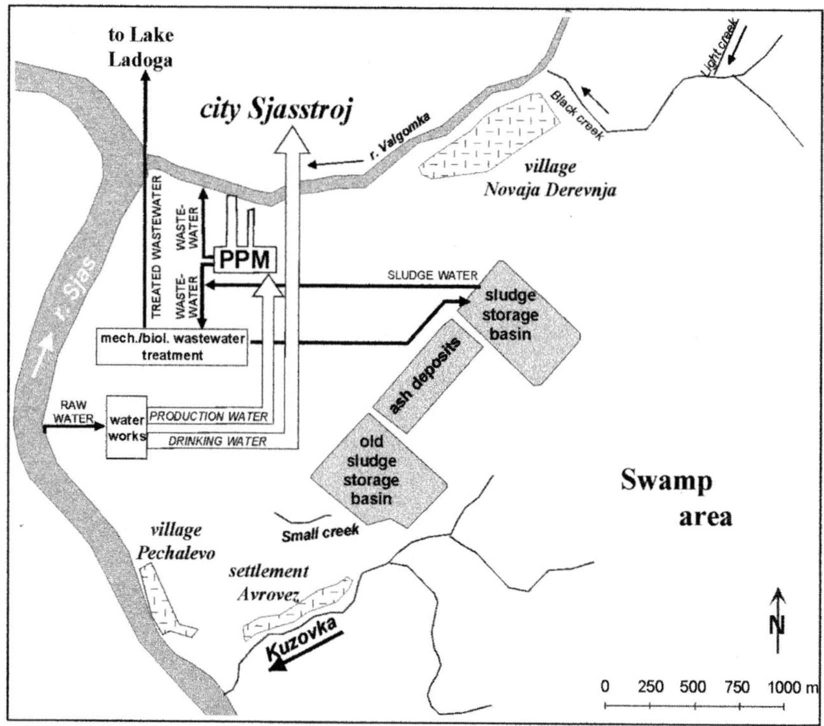

Fig. 5. Wastewater distribution PPM Sjasstroj

Fig. 5. Wastewater distribution at the Sjasstroj PPM.

west; some smaller creeks in the north-east; the Kuzovka creek in the south; and a large swamp in the east (Fig. 4). Furthermore, some villages (Novaja Derevnja to the north-east, Pechalevo to the south-west, Avrovez to the south) are located in the area.

Geologically, the investigation area is part of the north-east European platform. Limnic and limnic-glacial sediments lie on Wendian and Cambrian sandstones and limestones. These sediments are washed out along the river beds which are then filled with alluvial materials. There are two principal aquifers both consisting of Quaternary sandy to fine-sandy material. The upper aquifer with a thickness of about 5 m exists only locally underneath the sludge storage area and extends to the north-east of the investigation site. This aquifer is directly exposed to any potential pollutant input and is thus of main concern for a risk assessment evaluation. The lower aquifer is confined and is clearly separated from the upper aquifer by moraine silts and clays with a thickness of about 10 m. The groundwater flow in the upper aquifer is directed towards the River Valgomka in the north. The swamps to the south-east form the main groundwater recharge area (Fig. 4). Furthermore, given the main outcrop of the aquifer in the north-east of the investigation site, the aquifer also drains into Black creek.

Improved understanding of the hydrogeology of the site has been gained from the installation of eight groundwater monitoring wells, of which six are present in the upper aquifer and two in the lower aquifers.

The Sjasstroj PPM has been producing pulp since 1928 using sulphite technology. Bleaching with chlorine and hypochlorite started in 1932. Between 1965 and 1989, an average of 120 000 tonnes of pulp per year were produced. During the 1990s, the production capacity decreased continuously to a maximum of 15%, and occasionally 0%. In May 2000, the plant increased its production to about 50% and intends to return to its former capacities. The production water for the PPM of about 120 000 m d^{-1} at present (maximum capacity 270 000 m d^{-1}) is abstracted from the River Sjas.

Table 1. *Definition of source–pathway–target relationships*

#	Source	Flowpath	Target	Potential hazard for
1	Wastewater discharge into L. Ladoga	Mixing by flow, wind, and wave actions	Surface water of L. Ladoga	Lake ecosystem, drinking water abstraction from L. Ladoga
2	Wastewater discharge into R. Valgomka	Mixing by flow action	Surface water of R. Sjas	River ecosystem, drinking water abstraction from R. Sjas
3	Sludge discharge into active storage basin	Upper aquifer	Groundwater, draining creeks	Drinking water abstraction from domestic wells
4	Settled sludge in the old storage basin	Upper aquifer by rainwater infiltration	Groundwater, draining creeks	Drinking water abstraction from domestic wells
5	Deposits of solid PPM waste	Upper aquifer by rainwater infiltration	Groundwater, draining creeks	Drinking water abstraction from domestic wells

Source-pathway-target identification

Three pollution sources by direct wastewater discharges were distinguished at the site (see Fig. 5): (1) untreated wastewater from production processes such as wood preparation and cooling, discharged through pipes into the River Valgomka; (2) mechanically and biologically treated wastewater, discharged through a channel into Lake Ladoga; and (3) excess sludge water, discharged into sludge storage basins to separate sludge by settling. The separated sludge water flows back to the treatment plant. There are two sludge basins: an old, inoperative sludge basin in the south of the site, containing a large volume of settled sludge, and an operational basin. Supposedly the sludge storage basins have an impermeable lining.

Indirect potential pollution sources are solid waste deposits found over the PPM site, for example, ash, pyrite, and sulphur deposits that might be washed out by rainwater.

The drinking water supply for the town of Sjasstroj is abstracted from the River Sjas, the source of production water for the PPM. The waterworks treats the drinking water in various ways including chlorination. However, the population of Novaja Derevnja and Avrovez obtains its drinking water from domestic groundwater wells, while residents of Pechalevo take surface water direct from the River Sjas.

During the initial field investigation in September 1998, water samples were taken from both the wastewater and the excess sludge discharge of the treatment plant. The main parameters measured for the first set of samples included pH, conductivity, dissolved organic carbon, adsorbable organic halogens, ion-pair extractable organic sulphur compounds (IOS), sulphate, nitrate, chloride, ammonium, calcium, potassium, sodium, magnesium, total iron, manganese, cadmium, copper, lead and mercury. From the results obtained, it was apparent that due to the low PPM production capacity at that time; the samples analysed showed only slightly increased concentrations of AOX and DOC. As a result, it was decided that future sampling should focus on the determination of organic sum parameters and trace substances typical for PPM wastewaters such as resin acids, sulphonic acids and phenols. Assessment of the initial results led to the definition of source-pathway-target relations given in Table 1 and shown schematically in Figure 6.

Results of detailed field reconnaissance

For the second level of investigation, fieldwork focused on sampling groundwater, main wastewater discharges and sludge water outlets. Whereas treated wastewater is discharged directly into Lake Ladoga, sludge water storage in basins and disposal of solid materials are the main risk for groundwater pollution at the PPM site.

Dissolved organic carbon

The temporal DOC concentrations of biologically treated wastewater and excess sludge water are shown in Figure 7. The mean DOC concentration in wastewater was 264 mg l^{-1}. The low DOC concentration of 40 mg l^{-1} in September 1998 is not representative of PPM wastewater and occurred at a time of plant closure. The mean DOC concentration of excess sludge during the observation period was 245 mg l^{-1}, neglecting values from April 1999 and December 1999, which were probably diluted by rainwater or discharge of water from the public supply.

With respect to the high DOC concentrations of the treated wastewater discharged into Lake Ladoga, four test filter experiments were carried out to determine the biodegradability of the DOC using PPM wastewater mixed with Elbe

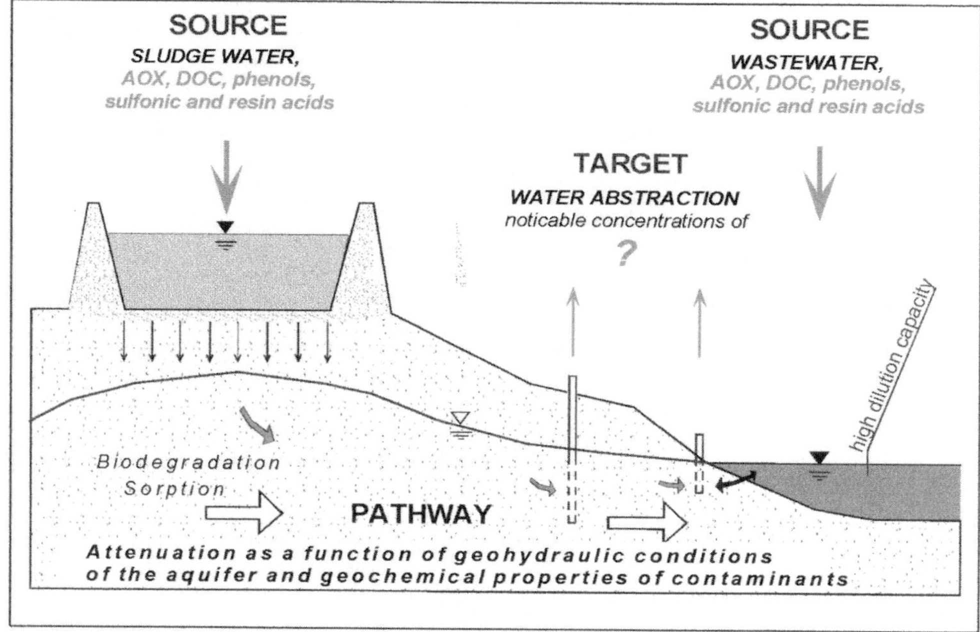

Fig. 6 Schematic diagram showing the problem formulation

Fig. 6. Schematic diagram showing the problem formulation.

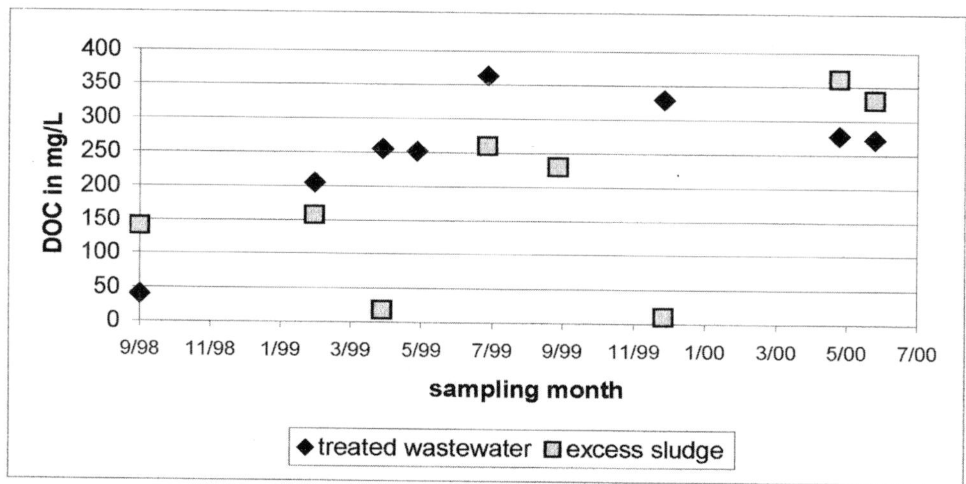

Fig. 7. DOC concentrations in wastewater and sludge water at PPM Sjasstroj

Fig. 7. DOC concentrations in wastewater and sludge water at the Sjasstroj PPM.

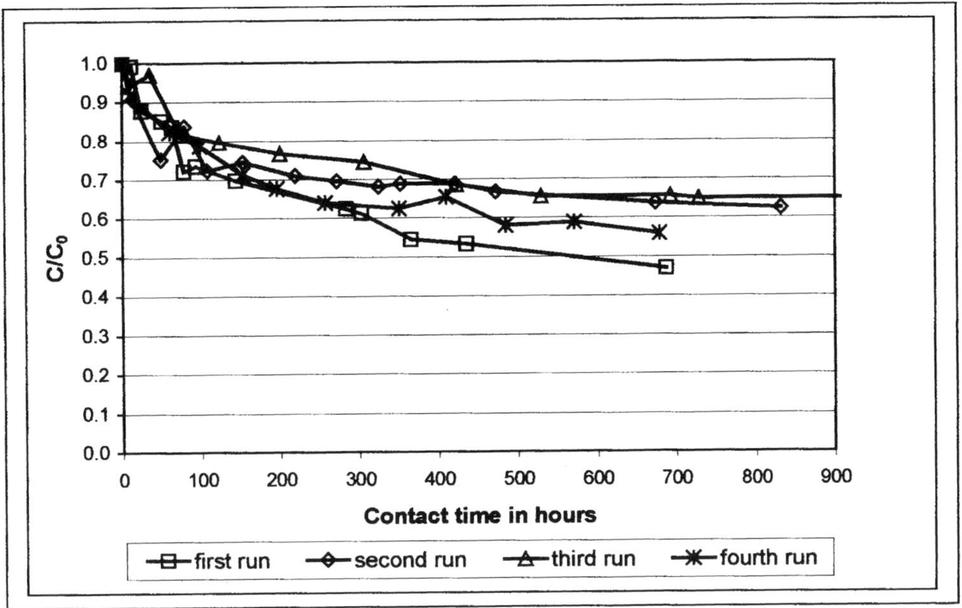

Fig. 8. DOC-biodegradation of test filter B.

river water. To assure reliability of the results, each experiment consisted of two test filter set-ups A and B, that were operated under the same conditions. Figure 8 displays the results for test filter B and shows that DOC concentrations decreased by up to 30% within the first 5 d, followed by a further slow decrease to a final concentration of 50 to 60% of the starting concentration.

After a contact time of approximately 500 h, the DOC concentration reached a nearly constant level. Thus, about half of the DOC in treated PPM wastewater is still biodegradable under aerobic conditions. Thus, direct wastewater discharge to Lake Ladoga results in oxygen consumption in the lake which can affect the local ecosystem. The poorly biodegradable portion of the DOC results in a high input of residual organic carbon to the lake. It is assumed that this portion is partly accumulating in the lake sediments, thereby having a limited impact on the oxygen content of the lake water. Potentially, an increase in DOC of the lake water will result in the formation of disinfection by-products during water treatment with chlorine, if lake water is used as the raw water resource. Consequently, a risk from organics discharged to Lake Ladoga could be expected if raw water is abstracted from wells at the banks of Lake Ladoga that are in hydraulic contact with the lake water.

The DOC concentrations of groundwater vary between 2 and 60 mg l^{-1} (Fig. 9). High concentrations of 28 mg l^{-1} and 44 mg l^{-1} occur in the immediate western vicinity of the new and old sludge storage basins respectively. As a result, groundwater downgradient of the basins is influenced by infiltrating sludge water with a noticeably higher DOC concentration. The higher DOC concentration of 60 mg l^{-1} to the north-east of the basins compared with that close to the basins is associated with a municipal landfill between the sludge basins and the measurement point. The lower DOC concentrations to the north of the basins, measured in domestic wells, might be the consequence of attenuation processes occurring along the groundwater flowpath. The only low DOC concentrations of 9 mg l^{-1} and 2 mg l^{-1} in groundwater to the south and to the south-east of the old sludge basin show no influences of infiltrated sludge water. The infiltrating sludge water does not cause changes in groundwater levels underneath the basins and, consequently, there is no change in the general groundwater flow direction in the area surrounding the basins. The DOC concentration of 2 mg l^{-1} in the lower aquifer also indicates no impact of

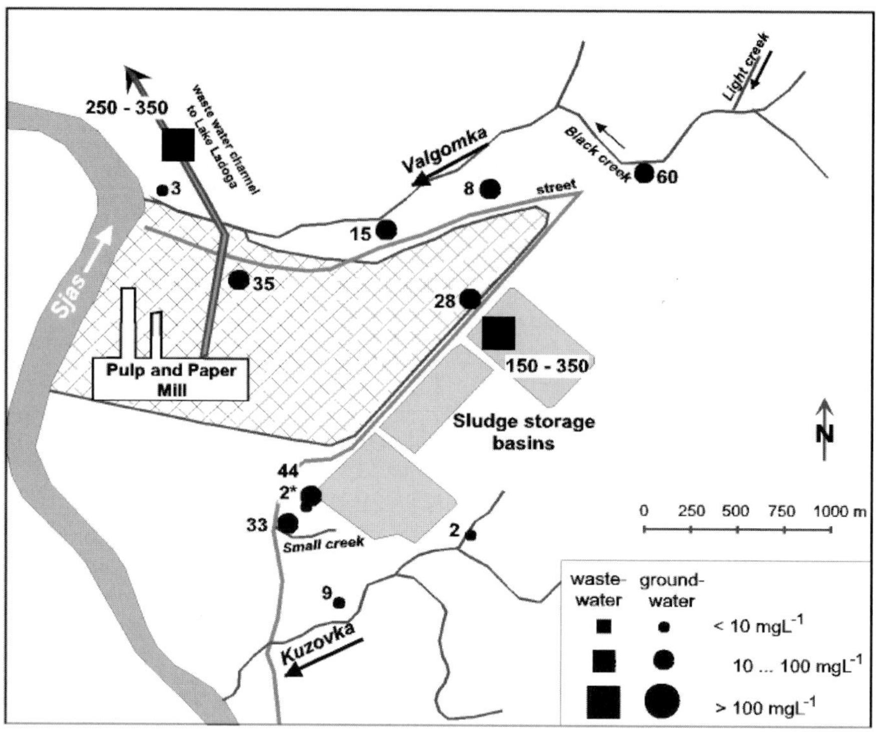

Fig. 9. DOC concentrations in mg l^{-1} (* characterizes the lower aquifer).

sludge water. This supports the assumption that there is no hydraulic connection between the upper and the lower aquifers.

For the drinking water supply to Sjasstroj, raw water is taken from the River Sjas. The raw water has a high DOC concentration of more than 10 mg l^{-1} and a high potential for the formation of disinfection by-products. Drinking water DOC was found at concentrations between 4 and 6 mg l^{-1}. Compared with the raw water, the drinking water treatment showed about 50% DOC removal. The residual DOC in the Sjas drinking water is assumed to be persistent natural organic compounds. Most surface water samples in the area contain high concentrations of organic carbon, which is typical for peat bogs. Consequently, there is no indication as yet that the PPM effluents affect the DOC of the raw water.

SAC-245nm

The UV absorbance at a wavelength of 254 nm for wastewater and groundwater downstream of the basins (W1, W5a) was between 80 and 400 m^{-1}. The groundwater in the south of the basins (W3) and the lower aquifer (W5) had very low absorbance (between 0.9 and 4.1 m^{-1}), thus we assume it is unaffected by infiltration processes from the basins.

Specific UV absorbance coefficient

Specific UV absorbance (SUVA) was calculated from the ratio of SAC-254 nm and DOC. The SUVA of wastewater after biological treatment was low (0.2–1.0 l mg^{-1} m^{-1}) indicating low aromaticity and a high content of sewage fulvic acids. Groundwater and surface water samples had higher SUVA values (2–3 l mg^{-1} m^{-1}) indicating aquagenic fulvic acids (Huber 2000).

Adsorbable organic halogens

The concentrations of AOX in wastewater and sludge water fluctuated within the range 78–2880 μg l^{-1}. Since May 2000, high values of 2420 and 2880 μg l^{-1} have been measured in wastewater and coincide with the restarting of the pulp and paper production, including the

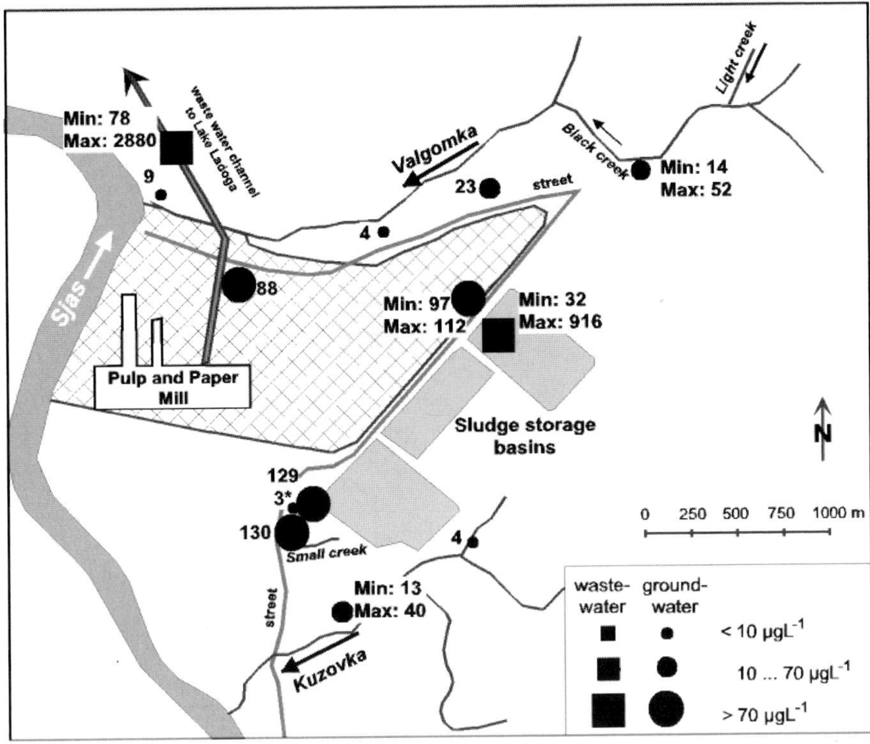

Fig. 10. AOX concentrations in mg l^{-1} (* characterizes the lower aquifer).

bleaching processes, at the Sjasstroj PPM. Concentrations of about 900 μg l^{-1} found in the sludge water coincide with the wastewater measurements. As with the DOC measurements, low AOX concentrations of 32 μg l^{-1} (April 1999) and 25 μg l^{-1} (December 1999) in the sludge water samples underline the probable occurrence of dilution processes during these months.

AOX concentrations of groundwater were found to be in the range 4 to 130 μg l^{-1} (Fig. 10). The highest impact is again observed to the north-west of the sludge water basins. Concentrations below 30 μg l^{-1} are not necessarily caused by PPM impacts. AOX concentrations greater than 100 μg l^{-1} at the groundwater monitoring wells to the west of the basins indicate the impact of infiltrating sludge water with a high proportion of poorly biodegraded chlorinated compounds. Other wells appear not to be affected, or biodegradation of AOX has occurred along the groundwater flowpaths.

The AOX concentration (about 30 μg l^{-1}) in the raw water for drinking water production for Sjas is low and does not indicate an impact from PPM effluents. However, due to chlorine disinfection of the raw water with a high DOC content, disinfection by-products were formed resulting in high AOX concentrations of up to 619 μg l^{-1} in drinking water. Further investigation is necessary to identify single chlorinated organic compounds in the drinking water.

Results of GC–MS screenings

In treated wastewater from the Sjasstroj PPM, phenols and chlorophenols, e.g. 4,5-dichloro-2-methoxyphenol, 2,4-dichlorophenol and 2,4,6-trichlorophenol, have been identified.

Sulphonic acids

There is widespread pollution of the aquatic environment in Europe by aromatic sulphonic acids (Hooper 1994; Neitzel *et al.* 1998). Given their anthropogenic origin, aromatic sulphonic acids can be used as indicators for human impacts on water resources. In general, the toxicity of sulphonic acids is quite low. The substances naphthalene-2,6-disulphonic acid,

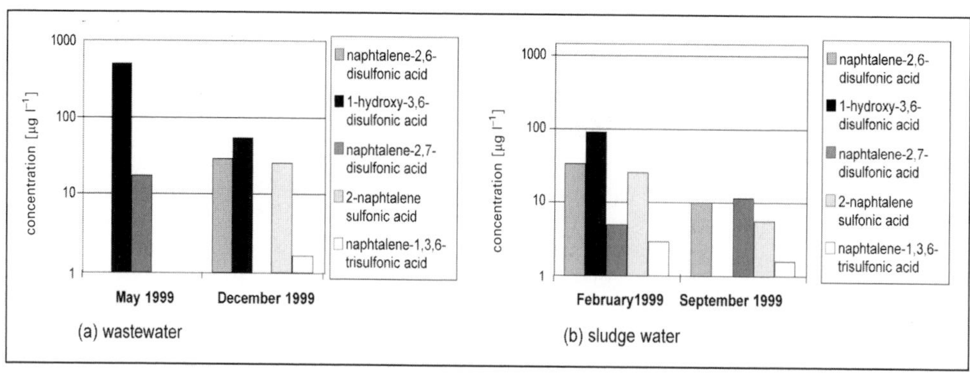

Fig. 11. Sulphonic acid concentrations (µg l^{-1}) in wastewater (**a**) and sludge water (**b**).

1-hydroxy-3,6-disulphonic acid, naphthalene-2,7-disulphonic acid and 2-naphthalene sulphonic acid were found in wastewater, surface water and groundwater samples. The measured concentrations were very variable, with the highest concentrations being found in wastewater (Fig. 11). Of all the sulphonic acids investigated, 1-hydroxy-3,6-disulphonic acid was found to be the most significant, with measured concentrations of between 34 and 481 µg l^{-1} in wastewaters. This acid was also found in groundwater at concentrations between 5 and 7 µg l^{-1} and might therefore be a useful indicator of PPM-related impacts and the determination of groundwater flow behaviour.

Resin acids

Resin acids present a high toxicity potential for aquatic organisms. They accumulate in sediments and bioaccumulate in fish. The minimum effective concentration of dehydroabietic acid to rainbow trout was found to be toxic at 20 µg l^{-1} (Oikari *et al.* 1983).

Dehydroabietic acid was found in all wastewater samples analysed at the Sjasstroj PPM. Maximum concentrations of resin acids found in PPM effluents were: dehydroabietic acid, 59 µg l^{-1}; isopimaric acid, 11.8 µg l^{-1}; and neoabietic acid, 3.9 µg l^{-1}. The highest concentration of dehydroabietic acid was determined in sludge water in September 1999 (Table 2). In the well near Avrovez, 7-oxo-dehydroabietic acid was also found.

Batch experiments were carried out to investigate the stability of dehydroabietic acid. Samples of ultrapure water (500 ml) and Elbe river water (600 ml) were each spiked with 50 µg dehydroabietic acid and stored at 4°C. Samples were taken after 1, 2, 8 and 17 days. Whereas the resin acid was nearly stable in ultrapure water, it was attenuated in river water by more than 90% within 17 days. These first results indicate that dehydroabietic acid is likely to be easily biodegradable in surface and groundwaters. Given the degradability of dehydroabietic acid, it is unlikely that this resin acid will exceed the minimum effective concentration of 20 µg l^{-1} in groundwater.

Table 2. *Resin acid concentrations (µg l^{-1}), Sjasstroj, September 1999*

Sampling point	Dehydroabietic acid	7-oxo-dehydroabietic acid
Sludge basin	67	21
Well Avrovez	–	19

Inorganic constituents

The maximum concentrations of inorganic constituents measured at different points and times by the Northwest Polytechnical Institute, St Petersburg, are presented in Table 3, together with the Russian admissible concentration limits for comparison. The results presented correspond to the period of low production capacity at the Sjasstroj PPM.

The high values for iron, manganese and heavy metals were mainly measured close to pyrite deposits. With one exception, the nitrate concentration in wastewater, sludge water and groundwater was below 20 mg l^{-} or not detectable. The aquifer around the sludge basins can be characterized by anaerobic conditions. Low chloride concentrations in wastewater, surface

Table 3. *Measured maximum concentrations in mg l^{-1} between February 1999 and April 2000*

Parameter		Russian admissible concentration limits	Maximum concentrations Wastewater	Surface water	Groundwater
Sulphate	SO_4^{2-}	500	191	756	24000*
Nitrate	NO_3^-	45	248	6	342
Chloride	Cl^-	350	112	33.1	795
Ammonium	NH_4^+	2	5.6	42.5	48.0
Aluminium	Al	0.5	100	10.9	103
Manganese	Mn	0.1	–	4.6	3.95
Dissolved iron	Fe	0.3	5.8	1280	8260*
Cadmium	Cd	0.001	0.07	0.011	0.002
Zinc	Zn	5	–	9.6	0.16
Mercury	Hg	0.0005	0.0002	0.0038	–

*beneath pyrite deposit.

water and groundwater samples do not permit use of chloride as a natural tracer for identification of flow paths in the aquifer.

Risk characterization by the Hazard Quotient (HQ) method

For application at contaminated industrial sites, the Russian risk assessment method, known as the Limit Concentration Factor (LCF) method, focuses on risk calculation by the determination of a toxicological index in the form of HQ values. This toxicological index, I_{tox} , indicates a potential risk if it is greater than 1, and is calculated as follows:

$$I_{tox} = \sum_{i=1}^{n} \frac{c_i}{LAC_i} \quad (1)$$

where i represents the constituents, $i = 1 \ldots n$; c_i is the measured concentration of constituent i; and LAC_i is the Russian limit for admissible concentration of constituent i.

In the calculation of I_{tox}, considered constituents are chosen arbitrarily, mainly as a function of laboratory and financial capabilities. The importance of different compounds in terms of their hazard potential is not evaluated. The derived results allow a comparative, quantitative assessment of analytical results for different measurement points, but the method incorporates no qualitative assessment. Also, the assessment is neither source- nor target-related. Table 4 gives an example for three different water samples from the Sjasstroj PPM.

For the calculations shown (Table 4), it is obvious that all three samples, including groundwater from the lower aquifer (#III), which is not considered to be connected to the upper aquifer, are predicted to be at high potential risk given the values of I_{tox} in excess of 1. The variation between the two measurements made for sample #II for November 1999 and February 2000 is mainly due to the high iron concentration recorded for November 1999. The higher concentration of the more critical phenols determined for February 2000 is not reflected in the toxicological index. Furthermore, substances with no toxicological potential, for example the COD value for sample #II, can determine the outcome of the toxicological index calculation. Another problem in this example calculation of I_{tox} is that concentrations of phenols, surfactants and the biochemical oxygen demand were not always available.

Conclusions

The situation of pulp and paper mill factories around Lakes Ladoga and Onega in north-west Russia presents an environmental hazard. The contamination of land over wide areas and the large volumes of wastewater requiring treatment and disposal threaten the sustainability of surface and domestic well supplies and lake ecosystems. In this study, the PPM site at Sjasstroj represented an example of a plant using sulphite cooking technology and bleaching processes with chlorine and hypochlorite to produce pulp and paper.

This study has demonstrated the value of a tiered approach to risk assessment at a complex PPM site, beginning with a survey of historical data and an initial field reconnaissance, followed by more detailed, targeted hydrogeological and hydrochemical investigations. Contamination of shallow groundwater from the impact of sludge storage basins was shown from analyses of DOC, AOX, and sulphonic and resin acids.

Table 4. *Calculation of the toxicological index* I_{tox} *based on results of analyses (mg* l^{-1}*)*

	Constituent, *i*	BOD	COD	SO_4^{2-}	Cl^-	NH_4^+	NO_2^-	NO_3^-	Fe	Al	Phenols	Surfactants	I_{tox}
	LAC_i	3	30	500	350	2	3	45	0.3	0.5	0.001	0.1	1
#	*Sampling time*												
I	09/99		3044	160	15	5.6	0.04	1.0	5	90			301
II	11/99	5.4	61	13	42.7	3.5	0.005	0.45	53	0.2	0.001	0.1	185
	02/00	0.9	44	6.9	83	5	0.002	0.15	2.3	4.6	0.003	0.13	25
III	02/00	0.6	7.7	5.8	4	1.45	0.005	0.0	0.6	4.5	0.002	0	14

I, excess sludge; II, groundwater close to the active sludge basin; III, groundwater lower aquifer; LAC_i, Russian limit for admissible concentration of constituent, *i*. BOD, biochemical oxygen demand; COD, Chemical oxygen demand.

Concentrations for DOC as high as 44 mg l^{-1} and for AOX as high as 130 μg l^{-1} were found in groundwater samples.

The described Russian LCF method as an example of the Hazard Quotient method of risk characterization is attractively simple. Its main disadvantages are that it can only be applied to compounds for which a limit for admissible concentration exists and, secondly, it can result in the masking of some important contaminants that are above the admissible limit. For example, looking at sample III in Table 4, there are only three significant contributors to I_{tox}: aluminium (64%), iron (14%) and phenols (14%). Consequently, it could be assumed that metals rather than organics are harmful, which is not the case generally. This shows that the LCF method should not be used blindly and that individual constituents of interest should, at the same time, be clearly identified. The method does, however, consider the cumulative effect of the presence of multiple contaminants and allows the comparison of environmental impacts at multiple targets across a large industrial complex such as a PPM.

In comparison to the LCF method, the detailed field reconnaissance survey contradicted the calculated toxicity index values for the wastewater, sludge and groundwater samples, which would otherwise predict a high potential environmental pollution risk. In reality, there is good evidence for natural attenuation of organic compounds in groundwater, e.g. the resin acid dehydroabietic acid was found in concentrations of 67 μg l^{-1} in sludge water but was not detected in groundwater samples. Biodegradation has a further cleaning effect on the water with up to 40–50% DOC being removed. However, the residual DOC can be problematic when chlorination of raw surface and groundwaters creates organo-chlorine by-products in treated water.

For the future, a combined framework linking an assessment of the technological hazards at a factory site to approaches adopting source-pathway-target models, as well as probabilistic tools of risk assessment, may provide an overall risk assessment methodology for application in Russia and elsewhere.

This work was supported financially by the EU INCO-Copernicus Programme (grant IC15 CT98-0134). We thank H. Börnick, L. Cornelius, A. Foley, H. Franz, C. Grützner, I. Klemm and B. Wiemer for undertaking analyses and assistance with laboratory experiments and fieldwork.

References

BIKSEY, T. M. & BERNHARDT, A. M. 2000. The right tool for the job. *Water, Environment & Technology*, **12**, 30–35.

DRABKOVA, V. G., RUMYANTSEV, V. A., SERGEEVA, L. V. & SLEPUKHINA, T. D. 1996. Ecological problems of Lake Ladoga: causes and solutions. *Hydrobiologia*, **322**, 1–7.

FRUMIN, G. T., CHERNYKH, O. A. & KRYLENKOVA, N. L. 1996. Lake Ladoga: chemical pollution and biochemical self-purification. *Hydrobiologia*, **322**, 143–147.

HOOPER, S. W. 1994. Biodegradation of sulphonated aromatics. *In*: CHAUDHRY, G. (ed.) *Biological Degradation and Biomediation of Toxic Chemicals.* Chapman & Hall, London, 169–182.

HUBER, S. 2000. *Humic substances characterization.* World Wide Web Address: http://www.doc-labor.de.

NEITZEL, P. L., ABEL, A., GRISCHEK, T., NESTLER, W. & WALTHER, W. 1998. Behaviour of aromatic sulphonic acids during bank infiltration and under laboratory conditions. *Vom Wasser*, **90**, 245–271 (in German).

OIKARI, A., LÖNN, B.-E., CAS-I-RÜN, M., NAKKRI, T., SNICKARS-NIKINMAA, B., BISTER, H. & VIRTANEN, E. 1983. Toxicological effects of dehydroabietic acid (DHAA) on the trout, *Salmo-gairdneri* Richardson, in fresh-water. *Water Research*, **17**, 81–89.

SONTHEIMER, H. 1988. The test filter concept: a method for the characterization of waters. *DVGW-Schriftenreihe Wasser*, **60**, 27–50.

VOROBIEVA, L. V., SELYUZHITSKII, G. V. & CHERNOVA, G. I. 1996. Ecologic and hygienic evaluation of Lake Ladoga as a source of drinking water. *Hydrobiologia*, **322**, 137–141.

Hydrocarbon contamination of groundwater at Ploiesti, Romania

M. A. ALBU[1], L. M. MORRIS[2], H. NASH[2] & M.O. RIVETT[3]

[1]*Department of Geology and Geophysics, University of Bucharest, Romania*

[2]*Wardell Armstrong, Lancaster Building, Newcastle-under-Lyme, Staffordshire ST5 1PQ, UK (e-mail: hnash@wardell-armstrong.com)*

[3]*School of Earth Sciences, University of Birmingham, Edgbaston, Birmingham B15 2TT, UK*

Abstract: The area of Ploiesti in Romania was among the first oil producing regions in the world. The refineries and oil pipelines have leaked petroleum products over many years. An area of at least 500 ha is contaminated by free product light non-aqueous-phase liquid (LNAPL), affecting one third of Ploiesti's water supplies. To date it has been possible to compensate by extending the well fields and using surface water, but the long-term sustainability of these measures has not been investigated. The Prahova and Teleajan rivers, tributaries of the River Danube, are also at risk. A conceptual map is presented to show the evolution of LNAPL contamination between 1980 and 2000, and the quantity of oil in each plume is estimated. Potential recovery of oil using a conventional pump-and-treat system is also calculated, with costs and possible revenue. An outline programme for investigation and remediation is presented, part of which is being implemented with a European Community grant. Although the severity of contamination was at least partly realised in the 1970s, only the recent move towards privatization could mobilize sufficient financial resources to address this problem and protect groundwater resources for future use.

Ploiesti (Fig. 1) has been the major centre for petroleum refining in Romania since the start of commercial oil production in the 1850s. Four oil refineries (Astra, Vega, Brazi and Teleajen; see Fig. 2) are situated between the Prahova and Teleajen tributaries of the River Ialomita, which flows to the River Danube. They are sited on extensive deposits of an alluvial fan (the Prahova–Teleajen fan), which provides the main source of drinking water for the urban and rural population, estimated at 400 000 in the city of Ploiesti and 700 000 for the whole of Prahova County. Industrial and agricultural supplies are also drawn from the aquifer.

Bombing by the Allies during the Second World War resulted in damage to the Astra and Vega refineries and associated pipework. The Brazi and Teleajen refineries together with the associated pipework and storage tanks have developed leaks since their construction in the 1960s. Earthquakes in 1977 and 1986 caused additional damage. Based on the extent of free-phase product identified to date, more than 105 000 tonnes of petroleum products, including crude oil, petrol and diesel oil, have leaked to the ground and are present as a light non-aqueous-phase liquid (LNAPL) floating on the water able. The LNAPL extends over at least 500 ha. In places oil is ponded at the ground surface and collected by residents for their own use.

The most immediate risk is clearly to human health, due to explosions, inhalation of vapour, dermal contact and possibly drinking contaminated water. Hydrocarbon vapours have resulted in explosions such as occurred in 1987 at a school, and in deaths by asphyxiation. Significant groundwater resources are impacted and additional resources threatened, by both

From: HISCOCK, K. M., RIVETT, M. O. & DAVISON, R. M. (eds) *Sustainable Groundwater Development*. Geological Society, London, Special Publications, **193**, 293–301. 0305-8719/02/ $15.00 The Geological Society of London 2002.

Fig. 1. Location map.

LNAPL and dissolved hydrocarbons. Although contamination of groundwater by petroleum products is commonplace worldwide (Environment Agency 1997; Mather *et al.* 1998), the example presented in this paper is probably one of the most extensive contamination cases reported to date.

This paper provides an example of a lack of environmental or sustainable groundwater management resulting in unacceptable, major impacts on groundwater quality. This lack of historical management has jeopardized both the short- and long-term sustainability of the groundwater resource. This type of problem is not uncommon in a country where historic industrial activity has often been located on phreatic aquifers of high vulnerability, with no regard for the potential pollution of either groundwater or surface water (Albu *et al.* 1994). Although the case study presented is largely qualitative due to the lack of environmental data, it nevertheless outlines the extent of LNAPL contamination and impacts on public water supplies from the alluvial aquifer. Measures that have been considered over a number of years to limit additional impacts on water supplies and other receptors, as well as to remediate existing contaminated areas are outlined. Sufficient resources to initiate action were only made available recently, when it was planned to privatize the Brazi refinery. Since privatization of the petroleum industry is an

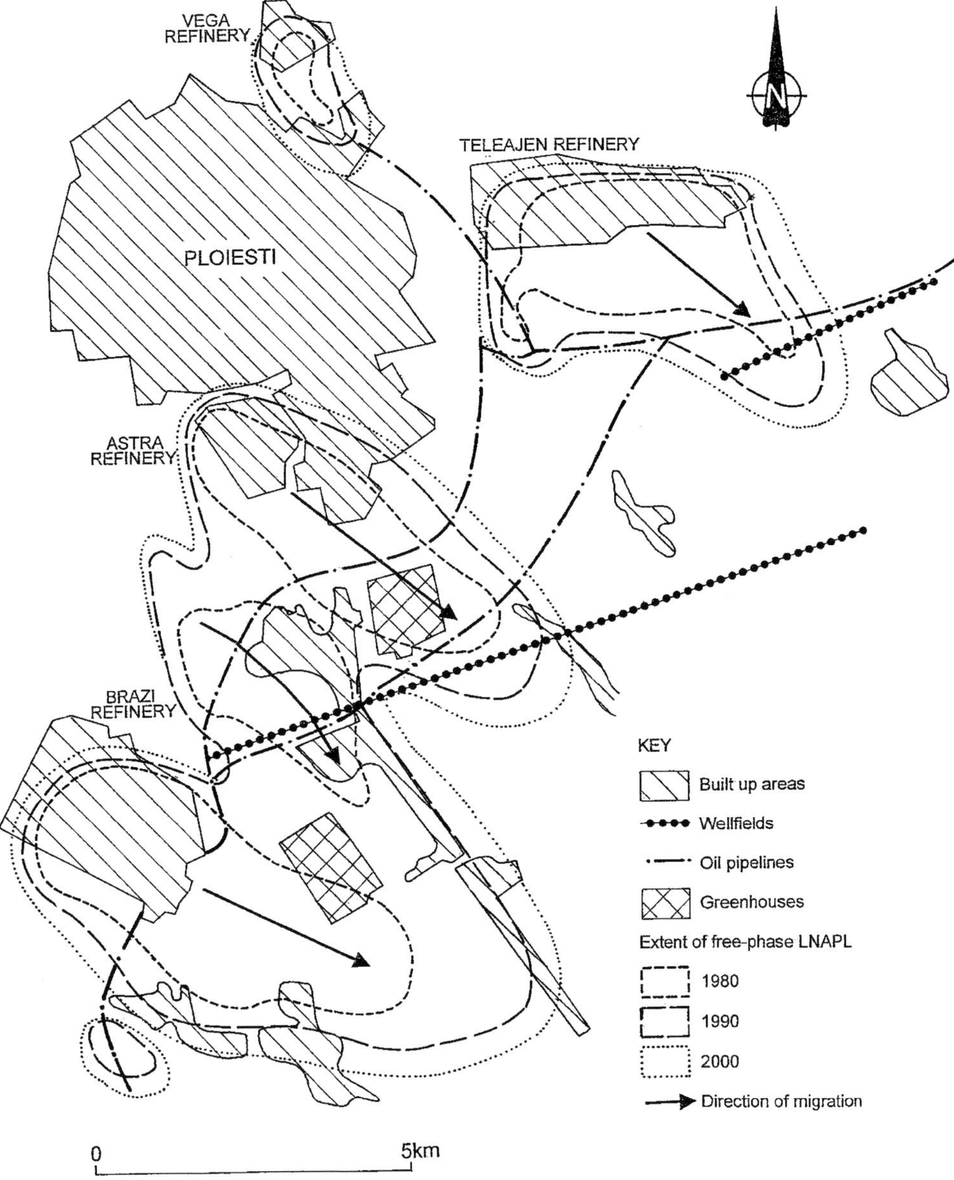

Fig. 2. Extent of LNAPL plumes.

essential step to Romania meeting requirements to join the European Community (EC), an EC grant has been made available for the initial stages of investigating the remediation of the area. The paper therefore provides an illustration of how the availability of financial resources may significantly influence achievement of a remediated, more sustainable scenario.

Hydrogeology

The alluvial deposits of the Prahova–Teleajen fan are more than 160 m thick and comprise sands and gravels interbedded with clay horizons, as indicated in Figure 3. This aquifer system constitutes one of the major groundwater units in Romania (Manescu *et al.* 1994).

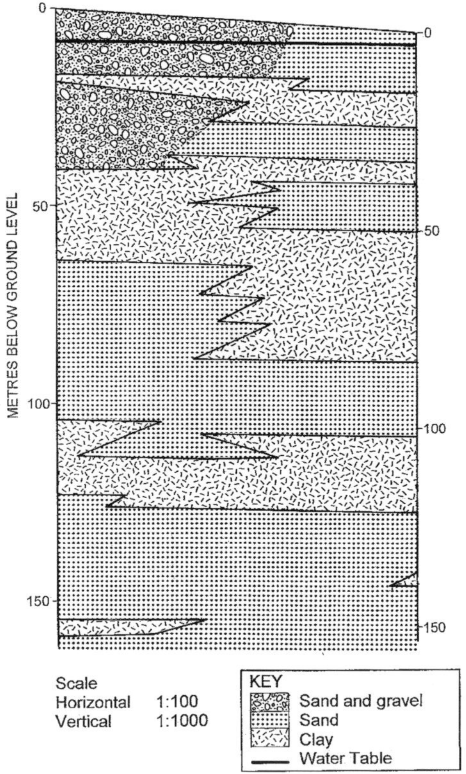

Fig. 3. Schematic section of alluvial aquifer.

Investigations were carried out in the period 1961–1969 by the Institute of Meteorology and Hydrology to provide more water to sustain industrial development (Avramescu *et al.* 1971). Tens of monitoring boreholes were installed as part of the national network, and aquifer tests and geophysical surveys were carried out. This work provided the basis for hydrogeological interpretation of the area. Further research was carried out for Petrobrazi SA in 1973 and between 1986 and 1993 to establish the extent of the hydrocarbon plumes. As part of these studies, groundwater flow and water balance modelling were carried out in 1990 at the University of Bucharest.

The sands and gravels have hydraulic conductivities ranging from 40 to 100 m d^{-1} (pump test data). Groundwater flow is to the south-east, in the same direction as surface water flow. A MODFLOW model developed by the University of Bucharest indicates that groundwater flow rates range from 100 m per annum in the vicinity of the Teleajen refinery to 200 m per annum in the vicinity of Brazi. The water table is shallow, ranging from close to ground level in the south-east to greater than 10 m below ground level in the north-west, where the seasonal fluctuation is up to 8 m. The low depths to groundwater and limited occurrence of near-surface clay horizons mean that the upper aquifer unit, from which the majority of abstraction occurs, is highly vulnerable to pollution.

The aquifer system is exploited by nine linear wellfields (Fig. 2). Before contamination became apparent, the wellfields operated at their installed capacity of about 300 Ml d^{-1}, from an estimated renewable resource of 430–600 Ml d^{-1}. Approximately 90% of the abstraction was from the shallow aquifer (2–40 m below ground level), and 10% from deeper aquifers (40–150 m below ground level). Four of the wellfields are located upgradient of the areas of contamination. The others either intersect areas with free-phase LNAPL (as shown in Fig. 2) or lie close to, or downgradient of, the area of contamination.

Contamination

The pollution of groundwater has not yet been fully assessed since investigations have generally been local and short term. The distribution of free-phase product (as opposed to emanating dissolved-phase plumes) is somewhat better known, and the sources of the hydrocarbon pollution are considered to be the four refineries (tanks, lagoons, processing towers and associated pipework) and leaking storage tanks and pipelines. This has resulted in the formation of four main 'plumes' of LNAPL originating from the refineries, as depicted in Figure 2. Each of these plumes is in the order of 5–9 km long. Smaller LNAPL plumes are also present to the south and north-east of the Brazi refinery, where major pipeline leaks have occurred. Such leaks have been exacerbated by holes in the pipeline made intentionally in order to steal the oil. In partial mitigation, the use of ponded oil from other leaks, for fuel or running vehicles could account for a significant volume of product, reducing the overall source of contamination.

The LNAPL spread relatively quickly in the period 1980 to 1990, as shown in Figure 2; plume extension was most apparent at the Brazi refinery, where the plume had extended approximately 2 km to the south-east over the 10-year period. More recently, the spread of LNAPL has slowed as indicated by the estimated extent in 2000. However, the latter is based on data from the early 1990s, and the actual position may be different. The thickness of LNAPL was measured in fully screened boreholes during the

Table 1. *Characteristics of pollution plumes*

Plume of contamination (estimated values)	Refinery Vega	Teleajen	Astra and Pipeline	Brazi	Total
Length (m)	5000	7000	9500	8300	
Width (m)	130	210	240	230	
Area (m^2)	500 000	1 300 000	1 800 000	1,500 000	5 100 000
Average thickness (m)	0.15	0.20	0.25	0.20	
Volume (m^3)*	37 500	130 000	225 000	150 000	542 500
Porosity	0.30	0.35	0.35	0.40	
Oil/water ratio	0.45	0.50	0.65	0.70	
Volume of oil (m^3)	5063	22 750	51 188	42 000	121 000
Mass of oil (tonne)	4455	20 020	45 045	36 960	106 480

*Volume = area x thickness / 2 (see text).

investigations in the 1970s and early 1990s, with the maximum thickness of 5 m being measured at Brazi. The geophysical investigations (Manescu *et al.* 1994) found that the resistivity contrast between contaminated and uncontaminated ground (generally more than 200 ohm m) was sufficient to delineate areas of contamination within the unsaturated zone as well as areas of ponding on the water table.

Obtaining reliable estimates of both the quantities of hydrocarbon LNAPL in aquifers from the thickness measured in monitoring wells and the proportion of LNAPL that is actually mobile and recoverable is difficult (Abdul *et al.* 1989; Wiedemeier *et al.* 1999). In particular, the thickness of free product measured in monitoring wells is generally greater than that in the aquifer. The extent of free LNAPL at the water table, oil/water ratios and porosity of the sands and gravels estimated from the field investigations are given in Table 1. The 'gross' volume of oil has been reduced by 50% to account for errors in the estimation of product thickness, in particular capillary effects, which will be relatively minor in this coarse-grained aquifer, and density effects. When this is taken into account it appears that 121 000 m^3 of LNAPL could be present. It must, however, be emphasized that this estimate is very approximate due to variations in the oil/water ratio, porosity and other factors, and because recent measurements of the extent of the LNAPL have only been estimated, not measured.

The Water Authority of Ploiesti has indicated that 170 Ml d^{-1}, more than 50% of the total wellfield abstraction, is contaminated with hydrocarbons. In addition to affecting the main wellfield, private domestic wells in the city and in several surrounding villages are contaminated. Approximately one-third of the private wells abstracting water from the shallow aquifer have become contaminated and have been abandoned. It appears that this has been compensated for by extending the affected wellfields into uncontaminated areas, by the use of surface water, and by extending the mains supply.

Groundwater currently used for public supply and agriculture is certainly contaminated over a wider area than the extent of LNAPL depicted in Figure 2. The high rates of groundwater flow indicate that dissolved hydrocarbons could migrate quickly and, if not attenuated, could already have reached the Prahova and Teleajen rivers, which are respectively 15 and 12 km downgradient of the LNAPL plumes at the Brazi and Teleajen refineries. If the LNAPL migrates further, groundwater downgradient will become more contaminated and the risks to river quality higher. One well field is less than 4 km downgradient of the LNAPL plume at Brazi, and is clearly at risk. The water table is closer to the ground surface in the direction of migration, and the area of contaminated soils will also increase if the LNAPL plume continues to extend, forming a source of continued groundwater contamination and affecting land use.

It is, however, anticipated that natural attenuation will limit the spread of dissolved plumes from the LNAPL. The migration of dissolved-phase plumes of soluble hydrocarbon components such as BTEX (benzene, toluene, ethylbenzene, xylenes) will be limited by natural attenuation processes such as sorption and biodegradation (Barker *et al.* 1987; McAllister & Chiang 1994). The recent 'plum-a-thon' studies of hundreds of sites (including sand and gravel aquifers, as in the area of Ploiesti) in the USA indicate that median lengths for BTEX plumes are in the order of 30–80 m down-

Table 2. *Budget costs for pilot project for LNAPL removal*

Stage	Activity	Consultancy fees (US$)	Expenses/ disbursements (US$)	Total (US$)
1	Consultations, collation and review of available data	31 000	8000	39 000
2	Preliminary investigation	40 000	28 000	68 000
3	Report and methodology for detailed investigation	16 000	4000	20 000
4	Detailed investigation	144 000	373 000	517 000
5	Selection and design of pilot remediation scheme	45 000	21 000	66 000
6	Field trials	119 000	400 000	519 000
7	Feasibility study of large-scale remediation	35 000	19 000	54 000
	TOTALS	430 000	853 000	1 283 000

gradient of LNAPL sources (Newell & Connor 1998). The majority (about 85%) are less than 100 m long with only 2% of plumes exceeding 250 m. The relatively low extension of plumes is ascribed to natural attenuation processes, particularly biodegradation. The scale of the LNAPL contamination at Ploiesti, and hence hydrocarbon loading, is significantly greater than the typical USA survey sites, giving an increased potential for extensive areas of anaerobic conditions to develop underlying and potentially downgradient of the LNAPL plumes. Anaerobic biodegradation of hydrocarbons is a complex process (Baedecker *et al.* 1993), but the degradation rates are typically much slower than aerobic rates, with benzene in particular being relatively persistent (Davis *et al.* 1999). It is therefore probable that dissolved plumes around the Ploiesti contamination are more extensive than those observed in the USA survey.

Proposed remediation

Despite the paucity of data, a proposal to start to investigate and deal with the problem was developed in 1997 to 1998 as the basis for the Romanian government to apply for international funds, and for one of the local authorities to raise internal funds. However, no action could be taken locally or by the Ministry of Industries in Bucharest or the EC, until privatization was firmly on the agenda.

The proposed remedial works aimed to protect groundwater resources as far as possible by reducing the levels of contamination and minimizing the potential for extension of the contaminant plumes. It was proposed to confirm the individual or joint sources of pollution and to define the current extent of LNAPL and affected soils, wells and watercourses within one of the main plumes. In particular, the extent and severity of groundwater contamination were to be defined, and additional information collected regarding the nature of hydrocarbon contamination and the mechanical, physical, chemical and hydrogeological properties of the aquifer.

A phased approach was suggested, with a preliminary investigation involving an assessment based on existing boreholes and data sources, with limited laboratory analysis of soils, water and product. A more detailed investigation would then be designed with provision for up to 100 investigation and monitoring boreholes, more extensive field and laboratory analyses and resistivity surveys.

An assessment would be made of groundwater pollution regarding the causes and effects of contamination; the location and evolution of the sources and plumes of contamination; the volume and nature of the contaminants and the risks associated with migration of the contamination. On completion of the assessment, appropriate methods and costs of remediation would be identified. Priority would be given to stopping the leaks, followed by removal and containment of LNAPL and then possibly by remediation or containment of the dissolved plume. To evaluate the effectiveness of the proposed remediation, and in particular the amount of oil that could be recovered and utilized, a field trial or pilot scheme would be undertaken.

Outline costs for the above activities for one of the main plumes, excluding the repair of leaks, are given in Table 2. Assuming the majority of the work would be undertaken by Romanian enterprises and consultants, funding in the order of US$1.3 million would be required for the suggested investigation.

In addition to reducing the risks to human health from contamination of the ground, water supply and agricultural environment, it is considered that a significant proportion of LNAPL could be recovered, initially by pump-and-treat methods, including dual-phase extraction of groundwater–oil and vapour contamina-

Table 3. *Estimated costs of shallow aquifer remediation*

	Vega	Teleajen	Refinery Astra and Pipeline	Brazi	Total
Oil recovery using conventional pump-and-treat system, comprising abstraction wells and air stripper with granular activated carbon units for off-gas treatment. 50 % recovery.					
Time (year)	8	24	38	23	
Cost (US$ million)	1.50	4.55	7.00	4.20	17.25
Estimated revenue from sale of recovered petroleum products (at US$15 per barrel)					
Volume of oil (m^3)	2531	11 375	25 594	21 000	60 500
Volume of oil (bbl)	15 900	71 500	160 900	132 000	380 300
Revenue (US$ million)	0.239	1.072	2.413	1.980	5.704
% revenue contribution to remediation costs	16	24	34	47	33

Table 4. *Additional revenue from other commodities*

	1 Year (US$)	50 Years (US$)
Revenue associated with continuing use of current water resource	16 000 to 32 000	800 000 to 1 600 000
Revenue associated with 50 hectares crop production, at 3 tonnes of grain per hectare per year	19 000	950 000

tion. The revenue potentially generated by recovery is seen as a key issue in providing funds to continue remedial works. The high permeability of the shallow aquifer, the thick zone of LNAPL in some areas and the relatively high water table are factors that would enhance the potential for LNAPL recovery, while the heterogeneity of the aquifer, including the presence of clays, would not. Until the investigations outlined above are completed, estimates of the volume of LNAPL, treatment costs and recovery rates are necessarily uncertain. An initial estimate, based on Holden *et al.* 1998, is presented in Table 3 assuming a recovery rate of 50%, which takes into account that a proportion of residual oil will be left in the ground as well as allowing for some of the uncertainties discussed above. For this scenario, some 60 000 m^3 of LNAPL would be recovered, and approximately US$20 million would be required for large-scale remediation by pump-and-treat (LNAPL recovery) methods.

This remediation would operate over an uncertain time scale, probably of some decades, since it would be phased geographically to address priority areas first. The residual contamination, which is expected to represent a significant proportion of the LNAPL, would be best removed by other in situ technologies such as soil vapour extraction, high-vacuum extraction, air(bio) sparging, and bioremediation depending on the specific contaminant and hydrogeological circumstance. Ex situ bioremediation, for example, may also be possible for excavatable, shallow contaminated material. Such in situ/ex situ methods, in view of their expense, would be implemented primarily at sensitive locations (e.g. to protect key receptors such as currently unaffected wellfields).

The revenue from the sale or use of recovered oil is difficult to quantify precisely, since recovery volumes are difficult to estimate and will probably vary across the area. The revenue is also dependent on the price of oil and the cost of any refining needed before sale. Based on the recovery of 60 500 m^3 of product and an oil price of US$15 per barrel, the revenue could potentially provide approximately a third of the costs of remediation (Table 3). In addition to revenue realised from the sale of the recovered oil, there would be lesser savings related to water supply and to crop production (Table 4).

The proposal was submitted to the local authorities in 1998 but no action was taken at that time due to the difficulty of finding the initial cash injection. The Romanian Ministry of

Industries faced the same constraint, while the EC advised that such works would only be considered as part of a privatization programme. Such funding delays have clearly exacerbated the extent of the contamination scenario and increased the ultimate remedial effort required.

EC-funded project to decrease environmental liability

In support of efforts to privatize the oil industry, the EC has now allocated funds for a project entitled 'Facilitating the Privatization and Reorganization of the Oil Plants in the Ploiesti Area by Decreasing their Environmental Liability'. This is targeted at the Brazi and Astra refineries and the main pipeline, and aims to evaluate soil and groundwater pollution, to develop a pilot project for recovery of hydrocarbons and to implement monitoring systems. The requirements for, and costs of, emergency response actions are to be identified in the early stages of the project. However, stopping leaks would not be addressed directly but would be included in the design of the pilot project for remediation. The estimate of recoverable oil will be refined during the pilot recovery project, and will be an important factor in assessing the feasibility of large-scale remediation.

The budget for technical assistance from EC funds is approximately US$500 000 (500 000 Euros), with an additional US$250 000 from in-kind contributions from the beneficiaries. The budget for investment and equipment in the regional monitoring network and pilot remediation is approximately US$3.25 million. The project is more broad ranging than that previously outlined for the local authority, but the allocation of funds is similar when the additional scope of works, including updating the regional monitoring system, is included.

Implications for sustainable groundwater development

Privatization of a number of urban water supply systems is also underway in Romania, funded partly by international loans. The emphasis has been strongly on upgrading existing infrastructure, repairing leaks etc. There has been little surplus funding to address the problems of pollution of groundwater and its impact on water supplies, in Ploiesti or elsewhere. However, the experience at Ploiesti shows that in the absence of environmental controls, alternative supplies may be required. It is possible that the situation has been accepted due to the fact that once contaminated, it may not be possible to clean up groundwater to drinking water standards within one generation. Alternative supplies, although more distant, have generally been available despite increased costs. However, as environmental controls increase, it is possible that as in the UK and other parts of Europe, more contamination of groundwater will be identified. Development of urban supplies from groundwater will therefore have to take into account historic and current industrial activity, and the costs of additional treatment of groundwater will have to be balanced against the costs of additional wells and longer pipelines from unaffected areas.

The long-term sustainable use of groundwater in the Ploiesti area clearly depends upon the success of plume containment and removal, the former being significantly easier to achieve. The nature of the remedial approach taken may also have a bearing on the sustainability of the groundwater resource. Consumptive use (e.g. by industry) or re-injection of treated groundwater to the aquifer is clearly desirable, as practised in the remediation of the 5 km-scale Tucson (Arizona, USA) dissolved solvent plume (Zhang & Brusseau 1999). Otherwise, large quantities of groundwater may be disposed of to surface water / sewer during the remediation effort. If clean-up of LNAPL-contaminated areas is not realisable in the short to medium term, the only possible use of affected groundwater would be for industrial purposes that are less quality sensitive. Further exploitation of the deeper, non-impacted aquifer units may be possible, although this may induce the downward migration of contamination. LNAPL buoyancy properties, natural attenuation and the probability of high sorption in the intervening clay horizons would, however, limit the migration of the contaminants and provide some protection of the deeper resources.

Conclusions

The contamination by petroleum hydrocarbons in the area of Ploiesti is a direct threat to the area's urban, rural and industrial water supply, affecting about one-third of the potable groundwater resources. Only with the recent moves to privatize the oil refineries have concerted efforts been made to address the problems, even with regard to stopping the leaks. It is not expected that the contamination problems of the area can be solved in the short lifetime of the EC project (15 months). However, the installation of monitoring systems and a pilot project for remediation should establish what measures are feasible to improve the situation and their costs. The extent of remediation, and even the protection of currently unaffected water supplies, will probably depend on the generation of a significant income from the recovery of oil. If all the refineries, pipeline operators and regulatory agencies work together, the project could provide the baseline of knowledge and expertise and that could eventually make significant improvements to the area of Ploiesti and to Romania's standing as it works towards meeting the environmental conditions required to join the European Union.

Providers of urban water supplies can expect additional costs due to problems of groundwater contamination for the foreseeable future, and this factor that will have to be considered when designing new supplies. The supply situation, does not appear to be critical at present, this is at least partly because industrial activity and industry's water usage has fallen sharply in recent years. However, this could change, and membership of the EC would hopefully entail a more productive economic activity than at present. In addition, many villages currently using private wells may wish to install a mains supply, increasing overall water usage. Timely identification and mitigation of groundwater pollution in areas used for water supply should therefore be given a high priority. Continued delays in remedial action only serve to exacerbate the existing, already major contamination problem and increasingly jeopardize prospects for a more sustainable future use and management of the Ploiesti groundwater resource.

The authors thank Phil Aldous and Brain Morris for their suggestions and for helping to place the case study in a wider context.

References

ABDUL, A. S., KIA, S. F. & GIBSON, T. L. 1989. Limitations of monitoring wells for the detection and quantification of petroleum products in soils and aquifers. *Ground Water Monitoring Review*, **9**, 90–99.

ALBU, M., TOMESCU, G., BRETOTEAN, M. & PANE, R. 1994. Types of industrial impact on groundwater resources in Romania. *Impact of Industrial Activities on Groundwater, Proceedings of the International Hydrogeological Symposium*, Constantza, Romania, 1994, 1–7.

AVRAMESCU, E., TOMESCU, G. & ALBU, M. 1971. Recherches Hydrologiques dans la zone du cone alluvionnaire de Prahova-Teleajen. *Meteorology and Hydrology* (Institutul de Meteorologie si Hidrologie), **1**, 62–75.

BAEDECKER, M. J., COZZARELLI, I. M., EGANHOUSE, R. P., SIEGEL, D. I. & BENNET, P.C. 1993. Crude oil in a shallow sand and gravel aquifer III Biogeochemical reaction and mass balance modelling in anoxic groundwater. *Applied Geochemistry*, **8**, 569–586.

BARKER, J. F., PATRICK, G. C. & MAJOR, D. 1987. Natural attenuation of aromatic hydrocarbons in a shallow sand aquifer. *Ground Water Monitoring Review*, **7**, 64–71.

DAVIS, G. B., BARBER, C., POWER, T. R., THIERRIN, J., PATTERSON, B. M., RAYNER, J. L. & WU, Q. 1999. The variability and intrinsic remediation of a BTEX plume in anaerobic sulphate-rich groundwater. *Journal of Contaminant Hydrology*, **36**, 265–290.

ENVIRONMENT AGENCY. 1997. *Groundwater pollution - Evaluation of the extent and character of groundwater pollution from point sources in England and Wales*. Environment Agency, Bristol.

HOLDEN, J. M. W., JONES, M. A., MIRALES-WILHELM, F. & WHITE, C. 1998. *Hydraulic measures for the treatment and control of groundwater pollution*. CIRIA, London, Report 186.

MCALLISTER, P. M. & CHIANG, C. Y. 1994. A practical approach to evaluating natural attenuation of contaminants in groundwater. *Ground Water Monitoring and Remediation*, **14**, 161–173.

MANESCU, M., BICA, I. & STAN, I. 1994. La pollution d'eau souterraine avec des produits petroliers dans la zone de Ploiesti. *Impact of Industrial Activities on Groundwater, Proceedings of the International Hydrogeological Symposium*, Constantza, Romania, 1994. 356–368.

MATHER, J., BANKS, D., DUMPLETON, S. & FERMOR, M. (eds) 1998. Section 4: Groundwater pollution by hydrocarbons. *In*: MATHER, J., BANKS, D., DUMPLETON, S. & FERMOR, M. (eds) *Groundwater Contaminants and their Migration*. Geological Society, London, Special Publications, **128**, 121-180.

NEWELL, C. J. & CONNOR, J. A. 1998. *Characteristics of dissolved petroleum hydrocarbon plumes: results from four studies*. API technical transfer bulletin, American Petroleum Institute, Washington, DC.

WIEDEMEIER, T. H., RIFAI, H. S, NEWELL, C. J. & WILSON, M. 1999. *Natural Attenuation of Fuels and Chlorinated Solvents in the Subsurface*. John Wiley & Sons, New York.

ZHANG, Z. & BRUSSEAU, M. L. 1999. Nonideal transport of reactive solutes in heterogeneous porous media. 5. Simulating regional-scale behaviour of a trichloroethene plume during pump-and-treat remediation. *Water Resources Research*, **35**, 2921–2935.

Assessments of the sensitivity to climate change of flow and natural water quality in four major carbonate aquifers of Europe

P. L. YOUNGER[1], G. TEUTSCH[2], E. CUSTODIO[3], T. ELLIOT[1,4], M. MANZANO[3] & M. SAUTER[2,5]

[1]*Water Resource Systems Research Laboratory, Department of Civil Engineering, University of Newcastle Upon Tyne, UK (e-mail: p.l.younger@ncl.ac.uk)*
[2]*Lehrstuhl für Angewändte Geologie, Eberhard-Karls Universität Tübingen, Germany*
[3]*Departamento de Ingeniería de Terreno, Universitat Politécnica de Catalunya, Spain*
[4]*Present address: School of Civil Engineering, Queen's University, Belfast*
[5]*Present address: Geologisches Institut, Universität Jena, Germany*

Abstract: A numerical modelling approach has been developed to predict the vulnerability of aquifers to future climate change. This approach encompasses changes in recharge regime, dynamics of flow and storage patterns within aquifers, and natural hydrochemical changes. An application of the approach has been made to four hypothetical spring catchments representative of major carbonate aquifers in three European climatic zones. Since prolific carbonate aquifers typically combine a high transmissivity with a low specific yield, they can be expected to be more sensitive than clastic aquifers to changes in recharge patterns. Simulations of the study systems to the middle of the 21st century predict different outcomes in the three different climate zones: (1) in the northern maritime zone (UK) recharge (and therefore discharge) is predicted to increase by as much as 21% in response to anticipated increases in precipitation; (2) in the continental zone (Germany) recharge in winter is predicted to remain approximately the same as at present, but summer recharge will decline dramatically (by as much as 32%), so that a net decrease in aquifer discharge is predicted; and (3) in the Mediterranean zone (Spain) recharge is predicted to decrease by as much as 16% of the present-day values. For all three systems, increases in water hardness in response to rising CO_2 are predicted, but are expected to be negligible in water resources terms.

Europe is highly dependent on its groundwater resources (Crampon *et al.* 1996). Even in countries like Britain and Germany, which are relatively rich in surface water resources, groundwater accounts for more than a third of all public water supply. In the Mediterranean countries of Europe, where surface water resources are limited, groundwater is often the sole source of public supply. Consequently, any major reductions in groundwater availability in such countries can be anticipated to have serious economic implications. Climate change clearly poses a threat to the long-term viability of European water resources (e.g. Arnell 1992; Price 1998). If current global warming prognoses are correct (Wigley *et al.* 1997; Arnell 1999), then serious consideration will need to be given to ways of meeting the supply-demand balance in future. While improvements in water conservation and demand management practices will have a considerable contribution to make, provision of further storage is virtually certain to become increasingly important in decades to come. With surface reservoirs becoming increasingly contentious in many parts of the world, groundwater storage is likely to assume even greater importance than at present in overall water resources management strategies (e.g. Price 1998).

Despite the likely increased importance of

From: Hiscock, K. M., Rivett, M. O. & Davison, R. M. (eds) *Sustainable Groundwater Development*. Geological Society, London, Special Publications, **193**, 303–323. 0305-8719/02/ $15.00 The Geological Society of London 2002.

groundwater resources under the changed climate conditions of the future, few assessments of the water resources implications of climate change have yet considered groundwater explicitly. To date, most assessments have focused on surface water quantity, principally through studies of changes in rainfall run-off behaviour which can be expected under possible future climatic conditions (e.g. Solomon *et al.* 1987; Arnell 1992, 1998). While an early attempt was made to predict possible changes in groundwater dynamics at the global scale in response to global warming (Zekster and Loaiciga 1993), relatively few aquifer-scale assessments have been made. Early studies were largely restricted to evaluations of changes in recharge (e.g. Vaccaro 1992; Cole *et al.* 1994; Sandstrom 1995). The implicit assumption in these studies was that total groundwater discharge would simply equal the total recharge under changed climate conditions (e.g. Bouraoui *et al.* 1999). This is logical enough if one is concerned only with long-term average volumes of discharge. However, where inter-annual or sub-annual variations in groundwater discharge rate are of interest (which is nearly always the case in water resources evaluations), then the degree to which flow and storage within an aquifer transform a recharge time-series into a discharge time-series with different characteristics becomes very important. In other words, groundwater flow and storage processes effectively attenuate discrete incoming parcels of infiltration to produce a relatively smooth, continuous aquifer discharge. Groundwater storage also results in aquifers having much longer residence times than river systems. In some cases the impacts of a given period of extreme rainfall (low or high) on aquifer responses may persist for several years. Hence aquifer responses to climate change might be expected to show a considerable temporal lag which will not be picked up by simply equating total recharge to total discharge.

In recognition of these complications a number of climate change impact studies have included model representations of groundwater flow and storage processes. Most numerous are studies in which aquifer processes are represented in a 'lumped' manner within catchment-scale water balance models (e.g. Gellens 1991; Ferrier *et al.* 1993; Panagoulia & Dimou 1996; Eltahir & Yeh 1999; Kilsby *et al.* 1999; Rosenberg *et al.* 1999; Dehn *et al.* 2000; Limbrick *et al.* 2000). These models have the advantage that they generally demand input data at the same sort of spatial resolution as the outputs generated by atmospheric general circulation models (GCMs). However, spatial resolution is only one consideration when choosing an appropriate model for aquifer responses to climate change. The 'lumped' models listed above differ widely in their internal functioning, varying from simple linear storage models (e.g. Dehn *et al.* 2000) to relatively sophisticated models based on aggregation of physically based response modelling to large scales (e.g. Kilsby *et al.* 1999). Given that it is generally inadvisable to apply empirical input-output response models to conditions that fall outside their calibrated ranges, the physically based aggregation approach is the most defensible for climate change impact assessments (Kilsby *et al.* 1999). Where possible, it would be preferable to use a fully physically based model with spatial resolution on the typical scale of interest for water resources evaluations.

Although physically based, distributed numerical models are very widely used in hydrogeology they have found surprisingly few applications in climate change impact studies to date. Wilkinson & Cooper (1993) and Cooper *et al.* (1995) used finite difference models to examine groundwater–surface water dynamics in hypothetical, idealized systems representative of systems found in southern and central England. More recently, numerical modelling has been used to evaluate the possibility of exacerbated saline intrusion into coastal aquifers in Egypt and India (Sherif & Singh 1999; Bobba *et al.* 2000), although evaluations in that setting are complicated by possible sea-level rises in addition to changes in recharge patterns. A numerical modelling study of the karstic Edwards Aquifer in Texas, USA concluded that resources in the aquifer can be expected to come under severe stress in the next few decades in the likely event that predicted decreases in recharge are realised (Loaiciga *et al.* 2000). While the isolated studies listed above are interesting, the fact that they are restricted to single climate zones limits their utility in gaining an overall impression of the range of possible climate change impacts on public-supply aquifers. A study that applies the same evaluation criteria to a number of aquifers in a range of climate settings is therefore desirable.

There is a further aspect of groundwater–climate interactions that is receiving increasing attention, namely the role of the subsurface as a possible sink (or source) for atmospheric CO_2. Recent studies have illustrated that consumption of CO_2 in weathering reactions worldwide is of sufficient magnitude that it might feasibly influence global warming rates over the medium to long term (Kump *et al.* 2000; Liu & Zhao 2000). What has not yet been considered is whether increased weathering rates in response to rising atmospheric CO_2 have important consequences for groundwater quality.

All studies of aquifer responses to climate change have to overcome problems of coupling climate change prediction results to the small-scale soil–plant–atmosphere dynamics that ultimately govern recharge (e.g. Kilsby *et al.* 1998; Bouraoui *et al.* 1999). There are a number of possible approaches to climate change impact assessments. At their simplest, assessments can be made by simply modifying present-day rainfall totals in proportion to anticipated changes in atmospheric temperature (e.g. Panagoulia & Dimou 1996; Loaiciga *et al.* 2000). Such simple approaches might overlook significant changes in evapotranspiration which are expected to result from any change in climate. Another approach is to assume that the present latitudinal zonation of climate patterns will simply shift north or south, so that future conditions in England can be directly compared with, say, present-day conditions in central France. This approach ignores the important factor of day length, which is determined astronomically and is thus independent of climate.

Overall, the most scientifically defensible approach to climate change impact assessments is to use the output from physically based, process-oriented GCM simulations (Hulme *et al.* 1994; Kilsby *et al.* 1998). GCMs are generally formulated as finite difference (or more rarely, finite element) solutions of large systems of coupled partial differential equations which describe the movement of atmospheric air masses. Despite their pre-eminence in climate impact studies, GCM outputs are difficult to apply to hydrological change studies for at least two reasons (Gleick 1986 1989):

(1) Problems of discretization: the spatial resolution of GCM grids is generally too coarse to provide hydroclimatological information at a scale relevant to catchment or aquifer modelling.
(2) Problems of process representation: while many GCMs include representations of Earth surface hydrology (which represent important feedbacks to the atmosphere), the process representations used in the GCMs are usually very simple and do not provide information of sufficient detail (spatial or temporal) for standard water resources modelling purposes.

The down-scaling of GCM output for water resources management studies is an area of active research with a rapidly changing orthodoxy (see, for instance, Kilsby *et al.* 1998; Mavromatis & Jones 1998, 1999; Wilby *et al.* 1998). For this reason, specific climate impact predictions tend to have an extremely short 'shelf-life'.

This paper addresses a number of the issues identified above. In particular, we describe: (1) the use of a purpose-written, physically based numerical modelling code to simulate the responses of carbonate aquifers to climate change, in terms of both quantity and quality (primarily hardness) of water; and (2) a technique for using GCM output in aquifer studies, which has then been applied across a range of systems representative of the main climate zones of western Europe.

The work reported here was undertaken under the aegis of an European Commission research project entitled 'GRACE' (Groundwater Resources and Climate Change Effects). Full details of the GRACE project are available elsewhere (Clemens *et al.* 1996, 1997; Sauter *et al.* 1997; Younger *et al.* 1997; Elliot *et al.* 1998; Manzano *et al.* 1998; Sauter & Liedl 1998); this paper brings together key findings to provide an overview of possible climate change impacts on carbonate aquifers throughout western Europe.

Carbonate aquifers and climate change

Sensitivity of carbonate aquifers to climate change

As the first full analysis of its kind, GRACE focused on the 'worst-case scenario' represented by carbonate aquifers. Carbonate aquifers may be regarded as a 'worst case' for the following reasons:

(1) Those aquifers most responsive to changes in recharge will be those with low specific yields and (at least locally) high transmissivities; carbonate aquifers, being generally fracture-flow systems, generally fall into this category.
(2) Carbonate aquifers might be expected to show exacerbated lowering of the water table in the longer term if changes in atmospheric carbon dioxide concentration induce dissolutional enlargement of fracture apertures (and thus an increase in permeability).
(3) If dissolution of carbonate rocks becomes more vigorous over time, then the hardness of the groundwater can be expected to increase, possibly resulting in unacceptable water quality.

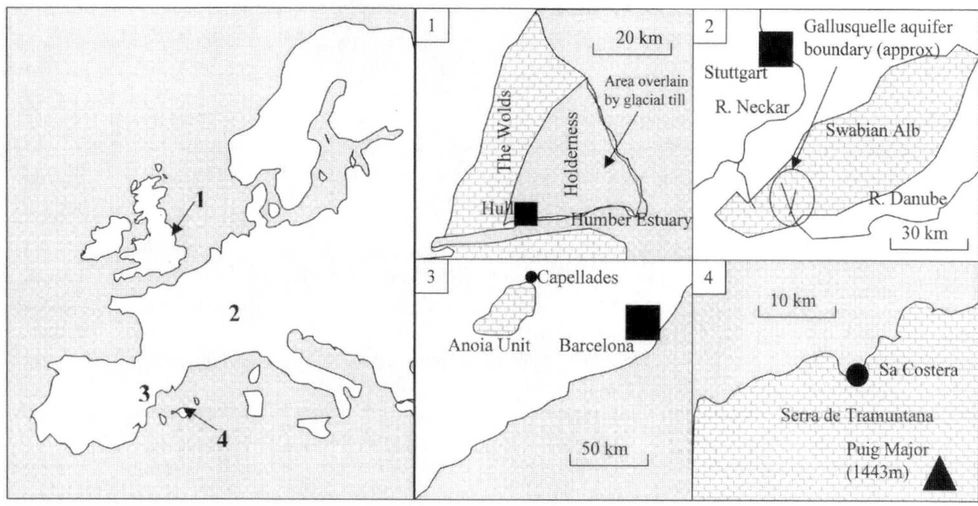

Fig. 1. Sketch map showing the locations of the four European carbonate aquifer systems discussed in the text. (Note that the right-hand edge of each box is oriented approximately north-south).

The first of these postulates is illustrated to some degree by the results obtained for the Edwards aquifer in Texas by Loaiciga *et al.* (2000), although their future climate scenarios were not rigorously based on GCM results. In this paper we provide inter-comparable, GCM-derived evaluations for four aquifers in three climatic settings. The latter two issues listed above have not been examined previously, and are therefore also considered in this study.

Carbonate aquifers in three European climatic zones were selected for detailed analysis (Fig. 1):

(1) The Yorkshire Chalk (UK; northern maritime climatic zone);
(2) The Gallusquelle aquifer (Germany; temperate continental climatic zone); and
(3) The Anoia Unit (Catalunya) and the Serra de Tramuntana (Mallorca) (Spain; Mediterranean climatic zone).

Comprehensive accounts of the hydrogeology of these aquifers are given elsewhere (Sauter 1992; Cardoso da Silva 1997; Younger *et al.* 1997; Lambán Jiménez 1998); hence only brief summaries are given below.

Yorkshire Chalk aquifer (UK) – northern maritime climate zone

The Yorkshire Chalk aquifer underlies an area of some 1800 km^2 north of the Humber Estuary in eastern England (study area 1 in Fig. 1). The hydrogeology of the aquifer has been described in detail (e.g. Foster & Milton 1974, 1976; Institute of Geological Sciences/Yorkshire Water Authority 1980; Elliot *et al.* 1998, 2001). Essentially the Yorkshire Chalk comprises two hydrogeological zones:

(1) A western zone of unconfined Chalk, in and adjoining the recharge area in the hills of the Yorkshire Wolds. Transmissivities in this zone vary from around 10^1 m^2 d^{-1} beneath interfluves to as much as 10^4 m^2 d^{-1} in the axes of the wider valleys. Unconfined storativities rarely exceed 0.02. Many natural springs drain this unconfined zone. Large well abstractions are also long-established. While some large conduits (diameters $\leq$0.5 m) have been observed in the Yorkshire Chalk at outcrop, these are rarely continuous over distances of more than a few metres. Hence, despite possessing some of the classical attributes of a karst system (such as a dense network of dry valleys), the Yorkshire Chalk is probably best regarded as proto-karstic.
(2) An eastern zone of confined Chalk, beneath the Holderness Plain, where the Chalk is overlain by Quaternary glacial sediments (predominantly low-permeability tills). Transmissivities in this area tend to fall in the range 10^1–10^2 m^2 d^{-1}, and confined storativities are usually in the range 10^{-5}–

10^{-7}. There are few natural discharges from this zone of the aquifer, though there are several significant well abstractions.

Although the total stratigraphic thickness of the Yorkshire Chalk exceeds 420 m, structural dip and weathering history have resulted in significant transmissivity development being restricted to the uppermost 80–100 m of the Chalk rock mass at any one point. Hydrochemical patterns in the aquifer coincide to some degree with the eastern and western hydrogeological zones (Pitman 1986). Elliot *et al.* (2001) have argued that hydrochemical patterns can be explained in terms of palaeohydrogeological events, the nature of which shed light on the sensitivity of the aquifer to climate change, inasmuch as the presence of bodies of saline water in parts of the aquifer are actually the result of previous changes in recharge regime and sea level rather than responses to artificial abstractions. The Yorkshire Chalk aquifer is an important source of public supply to the City of Hull and environs, both directly via well abstractions, and indirectly by providing most of the baseflow in the River Hull, from which water is also drawn for public supply. Within the last 5 years water levels in the Yorkshire Chalk have fallen to all-time low levels during a succession of dry summers, prompting fears that the aquifer may be a less reliable water resource than it had previously been considered (Elliot *et al.* 1998).

The Gallusquelle aquifer (Germany) – temperate continental climate zone

The Gallusquelle aquifer (study area 2 in Fig. 1) forms part of a regional-scale karstic mountain range, the Swabian Alb, in south-west Germany. The aquifer comprises limestones of Jurassic (Oxfordian and Kimmeridgian) age. (Note that these quintessentially English chronostratigraphic terms are also used in Germany.) The main aquifer units are the upper Kimmeridgian limestones. Within the catchment area, the whole limestone sequence dips south-east at about 1–2°. As in the Yorkshire Chalk, the effective aquifer base is not stratigraphically determined; rather it cuts discordantly across lithostratigraphic horizons according to the local degree of development of porosity and permeability (Sauter 1992). This discordance is a result of the genetic history of the aquifer and the petrography of the limestone rocks. Like the Yorkshire Chalk, the Gallusquelle aquifer has transmissivities in the range 10^1 to 10^4 m^2 d^{-1}, but in this case the higher values are associated with well-developed systems of dendritic, phreatic caves (Sauter 1992).

Most water resources in the Swabian Alb are exploited via natural springs; abstraction wells are rare, partly because of the difficulties of access in the mountainous terrain, and partly because the chances of hitting highly transmissive zones are rather slim. The Swabian Alb karst springs are also of immense ecological and emblematic importance, since they form the headwaters of the Danube, arguably Europe's greatest river.

The Anoia Unit, Catalunya (Spain) – Mediterranean climatic zone

The Anoia Unit (study area 3 in Fig. 1) is a tectonically complex hydrogeological system that lies in the Catalan Prelitoral Range, a range of mountains running south-west to north-east and rising to between 400 and 700 m asl some 60 km W of Barcelona. A recent comprehensive hydrogeological study of the area has been made by Lambán Jiménez (1998), from which the following summary has been compiled. The Anoia Unit underlies 160 km^2 and constitutes a multi-layered carbonate aquifer system of Triassic and Eocene age. The unit is surrounded on all sides by a variety of Caenozoic and Palaeozoic strata of relatively low permeability. Within the Anoia Unit are three carbonate aquifer horizons (Lower Muschelkalk, Upper Muschelkalk and Lower Eocene) which are fissured and lightly karstified. These are inter-bedded with two units which are generally presumed to be of lower permeability: a silty and clayey sand (Middle Muschelkalk) and a marly gypsiferous sequence (Keuper). Nevertheless, local karstification of the marls and gypsum results in the Keuper being highly permeable. The Lower Eocene aquifer is up to 60 m thick, while the Upper and Lower Muschelkalk aquifers are each about 100 m. However, folding produces greater apparent thicknesses. The whole Anoia Unit is smoothly folded, with the axes in a south-west to north-east disposition, and with all axes plunging to the north-east. Consequently, the older formations cropout to the south of the unit while the younger ones cropout in the north. These structural patterns control the hydraulic behaviour of the unit, with the degree of confinement of the aquifer layers varying over short distances.

Large areas of unsaturated limestone which overlie low permeability formations at altitude serve to conduct recharge to the lower-lying saturated areas. Discharge takes place mainly through three major spring zones (the largest being at Capellades; see Fig. 1), though there are also a number of public supply wells. Piezometric records over the last 23 years show a declining trend attributable to the onset of well abstractions in 1978. Data on the transmissivity of the aquifer system come from some pump tests carried out in the 1970s, from qualitative field observations, and from point dilution tracer tests performed in the 1970s and 1990s. In general, transmissivity values are higher in (or close to) fractured zones, ranging from 2000 to 3000 m^2 d^{-1}. Outside of the fracture zones values range from low (probably less than 10 m^2 d^{-1}) to medium transmissivity (Lambán Jiménez 1998).

Serra de Tramuntana, Mallorca (Spain) – Mediterranean climatic zone

The island of Mallorca lies approximately 160 km E of the Spanish mainland (Fig. 1), and has a total area of 3626 km^2. The island is subdivided into three geomorphological regions that coincide with the main hydrogeological systems: the high and abrupt mountain range called Serra de Tramuntana, which rises to 1400 m and effectively forms a 80 km long 'rampart' along the north-western coast of the island; the Central Plain, of relatively low relief; and the range called Serra de Llevant, in the south-east of the island.

The climate of the island is of Mediterranean type, with mean annual temperature around 17°C and mean annual rainfall quite variable due to orographic effects (ranging from 400 mm in low-lying zones of the island up to 1500 mm in the Tramuntana Range). Also, there is a wide variation in rainfall between dry and wet years, and between winter and summer in each year. Localized intense rainstorms are quite common.

The Serra de Tramuntana receives the highest rainfall of any area of Mallorca, which makes the aquifers of the area attractive as water resources, despite the difficulties inherent in groundwater development in mountainous terrain. The north-eastern and central sectors of the Serra de Tramuntana are predominantly composed of carbonate rocks of Liassic (Jurassic) and Raethian (Triassic) age, with limestones, dolomites and breccias predominating. These carbonates form aquifers that drain to three major springs, the most prolific of which is the Sa Costera Spring (study area 4 on Fig. 1). The catchment of Sa Costera includes the Puig Major, which is the highest peak in Mallorca. The Sa Costera spring is the largest single discharge of natural freshwater in Mallorca, and is therefore of considerable interest as a potential water resource. Unfortunately, the natural drainage of the Sa Costera is straight into the sea from a very remote cliff line (Fig. 1), and major investment is required to divert the waters by submarine pipeline to the major demand centres around the island's capital, Palma. The susceptibility of the Sa Costera to climate change is therefore a material consideration in the cost-benefit balance for its exploitation.

Recent geological and hydrogeological field investigations (Cardoso da Silva 1997) confirm that the catchment of the Sa Costera is structurally complex, with imbricate thrust sheets giving rise to a 'multi-reservoir' aquifer system underlying a total area of some 10 to 17 km^2. Definition of the aquifer hydrodynamics is not an easy task, and it is made no easier by a shortage of reliable data. The range of discharge rates of the spring is subject to considerable uncertainty, with two earlier unpublished studies giving alternative ranges of 4.9–22.7 $\times 10^6$ m^3 a^{-1} (average of 10.4 $\times 10^6$ m^3 a^{-1}) and 2.6–26.3 $\times 10^6$ m^3 a^{-1} (average of 8.2 $\times 10^6$ m^3 a^{-1}). New gauging facilities were installed at the start of the GRACE project to address these discrepancies. Data obtained to date reveal the Sa Costera to have a rather steady baseflow in the order of 17 Ml d^{-1}, with peak flows reaching 170 Ml d^{-1} in wet winter periods. These figures suggest that the smaller of the two annual flow ranges quoted above is probably the more reasonable.

Climate change assessment methodology

Aquifer simulation methods

Two groundwater modelling codes were developed within the GRACE project to support the climate change assessments: CAVE (*C*arbonate *A*quifer *V*oid *E*volution), and BALDOS.

CAVE simulates coupled flow and non-equilibrium carbonate dissolution in dual-porosity carbonate aquifer systems. CAVE is based upon the well-known USGS MODFLOW code (McDonald & Harbaugh 1984). MODFLOW has previously been adapted to simulate two-domain (= dual porosity) groundwater flow in carbonate aquifers by means of setting up two standard MODFLOW grids in parallel, with water being exchanged between them according to the heads in the corresponding cells of the two grids (Teutsch & Sauter 1991). In other words, the diffuse domain and the conduit domain within a carbonate aquifer are each represented by separate, but inter-communicating, equivalent porous media. This approach gives good results where only quantities of water are of interest (Sauter 1992). However, where carbonate dissolution needs to be considered, it is necessary to define geometrically the surface on which the dissolution takes place. In significantly fractured carbonate aquifers, recharge waters which are under-saturated with respect to calcite will penetrate the aquifer primarily in the conduit flow system. It therefore makes sense to define the geometry of the dissolving surface as being the walls of the conduit system. To achieve this goal, CAVE was configured so that the conduit domain is represented by a pipe network model of readily defined geometry, rather than by an equivalent porous medium as in previous studies. The algorithmic approach necessary to attain this model representation is documented in detail by Younger *et al.* (1997), and has also been published in summary form by Clemens *et al.* (1996, 1997) and Sauter *et al.* (1997). It should be noted here that the GRACE project only considered 2-D regional flow, using a single MODFLOW layer for the diffuse domain. Further generalization of the approach to three dimensions, and to the case where the diffuse domain is unsaturated, have recently been described by Adams & Younger (2001). Within the GRACE project, the CAVE code was used to simulate the German and UK study systems. The Spanish study systems had insufficient data to support the use of fully distributed modelling using CAVE. For these systems BALDOS was developed and applied.

BALDOS is a semi-distributed recharge-routing model that simulates both direct and indirect groundwater recharge (*sensu* Lerner *et al.* 1990). BALDOS was developed from a pre-existing code (BALAN), which calculates classic Penman-Grindley based infiltration and closes the water balance with measured aquifer discharges by the use of empirical time-lag coefficients (Samper & García-Vera 1992). As at least some of the recharge in carbonate terrains is indirect (for instance via dolines), the original BALAN code was updated to include process representations for recharge via dolines, either directly from surface run-off or via drainage in the epikarstic zone. As for CAVE, full algorithmic details for BALDOS are given by Younger *et al.* (1997). BALDOS was applied to the Anoia Unit and the Serra de Tramuntana aquifers (Spain).

Both CAVE and BALDOS were thoroughly tested for their ability to reproduce observed flow and carbonate chemistry patterns in the study aquifer systems (Younger *et al.* 1997). This engendered confidence in their applicability for the simulations of possible future climate change scenarios. Nevertheless, it must always be borne in mind that the use groundwater models to predict circumstances which necessarily fall outside of the range of observed conditions is fraught with uncertainties. To avoid giving the misleading impression that definitive predictions of the future had been made, rather than the tentative numerical experiments which the authors consider the simulations to constitute, all climate change impact assessments were undertaken for 'fictitious' catchments, which closely resembled real catchments but were not explicitly identified with them. Thus the models reported here may be termed 'scoping models' inasmuch as the predictions that they yield are meant to be indicative of the possible ranges of aquifer responses that might be encountered, providing a basis for further discussion rather than for site-specific management interventions.

Using GCM outputs to define recharge scenarios

The problems of spatial resolution and inadequate process representation which beset the application of GCM models in hydrological impact studies (Gleick 1986, 1989; Mavromatis & Jones 1998; Wilby *et al.* 1998) have been discussed above. To overcome problems of spatial resolution an approach developed by Viner & Hulme (1994) was adopted and modified. In this approach mesoscale (or even more localized) climatic patterns are related to the appropriate GCM values, which are regarded as grid-averaged values (Skelly & Henderson-Sellers 1996). With respect to problems of process representation, the starting point for the GRACE simulations was to recognize that GCM predictions of changes in primary climatological variables (e.g. temperature and precipitation) are generally regarded as being better than their predictions of secondary (derived) parameters such as evapotranspiration and surface run-off (e.g. Reed 1986; Gregory & Mitchell 1995; Kilsby *et al.* 1998). Recharge was therefore calculated using GCM predictions of temperature and precipitation (in lieu of any credible alternatives), utilizing well-known formulae to calculate potential evapotranspiration [e.g. Thornthwaite and/or Blaney-Criddle; see Palutikof *et al.* (1994) for a discussion of their applicability in climate impact studies], and standard (Penman-Grindley type) soil–water budgeting methods. To this end, GCM grid-point estimates for mean surface air temperature

(T) and mean precipitation (P) for the GRACE study areas were obtained from the UK Climate Impacts-Link project (coordinated by the University of East Anglia, UK).

GCM modelling is a rapidly developing field of research. Until the mid-1990s, GCM-based simulations of possible future climate scenarios were necessarily based on the use of so-called 'equilibrium models', in which the dynamics of the atmosphere with a fixed CO_2 concentration were simulated. Concentrations are normally expressed as '$1 \times CO_2$' for 'pre-warming' atmospheric CO_2 concentrations (323 ppm by volume, representing conditions in about 1968), and thus for various increments up to the predicted eventual doubling of pre-industrial CO_2 (to 646 ppm by volume) which is designated '$2 \times CO_2$'. Intermediate states are definable only by interpolation between the predictions of 'dynamic equilibrium' conditions represented by each of the models. By the mid-1990s a new generation of GCM models, termed 'transient GCMs', was becoming available (Viner *et al.* 1994). Transient GCMs simulate progressive changes in atmospheric circulation over simulation periods of many years in response to gradually rising atmospheric CO_2. Clearly this is conceptually more realistic than assuming that atmospheric circulation reaches a quasi-equilibrium at all stages of the increase in CO_2 concentration. However, transient GCMs suffer from the same problems of 'cold start' which bedevil transient groundwater flow models, in that initial conditions prescribed *a priori* by the modeller may not be consistent with the internal parameterization of the model at the start of the simulations. At the time of the GRACE study, serious 'cold start' problems with transient GCMs were still being tackled, and only a limited range of greenhouse gas emission scenarios had been incorporated into transient GCM runs (Viner & Hulme 1994). Consequently, the GRACE predictions were developed by an interpolation method based on equilibrium GCM output which was considered the most robust data available at the time.

The equilibrium GCM output used in GRACE originated in an equilibrium experiment named 'UKHI' (Viner & Hulme 1993*a*), which was obtained using a second-generation GCM code, configured to run with 10 min time steps, with radiation updated every 3 h, and assuming the oceanic heat reservoir to be a 50 m single mixed-layer 'slab-ocean'. Simulations were then obtained for 'Control' ($1 \times CO_2$) and 'Perturbed' ($2 \times CO_2$) cases. For each case, the simulations were performed until the climate had reached a quasi-equilibrium state. Thereafter, 10 years of data were obtained for each scenario. The output of these simulations provides spatial fields of the climate parameters of interest with cell-average values provided at grid-centre points. The resultant model spatial temperature change field at each GCM grid-point i as generated in these equilibrium experiments can then be expressed as grid-point changes in T per °C of global warming, i.e. the standardized change for a grid-point i value ($\Delta T_i{}^*$) from a GCM experiment:

$$\Delta T_i{}^* = (T_i(2 \times CO_2) - T_i\,(1 \times CO_2))/(\Delta T_{2x}) \quad (1)$$

where $T_i(2 \times CO_2)$ and $T_i(1 \times CO_2)$ are the 10-year monthly average temperature values at GCM grid-point i for the Perturbed and Control experiment fields respectively; and ΔT_{2X} is the GCM model climate sensitivity, which is defined as the global annual equilibrium temperature change for a doubling of CO_2 concentration (within the scenario represented by the particular GCM experiment) (e.g. Cess & Potter 1988). The UKHI experiment gives a ΔT_{2X} of 3.5°C, and this value is used to standardize the parameter change fields. (Ten-year monthly average values at a grid-point were used to define climatically relevant change values in GRACE.)

For precipitation, standardized changes between Control and Perturbed runs were defined in a similar manner, save that changes could here be expressed as percentages:

$$\Delta P_i{}^* = 100\ \{(P_i(2 \times CO_2) - P_i\,(1 \times CO_2))/(P_i\,(1 \times CO_2))\} \quad (2)$$

where $P_i(2 \times CO_2)$ and $P_i\,(1 \times CO_2)$ are the 10-year monthly average precipitation values at GCM grid-point i for the Perturbed and Control experiment fields respectively, and $\Delta P_i{}^*$ is the percentage change in the precipitation value at grid-point i for the climate perturbation experiment.

An assessment of percentage changes is used in equation (2) because precipitation is highly variable from month to month and place to place, and absolute values are difficult to interpret due to discrepancies between control climate data and direct observations.

Equations (1) and (2) do not directly provide for assessment of temporal changes in climate. Without resort to transient GCM output, this is best achieved by linking equilibrium GCM output to estimates of the effect of transient emissions of greenhouse gases on global temperature rise, assuming that changes in the T and

P variables directly follow this trend (e.g. Viner & Hulme 1993*b*). A climate model named 'MAGICC' (Model for the Assessment of Greenhouse gas Induced Climate Change; Hulme *et al.* 1995) has been used for this purpose. MAGICC provides mutually consistent estimates of global-mean CO_2 concentration and temperature change at yearly intervals over the future period 1990 to 2100, as functions of a range of user-selected scenarios of anthropogenic emissions of carbon dioxide, methane, nitrous oxide and halocarbons. MAGICC is conveniently implemented within the software package SPECTRE (Barrow *et al.* 1994). Although there are a number of uncertain parameters in MAGICC, by far the most important is the climate sensitivity (ΔT_{2X}). The Inter-Governmental Panel on Climate Change (IPCC) best estimate of ΔT_{2X} is 2.5°C, with a range of 1.5 to 4.5°C. In the SPECTRE software, MAGICC has been run for eight prescribed greenhouse gas emissions scenarios for each of these three values, generating a range of uncertainty for the resulting estimates of global mean temperature rise from each scenario (termed LOW, MID and HIGH estimates respectively). The above protocol allows scaling of the GCM climate variables (*T*, *P*) change scenario at any grid-point both in time (following defined transient emissions scenarios for greenhouse gases), and also with respect to different GCM model climate sensitivities.

From the plethora of alternative greenhouse gas emission scenarios available in MAGICC, three were selected for detailed aquifer response simulations using CAVE and BALDOS (Fig. 2). At the optimistic end of the spectrum is the 'fossil-fuel free energy future' (FFEF) scenario (sometimes called the 'Greenpeace scenario'), assuming a low climate sensitivity (ΔT_{2X} = 1.5°C). This scenario is referred to hereafter as 'FFEF LOW'. The mid-range and upper-bound estimates of likely climate change are both derived from the 1992 IPCC Scenario A (Houghton *et al.* 1992), hereafter referred to as IS92a, which basically assumes that 'business-as-usual' prevails (i.e. CO_2 controls, limitations on deforestation and other important controls on global warming do not become significantly more efficient than at present). The mid-range scenario is obtained by coupling the IS92a emissions scenario to the median estimate of climate sensitivity (ΔT_{2X} = 2.5°C), producing the 'IS92a MID' scenario. The upper-bound ('worst-case') scenario ('IS92a HIGH') couples the IS92a emissions prognosis with the maximum climate sensitivity value (ΔT_{2X} = 4.5°C). The long-term global temperature rises predicted

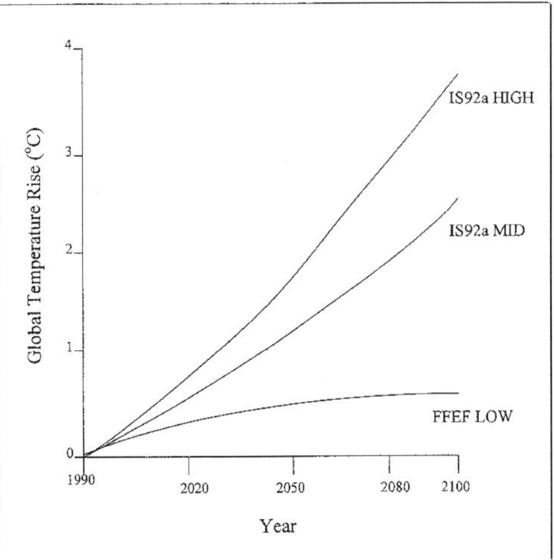

Fig 2. Global average temperature changes for the three scenarios selected for the climate change impact simulations.

for each of these three scenarios are shown in Figure 2. The implications of the three scenarios for the study aquifers (over 50 years from 1996 to 2045) can now be examined.

Climate change predictions to 2045 AD.

Timescale for predictions

Figure 2 illustrates that under the entire range of future climate scenarios, global temperatures are predicted to rise more or less steeply over the next century (and, by implication, beyond). A number of factors (which were not included in the climate models upon which Figure 2 is based) may in the event serve to slow down the temperature rise, such as the possibility of algal blooms in the world's oceans acting as a major sink for atmospheric CO_2, a process which appears to have contributed to the abrupt halt in global warming during earlier periods of the Earth's history (e.g. Bains *et al.* 2000). Nevertheless, the current consensus is that global warming is a reality, which will certainly give rise to real problems over the coming decades. As water resources managers are rarely able to justify discounting cost-benefit analyses over time-scales in excess of 50 years, predictions using the methodology outlined above have been restricted to the 50-year interval 1996–2045. Nevertheless, when examining the predictions presented below, it should be borne in mind that

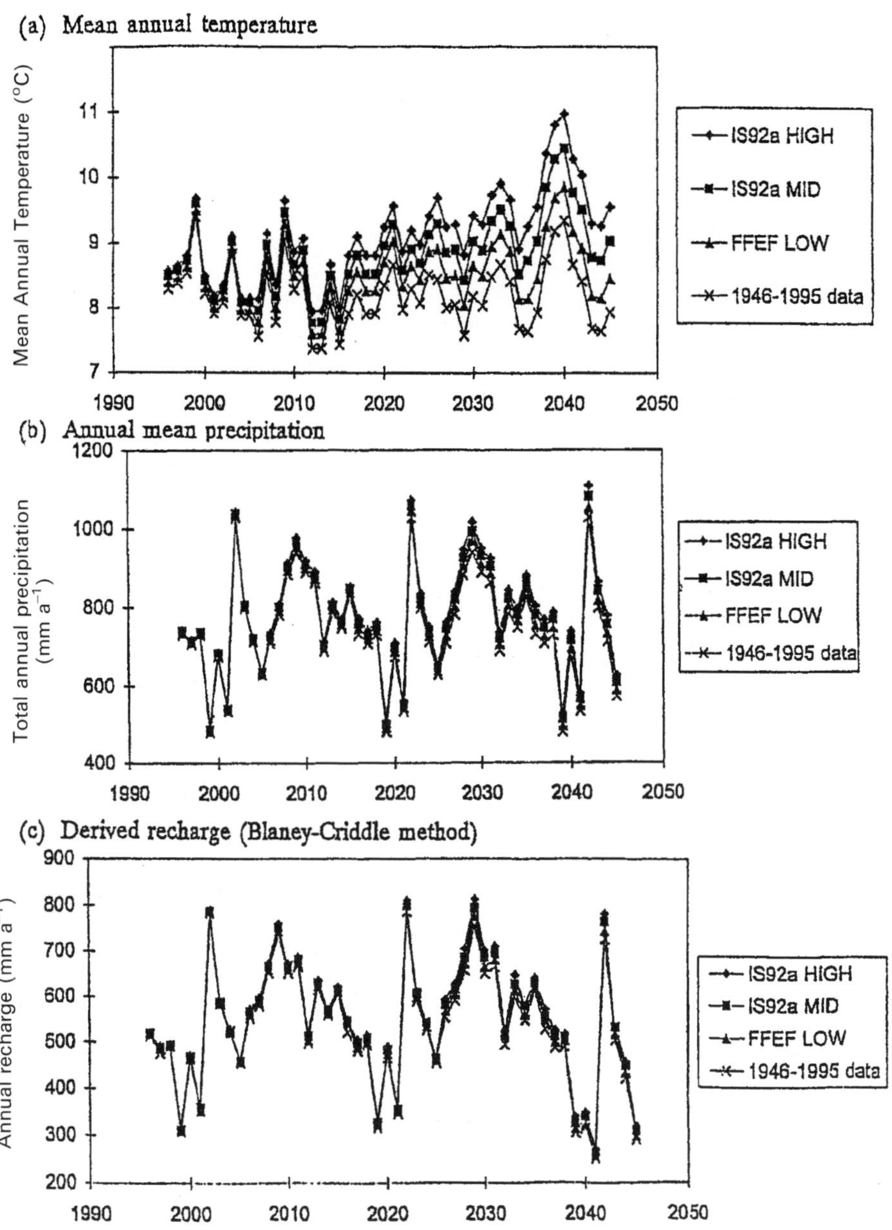

Fig. 3. Temperature (**a**), precipitation (**b**) and recharge (**c**) predictions for the 50 years to 2045 AD, for the Yorkshire Chalk study system.

the current consensus is that global temperature rises are likely to continue beyond 2045 AD (Hadley Centre 2000). Hence, even where predictions of changes to 2045 AD are relatively benign (as in the case of the Yorkshire Chalk), there is no guarantee that more negative water resources impacts will not ensue after 2045 AD.

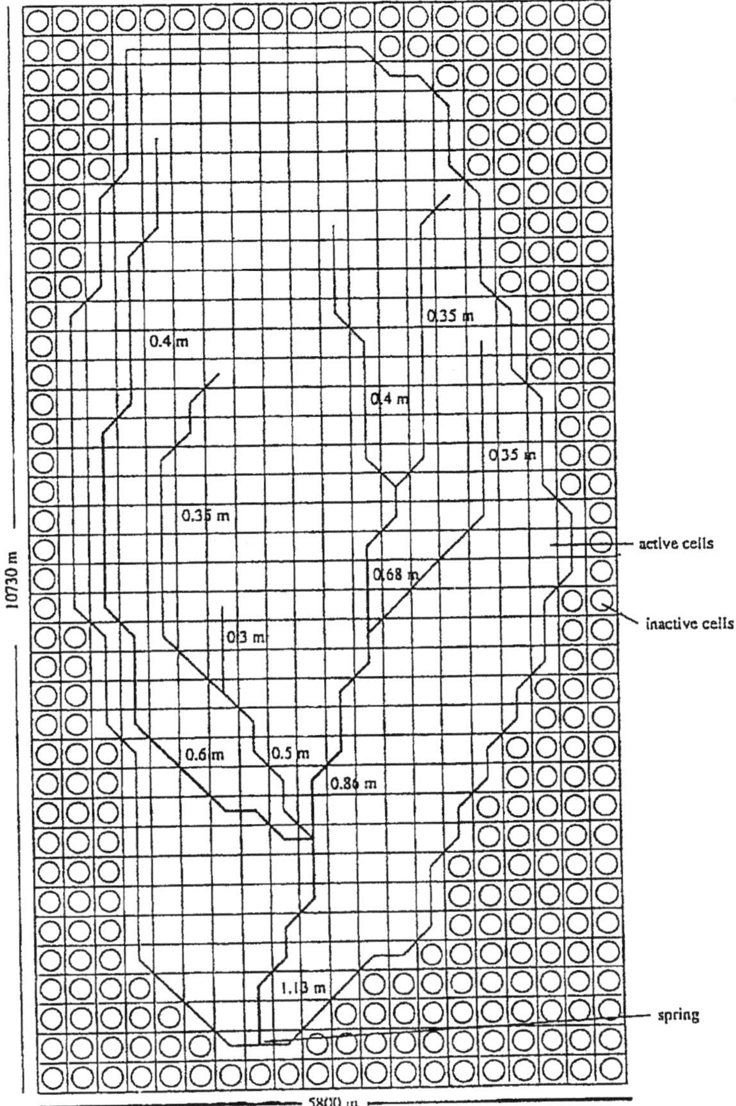

Fig. 4. Model grid for the hypothetical catchment representative of the Gallusquelle Aquifer study system in Germany.

Experimental approach

To facilitate comparisons between the four study systems in three quite different climate zones, it is necessary to ensure that we are comparing like with like. To this end, all of the simulations reported below were developed in accordance with the following experimental plan:

(1) All predictions are based on the temperature and precipitation estimates for future climate scenarios FFEF LOW, IS92a MID and IS92a HIGH (as described above) specific to GCM grids corresponding to each of the study systems. The temperature and precipitation data were used to derive

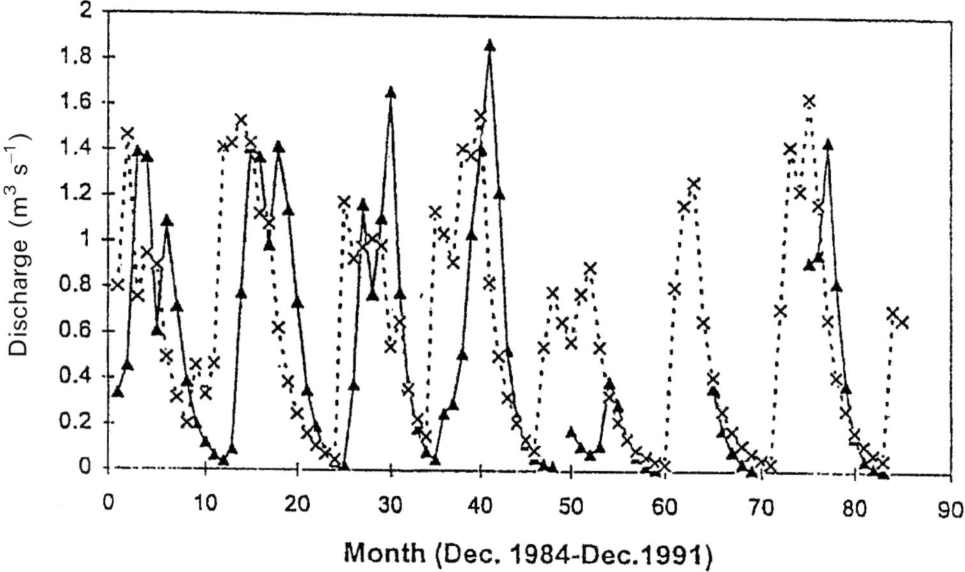

Fig. 5. A comparison between modelled (dashed line) flow patterns for the hypothetical catchment in the Yorkshire Chalk, and observed (solid lines) flow patterns in the real Tibthorpe spring catchment, obtained without site-specific calibration, illustrating that the magnitudes and general durations of periods of high flow and baseflow are captured in the hypothetical catchment model.

50-year time series of estimates of future recharge rates, using site-specific 'transfer functions' (such as a locally calibrated version of the well-known Blaney-Criddle formula; e.g. Palutikof *et al.* 1994). These transfer functions were developed by relating observed mean monthly values of temperature and precipitation to accepted actual recharge figures (obtained for present day conditions using Penman or similar methods, and checked by incorporation in regional groundwater flow models). An example of the temperature, precipitation and recharge predictions for the 50 years to 2045 AD is shown in Figure 3 for the Yorkshire Chalk study system. Sauter and Liedl (1998) and Manzano *et al.* (1998) have published the analogous predictions for the German and Spanish study systems respectively.

Adoption of this approach necessarily assumes that the functional relationship between temperature, precipitation and actual recharge will remain stationary as climate changes. While this is arguably a bold assumption, there is no practical alternative given the current state-of-the-art in predicting soil and vegetation changes in response to global warming.

(2) A 'hypothetical catchment' was developed as the basis for climate change experiments for each study area. The properties of the hypothetical catchments were defined in accordance with the generalized hydrogeological properties of the study aquifers identified during the individual regional investigations. Each hypothetical catchment was established to have the following properties:

(a) it will be a well-defined catchment area feeding a spring, to avoid the need to arbitrarily define well pumping regimes etc.
(b) it will be bounded for the most part by no-flow boundaries, except where fixed heads are needed to avoid singularities. This will ensure that the response is dominated by changes in recharge dynamics, without allowing arbitrary boundary inflows to compensate for reductions in recharge. Figure 4 shows the model grid for the hypothetical catchment representative of the Gallusquelle study system in Germany. (In this particular catchment the conduit system is based on well-characterized dendritic cave systems in the region, which have

been the subject of underground exploration and tracer testing (Sauter 1992). The equivalent grid for the Yorkshire case lacks this interesting feature.)

(c) for modern values of recharge, the catchment will exhibit an areally normalized flow distribution (e.g. flows expressed in terms of $m^3 s^{-1} km^{-2}$ of catchment area) which resembles that of a similar real system in the study area. Acceptable resemblance will be to order-of-magnitude precision and will be assessed in terms of the objective functions listed under point 4 below. A comparison between modelled and observed flow patterns obtained during model development for the hypothetical catchment in the Yorkshire Chalk is shown in Figure 5.

(3) Unless there were compelling reasons to the contrary (which proved to be the case for the Spanish catchments; Manzano *et al.* 1998), all simulations were performed using the CAVE software with the carbonate dissolution mode enabled. The maximum time-step for flow modelling purposes was monthly.

(4) The overall objective function for the flow simulations is the distribution of spring discharge rates. Because monthly time-steps were used, it is not feasible to use flow distribution descriptors which are most appropriately derived from daily data (such as Q_{95} etc.). It was therefore decided that the following descriptors would be logged throughout the prediction periods: total annual flow; minimum, maximum and mean flows for January and June; the maximum and minimum flow in each year; and the percentage deviation of the above from the 10-year average values for a simulation using modern temperature and precipitation values.

(5) The moving average of the calcium content (mg l^{-1}) of the spring discharge water was calculated throughout the 50-year simulation period, to allow evaluation of whether an increase (or indeed a decrease) in the total hardness of water supplies might be anticipated under changing climate conditions. Changes in conduit diameter at no less than two user-selected positions within the catchment were also logged at yearly intervals throughout the 50-year simulation period, to enable an evaluation of the possibility that enhanced dissolution rates serve to increase aquifer permeability.

Simulation results - flows

Tables 1–3 summarize the principal findings of the study for each of the four study areas. The tables compare the 10-year average behaviour at the end of the 50-year simulation period (i.e. the modelled results for 2036–2045) with the observed values for the 10-year period which preceded the modelled period (i.e. 1986–1995).

In assessing the predicted changes in groundwater recharge and discharge, the best way to proceed is to compare the magnitude of predicted changes with the magnitude of inter-annual variability (e.g. Hulme *et al.* 1999). Using this sort of approach, any change of less than about 10% is unlikely to be considered significant in water resources terms. Using a threshold of this magnitude, the results presented in Tables 1–3 do predict significant reductions (i.e. negative percentage values >10) in one or more of the indicator variables for the hypothetical catchments in Germany and Mallorca. By contrast, the magnitudes of predicted changes in flow rates in the Anoia Unit and the Yorkshire Chalk catchments are unlikely to exceed the magnitude of inter-annual variability. Nevertheless, the predicted trend in the Anoia Unit is towards decreasing groundwater recharge and discharge. In the case of the Yorkshire Chalk, the IS92a MID and HIGH scenarios (Tables 2 and 3 respectively) suggest that year-round increases in flow are likely. Only under the FFEF LOW scenario (Table 1) are any reductions in flow predicted for the Yorkshire Chalk, and even in that case the total annual flow is still predicted to rise slightly. On the face of it, therefore, global warming potentially represents local 'good news' for users of the Yorkshire Chalk aquifer. However, there is little scope for complacency as the global temperature trends shown in Figure 2, taken together with the latest predictions of climate change (Hadley Centre 2000), strongly suggest that extrapolation of predictions beyond 2045 might well lead to increases in evapotranspiration exceeding increases in rainfall, so that recharge will eventually decline later in the 21st century.

Turning to the results obtained for the Gallusquelle catchment in Germany, although rises are consistently predicted for the mean and maximum January flows (which coincide with periods of generally low demand), all other flows are predicted to be significantly in deficit compared with pre-1995 conditions (Tables 1–3). Nevertheless, as Sauter and Liedl (1998) have pointed out, the amplitude of inter-annual

Table 1. - *Summary of 'best-case' climate change impact predictions for the four GRACE study systems.*

Study System	δQ_{total} Annual	δQ_{min} Annual	δQ_{max} Annual	δQ_{min} January	δQ_{mean} January	δQ_{max} January	δQ_{mean} June	δQ_{min} June	δQ_{max} June
Yorkshire Chalk, United Kingdom	+5	–1	–0.5	–1	–0.1	–0.5	–0.1	–0.1	–0.5
Gallusquelle Aquifer, Germany	–12	–16	–10	–12	+10	+23	–31	–25	–26
Anoia Unit, Catlunya, Spain	–1	–3	+2	–1	+1	+2	–2	–1	–0.5
Serra de Tramuntana, Mallorca, Spain	–4	–6	–2	–	–	–	–	–	–

Results generated using GCM output for the optimistic scenario 'FFEF LOW', as described in the text. All values are in percentages, as the changes in the 10-year mean values between the periods 1986–1995 and 2036–2045.

δQ_{total} Annual, % change in the average total annual flow; δQ_{min} Annual and δQ_{max} Annual, % changes in the minimum and maximum annual flows respectively; δQ_{min} January, dQ_{mean} January, δQ_{max} January, and δQ_{min} June, δQ_{mean} June, δQ_{max} June are the corresponding % changes in minimum, mean and maximum flows in January and June as appropriate.

Table 2. *Summary of 'best-estimate' climate change impact predictions for the four GRACE study systems*

Study System Annual	δQ_{total} Annual	δQ_{min} Annual	δQ_{max} January	δQ_{min} January	δQ_{mean} January	δQ_{max} June	δQ_{min} June	δQ_{mean} June	δQ_{max}
Yorkshire Chalk, United Kingdom	+9	+1	+2	+1	+0.2	+1	+0.1	+0.1	+0.5
Gallusquelle Aquifer, Germany	−13	−17	−8	−10	+9	+21	−32	−27	−27
Anoia Unit, Catlunya, Spain	−3	−6	+2	−2	+1	+3	−4.5	−2	−1
Serra de Tramuntana, Mallorca, Spain	−7	−11	−5	–	–	–	–	–	–

Results generated using GCM output for the mid-range scenario 'IS92a MID', as described in the text. All values are in percentages, as the changes in the 10-year mean values between the periods 1986–1995 and 2036–2045.

δQ_{total} Annual, % change in the average total annual flow; δQ_{min} Annual and δQ_{max} Annual, % changes in the minimum and maximum annual flows respectively; δQ_{min} January, δQ_{mean} January, δQ_{max} January, and δQ_{min} June, δQ_{mean} June, δQ_{max} June, are the corresponding % changes in minimum, mean and maximum flows in January and June as appropriate.

Table 3. *Summary of 'worst-case' climate change impact predictions for the four GRACE study systems.*

Study System	δQ_{total} Annual	δQ_{min} Annual	δQ_{max} Annual	δQ_{min} January	δQ_{mean} January	δQ_{max} January	δQ_{min} June	δQ_{mean} June	δQ_{max} June
Yorkshire Chalk, United Kingdom	+21	+2	+4	+1.5	+0.5	+2	+0.1	+0.5	+1
Gallusquelle Aquifer, Germany	–13	–17	–8	–10	+8	+20	–32	–28	–27
Anoia Unit, Catalunya, Spain	–3	–8.5	+2	–2.5	+0.5	+3	–7	–2	–1.5
Serra de Tramuntana, Mallorca, Spain	–11	–16	–7	–	–	–	–	–	–

Results generated using GCM output for the optimistic scenario "IS92a HIGH", as described in the text. All values are in percentages, as the changes in the 10-year mean values between the periods 1986–1995 and 2036–2045.

δQ_{total} Annual, % change in the average total annual flow;. δQ_{min} Annual and δQ_{max} Annual, % changes in the minimum and maximum annual flows respectively; δQ_{min} January, δQ_{mean} January, δQ_{max} January, and δQ_{min} June, δQ_{mean} June, δQ_{max} June, corresponding % changes in minimum, mean and maximum flows in January and June as appropriate.

variability exceeds the magnitudes of the declines in flow up to 2045. Similar results have been reported in other climate change studies (e.g. Hulme *et al.* 1999). Consequently, unambiguous proof of any systematic decline in flows is only likely to be demonstrable in the second half of the 21st century. One remarkable aspect of the Gallusquelle predictions is that the results for all three scenarios are tightly clustered, with no statistically significant difference between FFEF LOW (Table 1) on the one hand and IS92a HIGH (Table 3) on the other. This somewhat surprising result prompts one to question whether the GCM output for the grid point relevant to Gallusquelle is subject to some mathematical singularity related to GCM configuration. For instance, the proximity of this GCM grid point to those which correspond to the Alps might be responsible for local instability in model performance. This possibility cannot be resolved using the data available to the GRACE team, but may well be worthy of re-examination using the next generation of GCM output.

The results for the Anoia Unit reveal patterns that are similar to, but less extreme than, those obtained for Gallusquelle, with consistent increases in mean and maximum January flows, but decreases in virtually all other indicators. The maximum decrease in recharge that can be expected in the region of the Anoia Unit over the next 50 years or so amounts to about 8% of the total present-day recharge, while the average decrease is about 3%. As the average present-day recharge rate for the period 1942–1991 was estimated to be some 80×10^6 m^3 a^{-1} (over a recharge area of 156 km^2; Lambán Jiménez 1998), the predicted decrease does not represent a particularly dramatic reduction of groundwater reserves. However, when the high inter-annual variability of recharge in the area is taken into account (with an estimated minimum in the past 50 years of 40×10^6 m^3 a^{-1}) the predicted decline in recharge suggests that droughts in the area may become somewhat more severe. Towards the end of the simulated period, recharge becomes increasingly more irregular and variable from year to year, even producing discrete recharge values higher than those of the present day (Younger *et al.* 1997; Manzano *et al.* 1998). This highlights the concern that increases in inter-annual variability may be a more serious general result of global warming than decreases in total effective rainfall (C. Kilsby, University of Newcastle, pers. comm., 1999).

Given the shortage of pre-1995 data for the Sa Costera system, it proved possible to calculate only a few of the indicators in Tables 1–3 to sufficient precision to warrant their inclusion. The figures given suggest that the Sa Costera catchment is the most vulnerable of all four systems studied, with a maximum decrease in recharge (relative to pre-1995 patterns) of as much as 16%. Simulated time series (Manzano *et al.* 1998) suggest an accelerated rate of decline in recharge towards the end of the simulated period. As for the Anoia Unit, the inter-annual variability of recharge increases during the latter part of the 50-year period. However, unlike the Anoia Unit, at no point does the predicted recharge exceed pre-1995 values.

The contrasting predictions for the four study systems clearly demonstrate the general sensitivity of these carbonate aquifers to climatic change. Where maritime climatic influences are strong (as in the UK), increased oceanic evaporation induced by global warming may well result in overall increases in available resources. Even in the continental climatic zone (e.g. southern Germany), atmospheric moisture derived from increased oceanic evaporation may sustain winter rains at their present level, but declines in summer rainfall (coupled with more intense evapotranspiration in a warmer climate), can be expected to result in an overall decrease in available groundwater resources. The inter-annual variability of rainfall in the Mediterranean climatic zone is well known (though its meteorological causes are complex). If this variability is exacerbated as a consequence of global warming (as we predict), then the need for major investment in storage systems (surface reservoirs or – to minimize evaporation losses – the use of artificial groundwater recharge) will become ever more pressing.

It must here be re-emphasized that global warming impacts are expected to accelerate beyond the 50-year investigation horizon of these simulations (Hadley Centre 2000), which is likely to have important implications for recharge variability. In particular, further work is necessary to evaluate whether the predicted rise in recharge in the Yorkshire Chalk system might prove to be followed by a decline in recharge in the decades beyond 2045 AD.

Simulation results – water quality (hardness)

Predictions of increases in the hardness of the groundwaters were obtained for the UK and German systems only, since the CAVE code was not applied to the Spanish systems. Figure 6 shows the predicted increases in dissolved calcium in the Gallusquelle aquifer over the 50-

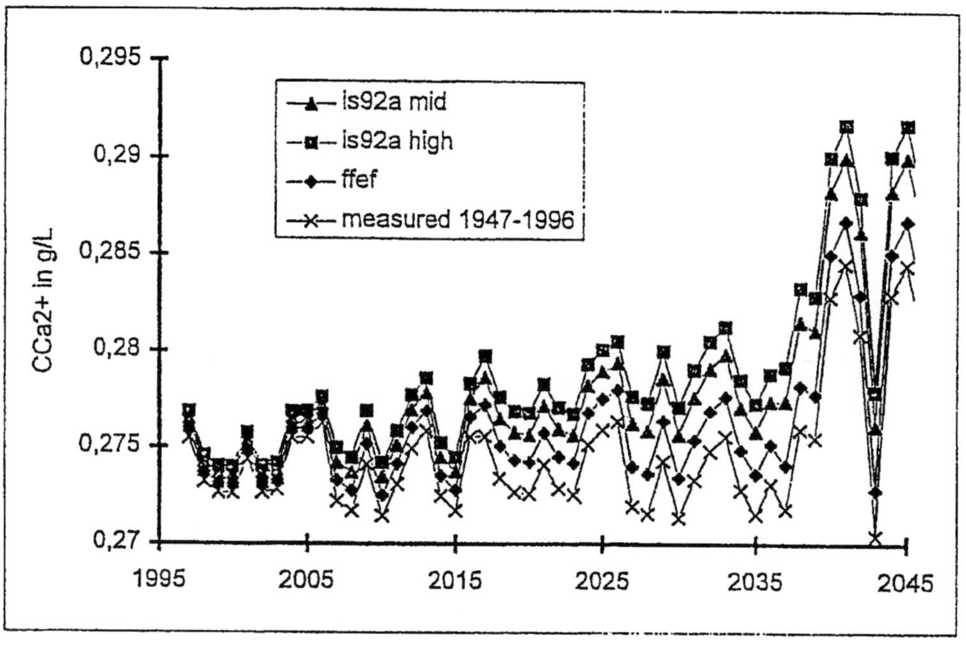

Fig. 6. Predicted climate change impacts on natural water quality: dissolved calcium concentrations('CCa^{2+}') in water flowing from the hypothetical catchment representing the Gallusquelle aquifer, Germany, compared with measured values.

year simulation period to 2045 AD. The overall rising trend in response to increases in atmospheric CO_2 concentrations is clearly evident for all three climate change scenarios. The noticeable peaks in the final decade of the simulations are simply results of a lack of dilution, since they correspond precisely to two episodes of low winter recharge rates. Despite the clear increases in dissolved calcium shown in Figure 6, it is important to examine the *y*-axis labels for the maximum increase in dissolved calcium in the Gallusquelle to 2045 AD amounts to no more than 8 mg l^{-1} (Sauter & Liedl 1998). The results obtained for the Yorkshire Chalk aquifer are very similar, with a maximum increase in dissolved calcium $\leq$10 mg l^{-1}. These increases are negligible in water resources terms, and therefore have little more than curiosity value. Nevertheless, steady increases in hardness in the world's carbonate aquifers represent a significant sink for atmospheric CO_2, which may make a non-negligible contribution to slowing global warming over long time-scales (cf. Kump *et al.* 2000; Liu & Zhao 2000).

Conclusions

Taken at face value, the flow predictions reported here suggest that considerable cause for concern exists in relation to the long-term viability of a number of carbonate aquifers currently used for public supply in central and southern Europe, even when predictions are made using the most optimistic of all future climate scenarios (FFEF LOW). Needless to say, these results are surrounded by sufficient uncertainty that they cannot be used as definitive predictions of future aquifer behaviour at particular points in time and space. A more probabilistic approach to the simulations might have resulted in formal confidence limits on these predictions, but would have been unlikely to change the fundamental conclusion that significant reductions in available resources can be expected in low- and mid-latitude carbonate aquifers in Europe by the middle of the present century.

The work reported here was funded by the European Commission, grant number Project EV5V - CT94 -

0471. We are grateful to Dirk Hückinghaus, Torsten Clemens, Javier Lambán and C. Tore for their contributions to the project.

References

ADAMS, R. & YOUNGER, P. L. 2001. A strategey for modeling ground water rebound in abandoned deep mine systems. *Ground Water*, **39**, 249–261.

ARNELL, N. W. 1992. Impacts of climatic change on river flow regimes in the UK. *Journal of the Institution of Water and Environmental Management*, **6**, 432–442.

ARNELL, N. W. 1998. Climate change and water resources in Britain. *Climatic Change*, **39**, 83–110.

ARNELL, N. W. 1999. Climate change and global water resources. *Global Environmental Change – Human and Policy Dimensions*, **9**, S31–S49.

BAINS, S., NORRIS, R. D., CORFIELD, R. M. & FAUL, K. L. 2000. Termination of global warmth at the Palaeocene/Eocene boundary through productivity feedback. *Nature*, **407**, 171–174.

BARROW, E., HULME, M. & JIANG, T. 1994. *SPECTRE – Spatial and Point Estimates of Climate Change due to Transient Emissions*. Software Manual, Version 1.1, June 1994. Climate Research Unit, University of East Anglia, UK.

BOBBA, A. G., SINGH, V. P., BERNDTSSON, R. & BENGTSSON, L. 2000. Numerical simulation of saltwater intrusion into Laccadive Island aquifers due to climate change. *Journal of the Geological Society of India*, **55**, 589–612.

BOURAOUI, F., VACHAUD, G., LI, L. Z. X., LE TREUT, H. & CHEN, T. 1999. Evaluation of the impact of climate changes on water storage and groundwater recharge at the watershed scale. *Climate Dynamics*, **15**, 153–161.

CARDOSO DA SILVA, G. 1997. *Comportamiento de los manantiales del karst nororiental de la Serra de Tramuntana, Mallorca* (Behaviour of the springs of the northwest karst of the Serra de Tramuntana, Mallorca). PhD thesis, Universitat Politécnica de Catalunya, Barcelona (In Spanish).

CESS, R. D. & POTTER, G. L. 1988. A methodology for understanding and intercomparing atmospheric climate feedback processes in General Circulation Models. *Journal of Geophysical Research*, **93**, 8305–8314.

CLEMENS, T. HÜCKINGHAUS, D., SAUTER, M., LIEDL, R. & TEUTSCH, G. 1996. A combined continuum and discrete network reactive transport model for the simulation of karst development. *International Association of Hydrological Sciences*, **237**, 309–318.

CLEMENS, T., HÜCKINGHAUS, D., SAUTER, M., LIEDL, R. & TEUTSCH, G. 1997. Simulation of the evolution of maze caves. *In: Proceedings of the 6th Conference on Limestone Hydrology and Fissured Media and the 12th International Congress of Speleology*, Neûchatel, Volume 2, 65–68.

COLE, J. A., OAKES, D. B., SLADE, S. & CLARK, K. J. 1994. Potential impacts of climatic change and of sea-level rise on the yields of aquifer, river and reservoir sources. *Journal of the Institution of Water and Environmental Management*, **8**: 591–606.

COOPER, D. M., WILKINSON, W. B. & ARNELL, N. W. 1995. The effects of climate changes on aquifer storage and river baseflow. *Hydrological Sciences Journal*, **40**, 615–631.

CRAMPON, N., CUSTODIO, E. & DOWNING, R. A. 1996. The hydrogeology of western Europe: A basic framework. *Quarterly Journal of Engineering Geology*, **29**, 163–180.

DEHN, M., BURGER, G., BUMA, J. & GASPARETTO, P. 2000. Impact of climate change on slope stability using expanded downscaling. *Engineering Geology*, **55**, 193–204.

ELLIOT, T., CHADHA, D. S. & YOUNGER, P. L. 2001. Water quality impacts and palaeohydrogeology in the Yorkshire Chalk Aquifer, UK. *Quarterly Journal of Engineering Geology and Hydrogeology*, **34**, 385–398.

ELLIOT, T., YOUNGER, P. L. & CHADHA, D. S. 1998. The future sustainability of groundwater resources in East Yorkshire: past and present perspectives. *In* WHEATER, H. & KIRBY, C. (eds), *Hydrology in a Changing Environment*. Wiley, Chichester, Volume II, 21–31.

ELTAHIR, E. A. B. & YEH, P. A. J. F. 1999. On the asymmetric response of aquifer water level to floods and droughts in Illinois. *Water Resources Research*, **35**, 1199–1217.

FERRIER, R. C., WHITEHEAD, P. G. & MILLER, J. D. 1993. Potential impacts of afforestation and climate-change on the stream water chemistry of the Monachyle Catchment. *Journal of Hydrology*, **145**, 453–466.

FOSTER, S. S. D. & MILTON, V. A. 1974. The permeability and storage of an unconfined Chalk aquifer. *Hydrological Sciences Bulletin*, **19**, 485–500.

FOSTER, S. S. D. & MILTON, V. A. 1976. *Hydrological basis for large-scale development of groundwater storage in the East Yorkshire Chalk*. Report of the Institute of Geological Sciences No 76/3.

GELLENS, D. 1991. Impact of a CO_2-induced climatic-change on river flow variability in 3 rivers in Belgium. *Earth Surface Processes and Landforms*, **16**, 619–625.

GLEICK, P. H. 1986. Methods for evaluating the regional hydrologic impacts of global climatic changes. *Journal of Hydrology*, **88**, 97–116.

GLEICK, P. H. 1989. Climate, hydrology, and water resources. *Reviews in Geophysics*, **27**, 329–344.

GREGORY, J. M. & MITCHELL, J. F. B. 1995. Simulation of daily variability of surface temperature and precipitation over Europe in the current and $2\times CO_2$ climates using the UKMO climate model. *Quarterly Journal of the Royal Meteorological Society*, **121**, 1451–1476.

HADLEY CENTRE. 2000. *COP6 – Climate Change. An update of recent research from the Hadley Centre*. UK Meteorological Office, Bracknell.

HOUGHTON, J. T., CALLANDER, B. A. & VARNEY, S. K. 1992. *Climate Change 1992: The Supplementary Report to the IPCC Scientific Assessment*. Cambridge University Press, Cambridge.

HULME, M., BARROW, E. M., ARNELL, N. W., HARRISON, P. A., JOHNS, T. C. & DOWNING, T. E.

1999. Relative impacts of human-induced climate change and natural climate variability. *Nature*, **397**, 688–691.

HULME, M., CONWAY, D., BROWN, O. & BARROW, E. 1994. *A 1961–90 baseline climatology and future climate change scenarios for Great Britain and Europe, Part III: Climate change scenarios for Great Britain and Europe (December 1994)*. A report accompanying the datasets prepared for the 'Landscape Dynamics and Climate Change' TIGER IV3.a Consortium. Climatic Research Unit, University of East Anglia, UK.

HULME, M., RAPER, S. C. B. & WIGLEY, T. M. L. 1995. An integrated framework to address climate changes (ESCAPE) and further developments of the global and regional climate modules (MAGICC). *Energy Policy*, **23**, 347–355.

INSTITUTE OF GEOLOGICAL SCIENCES/YORKSHIRE WATER AUTHORITY. 1980. *Hydrogeological Map of East Yorkshire* (1 : 100 000). Institute of Geological Sciences, London.

KILSBY, C. G., COWPERTWAIT, P. S. P., O'CONNELL, P. E. & JONES, P. D. 1998. Predicting rainfall statistics in England and Wales using atmospheric circulation variables. *International Journal of Climatology*, **18**, 523–539.

KILSBY, C. G., EWEN, J., SLOAN, W. T., BURTON, A., FALLOWS, C. S. & O'CONNELL, P. E. 1999. The UP modelling system for large scale hydrology: simulation of the Arkansas – Red River basin. *Hydrology and Earth System Sciences*, **3**, 137–149.

KUMP, L. R., BRANTLEY, S. L. & ARTHUR, M. A. 2000. Chemical weathering, atmospheric CO_2 and climate. *Annual Review of Earth and Planetary Sciences*, **28**, 611–667.

LAMBÁN JIMÉNEZ, L. J. 1998. *Estudio de la recarga y del funcionamiento hidrogeológico de la Unidad Nioa (Cordillera Prelitoral Catalana)*. [A study of the recharge and hydrogeological functioning of the Anoia Unit (Cordillera Prelitoral of Catalonia)]. PhD thesis, Universitat Politécnica de Catalunya, Barcelona. (in Spanish).

LERNER, D. N., ISSAR, A. S. & SIMMERS, I. 1990. *Groundwater recharge. A guide to understanding and estimating natural recharge*. International Contributions to Hydrogeology Volume 8. International Association of Hydrogeologists. Verlag Heinz Heise, Hannover.

LIMBRICK, K. J., WHITEHEAD, P. G., BUTTERFIELD, D. & REYNARD, N. 2000. Assessing the potential impacts of various climate change scenarios on the hydrological regime of the River Kennet at Theale, Berkshire, south-central England, UK: an application and evaluation of the new semi-distributed model, INCA. *Science of the Total Environment*, **251**, 539–555.

LIU, Z. & ZHAO, J. 2000. Contribution of carbonate rock weathering to the atmospheric CO_2 sink. *Environmental Geology*, **39**, 1053–1058.

LOAICIGA, H. A., MAIDMENT, D. R. & VALDES, J. B. 2000. Climate-change impacts in a regional karst aquifer, Texas, USA. *Journal of Hydrology*, **227**, 173–194.

MCDONALD, M. G. & HARBAUGH, A. W. 1984. *A modular three-dimensional finite-difference groundwater flow model*. United State Geological Survey Techniques of Water Resources Investigations. Reston, VA.

MANZANO, M., CUSTODIO, E., CARDOSO DA SILVA, G. & LAMBáN, J. 1998. *Modelación del efecto del cambio climático sobre la regarga en dos acuíferos carbonatados del área Mediterránea. (Modelling of the effects of climate change on the recharge of two carbonate aquifers in the Mediterranean region). Procedimientos del 4º Congreso Latinoamericano de Hidrología Subterránea* (Proccedings of the 4th Latin American Congress on Subsurface Hydrology). 322–333 (in Spanish).

MAVROMATIS, T. & JONES, P. D. 1998. Comparison of climate change scenario construction methodologies for impact assessment studies. *Agricultural and Forest Meteorology*, **91**, 51–67.

MAVROMATIS, T. & JONES, P. D. 1999. Evaluation of HadCM2 and direct use of daily GCM data in impact assessment studies. *Climatic Change*, **41**, 583–614.

PALUTIKOF, J. P., GOODESS; C. M. & GUO, X. (1994). Climate change, potential evapotranspiration and moisture availability in the Mediterranean Basin. *International Journal of Climatology*, **14**, 853–869.

PANNAGOULIA, D. & DIMOU, 1996. Sensitivities of groundwater - streamflow interaction to global climate change. *Hydrological Sciences Journal*, **41**, 781–796.

PITMAN, J. I. 1986. Chemical weathering of the East Yorkshire Chalk. In: Paterson, K. & Sweeting, M. M., (eds), *New Directions in Karst*. Norwich, GEO Books, 77–113.

PRICE, M. 1998. Water storage and climate change in Great Britain – the role of groundwater. *Proceedings of the Institution of Civil Engineers - Water, Maritime and Energy*, **130**, 42–50.

REED, D. N. 1986. Simulation of time series of temperature and precipitation over eastern England by an Atmospheric General Circulation model. *Journal of Climatology*, **6**, 233–253.

ROSENBERG, N. J., ESPSTEIN, D. J., WANG, D., VAIL, L., SRINIVASAN, R. & ARNOLD, J. G. 1999. Possible impacts of global warming on the hydrology of the Ogallala aquifer region. *Climatic Change*, **42**, 677–692.

SAMPER, J. & GARCÍA-VERA, M. A. 1992. *Programa BALAN. Versión 8.0. Manual de usuario*. Universitat Politécnica de Cataluña, Barcelona, Informe Interno 1–50. (BALAN Program Version 8.0, User's Manual; Internal report, in Spanish).

SANDSTROM, K. 1995. Modeling the effects of rainfall variability on groundwater recharge in semi-arid Tanzania. *Nordic Hydrology*, **26**, 313–330.

SAUTER, M. 1992. *Quantification and forecasting of regional groundwater flow and transport in a karst aquifer (Gallusquelle, Malm, SW. Germany)*. PhD thesis. Tübinger Geowissenschaftliche Arbeiten.

SAUTER, M. & LIEDL, R. 1998. The impact of climate change on water resources in a carbonate aquifer. *In*: WHEATER, H. & KIRBY, C. (eds), *Hydrology in a Changing Environment*. Wiley, Chichester. Volume II. 11–20.

SAUTER, M., LIEDL, R., CLEMENS, T. & HÜCKINGHAUS, D. 1997. Karst aquifer genesis – modelling approaches and controlling parameters. *Proceedings of the 6th Conference on Limestone Hydrology and Fissured Media and the 12th International Congress of Speleology*. Neûchatel, Volume 2, 107–110.

SHERIF, M. M. & SINGH, V. P. 1999. Effect of climate change on sea water intrusion in coastal aquifers. *Hydrological Processes*, **13**, 1277–1287.

SKELLY, W. C. & HENDERSON-SELLERS, A. 1996. Grid box or grid point: What type of data do GCMs deliver to climate impact researchers? *International Journal of Climatology*, **16**, 1079–1086.

SOLOMON, S. I., BERAN, M. & HOGG, W. 1987. *The influence of climate change and climatic variability on the hydrologic regime and water resources*. International Association of Hydrological Sciences Publication No **168**. (Proceedings of an International Symposium held at the International Union of Geodesy and Geophysics 19th General Assembly, Vancouver, Canada, August 1987). IAHS Press, Wallingford, UK.

TEUTSCH, G. & SAUTER, M. 1991. Groundwater modeling in karst terranes: scale effects, data aquisition and field validation. *Proceeding of the Third Conference on Hydrogeology, Ecology, Monitoring and Management of Groundwater in Karst Terranes*, Nashville, 1991.

VACCARO, J. J. 1992. Sensitivity of groundwater recharge estimates to cimate variability and change, Columbia Plateau, Washington. *Journal of Geophysical Research - Atmospheres*, **97**, 2821–2833.

VINER, D. & HULME, M. 1993*a*. *The UK Meteorological Office High Resolution GCM Equilibrium Experiment (UKHI)*. Technical Note No. 1 (February 1993) prepared for the UK Department of the Environment Climate Change Impacts/Predictive Modelling LINK (UK CI-LINK), Contract Reference No. **PECD 7/12/96**.

VINER, D. & HULME, M. 1993*b*. *Construction of climate change by linking GCM and STUGE output*. Technical Note No. 2 (March 1993) prepared for the UK department of the Environment Climate Change Impacts/Predictive Modelling LINK (UK CI-LINK), Contract Reference No. **PECD 7/12/96**.

VINER, D. & HULME, M. 1994. *The Climate Impacts LINK Project: Providing climate change scenarios for impacts assessment in the UK (February 1994)*. Prepared for the UK Department of the Environment Climate Change Impacts/Predictive Modelling LINK (UK CI-LINK), Contract Reference No. **PECD 7/12/96**.

VINER, D., HULME, M. & RAPER, S. C. B. 1994. *Construction of Climate Change Scenarios for the British Isles from GCM Transient Climate Change Experiments*. Technical Note No. 5 (April 1994) prepared for the UK Department of the Environment Climate Change Impacts/Predictive Modelling LINK (UK CI-LINK), Contract Reference No. **PECD 7/12/96**.

WIGLEY, T. M. L., JONES, P. D. & RAPER, S. C. B. 1997. The observed global warming record: What does it tell us? *Proceedings of the National Academy of Sciences of the USA*, **94**, 8314–8320.

WILBY, R. L., WIGLEY, T. M. L., CONWAY, D., JONES, P. D., HEWITSON, B. C., MAIN, J. & WILKS, D. S. 1998. Statistical downscaling of general circulation model output. *Water Resources Research*, **34**, 2995–3008.

WILKINSON, W. B. & COOPER, D. M. 1993. The response of idealized aquifer river systems to climate-change. *Hydrological Sciences Journal*, **38**, 379–390.

YOUNGER, P. L. & ADAMS, R. 1999. *Predicting mine water rebound*. Environment Agency R&D Technical Report **W179**. Bristol, UK.

YOUNGER, P. L., TEUTSCH, G., CUSTODIO, E., ELLIOT, T., SAUTER, M., MANZANO, M., LIEDL, R., CLEMENS, T., HÜCKINGHAUS, D., TORE, C. S., LAMBÁN, J. & CARDOSO DA SILVA, G. 1997. *Groundwater Resources and Climate change Effects – GRACE: Final Report*. European Commission Third Framework RTD Programme (Environment and Climate), Project **EV5V - CT94 - 0471**.

ZEKSTER, I. S. & LOAICIGA, H. A. 1993. Groundwater fluxes in the global hydrological cycle - past, present and future. *Journal of Hydrology*, **144**, 405–427.

Simulation of the impacts of climate change on groundwater resources in eastern England

I. YUSOFF[1,3], K. M. HISCOCK[1] & D. CONWAY[2]

[1]*School of Environmental Sciences, University of East Anglia, Norwich NR4 7TJ, UK (e-mail: k.hiscock@uea.ac.uk)*

[2]*School of Development Studies, University of East Anglia, Norwich NR4 7TJ, UK*

[3]*Present address: Department of Geology, University of Malaya, 50603, Kuala Lumpur, Malaysia*

Abstract: This study investigated the impacts of climate change on the Chalk aquifer in west Norfolk. A two-layer transient groundwater flow model of the aquifer system was calibrated and validated for the period 1980–1995 and provided the historic flow record for the climate change simulations. Two future scenarios were selected from the Hadley Centre's climate change experiments using HadCM2: (1) a medium-high (MH) emissions scenario; and (2) a medium-low (ML) emissions scenario of 'greenhouse' gases. Two future periods were considered: 2020–2035 and 2050–2065. Future recharge to the aquifer was estimated by adjusting the historic record of monthly precipitation and potential evapotranspiration by factors calculated from comparing control and future HadCM2-generated values. Impacts of climate change were evaluated by incorporating the monthly estimated recharge inputs within the flow model. The most noticeable and consistent result of the climate change impact simulations is the decrease in recharge expected in autumn for all scenarios (decreases ranging from 17 to 35%) as a consequence of the smaller amount of summer precipitation and increased autumn potential evapotranspiration. For the 2050MH scenario, these conditions lead to a 42% increase in autumn soil moisture deficit and a 26% reduction in recharge. Hence, west Norfolk can expect longer and drier summers that are predicted to have relatively little effect on summer groundwater levels (generally a 1 to 2% decrease) but will result in a decrease of up to 14% in autumn river baseflow volumes.

The hydrological cycle is an integral part of the climate system and is therefore involved in many of the interactions and feedback loops that give rise to the complexities of the system (Askew 1987). Climate change is expected to lead to an intensification of the global hydrological cycle and have major impacts on regional water resources (IPCC 1997). Figure 1 provides a summary of the likely impacts of climate change on natural hydrological systems. Changes in rainfall and evaporation as a result of higher 'greenhouse' gas concentrations are likely to result in changes in river flows and groundwater recharge. Changes in temperature, radiation, rainfall, evaporation, soil moisture and CO_2 concentrations all affect catchment ecosystems and land use, which in turn affect the catchment water balance. Water quality is affected by changes in temperature and rainfall that affect the volume of river flow and the degree of saline intrusion.

Many hydrologists prefer numerical models for the purpose of estimating the hydrological response to climate change. Models allow varying degrees of complexity for representing current and future climatic conditions. The use of a hydrological model offers important advantages over the direct use of hydrological output from global circulation models (GCMs) (Mearns *et al.* 1990). The linking of the two models enables hydrologists and water resource engineers to study a variety of climate change

From: Hiscock, K. M., Rivett, M. O. & Davison, R. M. (eds) *Sustainable Groundwater Development*. Geological Society, London, Special Publications, **193**, 325–344. 0305-8719/02/ $15.00 The Geological Society of London 2002.

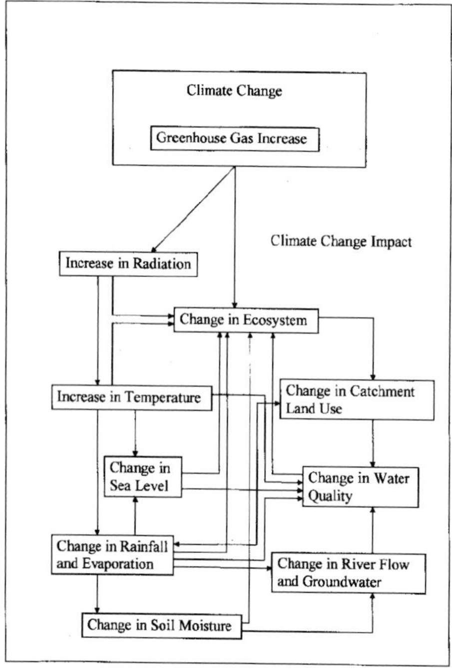

Fig. 1. Impacts of increasing 'greenhouse' gas concentrations on the natural hydrological cycle (after Arnell 1996).

effects, including both equilibrium and transient responses, and hypothetical responses. To apply a hydrological model in the estimation of climate change effects on water resources a few basic steps should be followed, as summarized in Table 1.

This paper presents a study of the effects of climate change on groundwater resources in part of the Great Ouse catchment, eastern England. The Anglian region is regularly affected by dry conditions in the warmer months of the year (the region regularly accounts for 80% of the total spray irrigation used in southern and eastern England; Palutikof *et al.* 1997). Sustained dry periods, rather than hot periods, as a result of future climate change, could greatly affect water provision in the region. Thus, the main aim of this study was to highlight the possible effects of climate change on groundwater resources in the selected aquifer, the Chalk aquifer system in west Norfolk. The approach adopted followed the procedures summarized in Table 1. The groundwater model simulations and the evaluation of the climate change impacts on groundwater resources, in terms of effects on recharge amounts, groundwater levels and baseflow volumes, are described.

Description of the study area

Geology

A location map of the study area showing the general solid geology is shown in Figure 2. The Upper Cretaceous Chalk, a white, fine-grained, fissured limestone, dips gently east at less than 1°. The western margin of the study area is bounded by the impermeable Gault Clay. The eastern half of the study area is overlain by late Pleistocene glacial deposits of variable thickness and extent. These deposits include boulder clay (Lowestoft Till), a brown calcareous clay with clasts of chalk and flint, and with occasional patches of glacial sands and gravels. The thickness of the boulder clay is generally between 15–30 m, with 60 m possible within sub-glacially eroded channels. In the main river valleys, alluvium and glacial sands and gravels cover the Chalk outcrop, especially in the River Nar valley.

Hydrogeology

The Chalk aquifer is the dominant hydrogeological unit, with the Chalk groundwater catchment area corresponding approximately to the

Table 1. *General method for estimating the impacts of climate change on water resources using a hydrological model and climate change scenarios (after Arnell 1992a)*

1	Develop a hydrological model for the study catchment and determine all relevant parameters. Use the current climatic input, observed groundwater levels and river flows for model validation
2	Perturb the historical time series of climatic data according to some climate change scenario (results from climate change experiments)
3	Simulate the hydrological characteristics of the catchment under the perturbed climate, using the calibrated hydrological model
4	Compare the model simulations of current and possible future hydrological characteristics to deduce the climate change impacts

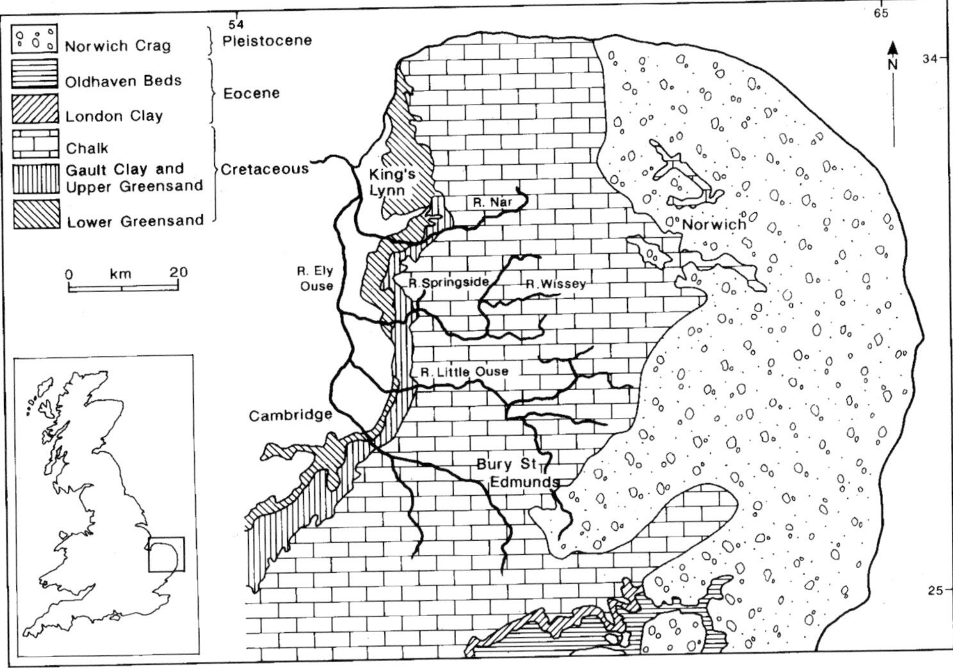

Fig. 2. Simplified map of the solid geology of East Anglia showing the study area.

surface water catchment area. Previous work in the study area has divided the Chalk aquifer into unconfined and semi-confined regions in the west and east, respectively, with the division corresponding to the Pleistocene boulder clay cover (Foster & Robertson 1977). In areas with thick boulder clay cover, the generally high piezometric surface produces confined aquifer characteristics. Unconfined or water table conditions can exist within the river valleys where there is no clay cover.

Based on the hydrochemical distribution of water types, recharge and groundwater flow in the Chalk aquifer system of East Anglia are shown by Lloyd *et al.* (1981) and Hiscock (1993) to occur predominantly in the valley areas where the Chalk permeability is well developed and the boulder clay cover is thin or absent. Groundwater divides are present between the river catchments but their positions are not static depending on seasonal head distributions in the aquifer.

Within the Chalk aquifer, flow zones are more likely to occur within an interval of 30 m below rest water level, and within an interval of 50 m from the top of the Chalk (Wooton 1994). A geophysical evaluation in the Rushall area, just to the east of the current study area (Foster & Robertson 1977), suggested that flow in the Chalk aquifer is mostly restricted to between −40 m above Ordnance Datum (m AOD) and above, with the most significant groundwater flow largely above −10 m AOD. In the Nar catchment, Chalk transmissivity values vary between 2000–3000 m^2 d^{-1} in the valley zone, to less than 100 m^2 d^{-1} in the interfluve areas where the boulder clay cover has restricted groundwater flow and aquifer development (Toynton 1983).

Hydrology

Average annual rainfall measured in the region from 1980–1990 is 582 mm with an average annual potential evapotranspiration of about 575 mm. Two major rivers drain the study area, the Rivers Nar and Wissey, and have estimated baseflow index values of between 0.7 and 0.9 (Owen 1969; Gustard *et al.* 1992). The River Stringside, a minor river, is a tributary of the River Wissey.

Water demand

The population of the Anglian region is predicted to be one of the fastest growing regions in

England and Wales in the next 25 years, with an increase in households of 0.6 million and an increase in population of 0.8 million (Environment Agency 2000). The demand for water in this period will change depending on a variety of societal factors as defined by the Foresight *Environmental Futures* scenarios (DTI 1999). For those scenarios where less sustainable water-use patterns prevail, and taking into account the increase in population, very large increases of up to 1487 Ml d^{-1} (>70% increase) in household water demand could occur by 2025. Alternatively, with more sustainable water use, an overall decline in water use could occur of up to 616 Ml d^{-1} (>25% decrease) (Environment Agency 2000). Furthermore, there is a large regional demand for water by agriculture with the demand for spray irrigation predicted to either decrease by up to 32 Ml d^{-1} by 2025, or increase by 83 Ml d^{-1} depending on which scenario is chosen based on customer preferences, international competition, crop varieties grown and efficiency of water use (Environment Agency 2000). In practice, it is unlikely that society will exclusively adopt one scenario, but given the relatively small amount of effective rainfall (about 140 mm a^{-1}) in the Anglian region it will be challenging to meet the higher demand forecasts even without the added susceptibility to future climate change.

Conceptual model

A two-layer conceptual groundwater model was designed to investigate the likely impact of climate change on groundwater resources in the study area (Fig. 3). An upper, semi-confining layer of glacial deposits represents the glacial boulder clay, silts, sands and gravels. A lower layer, with an assumed effective thickness of 50 m, and bottom elevation of −20 m AOD, represents the Chalk aquifer. The upper, boulder clay layer was estimated from borehole logs, the hydrogeological map of northern East Anglia (IGS 1976) and statements from selected reports (Aspinwall & Co. 1992*a*, *b*; Boar *et al.* 1994) to be 30 m thick and to lie in semi-hydraulic contact with the lower Chalk layer.

The boulder clay receives direct recharge from effective precipitation with water routed predominantly as horizontal groundwater flow through lenses of more permeable sand deposits to contribute to surface run-off at interfluvial margins and within river valleys. Recharge to the Chalk aquifer is in the form of direct recharge in the outcrop area in the west and as leakage from the boulder clay and river bed deposits elsewhere.

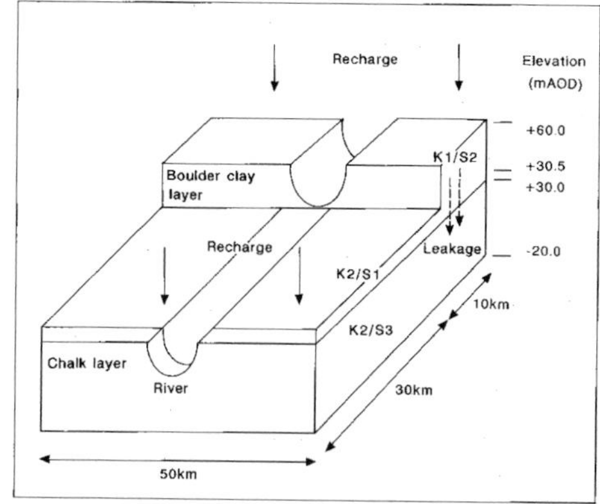

Fig. 3. Conceptual model of the hydrogeological system of west Norfolk. The upper layer represents the Quaternary glacial deposits (principally boulder clay) and the lower layer the Cretaceous Chalk aquifer. Zones of hydraulic conductivity (K) and storativity (S) are also shown.

Groundwater model design

The conceptual model was converted into a numerical groundwater flow model using the Groundwater Vistas MODFLOW package (McDonald & Harbaugh 1988; Rumbaugh & Rumbaugh 1996).

Model discretization

The conceptual model for the west Norfolk Chalk aquifer was translated into a numerical groundwater model with a total area of 2250 km^2. The model domain was discretized into grid dimensions of 1×1 km and 0.5×0.5 km square cells, with the smaller cell size assigned to the active model area where rivers, springs and abstractions are simulated. In total, the model contains 95 rows and 78 columns of cells, with 8474 active modelling cells, of which 521 are river cells and 11 are spring cells.

Model boundary conditions

All the external boundaries for the upper layer (boulder clay) are no flow boundaries as shown in Figure 4a. The upper reaches of the Rivers Nar and Wissey, which continue to flow in the upper layer, were modelled with the 'River Package' of MODFLOW and were assigned the same hydraulic conductivity as the underlying Chalk aquifer layer. Principal springs were modelled with the 'Drain Package'. There is no

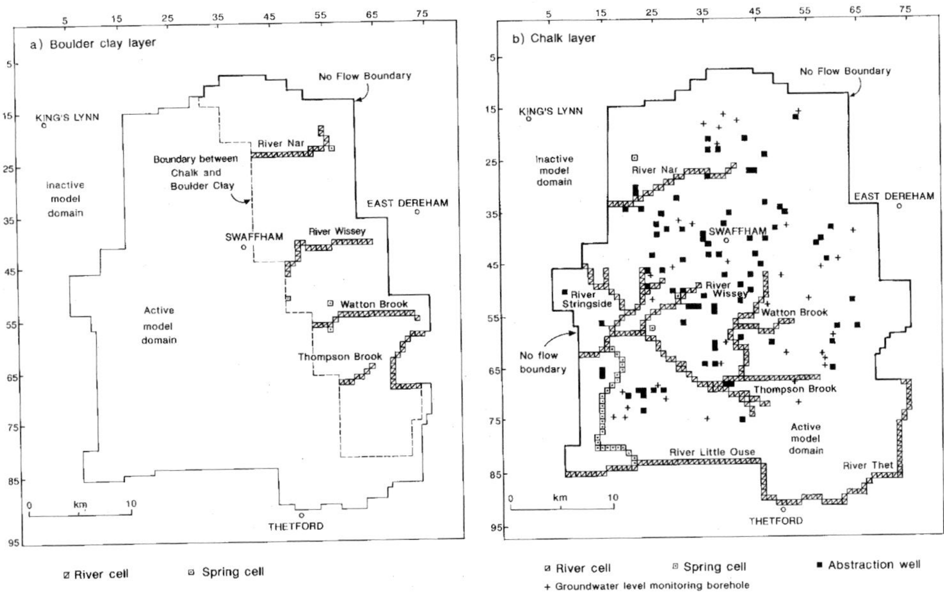

Fig. 4. Boundary conditions for the two-layer MODFLOW numerical model of the west Norfolk Chalk aquifer system. In (**a**) boundary conditions for the upper (boulder clay) layer of the model are shown, and in (**b**) boundary conditions for the lower (Chalk aquifer) layer are shown.

groundwater abstraction from the upper layer. The boundary conditions for the lower layer (Chalk aquifer) are shown in Figure 4b. No flow boundaries are used to represent the northern and western boundaries which, respectively, follow a flow line and the western limit of the Chalk. A further no flow boundary was used for the eastern boundary to represent the estimated regional groundwater divide. The remainder of the eastern boundary was modelled as a river boundary. The southern boundary was modelled as a river boundary representing the River Little Ouse. The model is able to simulate unconfined-confined Chalk aquifer conditions. The top elevation of the Chalk aquifer layer was assigned as 30 m AOD, and any groundwater levels which exceed this elevation are considered confined water levels.

Groundwater abstractions

Total licensed abstraction data for the area were obtained from the Environment Agency. To estimate actual abstraction quantities, 50% of the licensed abstraction amounts were used (Environment Agency staff, pers. comm.). However, actual abstraction data were requested and received for public water supplies operated by Anglian Water Services Ltd. Public water supplies account for the majority of groundwater abstractions in the study area.

Aquifer recharge

For the present west Norfolk Chalk aquifer model, it was decided, for ease of calculation in later climate change impacts modelling, to use the conventional method of Penman (Penman 1949, 1950) and a soil water balance calculation method to estimate recharge to the aquifer. An outline of the method is given by Howard & Lloyd (1979). The method calculates the actual evaporation, soil moisture deficit and effective rainfall or recharge using precipitation and potential evaporation input data. The timing of the decrease in evapotranspiration losses, as determined by the size of the soil moisture deficit, is controlled by the root constant value (a measure of the available soil water content) chosen for the vegetated surface. In this study, a root constant value of 70 mm resulted in the best fit ($R^2 = 0.931$) between the calculated effective precipitation values and comparable data produced by the Meteorological Office's MORECS system (Thompson *et al.* 1981) for the area. An average recharge, or effective precipitation, for

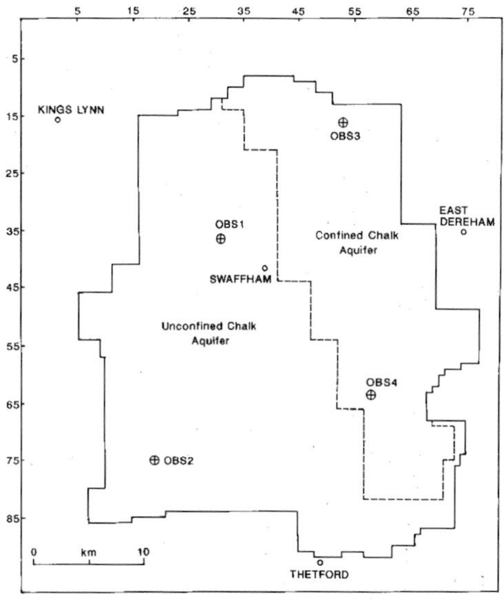

Fig. 5. Locations of the four selected Chalk observation boreholes used for calibration of the transient state MODFLOW numerical model.

1980–1990 of 3.43×10^{-4} m d^{-1} was calculated for the model area (Yusoff 2000).

Model calibration

The modelling strategy used in this study was to calibrate the model to the steady-state condition for the period 1980–1990, then to use the model output as the initial conditions for transient-state simulations. A calibrated transient model would then be used with the climate change scenarios. For the steady-state model, a 'trial and error' calibration process was used to match average field hydrogeological conditions (average groundwater levels and river baseflows) for the period 1980–1990. For the transient model, two calibration targets were chosen: groundwater levels for selected monitoring boreholes (two in the unconfined Chalk aquifer, two in the confined aquifer; see Fig. 5); and river baseflows of the Rivers Nar, Wissey and Stringside. In estimating the groundwater contribution to river flow, or baseflow, the separation method outlined by Ineson & Downing (1964) was used.

In the transient model, each month from 1980 to 1990 was treated as one stress period. Groundwater abstractions for each stress period remained unchanged, as monthly abstraction data were not available. The same MODFLOW solver package, the slice successive over-relaxation method, was used to solve the transient groundwater flow equation with an error criterion of 1% of steady-state recharge.

It was not possible to calibrate the model with a single value of hydraulic conductivity for each model layer, and so zones of different hydraulic conductivity were used, broadly validated by results from available pumping test and baseflow analyses. The hydraulic conductivity zones for each layer for the final transient model are shown in Figure 6. In general, the horizontal hydraulic conductivity of the Chalk aquifer is greater in the unconfined area and lower in the confined region, with a maximum value of 40 m d^{-1}. The vertical hydraulic conductivity of the Chalk was maintained as 10% of the horizontal value.

The three hydraulic conductivity zones in the boulder clay, K3, K5 and K9, each with a horizontal hydraulic conductivity value of 3 m d^{-1}, have small vertical hydraulic conductivity values (0.0003–0.003 m d^{-1}) which control the vertical leakage from this layer. The higher vertical hydraulic conductivity of zone K9 (0.8 m d^{-1}) represents an extensive cover of permeable sands and gravels within the mass of glacial deposits. For the final transient model, the water balance gave a total leakage to the confined Chalk aquifer from the boulder clay layer of about 30% and is similar in magnitude to the results obtained by Aspinwall & Co. (1992*a*, *b*). The remainder of the effective rainfall to the upper model layer is assumed to be directed horizontally as run-off to surface water courses.

With reference to Figure 3, three storage parameter zones are suggested by the conceptual model. As a result of model construction, a zero storage value, S1, is assigned to the 'transparent' layer overlying the Chalk aquifer (lower layer) in the west of the area, and therefore does not involve the transfer or storage of water. The distribution of the zones of different storage parameter values for the upper and lower layers used in the final transient model are shown in Figure 7. Depending on the presence or absence of the semi-confining boulder clay, Chalk storativity values range from 0.0001–0.03, and for the boulder clay itself, values are 0.01 in the north of the model area and 0.03 in the south.

Figures 8 and 9 illustrate, respectively, the graphical comparison between the observed field data and the final transient model simulated values for two of the observation boreholes and the river baseflow hydrographs. In general, the model performs well in simulating the observed groundwater hydrographs. A percentage mean absolute error of 24% in monthly baseflow for the River Nar is reasonable given the subjective

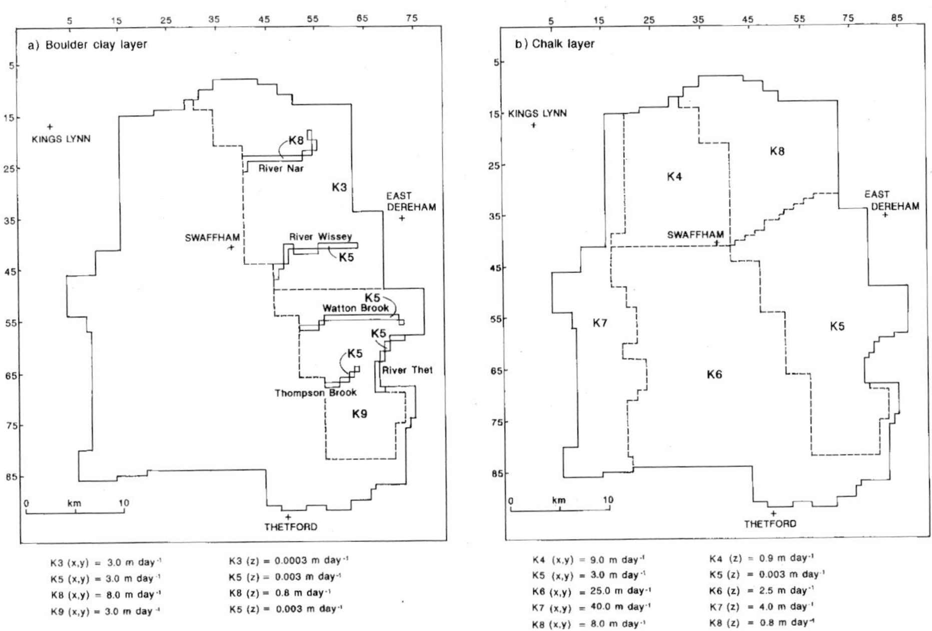

Fig. 6. Distribution of zones of hydraulic conductivity values for (**a**) the upper (boulder clay) layer and (**b**) the lower (Chalk aquifer) layer of the calibrated steady-state MODFLOW model of the west Norfolk Chalk aquifer system.

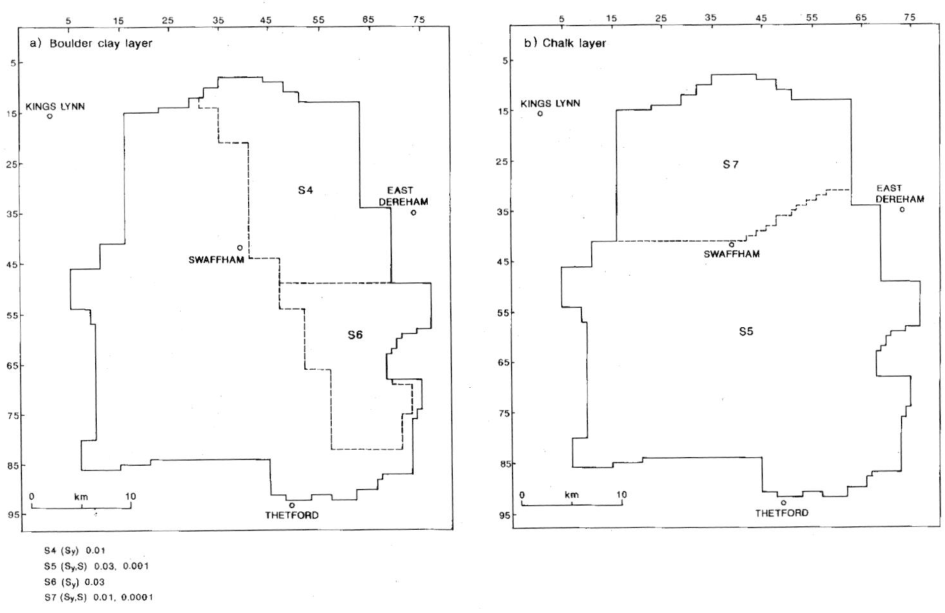

Fig. 7. Distribution of zones of specific yield and storage coefficient values for (**a**) the upper (boulder clay) layer and (**b**) the lower (Chalk aquifer) layer of the calibrated transient state MODFLOW model of the west Norfolk Chalk aquifer system.

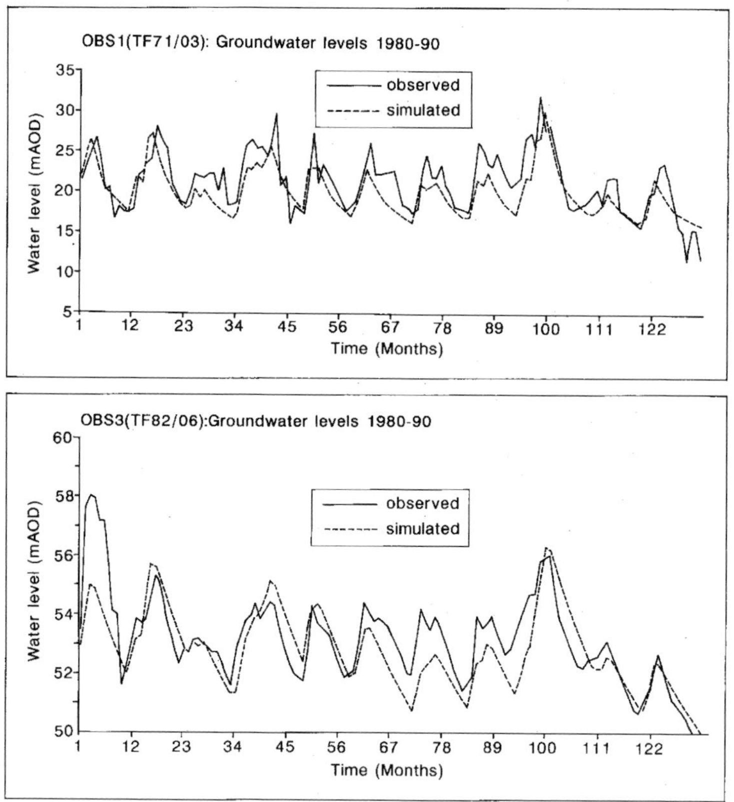

Fig. 8. Comparison between the observed and simulated groundwater levels for two selected Chalk observation boreholes (OBS1, unconfined; OBS3 confined) for the calibrated transient state MODFLOW model of the west Norfolk Chalk aquifer system.

method of baseflow separation. Compared with the River Nar, a poorer simulation was obtained for the Rivers Wissey and Stringside (errors of 42% and 104%, respectively). However, a good model simulation was achieved when comparing the mean annual baseflow for all three rivers (percentage difference between observed and model simulated mean annual flows of between 8 and 14%). The model captures the pattern of observed baseflow for the River Nar reasonably well, but for the Rivers Wissey and Stringside, it overestimates baseflow in periods of low flow. As a result, greater weight was given to the River Nar model when making future predictions of flow.

Validation

The final transient model was validated by extending the database used during the calibration process (1980–1990) by a further 5 years to 1995. In doing this, it is possible to study the effects of the initial conditions on model performance. In general, by starting the simulation period in 1980, the model provided a reasonable comparison to the independent dataset for 1991–1995, as shown by the selected groundwater and baseflow hydrographs given in Figure 10. As expected, the River Nar gave the best validation result using the calibrated model. At this point, the model was considered validated and ready for application in the prediction of hydrogeological conditions under future climate change scenarios.

Climate change scenarios

Global circulation models represent the most sophisticated method currently available for estimating the future effects of increasing green-

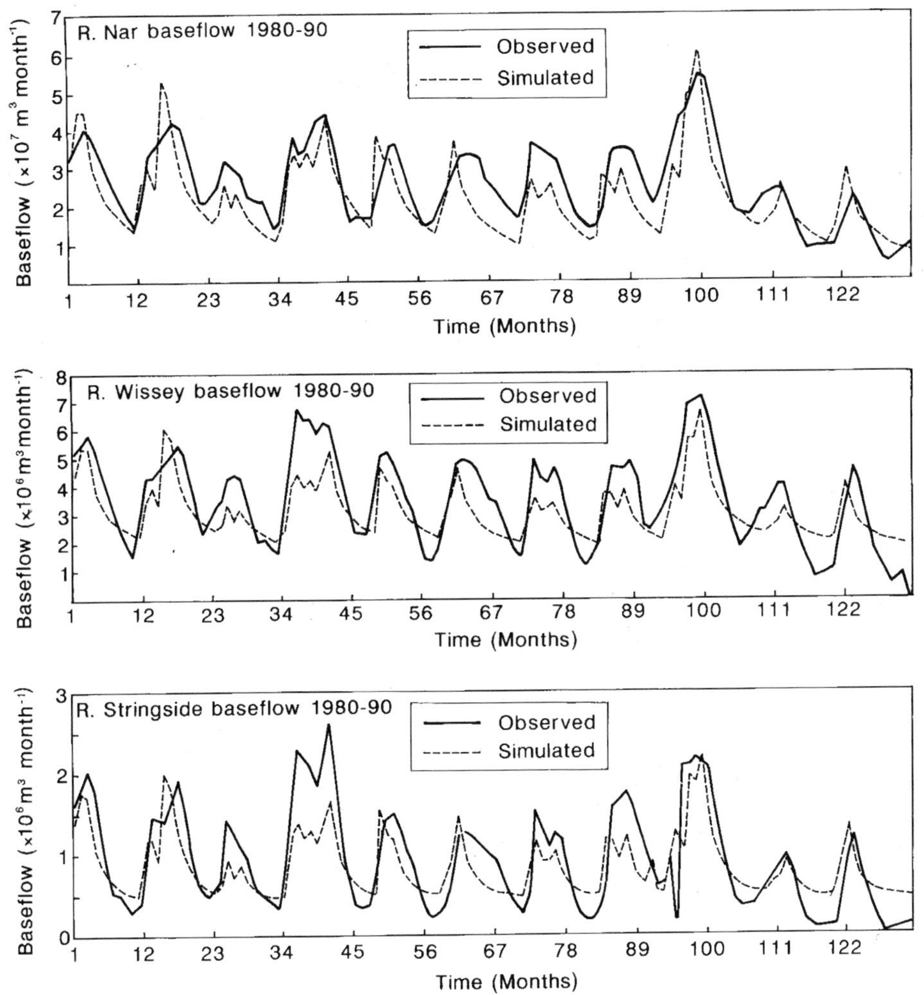

Fig. 9. Comparison between observed and simulated river baseflow run-off for the Rivers Nar, Wissey and Stringside for the calibrated transient state MODFLOW model of the west Norfolk Chalk aquifer system.

house gas concentrations on climate. Coupled ocean-atmosphere GCMs are mathematical representations of the time evolution of the global climate system. Model calculations are based on the laws of physics and are executed at widely spaced points of a three-dimensional grid at a resolution of approximately 5° latitude and 5° longitude (2.5° and 3.75°, respectively, for HadCM2). Transient experiments are conducted that involve GCMs which are run over a control period to simulate the climate without enhanced greenhouse gas forcing (and to ensure the ocean and atmosphere reach a steady state) and then forced with gradually increasing radiative forcing (representing increased concentrations of greenhouse gases and sulphate aerosols) over time to provide a time-dependent transient response of the climate system (Conway 1998).

Uncertainty is inherent in the development of climate scenarios due to unknown future emissions of greenhouse gases and aerosols, unknown global climate sensitivity and due to the difficulty of simulating the regional characteristics of climate change. It is because of these uncertainties that the regional changes in climate derived from GCM experiments are termed

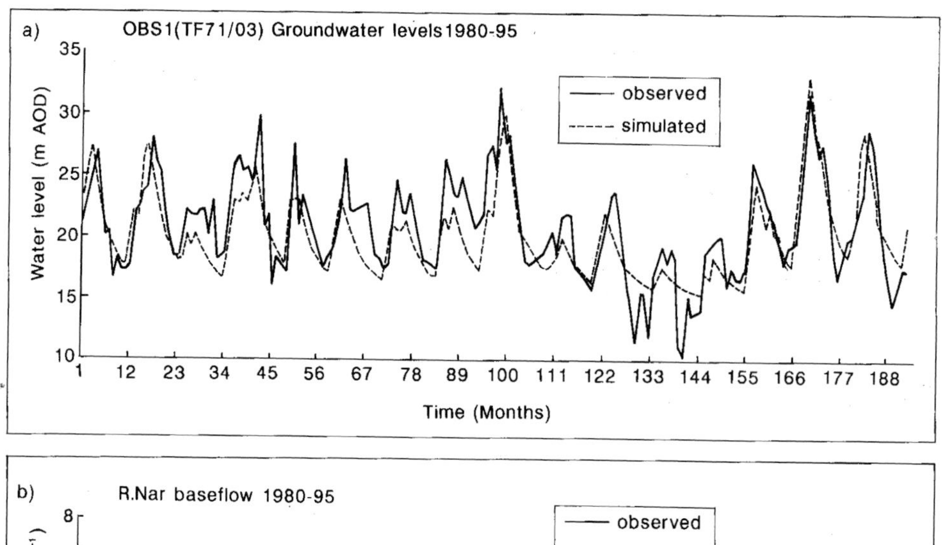

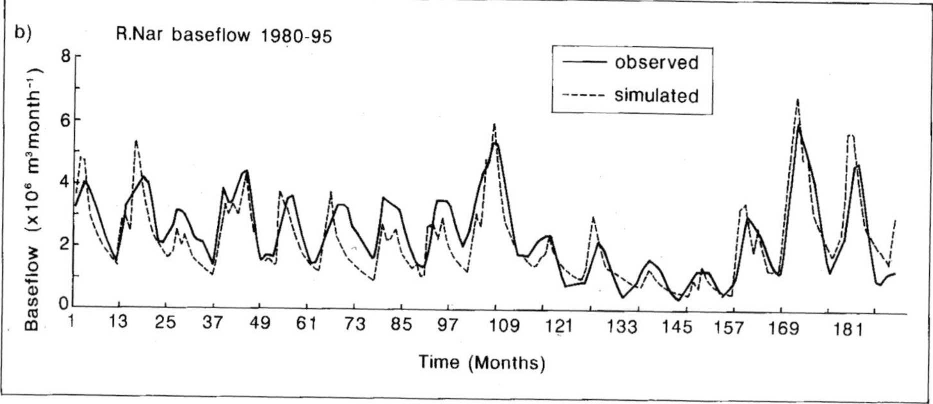

Fig. 10. Comparison between (**a**) the observed and simulated groundwater levels for unconfined Chalk observation borehole (OBS1) and (**b**) the observed and simulated river baseflow run-off for the River Nar for the validated transient state MODFLOW model of the west Norfolk Chalk aquifer system.

Table 2. *The four climate change scenarios of the UK Climate Impacts Programme (Hulme & Jenkins 1998)*

1	High (H): the HadCM2 GGa forced experiment; 1% per annum increase in equivalent carbon dioxide concentration with a high (4.5°C) climate sensitivity*
2	Medium-high (MH): the HadCM2 GGa forced experiment; 1% per annum increase in equivalent carbon dioxide concentration with a medium (2.5°C) climate sensitivity (used in this modelling study)
3	Medium-low (ML): the HadCM2 GGd forced experiment; 0.5% per annum increase in equivalent carbon dioxide concentration with a medium (2.5°C) climate sensitivity (used in this modelling study)
4	Low (L): the HadCM2 GGd forced experiment; 0.5% per annum increase in equivalent carbon dioxide concentration with a low (1.5°C) climate sensitivity

* The fully realized steady-state globally averaged surface air temperature change following the response of the climate system to a doubling of atmospheric CO_2 concentration

scenarios or projections and cannot be considered predictions. At best, these scenarios are plausible and physically consistent descriptions of a future climate. Scenarios serve to provide a framework from which the potential magnitude and nature of future climate change can be assessed and which provide input to process models and integrated assessments that try to quantify the potential impacts of climate change on the natural environment and society (Conway 1998).

Comprehensive reviews of climate change scenarios, including standards and construction, can be found in Carter & La Rovere (2001). In this study scenarios were generated to be consistent with the approach of the UK Climate Change Impacts Programme (UKCIP) (Hulme & Jenkins 1998). The scenarios represented the latest results from the HadCM2 climate change model experiments performed by the UK's Hadley Centre for Climate Prediction and Research. Given the difficulty of making firm predictions about future climate, the UKCIP approach is to present four alternative scenarios of climate change for the UK that span a reasonable range of possible future climates. The scenarios are labelled low (L), medium-low (ML), medium-high (MH) and high (H) and refer to global warming rates with different emission scenarios and climate sensitivities (low = 1.5°C; medium = 2.5°C; and high = 4.5°C). The details of the four proposed scenarios are given in Table 2. None of the scenarios include sulphate aerosol effects.

For the purpose of this study, only two climate change scenarios were selected, the medium-high (MH) and medium-low (ML) scenarios that represent the mid-range scenarios between the two extreme emissions scenarios. For each of these scenarios, two future, 15-year periods within the UKCIP time slices (and equal in length to the duration of the available hydrological records in this study) were considered for modelling climate change impacts on groundwater resources: the 2020s (2020–2035) and 2050s (2050–2065). Results were taken for the GCM grid box overlying eastern England. This approach results in four climate change scenarios: 2020MH, 2020ML, 2050MH and 2050ML. The global temperature increases predicted by the HadCM2 for these four scenarios range from 0.98°C (2020ML) to 2.11°C (2050MH) (Hulme & Jenkins 1998).

Viner *et al.* (2000) used the climate change scenarios presented by Hulme & Jenkins (1998) to describe likely future changes in climatic variables for the Anglian region. The change in mean annual temperature (with respect to the 1961–1990 average) is predicted to range from 1.0°C for the 2020ML scenario to 2.0°C for the 2050MH scenario. Summer temperatures are expected to warm faster than winter temperatures, contradicting the pattern across the UK as a whole. Changes in precipitation show marked seasonality in their patterns. Annually, there is a slight increase in precipitation (with respect to the 1961–1990 average) from 1% (2050MH scenario) to 3% (2020MH scenario). Most of this annual increase is borne out by large increases in winter precipitation which is as high as 13% in the 2050MH scenario. In contrast, summer precipitation decreases by as much as 19% for the 2050MH scenario. Potential evapotranspiration increases in all seasons, but by the greatest relative amount in the autumn and by the smallest amount in spring. By the 2050s, autumn potential evapotranspiration over the Anglian region is expected to be about 20–30% higher than at present.

The changes in summer precipitation and autumn potential evapotranspiration point to a marked decrease in water availability for the Anglian region. Water availability is likely to decrease throughout the year, compounded by a lengthening of the growing season leading to increased water uptake by plants and a shorter period of soil water retention. Viner *et al.* (2000) predict that this scenario will increase pressures upon the water available for domestic, agricultural and industrial consumption.

Modelling the impacts of climate change on groundwater resources

Changes in hydrological parameters

The main objective of the analysis of climate change impacts on groundwater resources was to develop scaling factors for changes in precipitation (P) and potential evapotranspiration (PE), using a method similar to that suggested by Arnell *et al.* (1997). The factors, i.e. percentage changes in P and PE, were defined for the future scenarios by comparing the monthly average of the control HadCM2 precipitation and potential evapotranspiration values with the monthly average values for the future climate scenarios (2020MH, 2020ML, 2050MH and 2050ML), thus: the precipitation or potential evapotranspiration factor = (Future/Control) × 100; where, Future is either the 2020s (2020–2035) or the 2050s (2050–2065) period, and Control is the HadCM2 control condition for the control (1975–1990) period.

Table 3. *Comparison of monthly and annual hydrological parameters for the observed present-day and perturbed future climate states (2020 MH, 2020 ML, 2050 MH, 2050 ML)*

Hydrological parameter	Parameter (measure)	Obs. (mm)	2020 MH	2020 ML	2050 MH	2050 ML
P	Monthly average (% change)	49.0	1.9	1.1	1.0	0.9
	Standard deviation (% change)	25.8	2.8	3.3	3.4	1.6
	Annual average (% change)	588.4	1.9	1.1	1.0	0.9
	Annual change (mm)		11.4	6.7	6.1	5.3
PE	Monthly average (% change)	48.9	11.4	9.0	18.7	12.8
	Standard deviation (% change)	34.4	8.4	13.5	22.1	9.6
	Annual average (% change)	586.6	11.4	9.0	18.7	12.8
	Annual change (mm)		66.9	52.9	109.8	75.2
AE	Monthly average (% change)	41.4	2.5	–0.9	–0.5	3.5
	Standard deviation (% change)	30.4	–1.5	5.7	–1.1	0.0
	Annual average (% change)	497.3	2.5	–0.9	–0.5	3.5
	Annual change (mm)		12.4	–4.6	–2.5	17.2
SMD	Monthly average (% change)	34.9	11.5	11.1	21.0	15.1
	Standard deviation (% change)	34.8	6.6	7.4	13.6	7.3
	Annual average (% change)	418.3	11.5	11.1	21.0	15.1
	Annual change (mm)		48.0	46.3	87.9	63.0
Recharge	Monthly average (% change)	10.8	–3.2	6.1	1.4	–10.4
	Standard deviation (% change)	17.8	1.4	9.3	10.0	–1.5
	Annual average (% change)	129.7	–3.2	6.1	1.4	–10.3
	Annual change (mm)		–4.1	7.9	1.8	–13.4

P, precipitation; PE, potential evapotranspiration; AE, actual evapotranspiration; SMD, soil moisture deficit; Obs., Observed present-day with averages based on 1980–1995

The derived percentage monthly factors represent the generalized effects of the two climate scenarios for the 2020s and 2050s. It was assumed that the monthly factors can be applied equally to each year of the observed historical data (1980–1995) to obtain future hydrological conditions. A serious limitation of this approach, which has been recognized and discussed by Arnell (1992*b*), is the fact that under the approach of perturbation to an historic time series, the frequency distribution of drought and flood occurrences does not change. Effectively, the same historic climate is being used as a model of future climate, with shifts being introduced only in the magnitude of each event in the series. However, the approach gives a good general indication of the possible range of changes in hydrological regimes.

The next stage of the analysis was to use these future conditions as input values in the conventional method of recharge calculation. Thus, monthly values of actual evapotranspiration, soil moisture deficit and recharge for the future climate scenarios were obtained. Table 3 presents a statistical analysis (at monthly and annual timescales) of the differences between the observed present-day and future hydrological states. Figure 11 shows a comparison of percentage changes in monthly average precipitation, potential evapotranspiration and recharge between the observed present-day and future states.

In general, and under the two future periods considered (2020s and 2050s), the precipitation, potential evapotranspiration and soil moisture deficit increase over monthly and annual timescales. The annual increase in precipitation (up to 11 mm) is small compared to the increase in potential evapotranspiration (up to 110 mm) and soil moisture deficit (up to 88 mm). Thus, the change in future potential evapotranspiration shows the greatest effect.

On the whole, the seasonal changes shown in Table 4 are greater than the monthly and annual changes which are almost equal in magnitude. Future precipitation increases in the winter (up to 13%) while potential evapotranspiration and soil moisture deficit increase in the autumn (by up to 33 and 42% respectively). However, summer precipitation decreases under all scenarios (up to 19%).

Even though constant climate change factors have been used, the month-to-month variability (standard deviation in Table 3) in future precipitation, potential evapotranspiration and soil moisture deficits increases slightly. The potential evapotranspiration shows the greatest variability and is increased by up to 22%, while

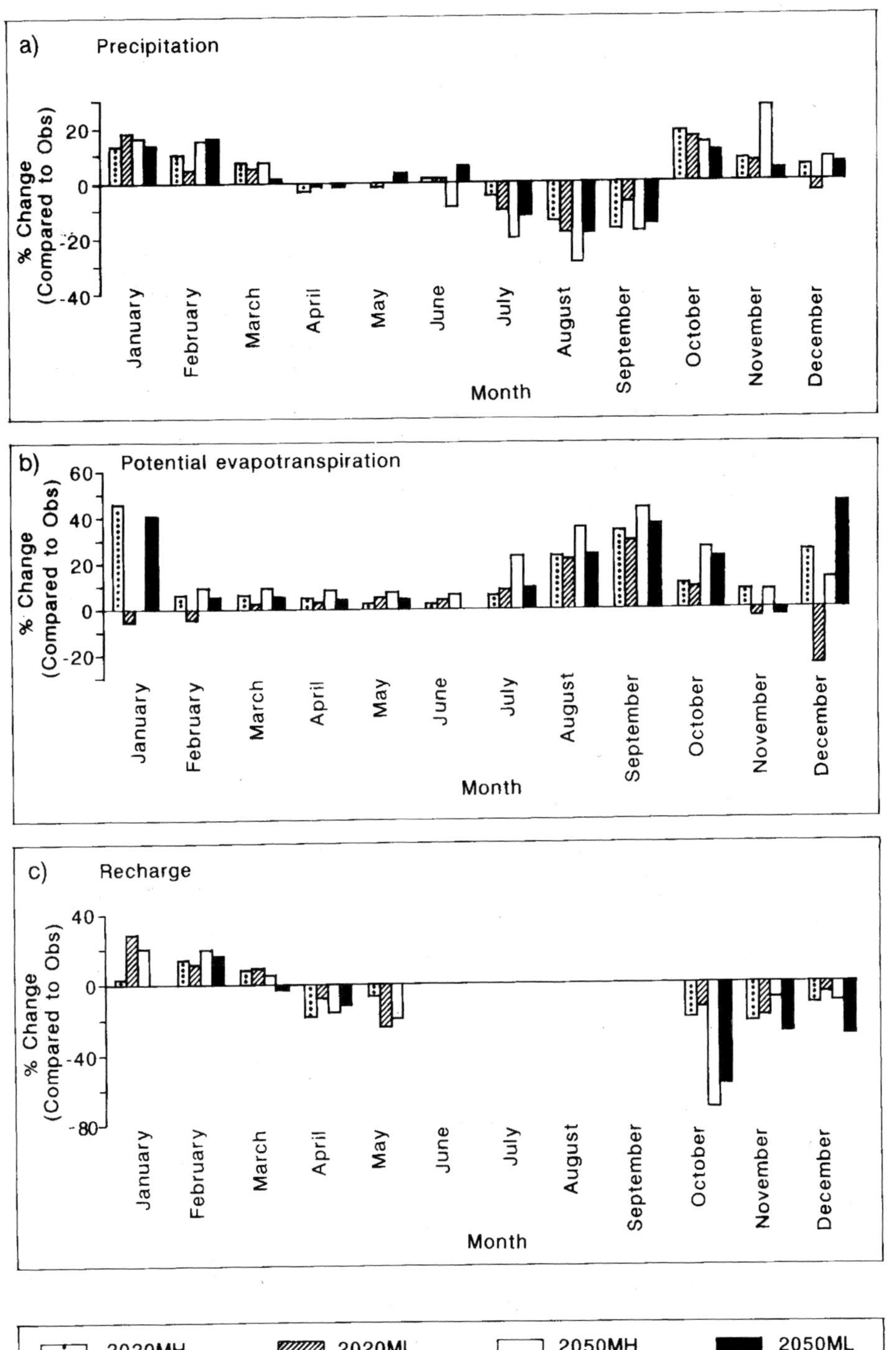

Fig. 11. Graphs comparing percentage changes in (**a**) monthly average precipitation, (**b**) monthly average potential evapotranspiration and (**c**) monthly average recharge between observed present-day climate and future climate predicted using output from the HadCM2 climate change model. Obs, observed present-day climate.

Table 4. *Comparison of seasonal hydrological parameters for the observed present-day and perturbed future climate states (2020 MH, 2020 ML, 2050 MH, 2050 ML)*

Hydrological parameter	Season	Obs. (mm)	2020 MH (% change)	2020 ML (% change)	2050 MH (% change)	2050 ML (% change)
P	Summer	48.1	-5.6	–8.4	–18.7	–7.3
	Autumn	57.0	2.8	5.3	7.1	–0.1
	Winter	44.6	9.3	6.4	12.7	11.0
	Spring	46.4	1.6	0.9	2.9	1.0
PE	Summer	92.3	9.9	11.0	21.2	10.6
	Autumn	35.4	22.9	18.8	33.3	26.8
	Winter	12.4	24.6	–10.3	7.8	28.4
	Spring	55.5	3.6	3.8	7.7	4.2
AE	Summer	67.1	–4.4	–7.0	–13.5	–2.5
	Autumn	31.4	4.2	6.3	9.9	6.7
	Winter	12.4	24.6	–10.3	7.8	28.4
	Spring	54.9	4.9	4.5	7.6	3.3
SMD	Summer	73.0	5.5	7.3	12.2	5.3
	Autumn	42.9	24.2	20.5	41.6	33.7
	Winter	1.7	6.3	–4.7	19.0	86.1
	Spring	21.8	6.7	6.4	10.2	5.6
Recharge	Summer	0.0	0.0	0.0	0.0	0.0
	Autumn	6.6	–21.3	–17.0	–25.7	–35.3
	Winter	25.7	0.1	13.7	10.1	–6.1
	Spring	11.0	0.2	2.3	–2.5	–5.1

P, precipitation; PE, potential evapotranspiration; AE, actual evapotranspiration; SMD, soil moisture deficit; Obs., observed present-day with averages based on 1980–1995, summer (June–August), autumn (September–November), winter (December–February) and spring (March–May).

the lowest variability is evident in the precipitation record (between 2 to 3%). The 2050MH scenario shows the most variability.

Future changes in actual evapotranspiration show a different trend (Table 3). The 2020MH and 2050ML scenarios result in similar increases in the actual evapotranspiration over monthly and annual timescales (up to 4%, or 17 mm), while the 2020ML and 2050MH scenarios result in relatively smaller decreases (up to 1% or 5 mm). The seasonal changes are greater than the monthly and annual changes. Seasonal actual evaporation (Table 4) is increased (up to 28%) in autumn, winter and spring for all scenarios except for the winter 2020ML scenario. The summer values of actual evapotranspiration are decreased (up to 14%) for all scenarios.

As a consequence of the increases in precipitation, potential evapotranspiration and soil moisture deficit in the future, there is a resultant change in the calculated recharge to the Chalk aquifer system. Given that recharge to the aquifer is generally a complex and non-linear process, two general trends in the future recharge are detected. The first trend is shown by the 2020MH and 2050ML scenarios that produce a decrease in recharge at monthly and annual timescales (Table 3). The second trend is shown by the 2020ML and 2050MH scenarios that produce an increase in recharge. The annual decrease for the first trend (up to 13 mm) is greater than the annual increase for the second trend (up to 8 mm). For the 2020ML scenario, the increase in recharge is due to a relatively large increase in precipitation coupled with the smallest increase in potential evapotranspiration and soil moisture deficit. For the 2050ML scenario, the smallest increase in precipitation coupled with a relatively large increase in potential evapotranspiration and soil moisture deficit cause a large reduction in potential recharge.

The seasonal change in future recharge is greater than the monthly and annual changes which are almost equal in magnitude. Winter generally represents the greatest increase in recharge by up to 14%, or 3.5 mm, for the 2020ML scenario. All the autumn recharge values show a notable reduction in seasonal recharge, with the greatest change of 35%, or 2.3 mm, recorded for the 2050ML scenario.

In terms of future hydrological states, the 2020s can generally be represented as having higher precipitation and groundwater recharge, i.e. a relatively wet period. On the other hand, the 2050s represent a period of higher potential

Table 5. *Comparison of monthly groundwater levels for the observed present-day and perturbed future climate states (2020 MH, 2020 ML, 2050 MH, 2050 ML)*

Observation borehole	Parameter (measure)	Obs (m)	2020 MH (m)	2020 ML (m)	2050 MH (m)	2050 ML (m)
OBS1	Change in monthly average groundwater level	20.1	−0.1	0.4	0.1	−0.6
	Standard deviation (% change)	3.3	−1.5	5.8	3.6	−4.9
OBS2	Change in monthly average groundwater level	2.8	0.0	0.1	0.0	−0.1
	Standard deviation (% change)	0.7	−1.5	5.9	4.4	−2.9
OBS3	Change in monthly average groundwater level	52.7	−0.1	0.4	0.2	−0.6
	Standard deviation (% change)	1.9	−3.2	0.0	−2.7	−4.8
OBS4	Change in monthly average groundwater level	37.3	−0.1	0.5	0.2	−0.7
	Standard deviation (% change)	1.6	−1.8	−0.6	−3.1	−1.8

OBS, Observation borehole; Obs., observed present-day with averages based on 1980–1995.

evapotranspiration and soil moisture deficit, i.e. a relatively dry period.

Changes in groundwater levels and baseflow volumes

Climate change impacts on groundwater resources were evaluated by incorporating the monthly estimated recharge inputs derived for the future climate scenarios into the validated transient model and comparing the relative changes (over monthly, seasonal and annual timescales) in groundwater levels and river baseflow volumes. The results of the effect of climate change on groundwater levels and baseflows for all scenarios considered are given in Tables 5–8. Although the model calibration process suggested that the baseflow of the River Nar is reasonably well simulated compared to that of the Rivers Wissey and Stringside, the output for the latter two rivers is still of interest in this study in order to look at relative changes in baseflow volumes, as opposed to absolute values. Figure 12 shows a comparison of percentage changes in monthly average groundwater levels in an unconfined Chalk observation borehole (OBS1) and baseflow volume for the River Nar between the observed present-day and future climate states.

In general, two future trends in groundwater levels are produced which, as expected, are similar in trend to the pattern of future recharge. The first trend is shown by the 2020MH and the 2050ML scenarios and the second trend by the 2020ML and 2050MH scenarios. The first trend shows a decrease in monthly groundwater levels of up to 70 cm (Table 5). The second, opposing trend shows a relatively smaller increase in monthly groundwater levels of up to 50 cm. The changes in groundwater levels are relatively greater for the confined section of the aquifer (OBS4) than the unconfined area. The seasonal change in groundwater levels is generally greater than the monthly changes (Table 6). For OBS1, the seasonal decrease is up to 97 cm (2050ML winter scenario) while the increase is up to 61 cm (2020ML spring scenario), the highest values for any of the four calibration target boreholes.

In terms of the likely impact of changes in groundwater levels on water resources, the change in the frequency of low groundwater levels in the future over an annual timescale and during the summer season are important (Table 6). In general, the 2020MH and 2050ML scenarios result in an increased frequency of low groundwater levels over an annual period (up to 12%) but with no change in the summer. Meanwhile, the 2020ML and 2050MH scenarios produce a decrease in the frequency of low groundwater levels over an annual period (up to 12%) and in the summer (up to 10%).

Overall, the 2020ML scenario predicts the prospect of an increase in groundwater levels in the future, while the 2050ML scenario predicts a decrease in groundwater levels compared with the present day. In terms of the two future periods modelled, the 2020s represent a period of relatively higher groundwater levels compared to the 2050s.

As with changes in groundwater levels, the river catchments in the study area show different magnitudes of response to the future climate change scenarios. Two future trends in baseflow

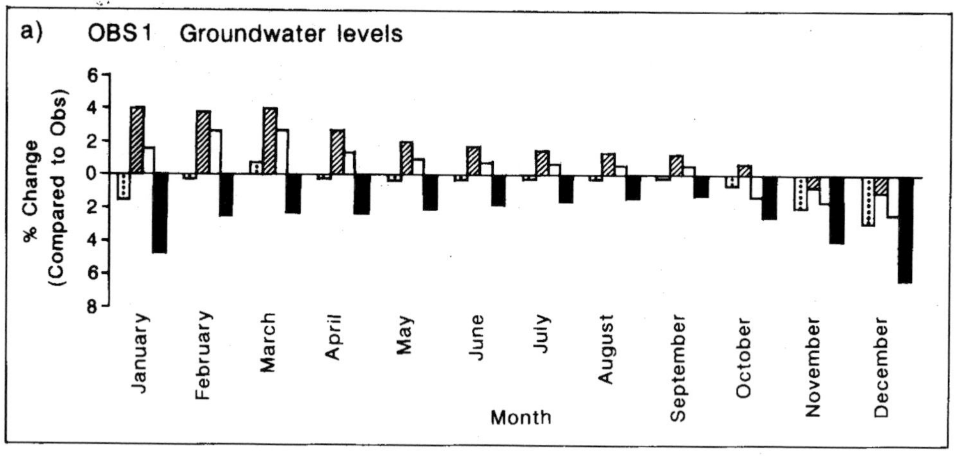

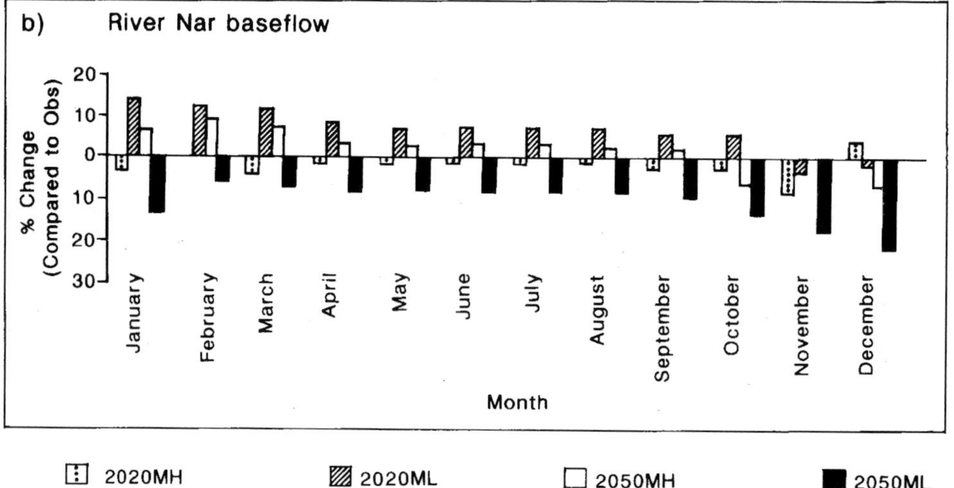

Fig. 12. Graphs comparing percentage changes in (**a**) monthly average groundwater levels at Chalk observation borehole OBS1 and (**b**) monthly average baseflow volume for the River Nar between observed present-day climate and future climate predicted using the calibrated transient state MODFLOW model. Obs., observed present-day climate.

volumes are detected using the monthly and annual output from the groundwater model. The first trend is shown by the 2020MH and 2050ML scenarios that produce monthly and annual decreases in baseflow volumes for all timescales (Table 7). The second trend is shown by the 2020ML and the 2050MH scenarios that produce monthly and annual increases in baseflow for all timescales.

Monthly and annual changes in baseflow volume are very similar for the Rivers Nar and Wissey (Table 7). The smallest catchment, the River Stringside, shows larger annual than monthly changes in baseflow. The River Nar has the highest monthly decrease (11% for 2050ML) and increase (7% for 2020ML) for all the proposed scenarios. The seasonal change is generally greater than the monthly and annual changes (Table 8). The seasonal decrease for the River Nar is up to 14% for the 2050ML autumn scenario while the increase is up to 9% for the 2020ML winter and spring scenarios.

In terms of the likely impact of changes in baseflow volumes on the in-stream ecology, the frequency of low baseflow volumes (expressed as the percentage of the year during which the

Table 6. *Comparison of seasonal groundwater levels for the observed present-day and perturbed future climate states (2020 MH, 2020 ML, 2050 MH, 2050 ML)*

Observation borehole	Parameter	Obs. (m)	2020 MH (m)	2020 ML (m)	2050 MH (m)	2050 ML (m)
OBS1	Summer (June–Aug)	19.0	−0.06	0.29	0.11	−0.30
	Autumn (Sep–Nov)	18.0	−0.18	0.07	−0.14	−0.47
	Winter (Dec–Feb)	21.6	−0.35	0.50	0.15	−0.97
	Spring (Mar–May)	21.9	0.00	0.61	0.35	−0.50
	%Year GWL ≤ Obs. ave.	0.5	12.0	−12.0	0.0	12.0
	%Summer GWL ≤ Obs. ave. summer	0.6	0.0	−10.4	−10.4	0.0

GWL, Groundwater level; OBS1, observation borehole 1; Obs., observed present-day, averages based on 1980–1995.

Table 7. *Comparison of monthly and annual river baseflow volumes for the observed present-day and perturbed future climate states (2020 MH, 2020 ML, 2050 MH, 2050 ML)*

River	Parameter	Obs. (m^3 per month)	2020 MH (% change)	2020 ML (% change)	2050 MH (% change)	2050 ML (% change)
River Nar	Monthly average	2.21×10^6	−2.1	7.3	2.4	−10.9
	Standard deviation	1.22×10^6	1.7	7.4	3.9	−6.7
	Annual average	2.65×10^7 m^3	−2.1	7.4	2.4	−10.7
River Wissey	Monthly average	3.12×10^6	−1.4	5.3	2.3	−6.8
	Standard deviation	1.16×10^6	−2.0	9.0	5.4	−7.3
	Annual average	3.74×10^7 m^3	−1.4	5.3	2.4	−6.7
River Stringside	Monthly average	8.48×10^5	0.0	0.1	0.0	−0.1
	Standard deviation	4.38×10^5	−3.1	15.1	6.9	−4.6
	Annual average	1.02×10^7 m^3	−2.3	5.2	1.5	−6.1

Obs., observed present-day with averages based on 1980–1995.

Table 8. *Comparison of seasonal river baseflow volumes for the observed present-day and perturbed future climate states (2020 MH, 2020 ML, 2050 MH, 2050 ML)*

River	Parameter	Obs. (m^3 per month)	2020 MH (% change)	2020 ML (% change)	2050 MH (% change)	2050 ML (% change)
River Nar	Summer (June–Aug)	1.72×10^6	−1.2	7.0	2.9	−8.1
	Autumn (Sep–Nov)	1.53×10^6	−4.6	2.6	−3.3	−13.7
	Winter (Dec–Feb)	2.85×10^6	−0.4	8.8	3.5	−13.3
	Spring (Mar–May)	2.73×10^6	−2.9	9.2	4.4	−8.1
	%Year BV ≤ Obs. ave.	50%	12.0	−12.0	0.0	24.0
	%Summer BV ≤ Obs. ave. summer	56%	0.0	0.0	0.0	11.6

BV, baseflow volume; Obs., observed present-day, averages based on 1980–1995.

average future river baseflow volumes are at or below the present-day average) over annual and summer season timescales is an important consideration. For the River Nar (Table 8), the 2020MH and 2050ML scenarios produce an increase in the percentage frequency (an annual increase up to 24% and a summer increase up to 12%). The 2020ML scenario produces a decrease in the low baseflow frequency by 12% while the summer frequency shows no change for this scenario.

Conclusions

The results of modelling the impacts of climate change on groundwater resources should be treated with caution given the uncertainty in modelling future climate change. Although a simple approach has been adopted in this study, with the application of climate change factors to the historic record to represent future climate conditions, the model outputs are nevertheless considered representative of potential climate change impacts on groundwater resources in the Chalk aquifer system in west Norfolk.

Two opposite trends are predicted from the modelling of climate change scenarios for the two future periods considered (2020s and 2050s). Of these two trends, the first produces larger changes, as follows:

(1) the 2050ML scenario predicts an overall annual decrease in recharge of 13 mm; a seasonal winter decrease in groundwater level of 97 cm; an increase in the annual frequency of low groundwater levels of 12%; a seasonal autumn decrease of 14% in the baseflow volume of the River Nar; and a 24% increase in the annual frequency of low river baseflow volumes in the River Nar;

(2) the 2020ML scenario predicts an overall annual increase in recharge of 8 mm; a seasonal spring increase in groundwater level of 61 cm; a decrease in the annual frequency of low groundwater levels of 12%; seasonal winter and spring increases of 9% in the baseflow volume of the River Nar; and a 12% decrease in the annual frequency of low river baseflow volumes in the River Nar.

From the point of view of managing future groundwater resources, the most optimistic outcome is produced by the 2020ML climate change scenario, where the annual recharge amount is predicted to be 6% greater than the present value (1980–1995 average), while the worst outcome is produced by the 2050ML scenario in which the annual recharge is predicted to be 10% less than the present value. A 10% decrease in recharge by the middle of this century compares with a projected increase in household water demand of over 70% by 2025 for the least sustainable water-use pattern.

The most noticeable and consistent result of the climate change impact simulations is the decrease in recharge expected in autumn for all scenarios (decreases ranging from 17 to 35%) as a consequence of the smaller amount of summer precipitation and increased autumn potential evapotranspiration. For the 2050MH scenario, these conditions lead to a 42% increase in autumn soil moisture deficit and a 26% reduction in recharge. Hence, the Anglian region can expect longer and drier summers and a delay in the start of groundwater recharge in the autumn and winter period. The drier conditions will have relatively little effect on summer groundwater levels (generally a 1–2% decrease) but the evident decrease of up to 14% in autumn baseflow volume for the 2050MH scenario and the associated increased frequency in summer low baseflow of 12% indicate that Chalk groundwater-fed rivers such as the River Nar may show environmental impacts, especially at a time when spray irrigation demand is likely to be high.

In comparing the magnitude of climate change impacts on river flows with the effects of spray irrigation demand from groundwater resources, Hiscock *et al.* (2001) showed that irrigation losses were greatest during dry growing seasons, particularly during the drought years 1989–1992 when the spray irrigation requirement was high. River flows in the River Nar experienced a maximum loss of 24% in the weekly mean flow and explained the major part of the gross and seasonal losses to flow in the catchment. Hence, in the future, and unless there is a change in the economic activity in the region, there are likely to be conflicting demands for water to serve domestic, agricultural and industrial requirements as well as to meet environmental requirements. Resolving these issues in the face of the compounding effects of climate change in reducing water availability in the drier months of the year will become one of the greater challenges for future sustainable water resources management in the Anglian region.

The University of Malaya is gratefully acknowledged for its financial support given to Ismail Yusoff under its Lectureship Training Scheme. This study benefited from the support of Environmental Simulations International in providing the Groundwater Vistas MODFLOW code and for helpful discussions in the use of the model. The GCM data were supplied by the Climate Impacts LINK Project (DEFRA contract EPG1/1/124) on behalf of the Hadley Centre and UK Meteorological Office. Anglian Water Services and the Environment Agency (Anglian Region) are thanked for their assistance in providing hydrogeological data. The authors thank Mike Carey and David Lister for their careful and constructive reviews of an earlier version of this paper.

References

ARNELL, N. W. 1992*a*. Factors controlling the effects of climate change on river flow regimes in a humid temperate environment. *Journal of Hydrology*, **132**, 321–342.

ARNELL, N. W. 1992*b*. Impact of climatic change on river flow regimes in the UK. *Journal of the Institution of Water & Environmental Management*, **6**, 432–442.

ARNELL, N. 1996. *Global Warming, River Flows and Water Resources*. John Wiley & Sons, Chichester.

ARNELL, N., REYNARD, N., KING, R., PRUDHOMME, C. & BRANSON, J. 1997. *Effects of climate change on river flows and groundwater resources: guidelines for resource assessment*. Report prepared for the Environment Agency and Water Industry Environment and Quality Committee by the UK Water Industry Research Limited, Report no. **97/CL/04/1**.

ASKEW, A. J. 1987. Climate change and water resources. *IAHS Publications*, **168**, 421–430.

ASPINWALL & COMPANY. 1992*a*. *A review of the possible impact of public water supply abstraction on surface water flow in the Watton area*. Unpublished report to the National Rivers Authority (Anglian Region).

ASPINWALL & COMPANY. 1992*b*. *A review of the water resources of the catchment of the River Wissey, Norfolk*. Unpublished report to the National Rivers Authority (Anglian Region).

BOAR, R. R., LISTER, D. H., HISCOCK, K. M. & GREEN, F. M. L. 1994. *The effects of water resources management on the Rivers Bure, Wensum and Nar in North Norfolk*. Unpublished report to the National Rivers Authority.

CARTER, T. R. & LA ROVERE, E. L. 2001. Developing and applying scenarios. *In*: MCCARTHY, J., CANZIANI, O. & LEARY, N. (eds) *Climate Change 2001: Impacts, Adaptation and Vulnerability*. Cambridge University Press, Cambridge.

CONWAY, D. 1998. Recent climate variability and future climate change scenarios for Great Britain. *Progress in Physical Geography*, **22**, 350–374.

DTI. 1999. *Environmental Futures*. Department of Trade & Industry, Foresight, Office of Science & Technology, London.

ENVIRONMENT AGENCY. 2000. *Water resources for the future. A strategy for Anglian Region*. Environment Agency, Peterborough.

FOSTER, S. S. D. & ROBERTSON, A. S. 1977. Evaluation of a semi-confined Chalk aquifer in East Anglia. *Proceedings of the Institution of Civil Engineers*, **63**, 803–817.

GUSTARD, A., BULLOCK, A. & DIXON, J. M. 1992 *Low flow estimation in the United Kingdom*. Institute of Hydrology, Wallingford, **Report No. 108.**

HISCOCK, K. M. 1993. The influence of pre-Devensian deposits on the hydrogeochemistry of the Chalk aquifer system of north Norfolk, UK. *Journal of Hydrology*, **144**, 335–369.

HISCOCK, K. M., LISTER, D. H., BOAR, R. R. & GREEN, F. M. L. 2001. An integrated assessment of long-term changes in the hydrology of three lowland rivers in eastern England. *Journal of Environmental Management*, **61**, 195–214.

HOWARD, K. W. F. & LLOYD, J. W. 1979. The sensitivity of parameters in the Penman evaporation equations and direct recharge balance. *Journal of Hydrology*, **41**, 329–344.

HULME, M. & JENKINS, G. J. 1998. *Climate change scenarios for the United Kingdom: scientific report*. UK Climate Impacts Programme, Climatic Research Unit, Norwich, **Technical Report No. 1**.

IGS. 1976. 1 : 125 000 Hydrogeological map of northern East Anglia. Institute of Geological Sciences, London.

INESON, J. & DOWNING, R. A. 1964. The groundwater component of river discharge and its relationship to hydrogeology. *Journal of the Institution of Water Engineers & Scientists*, **18**, 519–541.

IPCC. 1997. The regional impacts of climate change: an assessment of vulnerability. *In*: WATSON, R. T., ZINYOWERA, M. C. & MOSS, R. H. (eds) *A Special Report of the IPCC WGII*. Summary for Policymakers. Intergovernmental Panel on Climate Change. Cambridge University Press, Cambridge.

LLOYD, J. W., HARKER, D. & BAXENDALE, R. A. 1981. Recharge mechanisms and groundwater flow in the Chalk and drift deposits of southern East Anglia. *Quarterly Journal of Engineering Geology*, **14**, 87–96.

MCDONALD, M. G. & HARBAUGH, A. W. 1988. *A modular three-dimensional finite difference groundwater flow model*. Techniques of Water-Resources Investigations of the United States Geological Survey, **06-A1**.

MEARNS, L. O., GLEICK, P. H. & SCHNEIDER, S. H. 1990. Climate forecasting. In WAGGONER, P. E. (ed.) *Climate change and US Water Resources*. John Wiley & Sons, Toronto.

OWEN, M. 1969. *Some aspects of the hydrogeology of the Nar, Wissey, Little Ouse, Lark and Granta catchments*. PhD thesis, Imperial College, London.

PALUTIKOF, J. P., SUBAK, S. & AGNEW, M. D. (eds) 1997. *Economic Impacts of the Hot Summer and Unusually Warm Year of 1995*. Climatic Research Unit, University of East Anglia, Norwich.

PENMAN, H. L. 1949. The dependence of transpiration on weather and soil conditions. *Journal of Soil Sciences*, **1**, 74–89.

PENMAN, H. L. 1950. The water balance of the Stour catchment area. *Journal of the Institution of Water Engineers & Scientists*, **4**, 457–469.

RUMBAUGH, J. O. & RUMBAUGH, D. B. 1996. *Guide to Using Groundwater Vistas*. Environmental Simulations Inc., Herndon, Virginia.

THOMPSON, N., BARRIE, I. A. & AYLES, M. 1981. *The Meteorological Office Rainfall and Evaporation Calculation System: MORECS*. Meteorological Office, Bracknell. **Hydrological Memorandum No. 45.**

TOYNTON, R. 1983. The relation between fracture patterns and hydraulic anisotropy in the Norfolk Chalk, England. *Quarterly Journal of Engineering Geology*, **16**, 169–185.

VINER, D., JORDAN, A., LORENZONI, I. & FAVIS-MORTLOCK, D. 2000. *The potential impact of climate change in the Norfolk Arable Land Management Initiative (NALMI) area over the next 30-50 years*. Report prepared for the Countryside Agency. Climatic Research Unit, University of East Anglia,

Norwich.

WOOTON, N. A. 1994. *Are hardgrounds in the Chalk aquifer zones of enhanced groundwater flow*? MSc thesis, University of East Anglia, Norwich.

YUSOFF, I. 2000. *Modelling climate change impacts on groundwater resources in the West Norfolk Chalk aquifer*. PhD thesis, University of East Anglia, Norwich.

Index

Note: Page numbers in *italic* refer to illustrations, those in **bold** type to tables.